# The Astronomical Almanac (2021-2025)

## A Comprehensive Guide to Night Sky Events

### Richard J. Bartlett

Cover Images

Top image: Venus moving through the stars of the Pleiades star cluster in Taurus, April 2nd to 4th 2020. The images were taken using a 120 – 400mm telephoto lens using a Nikon D750 DSLR. The two images either side of closest encounter (the left and right-most exposures of Venus) were 30 second exposures. ISO 800 @ F25. The image taken on the day of conjunction (the middle exposure of Venus) was a 60 second exposure. ISO 800 @F29. A higher F setting produces enhanced diffraction spikes, produced by the iris in the lens.

Image taken by Dave Eagle FRAS. Dave has had a lifelong passion for astronomy and is a Mobile planetarium Operator, Author, Astrophotographer and Educator. You can visit his website at https://www.star-gazing.co.uk/

Bottom images: The phases of Venus, from May to August 2004.

Credit: Statis Kalyvas - VT-2004 programme

(Wikimedia, https://commons.wikimedia.org/wiki/File:Phases_Venus.jpg)

First Edition, May 2020

*For all the astronomers who bring the wonder of it all to all those who wonder.*

# Contents

# Introduction

# About this Book

## The Star Charts and Observing Lists

The book is broken up into two main parts. The first (and much smaller) part contains the star charts and observing lists. It would be impractical and redundant to produce star charts for every month of every year as the night sky in early May this year is essentially the same as the night sky in May next year – only the positions of the Moon and planets have changed. (Besides which, as I've broken each month into three, it would require the production of 180 star charts and additional pages.)

Instead I've produced 24 charts with suggested deep sky objects to observe with each chart. The idea is that you use the tables at the beginning of the section to look up the appropriate chart for the desired time of year and hour of night. The charts themselves depict the night sky for observers at a latitude of 45° north but don't show the planets. If you'd like to download a PDF with charts and observing lists for the United Kingdom you can do so here: http://tinyurl.com/ukstars

## Summary Tables

The second, largest part of the book is the data for the next five years. At the beginning of each year are several summary tables which show you, at-a-glance, the visibility of the planets as well as listing eclipses, major planetary events and meteor showers.

The first table details the visibility of the planets throughout the year. On the left side are the planets visible in the morning sky while on the right are the planets visible in the evening. The planets are abbreviated as follows:

| **Me** | Mercury | **Ve** | Venus | **Ma** | Mars | **Ju** | Jupiter |
|---|---|---|---|---|---|---|---|
| **Sa** | Saturn | **Ur** | Uranus | **Ne** | Neptune | | |

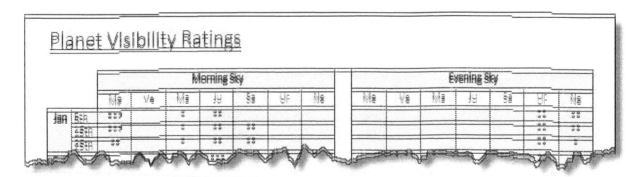

The * represent a visibility rating for each planet at that time, from * for poor to ***** for excellent. This is something I've added to give you an idea of the best times to observe the planet. The rating is based upon an average of the planet's magnitude, apparent diameter and elongation from the Sun. The rating for each planet is also based upon its own maximum and minimum for that criteria; for example, the brightness rating for Neptune is based upon its own minimum and maximum magnitude rather than the minimum and maximum possible brightness for planets as a whole. That way Neptune isn't rated lowly because it's nowhere near as bright as Venus.

For example, an outer planet at opposition will often score 5 stars as it's at its brightest, largest and appears furthest from the Sun in the sky at that time. It might sometimes only rate 4 stars as the distance between the Earth and the planet will vary with each opposition. If the planet is further away than usual then it might appear slightly smaller and fainter than at other oppositions and this may bring its rating down.

The ratings for Mercury and Venus may be a little different. When they're at their largest they're between the Earth and the Sun and not visible (they're also at their faintest at this time) so the highest rating for these two planets will typically be 4 stars. (I haven't found any circumstances yet where these planets rate 5 stars.)

Below the planet visibility table is a summary of eclipses. This gives you the date and time of maximum eclipse (in Universal Time) as well as where the eclipse might be visible from. It might be worth noting that, in the interests of formatting, I've had to remove some locations, specifically the oceans, Arctic and Antarctic, from some descriptions. (Otherwise the list of locations would require two lines and I don't have the space on the page for that.) Populated areas have always been listed.

Next we have a list of planetary highlights. This table is not all-inclusive; it doesn't include lunar events because there are too many and, again, there simply isn't enough space on the page. Instead it highlights the best conjunctions and events (ie, elongations for Mercury and Venus, oppositions for the outer planets) with a focus on the closest conjunctions and most easily observable events. Again, not every event is listed for formatting reasons but any omissions are included in the list of daily events for each month.

I've included the elongation from the Sun in the column labelled Elon. This is to give you an idea of how far away from the Sun the event occurs. (Where two planets are involved, I've listed the elongation of the brighter planet.)

Lastly, there is a table of major meteor showers. The table lists the start and end dates but is sorted by the shower maximum date. The Zenith Hourly Rate (ZHR) is listed, which describes the number of meteors you might expect to see under ideal conditions if the source (or radiant) were directly overhead. There is also a rating for speed and brightness. The brighter the meteor, the higher the rating but in terms of speed, the slower the meteor the higher the rating. (My thinking was that it would be easier to see.)

## Major Meteor Showers

| Shower Name | Start Date | End Date | Peak | ZHR | Speed | Brightness | Moon |
|---|---|---|---|---|---|---|---|
| Quadrantids | Dec 28 | Jan 12 | Jan 3 | 120 | *** | ***** | ○ |
| Lyrids | Apr 16 | Apr 25 | Apr 22 | 18 | *** | ***** | ◗ |
| Eta Aquariids | Apr 24 | May 19 | May 7 | 40 | * | **** | ◖ |
| | | | May 13 | | | | |

Each meteor shower also has a depiction of the lunar phase on the peak date. As almost every astronomer knows, the full Moon is simply not a good time for meteors as the light will brighten the sky and drown out the fainter shooting stars. As most meteor showers are best observed in the early hours of the morning, the best time during the lunar cycle for meteor spotting is between last quarter and first quarter (and ideally at the new Moon.)

## Monthly Pages

Each month is broken up into three segments of roughly ten days (e.g., the 1st to the 10th, 11 to the 20th, 21st to the 31st.) For each ten day period you'll find data and graphical depictions of the Moon and planets. The Moon graphics represent a field of view of 35 arc-minutes while the depictions of the planets represent a field of view just slightly more than 1 arc-minute (1.1 arc-minutes to be precise).

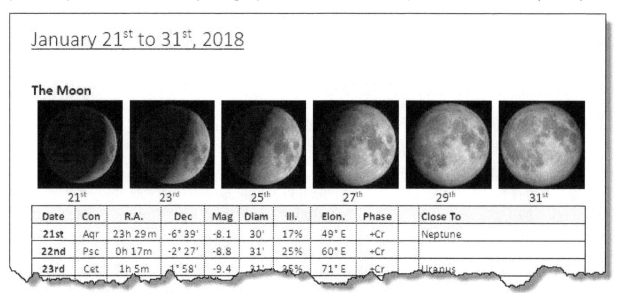

## January 21st to 31st, 2018

### The Moon

| Date | Con | R.A. | Dec | Mag | Diam | Ill. | Elon. | Phase | Close To |
|------|-----|------|-----|-----|------|------|-------|-------|----------|
| 21st | Aqr | 23h 29m | -6° 39' | -8.1 | 30' | 17% | 49° E | +Cr | Neptune |
| 22nd | Psc | 0h 17m | -2° 27' | -8.8 | 31' | 25% | 60° E | +Cr | |
| 23rd | Cet | 1h 5m | 1° 58' | -9.4 | 31' | 35% | 71° E | +Cr | Uranus |

The idea is that the planets are proportionally depicted as you'd see them if your eyepiece were to have that field of view. I haven't depicted Uranus and Neptune as the change in apparent diameter is minimal. Additionally, as those planets appear much smaller than the others the images might be difficult to clearly reproduce. (There's also the issue of formatting and layout to take into account.)

The Moon and planets all have data tables to list their positions and general properties (a key is provided below) and, again, there's a rating of the planet's visibility for that date. There's also a column to list which bright stars, star clusters and planets the object appears close to. Typically I've listed a nearby object if the distance between them is about five degrees or less.

Lastly, at the end of each ten day segment is a list of key astronomical events. With a few exceptions, these are all events that can be seen and observed. The exceptions are when a planet switches from retrograde to prograde (and vice versa), is in conjunction with the Sun or it's a New Moon. I've included those events as I would consider them to be of broad general interest to the amateur astronomer, even though they're not directly observable.

Angular separations between the planets and other objects are included but not for the Moon. As the Moon is much closer than the planets, the angle of separation will vary depending upon your location in the world.

Again, in a few instances where there are a lot of events during a ten day period I've had to omit a small number of more inconsequential events. For example, the Moon close to Pluto (making Pluto even harder to see) or an asteroid in conjunction with the Sun.

As always, the graphics and much of the data were generated using the *Mobile Observatory* app by Wolfgang Zima and I continue to be indebted to his generosity. It's available for Android devices on Google Play and you can get more information about the app from http://zima.co

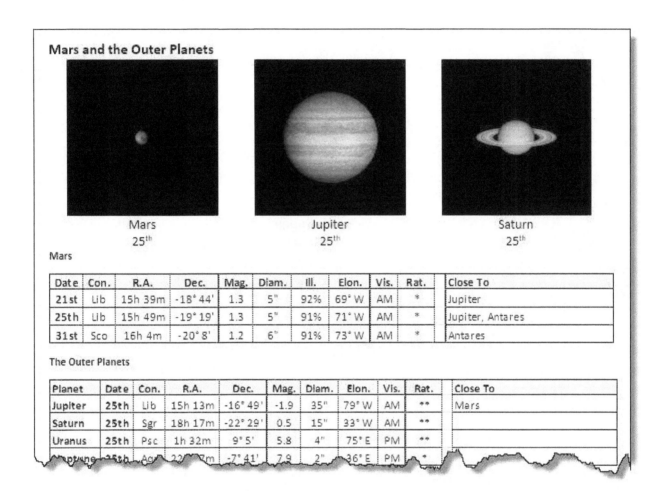

## Mars and the Outer Planets

|  | Mars 25th | Jupiter 25th | Saturn 25th |

### Mars

| Date | Con. | R.A. | Dec. | Mag. | Diam. | Ill. | Elon. | Vis. | Rat. | Close To |
|------|------|------|------|------|-------|------|-------|------|------|----------|
| 21st | Lib | 15h 39m | -18° 44' | 1.3 | 5" | 92% | 69° W | AM | * | Jupiter |
| 25th | Lib | 15h 49m | -19° 19' | 1.3 | 5" | 91% | 71° W | AM | * | Jupiter, Antares |
| 31st | Sco | 16h 4m | -20° 8' | 1.2 | 6" | 91% | 73° W | AM | * | Antares |

### The Outer Planets

| Planet | Date | Con. | R.A. | Dec. | Mag. | Diam. | Elon. | Vis. | Rat. | Close To |
|--------|------|------|------|------|------|-------|-------|------|------|----------|
| Jupiter | 25th | Lib | 15h 13m | -16° 49' | -1.9 | 35" | 79° W | AM | ** | Mars |
| Saturn | 25th | Sgr | 18h 17m | -22° 29' | 0.5 | 15" | 33° W | AM | ** | |
| Uranus | 25th | Psc | 1h 32m | 9° 5' | 5.8 | 4" | 75° E | PM | ** | |
| Neptune | 25th | Aqr | 22h 7m | -7° 41' | 7.9 | 2" | 36° E | PM | * | |

## Key to Lunar and Planetary Data Tables

These tables provide a key to the abbreviations and acronyms used for the Moon and planets. Please note that the position data (ie, the R.A. and declination) are accurate for 12:00 Universal Time.

| +Cr | Waxing Crescent Moon | FQ | First Quarter Moon | LQ | Last Quarter Moon |
|-----|---------------------|-----|-------------------|-----|------------------|
| -Cr | Waning Crescent Moon | +G | Waxing Gibbous Moon | NM | New Moon |
| FM | Full Moon | -G | Waning Gibbous Moon | | |

| AM | Morning visibility | Gem | Gemini | Psc | Pisces |
|---|---|---|---|---|---|
| AN | All Night | Ill | Illumination | R.A. | Right Ascension |
| Aqr | Aquarius | Leo | Leo | Rat | Visibility Rating |
| Ari | Aries | Lib | Libra | Sco | Scorpius |
| Cap | Capricornus | Mag | Magnitude | Sex | Sextans |
| Cet | Cetus | NV | Not Visible | Sgr | Sagittarius |
| Con | Constellation | Oph | Ophiuchus | Tau | Taurus |
| Dec | Declination | Ori | Orion | Vir | Virgo |
| Diam | Apparent Diameter | PM | Evening Sky | Vis | Visibility (AM/PM/AN) |
| Elong | Elongation from Sun | | | | |

## About the Author

Photo by my son, James Bartlett

I've had an interest in astronomy since I was six and although my interest has waxed and waned like the Moon, I've always felt compelled to stop and stare at the stars.

In the late 90's, I discovered the booming frontier of the internet, and like a settler in the Midwest, I quickly staked my claim on it. I started to build a (now-defunct) website called *StarLore*. It was designed to be an online resource for amateur astronomers who wanted to know more about the constellations – and all the stars and deep sky objects to be found within them. It was quite an undertaking.

After the website was featured in the February 2001 edition of *Sky & Telescope* magazine, I began reviewing astronomical websites and software for their rival, *Astronomy*. This was something of a dream come true; I'd been reading the magazine since I was a kid and now my name was regularly appearing in it.

Unfortunately, a financial downturn forced my monthly column to be cut after a few years but I'll always be grateful for the chance to write for the world's best-selling astronomy magazine.

I emigrated from England to the United States in 2004 and spent three years under relatively clear, dark skies in Oklahoma. I then relocated to Kentucky in 2008 and then California in 2013. I now live in the suburbs of Los Angeles; not the most ideal location for astronomy, but there are still a number of naked eye events that are easily visible on any given night.

## The Author Online

Amazon US: http://tinyurl.com/rjbamazon-us

Amazon UK: http://tinyurl.com/rjbamazon-uk

Facebook: http://tinyurl.com/rjbfacebook

Twitter: http://tinyurl.com/rjbtwitter

Instagram: richardjbartlett

Email: astronomywriter@gmail.com

Podcast: http://tinyurl.com/snspod

Facebook Group:
http://tinyurl.com/snsfacebookgroup

# Star Chart Tables

If observing during daylight savings time, first deduct one hour and then refer to the corresponding chart number. For example, for 10pm daylight savings time in early August, use chart 18.

|  | 6pm | 7pm | 8pm | 9pm | 10pm | 11pm |
|---|---|---|---|---|---|---|
| Early January | 1 | 2 | 3 | 4 | 5 | 6 |
| Late January | 2 | 3 | 4 | 5 | 6 | 7 |
| Early February | 3 | 4 | 5 | 6 | 7 | 8 |
| Late February | 4 | 5 | 6 | 7 | 8 | 9 |
| Early March | 5 | 6 | 7 | 8 | 9 | 10 |
| Late March | 6 | 7 | 8 | 9 | 10 | 11 |
| Early April | 7 | 8 | 9 | 10 | 11 | 12 |
| Late April | 8 | 9 | 10 | 11 | 12 | 13 |
| Early May | 9 | 10 | 11 | 12 | 13 | 14 |
| Late May | 10 | 11 | 12 | 13 | 14 | 15 |
| Early June | 11 | 12 | 13 | 14 | 15 | 16 |
| Late June | 12 | 13 | 14 | 15 | 16 | 17 |
| Early July | 13 | 14 | 15 | 16 | 17 | 18 |
| Late July | 14 | 15 | 16 | 17 | 18 | 19 |
| Early August | 15 | 16 | 17 | 18 | 19 | 20 |
| Late August | 16 | 17 | 18 | 19 | 20 | 21 |
| Early September | 17 | 18 | 19 | 20 | 21 | 22 |
| Late September | 18 | 19 | 20 | 21 | 22 | 23 |
| Early October | 19 | 20 | 21 | 22 | 23 | 24 |
| Late October | 20 | 21 | 22 | 23 | 24 | 1 |
| Early November | 21 | 22 | 23 | 24 | 1 | 2 |
| Late November | 22 | 23 | 24 | 1 | 2 | 3 |
| Early December | 23 | 24 | 1 | 2 | 3 | 4 |
| Late December | 24 | 1 | 2 | 3 | 4 | 5 |

If observing during daylight savings time, first deduct one hour and then refer to the corresponding chart number. For example, for 2am daylight savings time in early July, use chart 20.

|  | 12am | 1am | 2am | 3am | 4am | 5am | 6am |
|---|---|---|---|---|---|---|---|
| Early January | 7 | 8 | 9 | 10 | 11 | 12 | 13 |
| Late January | 8 | 9 | 10 | 11 | 12 | 13 | 14 |
| Early February | 9 | 10 | 11 | 12 | 13 | 14 | 15 |
| Late February | 10 | 11 | 12 | 13 | 14 | 15 | 16 |
| Early March | 11 | 12 | 13 | 14 | 15 | 16 | 17 |
| Late March | 12 | 13 | 14 | 15 | 16 | 17 | 18 |
| Early April | 13 | 14 | 15 | 16 | 17 | 18 | 19 |
| Late April | 14 | 15 | 16 | 17 | 18 | 19 | 20 |
| Early May | 15 | 16 | 17 | 18 | 19 | 20 | 21 |
| Late May | 16 | 17 | 18 | 19 | 20 | 21 | 22 |
| Early June | 17 | 18 | 19 | 20 | 21 | 22 | 23 |
| Late June | 18 | 19 | 20 | 21 | 22 | 23 | 24 |
| Early July | 19 | 20 | 21 | 22 | 23 | 24 | 1 |
| Late July | 20 | 21 | 22 | 23 | 24 | 1 | 2 |
| Early August | 21 | 22 | 23 | 24 | 1 | 2 | 3 |
| Late August | 22 | 23 | 24 | 1 | 2 | 3 | 4 |
| Early September | 23 | 24 | 1 | 2 | 3 | 4 | 5 |
| Late September | 24 | 1 | 2 | 3 | 4 | 5 | 6 |
| Early October | 1 | 2 | 3 | 4 | 5 | 6 | 7 |
| Late October | 2 | 3 | 4 | 5 | 6 | 7 | 8 |
| Early November | 3 | 4 | 5 | 6 | 7 | 8 | 9 |
| Late November | 4 | 5 | 6 | 7 | 8 | 9 | 10 |
| Early December | 5 | 6 | 7 | 8 | 9 | 10 | 11 |
| Late December | 6 | 7 | 8 | 9 | 10 | 11 | 12 |

# Chart 1

| Designation | Name | Con: | Type | R.A: | Dec: | Mag | Size/Sep |
|---|---|---|---|---|---|---|---|
| Gam And | Almach | And | MS | 02h 04m | +42° 20' | 2.1 | 10" |
| M 31 | Andromeda Galaxy | And | Gx | 00h 43m | +41° 16' | 4.3 | 156' |
| NGC 752 | Golf Ball Cluster | And | OC | 01h 58m | +37° 47' | 6.6 | 75' |
| Pi And | | And | MS | 00h 38m | +33° 49' | 4.4 | 36" |
| 107 Aqr | | Aqr | MS | 23h 47m | -18° 35' | 5.3 | 7" |
| 94 Aqr | | Aqr | MS | 23h 19m | -13° 28' | 5.2 | 13" |
| NGC 7293 | Helix Nebula | Aqr | PN | 22h 30m | -20° 50' | 6.3 | 16' |
| Zet Aqr | | Aqr | MS | 22h 29m | -00° 01' | 3.7 | 2" |

| Designation | Name | Con: | Type | R.A.: | Dec.: | Mag | Size/Sep |
|---|---|---|---|---|---|---|---|
| 41 Aqr | | Aqr | MS | 22h 14m | -21° 04' | 5.3 | 5" |
| 53 Aqr | | Aqr | MS | 22h 27m | -16° 45' | 5.6 | 3" |
| Lam Ari | | Ari | MS | 01h 59m | +23° 41' | 4.8 | 37" |
| Gam Ari | Mesarthim | Ari | MS | 01h 54m | +19° 22' | 4.6 | 8" |
| 30 Ari | | Ari | MS | 02h 37m | +24° 39' | 6.5 | 39" |
| Kemble 1 | Kemble's Cascade | Cam | Ast | 03h 57m | +63° 04' | 5.0 | 180' |
| Sig Cas | | Cas | MS | 23h 59m | +55° 45' | 4.9 | 3" |
| NGC 457 | Owl Cluster | Cas | OC | 01h 20m | +58° 17' | 5.1 | 20' |
| M 52 | | Cas | OC | 23h 25m | +61° 36' | 8.2 | 15' |
| Struve 163 | | Cas | MS | 01h 51m | +64° 51' | 6.5 | 35" |
| Struve 3053 | | Cas | MS | 00h 03m | +66° 06' | 5.9 | 15" |
| Eta Cas | Achird | Cas | MS | 00h 50m | +57° 54' | 3.6 | 13" |
| M 103 | | Cas | OC | 01h 33m | +60° 39' | 6.9 | 5' |
| NGC 281 | | Cas | OC | 00h 53m | +56° 38' | 7.4 | 4' |
| Iot Cas | | Cas | MS | 02h 29m | +67° 24' | 4.5 | 7" |
| NGC 7789 | Herschel's Spiral Cluster | Cas | OC | 23h 57m | +56° 43' | 7.5 | 25' |
| NGC 559 | | Cas | OC | 01h 30m | +63° 18' | 7.4 | 6' |
| NGC 659 | Ying Yang Cluster | Cas | OC | 01h 44m | +60° 40' | 7.2 | 5' |
| NGC 663 | | Cas | OC | 01h 46m | +61° 14' | 6.4 | 14' |
| Del Cep | | Cep | MS/Var | 22h 29m | +58° 25' | 3.5-4.4 | 41" |
| Xi Cep | Alkurhah | Cep | MS | 22h 04m | +64° 38' | 4.3 | 8" |
| Omi Cep | | Cep | MS | 23h 19m | +68° 07' | 4.8 | 3" |
| Gam Cet | Kaffaljidhma | Cet | MS | 02h 43m | +03° 14' | 3.5 | 3" |
| Omi Cet | Mira | Cet | Var | 02h 19m | -02° 59' | 2.0-10.1 | N/A |
| NGC 7243 | | Lac | OC | 22h 15m | +49° 54' | 6.7 | 29' |
| 8 Lac | | Lac | MS | 22h 36m | +39° 38' | 5.7 | 82" |
| Bet Per | Algol | Per | Var | 03h 08m | +40° 57' | 2.1-3.4 | N/A |
| Eps Per | | Per | MS | 03h 58m | +40° 01' | 2.9 | 9" |
| M 34 | | Per | OC | 02h 42m | +42° 46' | 5.8 | 35' |
| NGC 869/884 | Double Cluster | Per | OC | 02h 21m | +57° 08' | 4.4 | 18' |
| Eta Per | | Per | MS | 02h 51m | +55° 54' | 3.8 | 29" |
| NGC 1245 | | Per | OC | 03h 15m | +47° 15' | 7.7 | 10' |
| Melotte 20 | Alpha Persei Moving Cluster | Per | OC | 03h 24m | +49° 52' | 2.3 | 300' |
| TX Psc | | Psc | Var/CS | 23h 46m | +03° 29' | 4.5-5.3 | N/A |
| Alp Psc | Alrisha | Psc | MS | 02h 02m | +02° 46' | 3.8 | 2" |
| Psi1 Psc | | Psc | MS | 01h 06m | +21° 28' | 5.3 | 30" |
| Zet Psc | | Psc | MS | 01h 14m | +07° 35' | 5.2 | 23" |
| 55 Psc | | Psc | MS | 00h 40m | +21° 26' | 5.4 | 6" |
| 65 Psc | | Psc | MS | 00h 50m | +27° 43' | 7.0 | 4" |
| M 45 | Pleiades | Tau | OC | 03h 47m | +24° 07' | 1.5 | 120' |
| M 33 | Triangulum Galaxy | Tri | Gx | 01h 34m | +30° 40' | 6.4 | 62' |
| Alp UMi | Polaris | UMi | MS | 02h 51m | +89° 20' | 2.0 | 18" |

# Chart 2

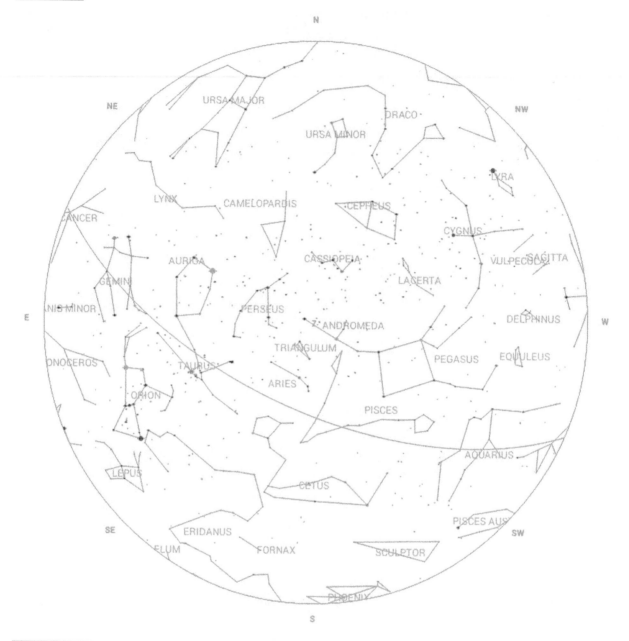

| Designation | Name | Con. | Type | R.A. | Dec. | Mag | Size/Sep |
|---|---|---|---|---|---|---|---|
| Gam And | Almach | And | MS | 02h 04m | +42° 20' | 2.1 | 10" |
| M 31 | Andromeda Galaxy | And | Gx | 00h 43m | +41° 16' | 4.3 | 156' |
| NGC 752 | Golf Ball Cluster | And | OC | 01h 58m | +37° 47' | 6.6 | 75' |
| NGC 7662 | Blue Snowball Nebula | And | PN | 23h 26m | +42° 32' | 8.6 | 17" |
| Pi And | | And | MS | 00h 38m | +33° 49' | 4.4 | 36" |
| Lam Ari | | Ari | MS | 01h 59m | +23° 41' | 4.8 | 37" |
| Gam Ari | Mesarthim | Ari | MS | 01h 54m | +19° 22' | 4.6 | 8" |
| 30 Ari | | Ari | MS | 02h 37m | +24° 39' | 6.5 | 39" |

| Designation | Name | Con. | Type | R.A. | Dec. | Mag | Size/Sep |
|---|---|---|---|---|---|---|---|
| U Cam | | Cam | Var/CS | 03h 42m | +62° 39' | 7.0-7.5 | N/A |
| ST Cam | | Cam | Var/CS | 04h 51m | +68° 10' | 7.0-8.4 | N/A |
| Kemble 1 | Kemble's Cascade | Cam | Ast | 03h 57m | +63° 04' | 5.0 | 180' |
| NGC 1502 | Jolly Roger Cluster | Cam | OC | 04h 08m | +62° 20' | 4.1 | 8' |
| NGC 457 | Owl Cluster | Cas | OC | 01h 20m | +58° 17' | 5.1 | 20' |
| Iot Cas | | Cas | MS | 02h 29m | +67° 24' | 4.5 | 7" |
| Sig Cas | | Cas | MS | 23h 59m | +55° 45' | 4.9 | 3" |
| Struve 163 | | Cas | MS | 01h 51m | +64° 51' | 6.5 | 35" |
| Struve 3053 | | Cas | MS | 00h 03m | +66° 06' | 5.9 | 15" |
| Eta Cas | Achird | Cas | MS | 00h 50m | +57° 54' | 3.6 | 13" |
| M 103 | | Cas | OC | 01h 33m | +60° 39' | 6.9 | 5' |
| M 52 | | Cas | OC | 23h 25m | +61° 36' | 8.2 | 15' |
| NGC 281 | | Cas | OC | 00h 53m | +56° 38' | 7.4 | 4' |
| NGC 559 | | Cas | OC | 01h 30m | +63° 18' | 7.4 | 6' |
| NGC 659 | Ying Yang Cluster | Cas | OC | 01h 44m | +60° 40' | 7.2 | 5' |
| NGC 663 | | Cas | OC | 01h 46m | +61° 14' | 6.4 | 14' |
| NGC 7789 | Herschel's Spiral Cluster | Cas | OC | 23h 57m | +56° 43' | 7.5 | 25' |
| Omi Cep | | Cep | MS | 23h 19m | +68° 07' | 4.8 | 3" |
| Gam Cet | Kaffajidhma | Cet | MS | 02h 43m | +03° 14' | 3.5 | 3" |
| Omi Cet | Mira | Cet | Var | 02h 19m | -02° 59' | 2.0-10.1 | N/A |
| 32 Eri | | Eri | MS | 03h 54m | -02° 57' | 4.4 | 7" |
| 40 Eri | Keid | Eri | MS | 04h 15m | -07° 39' | 4.4 | 83" |
| NGC 1528 | m & m Double Cluster | Per | OC | 04h 15m | +51° 13' | 6.4 | 16' |
| Bet Per | Algol | Per | Var | 03h 08m | +40° 57' | 2.1-3.4 | N/A |
| Eps Per | | Per | MS | 03h 58m | +40° 01' | 2.9 | 9" |
| NGC 1245 | | Per | OC | 03h 15m | +47° 15' | 7.7 | 10' |
| M 34 | | Per | OC | 02h 42m | +42° 46' | 5.8 | 35' |
| NGC 869/884 | Double Cluster | Per | OC | 02h 21m | +57° 08' | 4.4 | 18' |
| Eta Per | | Per | MS | 02h 51m | +55° 54' | 3.8 | 29" |
| Melotte 20 | Alpha Persei Moving Cluster | Per | OC | 03h 24m | +49° 52' | 2.3 | 300' |
| Alp Psc | Alrisha | Psc | MS | 02h 02m | +02° 46' | 3.8 | 2" |
| Psi1 Psc | | Psc | MS | 01h 06m | +21° 28' | 5.3 | 30" |
| Zet Psc | | Psc | MS | 01h 14m | +07° 35' | 5.2 | 23" |
| 55 Psc | | Psc | MS | 00h 40m | +21° 26' | 5.4 | 6" |
| 65 Psc | | Psc | MS | 00h 50m | +27° 43' | 7.0 | 4" |
| TX Psc | | Psc | Var/CS | 23h 46m | +03° 29' | 4.5-5.3 | N/A |
| M 45 | Pleiades | Tau | OC | 03h 47m | +24° 07' | 1.5 | 120' |
| Melotte 25 | Hyades | Tau | OC | 04h 27m | +15° 52' | 0.8 | 330' |
| NGC 1647 | Pirate Moon Cluster | Tau | OC | 04h 46m | +19° 07' | 6.2 | 40' |
| M 33 | Triangulum Galaxy | Tri | Gx | 01h 34m | +30° 40' | 6.4 | 62' |
| Alp UMi | Polaris | UMi | MS | 02h 51m | +89° 20' | 2.0 | 18" |

# Chart 3

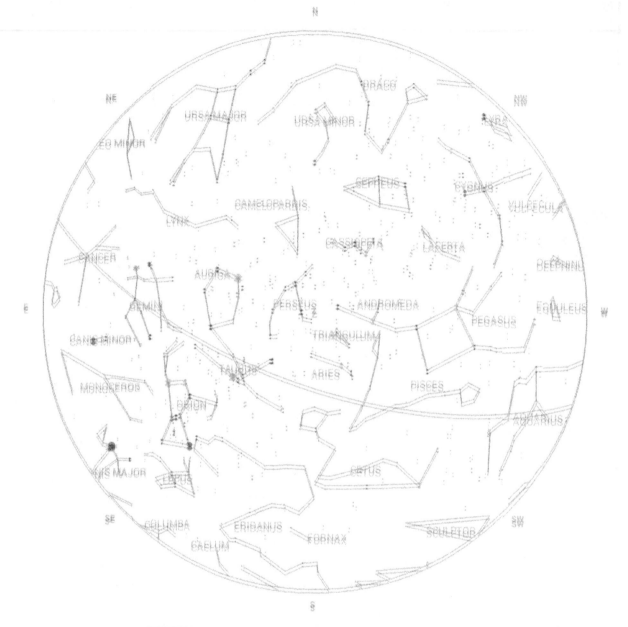

| Designation | Name | Con: | Type | R.A: | Dec: | Mag | Size/Sep |
|---|---|---|---|---|---|---|---|
| Gam And | Almach | And | MS | 02h 04m | +42° 20' | 2.1 | 10" |
| M 31 | Andromeda Galaxy | And | Gx | 00h 43m | +41° 16' | 4.3 | 156' |
| Pi And | | And | MS | 00h 38m | +33° 49' | 4.4 | 36" |
| Lam Ari | | Ari | MS | 01h 59m | +23° 41' | 4.8 | 37" |
| Gam Ari | Mesarthim | Ari | MS | 01h 54m | +19° 22' | 4.6 | 8" |
| 30 Ari | | Ari | MS | 02h 37m | +24° 39' | 6.5 | 39" |
| M 37 | | Aur | OC | 05h 52m | +32° 33' | 6.2 | 14' |
| M 36 | | Aur | OC | 05h 36m | +34° 08' | 6.5 | 10' |

| Designation | Name | Con: | Type | R.A.: | Dec.: | Mag | Size/Sep |
|---|---|---|---|---|---|---|---|
| 14 Aur | | Aur | MS | 05h 15m | +32° 41' | 5.0 | 15" |
| Kemble 1 | Kemble's Cascade | Cam | Ast | 03h 57m | +63° 04' | 5.0 | 180' |
| NGC 1502 | Jolly Roger Cluster | Cam | OC | 04h 08m | +62° 20' | 4.1 | 8' |
| NGC 457 | Owl Cluster | Cas | OC | 01h 20m | +58° 17' | 5.1 | 20' |
| Iot Cas | | Cas | MS | 02h 29m | +67° 24' | 4.5 | 7" |
| Struve 3053 | | Cas | MS | 00h 03m | +66° 06' | 5.9 | 15" |
| Eta Cas | Achird | Cas | MS | 00h 50m | +57° 54' | 3.6 | 13" |
| NGC 663 | | Cas | OC | 01h 46m | +61° 14' | 6.4 | 14' |
| Gam Cet | Kaffaljidhma | Cet | MS | 02h 43m | +03° 14' | 3.5 | 3" |
| Omi Cet | Mira | Cet | Var | 02h 19m | -02° 59' | 2.0-10.1 | N/A |
| 32 Eri | | Eri | MS | 03h 54m | -02° 57' | 4.4 | 7" |
| 40 Eri | Keid | Eri | MS | 04h 15m | -07° 39' | 4.4 | 83" |
| Sig Ori | | Ori | MS | 05h 40m | -02° 36' | 3.8 | 42" |
| Bet Ori | Rigel | Ori | MS | 05h 14m | -08° 12' | 0.2 | 10" |
| Alp Ori | Betelgeuse | Ori | Var | 05h 55m | +07° 24' | 0.4-1.3 | N/A |
| Eta Ori | | Ori | MS | 05h 25m | -02° 24' | 3.3 | 2" |
| Zet Ori | Alnitak | Ori | MS | 05h 41m | -01° 57' | 1.8 | 3" |
| 23 Ori | | Ori | MS | 05h 23m | +03° 33' | 5.0 | 32" |
| Iot Ori | Nair al Saif | Ori | MS | 05h 35m | -05° 55' | 2.8 | 11" |
| Collinder 70 | Epsilon Orionis Cluster | Ori | OC | 05h 36m | -01° 00' | 0.4 | 150' |
| Struve 747 | | Ori | MS | 05h 35m | -05° 55' | 4.8 | 36" |
| Collinder 72 | | Ori | OC | 05h 35m | -05° 55' | 2.5 | 20' |
| Del Ori | Mintaka | Ori | MS | 05h 33m | -00° 18' | 2.1 | 53" |
| Lam Ori | Meissa | Ori | MC | 05h 36m | +09° 56' | 3.4 | 4" |
| M 42 | Orion Nebula | Ori | Neb | 05h 35m | -05° 23' | 4.0 | 40' |
| NGC 1981 | Coal Car Cluster | Ori | OC | 05h 35m | -04° 26' | 4.2 | 28' |
| Collinder 69 | Lambda Orionis Cluster | Ori | OC | 05h 35m | +09° 56' | 2.8 | 70' |
| NGC 1528 | m & m Double Cluster | Per | OC | 04h 15m | +51° 13' | 6.4 | 16' |
| Bet Per | Algol | Per | Var | 03h 08m | +40° 57' | 2.1-3.4 | N/A |
| Eps Per | | Per | MS | 03h 58m | +40° 01' | 2.9 | 9" |
| M 34 | | Per | OC | 02h 42m | +42° 46' | 5.8 | 35' |
| NGC 869/884 | Double Cluster | Per | OC | 02h 21m | +57° 08' | 4.4 | 18' |
| Eta Per | | Per | MS | 02h 51m | +55° 54' | 3.8 | 29" |
| Melotte 20 | Alpha Persei Moving Cluster | Per | OC | 03h 24m | +49° 52' | 2.3 | 300' |
| Alp Psc | Alrisha | Psc | MS | 02h 02m | +02° 46' | 3.8 | 2" |
| Psi1 Psc | | Psc | MS | 01h 06m | +21° 28' | 5.3 | 30" |
| Zet Psc | | Psc | MS | 01h 14m | +07° 35' | 5.2 | 23" |
| 55 Psc | | Psc | MS | 00h 40m | +21° 26' | 5.4 | 6" |
| M 45 | Pleiades | Tau | OC | 03h 47m | +24° 07' | 1.5 | 120' |
| Melotte 25 | Hyades | Tau | OC | 04h 27m | +15° 52' | 0.8 | 330' |
| NGC 1647 | Pirate Moon Cluster | Tau | OC | 04h 46m | +19° 07' | 6.2 | 40' |
| 118 Tau | | Tau | MS | 05h 29m | +25° 09' | 5.5 | 5" |
| M 33 | Triangulum Galaxy | Tri | Gx | 01h 34m | +30° 40' | 6.4 | 62' |
| Alp UMi | Polaris | UMi | MS | 02h 51m | +89° 20' | 2.0 | 18" |

# Chart 4

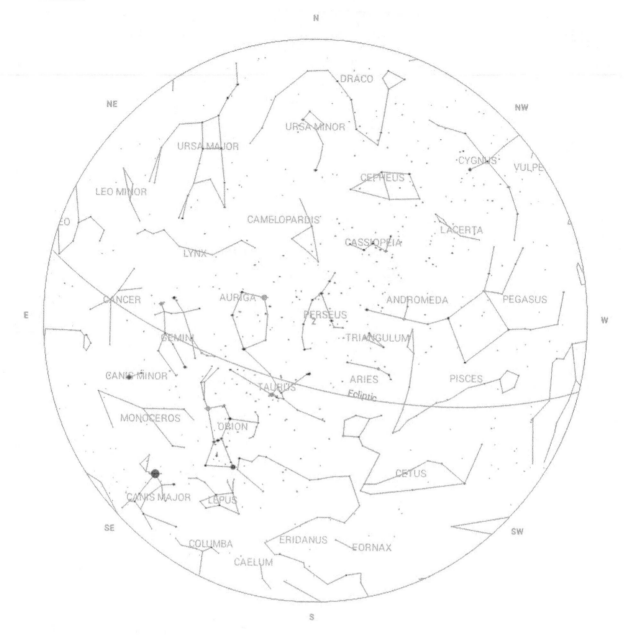

| Designation | Name | Con. | Type | R.A. | Dec. | Mag | Size/Sep |
|---|---|---|---|---|---|---|---|
| Gam And | Almach | And | MS | 02h 04m | +42° 20' | 2.1 | 10" |
| Lam Ari | | Ari | MS | 01h 59m | +23° 41' | 4.8 | 37" |
| Gam Ari | Mesarthim | Ari | MS | 01h 54m | +19° 22' | 4.6 | 8" |
| UU Aur | | Aur | Var/CS | 06h 36m | +38° 27' | 5.3-6.5 | N/A |
| The Aur | | Aur | MS | 06h 00m | +37° 13' | 2.7 | 4" |
| M 37 | | Aur | OC | 05h 52m | +32° 33' | 6.2 | 14' |
| 14 Aur | | Aur | MS | 05h 15m | +32° 41' | 5.0 | 15" |
| Kemble 1 | Kemble's Cascade | Cam | Ast | 03h 57m | +63° 04' | 5.0 | 180' |

| Designation | Name | Con. | Type | R.A. | Dec. | Mag | Size/Sep |
|---|---|---|---|---|---|---|---|
| NGC 1502 | Jolly Roger Cluster | Cam | OC | 04h 08m | +62° 20' | 4.1 | 8' |
| NGC 457 | Owl Cluster | Cas | OC | 01h 20m | +58° 17' | 5.1 | 20' |
| Iot Cas | | Cas | MS | 02h 29m | +67° 24' | 4.5 | 7" |
| NGC 663 | | Cas | OC | 01h 46m | +61° 14' | 6.4 | 14' |
| Gam Cet | Kaffajidhma | Cet | MS | 02h 43m | +03° 14' | 3.5 | 3" |
| Omi Cet | Mira | Cet | Var | 02h 19m | -02° 59' | 2.0-10.1 | N/A |
| 32 Eri | | Eri | MS | 03h 54m | -02° 57' | 4.4 | 7" |
| 40 Eri | Keid | Eri | MS | 04h 15m | -07° 39' | 4.4 | 83" |
| M 35 | | Gem | OC | 06h 09m | +24° 21' | 5.6 | 25' |
| 38 Gem | | Gem | MS | 06h 55m | +13° 11' | 4.7 | 7" |
| Gam Lep | | Lep | MS | 05h 45m | -22° 27' | 3.6 | 98" |
| R Lep | Hind's Crimson Star | Lep | Var/CS | 05h 00m | -14° 48' | 5.5-11.7 | N/A |
| 12 Lyn | | Lyn | MS | 06h 46m | +59° 27' | 4.9 | 9" |
| Sig Ori | | Ori | MS | 05h 40m | -02° 36' | 3.8 | 42" |
| Bet Ori | Rigel | Ori | MS | 05h 14m | -08° 12' | 0.2 | 10" |
| Alp Ori | Betelgeuse | Ori | Var | 05h 55m | +07° 24' | 0.4-1.3 | N/A |
| Eta Ori | | Ori | MS | 05h 25m | -02° 24' | 3.3 | 2" |
| Zet Ori | Alnitak | Ori | MS | 05h 41m | -01° 57' | 1.8 | 3" |
| 23 Ori | | Ori | MS | 05h 23m | +03° 33' | 5.0 | 32" |
| Iot Ori | Nair al Saif | Ori | MS | 05h 35m | -05° 55' | 2.8 | 11" |
| Collinder 70 | Epsilon Orionis Cluster | Ori | OC | 05h 36m | -01° 00' | 0.4 | 150' |
| Struve 747 | | Ori | MS | 05h 35m | -05° 55' | 4.8 | 36" |
| Collinder 72 | | Ori | OC | 05h 35m | -05° 55' | 2.5 | 20' |
| Del Ori | Mintaka | Ori | MS | 05h 33m | -00° 18' | 2.1 | 53" |
| Lam Ori | Meissa | Ori | MC | 05h 36m | +09° 56' | 3.4 | 4" |
| M 42 | Orion Nebula | Ori | Neb | 05h 35m | -05° 23' | 4.0 | 40' |
| NGC 1981 | Coal Car Cluster | Ori | OC | 05h 35m | -04° 26' | 4.2 | 28' |
| Collinder 69 | Lambda Orionis Cluster | Ori | OC | 05h 35m | +09° 56' | 2.8 | 70' |
| NGC 1528 | m & m Double Cluster | Per | OC | 04h 15m | +51° 13' | 6.4 | 16' |
| Bet Per | Algol | Per | Var | 03h 08m | +40° 57' | 2.1-3.4 | N/A |
| Eps Per | | Per | MS | 03h 58m | +40° 01' | 2.9 | 9" |
| M 34 | | Per | OC | 02h 42m | +42° 46' | 5.8 | 35' |
| NGC 869/884 | Double Cluster | Per | OC | 02h 21m | +57° 08' | 4.4 | 18' |
| Eta Per | | Per | MS | 02h 51m | +55° 54' | 3.8 | 29" |
| Melotte 20 | Alpha Persei Moving Cluster | Per | OC | 03h 24m | +49° 52' | 2.3 | 300' |
| Alp Psc | Alrisha | Psc | MS | 02h 02m | +02° 46' | 3.8 | 2" |
| Psi1 Psc | | Psc | MS | 01h 06m | +21° 28' | 5.3 | 30" |
| Zet Psc | | Psc | MS | 01h 14m | +07° 35' | 5.2 | 23" |
| M 45 | Pleiades | Tau | OC | 03h 47m | +24° 07' | 1.5 | 120' |
| Melotte 25 | Hyades | Tau | OC | 04h 27m | +15° 52' | 0.8 | 330' |
| NGC 1647 | Pirate Moon Cluster | Tau | OC | 04h 46m | +19° 07' | 6.2 | 40' |
| 118 Tau | | Tau | MS | 05h 29m | +25° 09' | 5.5 | 5" |
| M 33 | Triangulum Galaxy | Tri | Gx | 01h 34m | +30° 40' | 6.4 | 62' |
| Alp UMi | Polaris | UMi | MS | 02h 51m | +89° 20' | 2.0 | 18" |

# Chart 5

| Designation | Name | Con: | Type | R.A: | Dec: | Mag | Size/Sep |
|---|---|---|---|---|---|---|---|
| Gam And | Almach | And | MS | 02h 04m | +42° 20' | 2.1 | 10" |
| The Aur | | Aur | MS | 06h 00m | +37° 13' | 2.7 | 4" |
| 14 Aur | | Aur | MS | 05h 15m | +32° 41' | 5.0 | 15" |
| U Cam | | Cam | Var/ES | 03h 42m | +62° 39' | 7.0-7.5 | N/A |
| ST Cam | | Cam | Var/ES | 04h 51m | +68° 10' | 7.0-8.4 | N/A |
| Kemble 1 | Kemble's Cascade | Cam | Ast | 03h 57m | +63° 04' | 5.0 | 180' |
| NGC 1502 | Jolly Roger Cluster | Cam | OC | 04h 08m | +62° 20' | 4.1 | 8' |
| Iot Cas | | Cas | MS | 02h 29m | +67° 24' | 4.5 | 7" |

| Designation | Name | Con: | Type | R.A.: | Dec: | Mag | Size/Sep |
|---|---|---|---|---|---|---|---|
| 32 Eri | | Eri | MS | 03h 54m | -02° 57' | 4.4 | 7'' |
| 40 Eri | Keid | Eri | MS | 04h 15m | -07° 39' | 4.4 | 83'' |
| M 35 | | Gem | OC | 06h 09m | +24° 21' | 5.6 | 25' |
| 38 Gem | | Gem | MS | 06h 55m | +13° 11' | 4.7 | 7'' |
| Alp Gem | Castor | Gem | MS | 07h 35m | +31° 53' | 1.6 | 3'' |
| Del Gem | Wasat | Gem | MS | 07h 20m | +21° 59' | 3.5 | 6'' |
| Kap Gem | | Gem | MS | 07h 44m | +24° 24' | 3.6 | 7'' |
| Gam Lep | | Lep | MS | 05h 45m | -22° 27' | 3.6 | 98'' |
| R Lep | Hind's Crimson Star | Lep | Var/CS | 05h 00m | -14° 48' | 5.5-11.7 | N/A |
| 12 Lyn | | Lyn | MS | 06h 46m | +59° 27' | 4.8 | 9'' |
| 19 Lyn | | Lyn | MS | 07h 23m | +55° 17' | 5.8 | 215'' |
| NGC 2353 | | Mon | OC | 07h 15m | -10° 16' | 5.2 | 18' |
| Bet Mon | | Mon | MS | 06h 29m | -07° 02' | 3.8 | 7'' |
| NGC 2237 | Rosette Nebula | Mon | Neb | 06h 32m | +04° 59' | 5.5 | 70' |
| NGC 2244 | | Mon | OC | 06h 32m | +04° 57' | 5.2 | 29' |
| NGC 2264 | Christmas Tree Cluster | Mon | OC | 06h 41m | +09° 54' | 4.1 | 39' |
| Eps Mon | | Mon | MS | 06h 24m | +04° 36' | 4.3 | 13'' |
| Sig Ori | | Ori | MS | 05h 40m | -02° 36' | 3.8 | 42'' |
| Bet Ori | Rigel | Ori | MS | 05h 14m | -08° 12' | 0.2 | 10'' |
| Alp Ori | Betelgeuse | Ori | Var | 05h 55m | +07° 24' | 0.4-1.3 | N/A |
| Eta Ori | | Ori | MS | 05h 25m | -02° 24' | 3.3 | 2'' |
| Zet Ori | Alnitak | Ori | MS | 05h 41m | -01° 57' | 1.8 | 3'' |
| 23 Ori | | Ori | MS | 05h 23m | +03° 33' | 5.0 | 32'' |
| Iot Ori | Nair al Saif | Ori | MS | 05h 35m | -05° 55' | 2.8 | 11'' |
| W Ori | | Ori | Var/CS | 05h 05m | +01° 11' | 6.2-7.0 | N/A |
| Collinder 70 | Epsilon Orionis Cluster | Ori | OC | 05h 36m | -01° 00' | 0.4 | 150' |
| Struve 747 | | Ori | MS | 05h 35m | -05° 55' | 4.8 | 36'' |
| Collinder 72 | | Ori | OC | 05h 35m | -05° 55' | 2.5 | 20' |
| Del Ori | Mintaka | Ori | MS | 05h 33m | -00° 18' | 2.1 | 53'' |
| Lam Ori | Meissa | Ori | MC | 05h 36m | +09° 56' | 3.4 | 4'' |
| M 42 | Orion Nebula | Ori | Neb | 05h 35m | -05° 23' | 4.0 | 40' |
| NGC 1981 | Coal Car Cluster | Ori | OC | 05h 35m | -04° 26' | 4.2 | 28' |
| Collinder 69 | Lambda Orionis Cluster | Ori | OC | 05h 35m | +09° 56' | 2.8 | 70' |
| BL Ori | | Ori | Var/CS | 06h 26m | +14° 43' | 6.3-7.0 | N/A |
| Bet Per | Algol | Per | Var | 03h 08m | +40° 57' | 2.1-3.4 | N/A |
| Eps Per | | Per | MS | 03h 58m | +40° 01' | 2.9 | 9'' |
| M 34 | | Per | OC | 02h 42m | +42° 46' | 5.8 | 35' |
| NGC 869/884 | Double Cluster | Per | OC | 02h 21m | +57° 08' | 4.4 | 18' |
| Eta Per | | Per | MS | 02h 51m | +55° 54' | 3.8 | 29'' |
| Melotte 20 | Alpha Persei Moving Cluster | Per | OC | 03h 24m | +49° 52' | 2.3 | 300' |
| M 45 | Pleiades | Tau | OC | 03h 47m | +24° 07' | 1.5 | 120' |
| Melotte 25 | Hyades | Tau | OC | 04h 27m | +15° 52' | 0.8 | 330' |
| 118 Tau | | Tau | MS | 05h 29m | +25° 09' | 5.5 | 5'' |
| Alp UMi | Polaris | UMi | MS | 02h 51m | +89° 20' | 2.0 | 18'' |

# Chart 6

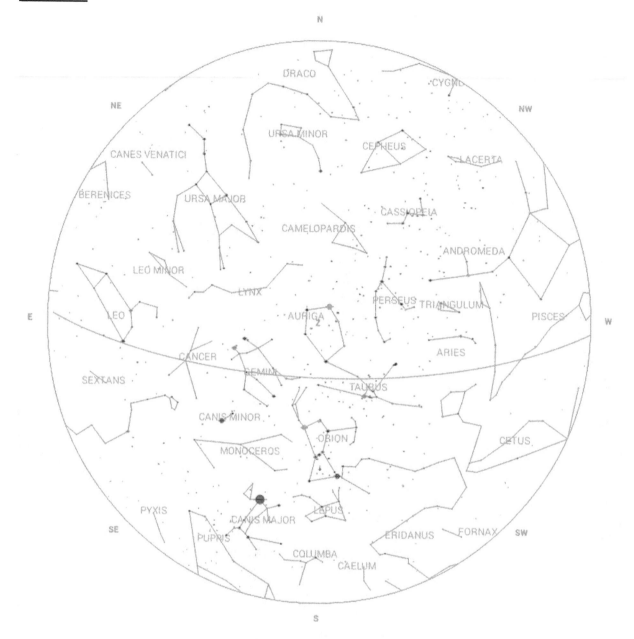

| Designation | Name | Con. | Type | R.A. | Dec. | Mag | Size/Sep |
|---|---|---|---|---|---|---|---|
| UU Aur | | Aur | Var/CS | 06h 36m | +38° 27' | 5.3-6.5 | N/A |
| The Aur | | Aur | MS | 06h 00m | +37° 13' | 2.7 | 4" |
| 14 Aur | | Aur | MS | 05h 15m | +32° 41' | 5.0 | 15" |
| Kemble 1 | Kemble's Cascade | Cam | Ast | 03h 57m | +63° 04' | 5.0 | 180' |
| NGC 1502 | Jolly Roger Cluster | Cam | OC | 04h 08m | +62° 20' | 4.1 | 8' |
| M 41 | | CMa | OC | 05h 46m | -20° 45' | 5.0 | 39' |
| Eps CMa | Adhara | CMa | MS | 06h 59m | -28° 58' | 1.5 | 8" |
| NGC 2362 | Tau Canis Majoris Cluster | CMa | OC | 07h 19m | -24° 57' | 3.8 | 5' |

| Designation | Name | Con. | Type | R.A. | Dec. | Mag | Size/Sep |
|---|---|---|---|---|---|---|---|
| Iot Cnc | | Cnc | MS | 08h 47m | +28° 46' | 4.0 | 30" |
| M 44 | Praesepe | Cnc | OC | 08h 40m | +19° 40' | 3.9 | 70' |
| Zet Cnc | Tegmen | Cnc | MS | 08h 12m | +17° 39' | 4.7 | 6" |
| Phi2 Cnc | | Cnc | MS | 08h 27m | +26° 56' | 5.6 | 5" |
| 57 Cnc | | Cnc | MS | 08h 54m | +30° 35' | 5.4 | 56" |
| X Cnc | | Cnc | Var/CS | 08h 55m | +17° 14' | 5.6-7.5 | N/A |
| 32 Eri | | Eri | MS | 03h 54m | -02° 57' | 4.4 | 7" |
| 40 Eri | Keid | Eri | MS | 04h 15m | -07° 39' | 4.4 | 83" |
| M 35 | | Gem | OC | 06h 09m | +24° 21' | 5.6 | 25' |
| 38 Gem | | Gem | MS | 06h 55m | +13° 11' | 4.7 | 7" |
| Alp Gem | Castor | Gem | MS | 07h 35m | +31° 53' | 1.6 | 3" |
| Del Gem | Wasat | Gem | MS | 07h 20m | +21° 59' | 3.5 | 6" |
| Kap Gem | | Gem | MS | 07h 44m | +24° 24' | 3.6 | 7" |
| Gam Lep | | Lep | MS | 05h 45m | -22° 27' | 3.6 | 98" |
| R Lep | Hind's Crimson Star | Lep | Var/CS | 05h 00m | -14° 48' | 5.5-11.7 | N/A |
| 12 Lyn | | Lyn | MS | 06h 46m | +59° 27' | 4.9 | 9" |
| 19 Lyn | | Lyn | MS | 07h 23m | +55° 17' | 5.8 | 215" |
| NGC 2353 | | Mon | OC | 07h 15m | -10° 16' | 5.2 | 18' |
| Bet Mon | | Mon | MS | 06h 29m | -07° 02' | 3.8 | 7" |
| NGC 2237 | Rosette Nebula | Mon | Neb | 06h 32m | +04° 59' | 5.5 | 70' |
| NGC 2244 | | Mon | OC | 06h 32m | +04° 57' | 5.2 | 29' |
| NGC 2264 | Christmas Tree Cluster | Mon | OC | 06h 41m | +09° 54' | 4.1 | 39' |
| Eps Mon | | Mon | MS | 06h 24m | +04° 36' | 4.3 | 13" |
| Sig Ori | | Ori | MS | 05h 40m | -02° 36' | 3.8 | 42" |
| Bet Ori | Rigel | Ori | MS | 05h 14m | -08° 12' | 0.2 | 10" |
| Alp Ori | Betelgeuse | Ori | Var | 05h 55m | +07° 24' | 0.4-1.3 | N/A |
| Eta Ori | | Ori | MS | 05h 25m | -02° 24' | 3.3 | 2" |
| Zet Ori | Alnitak | Ori | MS | 05h 41m | -01° 57' | 1.8 | 3" |
| 23 Ori | | Ori | MS | 05h 23m | +03° 33' | 5.0 | 32" |
| Iot Ori | Nair al Saif | Ori | MS | 05h 35m | -05° 55' | 2.8 | 11" |
| Collinder 70 | Epsilon Orionis Cluster | Ori | OC | 05h 36m | -01° 00' | 0.4 | 150' |
| Struve 747 | | Ori | MS | 05h 35m | -05° 55' | 4.8 | 36" |
| Collinder 72 | | Ori | OC | 05h 35m | -05° 55' | 2.5 | 20' |
| Del Ori | Mintaka | Ori | MS | 05h 33m | -00° 18' | 2.1 | 53" |
| Lam Ori | Meissa | Ori | MC | 05h 36m | +09° 56' | 3.4 | 4" |
| M 42 | Orion Nebula | Ori | Neb | 05h 35m | -05° 23' | 4.0 | 40' |
| NGC 1981 | Coal Car Cluster | Ori | OC | 05h 35m | -04° 26' | 4.2 | 28' |
| Collinder 69 | Lambda Orionis Cluster | Ori | OC | 05h 35m | +09° 56' | 2.8 | 70' |
| Bet Per | Algol | Per | Var | 03h 08m | +40° 57' | 2.1-3.4 | N/A |
| Eps Per | | Per | MS | 03h 58m | +40° 01' | 2.9 | 9" |
| Melotte 20 | Alpha Persei Moving Cluster | Per | OC | 03h 24m | +49° 52' | 2.3 | 300' |
| M 45 | Pleiades | Tau | OC | 03h 47m | +24° 07' | 1.5 | 120' |
| Melotte 25 | Hyades | Tau | OC | 04h 27m | +15° 52' | 0.8 | 330' |
| 118 Tau | | Tau | MS | 05h 29m | +25° 09' | 5.5 | 5" |

# Chart 7

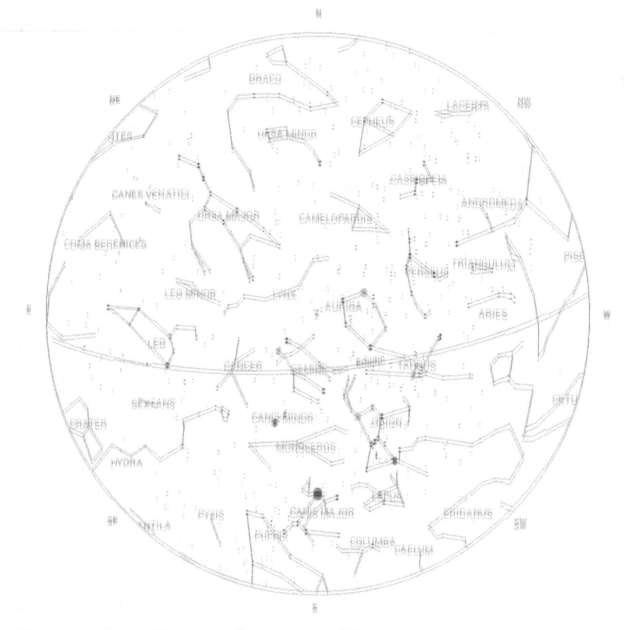

| Designation | Name | Con. | Type | R.A. | Dec. | Mag | Size/Sep |
|---|---|---|---|---|---|---|---|
| 14 Aur | | Aur | MS | 05h 15m | +32° 41' | 5.0 | 15" |
| The Aur | | Aur | MS | 06h 00m | +37° 13' | 2.7 | 4" |
| M 37 | | Aur | OC | 05h 52m | +32° 33' | 6.2 | 14' |
| UU Aur | | Aur | Var/ES | 06h 36m | +38° 27' | 5.3-6.5 | N/A |
| NGC 1502 | Jolly Roger Cluster | Cam | OC | 04h 08m | +62° 20' | 4.1 | 8' |
| Eps CMa | Adhara | CMa | MS | 06h 59m | -28° 58' | 1.5 | 8" |
| M 41 | | CMa | OC | 06h 46m | -20° 45' | 5.0 | 39' |
| NGC 2362 | Tau Canis Majoris Cluster | CMa | OC | 07h 19m | -24° 57' | 3.8 | 5' |

| Designation | Name | Con: | Type | R.A.: | Dec: | Mag | Size/Sep |
|---|---|---|---|---|---|---|---|
| Zet Cnc | Tegmen | Cnc | MS | 08h 12m | +17° 39' | 4.7 | 6" |
| Phi2 Cnc | | Cnc | MS | 08h 27m | +26° 56' | 5.6 | 5" |
| Iot Cnc | | Cnc | MS | 08h 47m | +28° 46' | 4.0 | 30" |
| 57 Cnc | | Cnc | MS | 08h 54m | +30° 35' | 5.4 | 56" |
| M 44 | Praesepe | Cnc | OC | 08h 40m | +19° 40' | 3.9 | 70' |
| X Cnc | | Cnc | Var/ES | 08h 55m | +17° 14' | 5.6-7.5 | N/A |
| 38 Gem | | Gem | MS | 06h 55m | +13° 11' | 4.7 | 7" |
| Del Gem | Wasat | Gem | MS | 07h 20m | +21° 59' | 3.5 | 6" |
| Alp Gem | Castor | Gem | MS | 07h 35m | +31° 53' | 1.6 | 3" |
| Kap Gem | | Gem | MS | 07h 44m | +24° 24' | 3.6 | 7" |
| M 35 | | Gem | OC | 06h 09m | +24° 21' | 5.6 | 25' |
| R Leo | Peltier's Variable Star | Leo | Var | 09h 48m | +11° 26' | 4.4-10.5 | N/A |
| Gam Lep | | Lep | MS | 05h 45m | -22° 27' | 3.6 | 98" |
| R Lep | Hind's Crimson Star | Lep | Var/ES | 05h 00m | -14° 48' | 5.5-11.7 | N/A |
| 12 Lyn | | Lyn | MS | 06h 46m | +59° 27' | 4.9 | 9" |
| 19 Lyn | | Lyn | MS | 07h 23m | +55° 17' | 5.6 | 215" |
| 38 Lyn | | Lyn | MS | 09h 19m | +36° 48' | 3.8 | 3" |
| Eps Mon | | Mon | MS | 06h 24m | +04° 36' | 4.3 | 13" |
| Bet Mon | | Mon | MS | 06h 29m | -07° 02' | 3.8 | 7" |
| NGC 2237 | Rosette Nebula | Mon | Neb | 06h 32m | +04° 59' | 5.5 | 70' |
| NGC 2244 | | Mon | OC | 06h 32m | +04° 57' | 5.2 | 29' |
| NGC 2264 | Christmas Tree Cluster | Mon | OC | 06h 41m | +09° 54' | 4.1 | 39' |
| NGC 2301 | Hagrid's Dragon | Mon | OC | 06h 52m | +00° 28' | 6.3 | 14' |
| NGC 2353 | | Mon | OC | 07h 15m | -10° 16' | 5.2 | 18' |
| Lam Ori | Meissa | Ori | MS | 05h 36m | +09° 56' | 3.4 | 4" |
| Bet Ori | Rigel | Ori | MS | 05h 14m | -08° 12' | 0.2 | 10" |
| 23 Ori | | Ori | MS | 05h 23m | +03° 33' | 5.0 | 32" |
| Eta Ori | | Ori | MS | 05h 25m | -02° 24' | 3.3 | 2" |
| Del Ori | Mintaka | Ori | MS | 05h 33m | -00° 18' | 2.1 | 53" |
| Iot Ori | Nair al Saif | Ori | MS | 05h 35m | -05° 55' | 2.8 | 11" |
| Struve 747 | | Ori | MS | 05h 35m | -05° 55' | 4.8 | 36" |
| Sig Ori | | Ori | MS | 05h 40m | -02° 36' | 3.8 | 42" |
| Zet Ori | Alnitak | Ori | MS | 05h 41m | -01° 57' | 1.8 | 3" |
| M 42 | Orion Nebula | Ori | Neb | 05h 35m | -05° 23' | 4.0 | 40' |
| Collinder 72 | | Ori | OC | 05h 35m | -05° 55' | 2.5 | 30' |
| NGC 1981 | Coal Car Cluster | Ori | OC | 05h 35m | -04° 26' | 4.2 | 28' |
| Collinder 69 | Lambda Orionis Cluster | Ori | OC | 05h 35m | +09° 56' | 2.8 | 70' |
| Collinder 70 | Epsilon Orionis Cluster | Ori | OC | 05h 36m | -01° 00' | 0.4 | 150' |
| Alp Ori | Betelgeuse | Ori | Var | 05h 55m | +07° 24' | 0.4-1.3 | N/A |
| M 47 | | Pup | OC | 07h 37m | -14° 29' | 4.3 | 25' |
| 118 Tau | | Tau | MS | 05h 29m | +25° 09' | 5.5 | 5" |
| Melotte 25 | Hyades | Tau | OC | 04h 27m | +15° 52' | 0.8 | 330' |
| NGC 1647 | Pirate Moon Cluster | Tau | OC | 04h 46m | +19° 07' | 6.2 | 40' |

# Chart 8

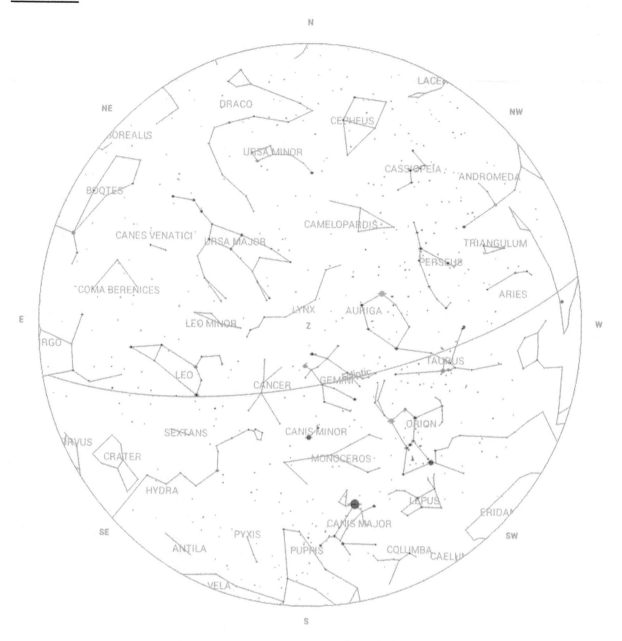

| Designation | Name | Con. | Type | R.A. | Dec. | Mag | Size/Sep |
|---|---|---|---|---|---|---|---|
| 14 Aur | | Aur | MS | 05h 15m | +32° 41' | 5.0 | 15" |
| The Aur | | Aur | MS | 06h 00m | +37° 13' | 2.7 | 4" |
| M 37 | | Aur | OC | 05h 52m | +32° 33' | 6.2 | 14' |
| UU Aur | | Aur | Var/CS | 06h 36m | +38° 27' | 5.3-6.5 | N/A |
| Eps CMa | Adhara | CMa | MS | 06h 59m | -28° 58' | 1.5 | 8" |
| M 41 | | CMa | OC | 05h 46m | -20° 45' | 5.0 | 39' |
| NGC 2362 | Tau Canis Majoris Cluster | CMa | OC | 07h 19m | -24° 57' | 3.8 | 5' |
| Zet Cnc | Tegmen | Cnc | MS | 08h 12m | +17° 39' | 4.7 | 6" |

| Designation | Name | Con. | Type | R.A. | Dec. | Mag | Size/Sep |
|---|---|---|---|---|---|---|---|
| Phi2 Cnc | | Cnc | MS | 08h 27m | +26° 56' | 5.6 | 5" |
| Iot Cnc | | Cnc | MS | 08h 47m | +28° 46' | 4.0 | 30" |
| 57 Cnc | | Cnc | MS | 08h 54m | +30° 35' | 5.4 | 56" |
| M 44 | Praesepe | Cnc | OC | 08h 40m | +19° 40' | 3.9 | 70' |
| X Cnc | | Cnc | Var/CS | 08h 55m | +17° 14' | 5.6-7.5 | N/A |
| 38 Gem | | Gem | MS | 06h 55m | +13° 11' | 4.7 | 7" |
| Del Gem | Wasat | Gem | MS | 07h 20m | +21° 59' | 3.5 | 6" |
| Alp Gem | Castor | Gem | MS | 07h 35m | +31° 53' | 1.6 | 3" |
| Kap Gem | | Gem | MS | 07h 44m | +24° 24' | 3.6 | 7" |
| M 35 | | Gem | OC | 06h 09m | +24° 21' | 5.6 | 25' |
| Eps Hya | | Hya | MS | 08h 47m | +06° 25' | 3.4 | 3" |
| M 48 | | Hya | OC | 08h 14m | -05° 45' | 5.5 | 30' |
| U Hya | | Hya | Var/CS | 10h 38m | -13° 23' | 4.8-6.5 | N/A |
| R Leo | Peltier's Variable Star | Leo | Var | 09h 48m | +11° 26' | 4.4-10.5 | N/A |
| Gam Leo | Algieba | Leo | MS | 10h 20m | +19° 50' | 2.0 | 5" |
| 54 Leo | | Leo | MS | 10h 56m | +24° 45' | 4.3 | 6" |
| 12 Lyn | | Lyn | MS | 06h 46m | +59° 27' | 4.9 | 9" |
| 19 Lyn | | Lyn | MS | 07h 23m | +55° 17' | 5.8 | 215" |
| 38 Lyn | | Lyn | MS | 09h 19m | +36° 48' | 3.8 | 3" |
| Eps Mon | | Mon | MS | 06h 24m | +04° 36' | 4.3 | 13" |
| Bet Mon | | Mon | MS | 06h 29m | -07° 02' | 3.8 | 7" |
| NGC 2237 | Rosette Nebula | Mon | Neb | 06h 32m | +04° 59' | 5.5 | 70' |
| NGC 2244 | | Mon | OC | 06h 32m | +04° 57' | 5.2 | 29' |
| NGC 2264 | Christmas Tree Cluster | Mon | OC | 06h 41m | +09° 54' | 4.1 | 39' |
| NGC 2301 | Hagrid's Dragon | Mon | OC | 06h 52m | +00° 28' | 6.3 | 14' |
| NGC 2353 | | Mon | OC | 07h 15m | -10° 16' | 5.2 | 18' |
| Lam Ori | Meissa | Ori | MC | 05h 36m | +09° 56' | 3.4 | 4" |
| Bet Ori | Rigel | Ori | MS | 05h 14m | -08° 12' | 0.2 | 10" |
| 23 Ori | | Ori | MS | 05h 23m | +03° 33' | 5 | 32" |
| Eta Ori | | Ori | MS | 05h 25m | -02° 24' | 3.3 | 2" |
| Del Ori | Mintaka | Ori | MS | 05h 33m | -00° 18' | 2.1 | 53" |
| Iot Ori | Nair al Saif | Ori | MS | 05h 35m | -05° 55' | 2.8 | 11" |
| Struve 747 | | Ori | MS | 05h 35m | -05° 55' | 4.8 | 36" |
| Sig Ori | | Ori | MS | 05h 40m | -02° 36' | 3.8 | 42" |
| Zet Ori | Alnitak | Ori | MS | 05h 41m | -01° 57' | 1.8 | 3" |
| M 42 | Orion Nebula | Ori | Neb | 05h 35m | -05° 23' | 4.0 | 40' |
| Collinder 72 | | Ori | OC | 05h 35m | -05° 55' | 2.5 | 20' |
| NGC 1981 | Coal Car Cluster | Ori | OC | 05h 35m | -04° 26' | 4.2 | 28' |
| Collinder 69 | Lambda Orionis Cluster | Ori | OC | 05h 35m | +09° 56' | 2.8 | 70' |
| Collinder 70 | Epsilon Orionis Cluster | Ori | OC | 05h 36m | -01° 00' | 0.4 | 150' |
| Alp Ori | Betelgeuse | Ori | Var | 05h 55m | +07° 24' | 0.4-1.3 | N/A |
| M 47 | | Pup | OC | 07h 37m | -14° 29' | 4.3 | 25' |
| 118 Tau | | Tau | MS | 05h 29m | +25° 09' | 5.5 | 5" |

Chart 9

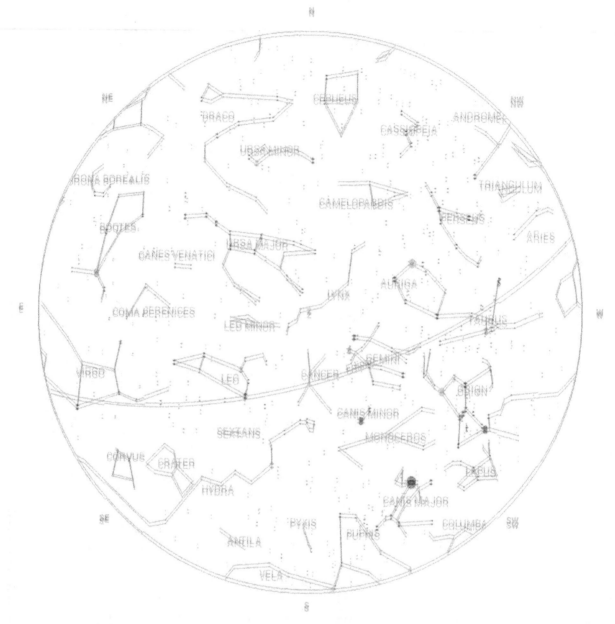

| Designation | Name | Con: | Type | R.A: | Dec: | Mag | Size/Sep |
|---|---|---|---|---|---|---|---|
| The Aur | | Aur | MS | 06h 00m | +37° 13' | 2.7 | 4" |
| UU Aur | | Aur | Var/CS | 06h 36m | +38° 27' | 5.3-6.5 | N/A |
| NGC 2403 | | Cam | Gx | 07h 37m | +65° 36' | 8.8 | 20' |
| Zet Cnc | Tegmen | Cnc | MS | 08h 12m | +17° 39' | 4.7 | 6" |
| Phi2 Cnc | | Cnc | MS | 08h 27m | +26° 56' | 5.6 | 5" |
| Iot Cnc | | Cnc | MS | 08h 47m | +28° 46' | 4.0 | 30" |
| 57 Cnc | | Cnc | MS | 08h 54m | +38° 35' | 5.4 | 56" |
| M 44 | Praesepe | Cnc | OC | 08h 40m | +19° 40' | 3.9 | 70' |

| Designation | Name | Con: | Type | R.A. | Dec. | Mag | Size/Sep |
|---|---|---|---|---|---|---|---|
| M 67 | | Cnc | OC | 08h 51m | +11° 48' | 7.4 | 25' |
| X Cnc | | Cnc | Var/CS | 08h 55m | +17° 14' | 5.6-7.5 | N/A |
| 38 Gem | | Gem | MS | 06h 55m | +13° 11' | 4.7 | 7" |
| Del Gem | Wasat | Gem | MS | 07h 20m | +21° 59' | 3.5 | 6" |
| Alp Gem | Castor | Gem | MS | 07h 35m | +31° 53' | 1.6 | 3" |
| Kap Gem | | Gem | MS | 07h 44m | +24° 24' | 3.6 | 7" |
| M 35 | | Gem | OC | 06h 09m | +24° 21' | 5.6 | 25' |
| NGC 2392 | Eskimo Nebula | Gem | PN | 07h 29m | +20° 55' | 8.6 | 47" |
| NGC 3242 | Ghost of Jupiter | Hya | PN | 10h 25m | -18° 39' | 8.6 | 40" |
| Eps Hya | | Hya | MS | 08h 47m | +06° 25' | 3.4 | 3" |
| M 48 | | Hya | OC | 08h 14m | -05° 45' | 5.5 | 30' |
| U Hya | | Hya | Var/CS | 10h 38m | -13° 23' | 4.8-6.5 | N/A |
| R Leo | Peltier's Variable Star | Leo | Var | 09h 48m | +11° 26' | 4.4-10.5 | N/A |
| Gam Leo | Algieba | Leo | MS | 10h 20m | +19° 50' | 2.0 | 5" |
| 54 Leo | | Leo | MS | 10h 56m | +24° 45' | 4.3 | 6" |
| M 66 | | Leo | Gx | 11h 20m | +13° 00' | 8.7 | 9' |
| 12 Lyn | | Lyn | MS | 06h 46m | +59° 27' | 4.9 | 9" |
| 19 Lyn | | Lyn | MS | 07h 23m | +55° 17' | 5.8 | 215" |
| 38 Lyn | | Lyn | MS | 09h 19m | +36° 48' | 3.8 | 3" |
| Eps Mon | | Mon | MS | 06h 24m | +04° 36' | 4.3 | 13" |
| NGC 3521 | | Leo | Gx | 11h 06m | -00° 02' | 9.0 | 10' |
| Bet Mon | | Mon | MS | 06h 29m | -07° 02' | 3.8 | 7" |
| Iot Leo | | Leo | MS | 11h 24m | +10° 32' | 3.9 | 2" |
| NGC 2237 | Rosette Nebula | Mon | Neb | 06h 32m | +04° 59' | 5.5 | 70' |
| NGC 2244 | | Mon | OC | 06h 32m | +04° 57' | 5.2 | 29' |
| NGC 2264 | Christmas Tree Cluster | Mon | OC | 06h 41m | +09° 54' | 4.1 | 39' |
| NGC 2301 | Hagrid's Dragon | Mon | OC | 06h 52m | +00° 28' | 6.3 | 14' |
| M 50 | | Mon | OC | 07h 03m | -08° 23' | 7.2 | 14' |
| NGC 2353 | | Mon | OC | 07h 15m | -10° 16' | 5.2 | 18' |
| NGC 2506 | | Mon | OC | 08h 00m | -10° 46' | 8.9 | 12' |
| NGC 2175 | | Ori | Neb | 06h 10m | +20° 29' | 6.8 | 22' |
| NGC 2169 | 37 Cluster | Ori | OC | 06h 09m | +13° 58' | 7.0 | 5' |
| NGC 2467 | | Pup | Neb | 07h 52m | -26° 26' | 7.1 | 14' |
| M 47 | | Pup | OC | 07h 37m | -14° 29' | 4.3 | 25' |
| M 46 | | Pup | OC | 07h 42m | -14° 48' | 6.6 | 20' |
| M 93 | | Pup | OC | 07h 45m | -23° 51' | 6.5 | 10' |
| NGC 2539 | Dish Cluster | Pup | OC | 08h 11m | -12° 49' | 8.0 | 9' |
| M 97 | Owl Nebula | UMa | PN | 11h 15m | +55° 01' | 9.7 | 3' |
| Xi UMa | Alula Australis | UMa | MS | 11h 18m | +31° 32' | 4.4 | 2" |
| VY UMa | | UMa | Var/CS | 10h 45m | +67° 25' | 5.9-6.5 | N/A |
| M 81 | Bode's Galaxy | UMa | Gx | 09h 56m | +69° 04' | 7.8 | 22' |
| M 82 | Cigar Galaxy | UMa | Gx | 09h 56m | +69° 41' | 9.0 | 9' |

# Chart 10

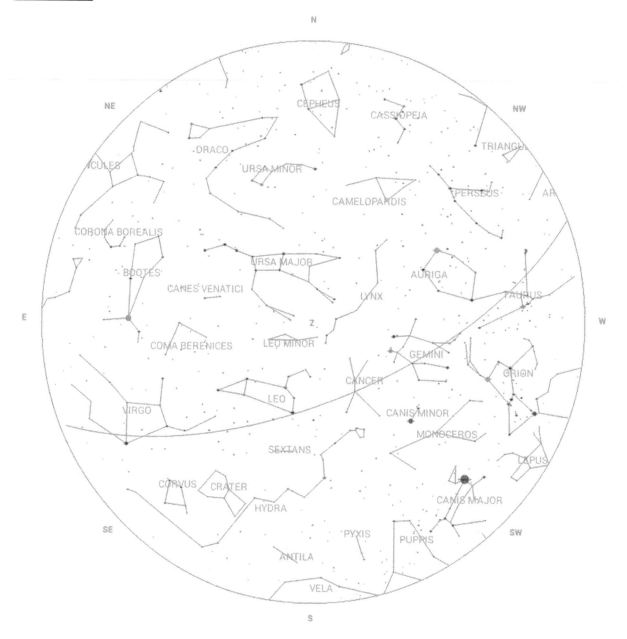

| Designation | Name | Con. | Type | R.A. | Dec. | Mag | Size/Sep |
|---|---|---|---|---|---|---|---|
| NGC 2403 | | Cam | Gx | 07h 37m | +65° 36' | 8.8 | 20' |
| Zet Cnc | Tegmen | Cnc | MS | 08h 12m | +17° 39' | 4.7 | 6" |
| Phi2 Cnc | | Cnc | MS | 08h 27m | +26° 56' | 5.6 | 5" |
| Iot Cnc | | Cnc | MS | 08h 47m | +28° 46' | 4.0 | 30" |
| 57 Cnc | | Cnc | MS | 08h 54m | +30° 35' | 5.4 | 56" |
| M 44 | Praesepe | Cnc | OC | 08h 40m | +19° 40' | 3.9 | 70' |
| M 67 | | Cnc | OC | 08h 51m | +11° 48' | 7.4 | 25' |
| X Cnc | | Cnc | Var/CS | 08h 55m | +17° 14' | 5.6-7.5 | N/A |
| M 64 | Black Eye Galaxy | Com | Gx | 12h 57m | +21° 41' | 9.3 | 10' |

| Designation | Name | Con. | Type | R.A. | Dec. | Mag | Size/Sep |
|---|---|---|---|---|---|---|---|
| Melotte 111 | Coma Star Cluster | Com | OC | 12h 25m | +26° 06' | 2.9 | 120' |
| NGC 4725 | | Com | Gx | 12h 50m | +25° 30' | 9.9 | 10' |
| 24 Com | | Com | MS | 12h 35m | +18° 23' | 5.0 | 20" |
| M 106 | | CVn | Gx | 12h 19m | +47° 18' | 9.1 | 17' |
| Alp CVn | Cor Caroli | CVn | MS | 12h 56m | +38° 19' | 2.9 | 19" |
| M 94 | | CVn | Gx | 12h 51m | +41° 07' | 8.7 | 10' |
| 2 CVn | | CVn | MS | 12h 16m | +40° 40' | 5.7 | 11" |
| NGC 4656 | Hook Galaxy | CVn | Gx | 12h 44m | +32° 10' | 9.7 | 9' |
| NGC 4449 | | CVn | Gx | 12h 28m | +44° 06' | 9.5 | 5' |
| Y CVn | La Superba | CVn | CS | 12h 45m | +45° 26' | 5.2-5.5 | N/A |
| NGC 4490 | Cocoon Galaxy | CVn | Gx | 12h 31m | +41° 39' | 9.8 | 6' |
| NGC 4631 | Whale Galaxy | CVn | Gx | 12h 42m | +32° 33' | 9.5 | 13' |
| RY Dra | | Dra | CS | 12h 56m | +66° 00' | 6.0-8.0 | N/A |
| Del Gem | Wasat | Gem | MS | 07h 20m | +21° 59' | 3.5 | 6" |
| Alp Gem | Castor | Gem | MS | 07h 35m | +31° 53' | 1.6 | 3" |
| Kap Gem | | Gem | MS | 07h 44m | +24° 24' | 3.6 | 7" |
| NGC 2392 | Eskimo Nebula | Gem | PN | 07h 29m | +20° 55' | 8.6 | 47" |
| M 68 | | Hya | GC | 12h 39m | -26° 45' | 7.3 | 11' |
| NGC 3242 | Ghost of Jupiter | Hya | PN | 10h 25m | -18° 39' | 8.6 | 40" |
| Eps Hya | | Hya | MS | 08h 47m | +06° 25' | 3.4 | 3" |
| M 48 | | Hya | OC | 08h 14m | -05° 45' | 5.5 | 30' |
| U Hya | | Hya | Var/CS | 10h 38m | -13° 23' | 4.8-6.5 | N/A |
| R Leo | Peltier's Variable Star | Leo | Var | 09h 48m | +11° 26' | 4.4-10.5 | N/A |
| Gam Leo | Algieba | Leo | MS | 10h 20m | +19° 50' | 2.0 | 5" |
| 54 Leo | | Leo | MS | 10h 56m | +24° 45' | 4.3 | 6" |
| M 66 | | Leo | Gx | 11h 20m | +13° 00' | 9.7 | 9' |
| NGC 3521 | | Leo | Gx | 11h 06m | -00° 02' | 9.9 | 10' |
| Iot Leo | | Leo | MS | 11h 24m | +10° 32' | 3.9 | 2" |
| 19 Lyn | | Lyn | MS | 07h 23m | +55° 17' | 5.8 | 215" |
| 38 Lyn | | Lyn | MS | 09h 19m | +36° 48' | 3.8 | 3" |
| M 40 | Winnecke 4 | UMa | MS | 12h 22m | +58° 05' | 9.6 | |
| M 97 | Owl Nebula | UMa | PN | 11h 15m | +55° 01' | 9.7 | 3' |
| Xi UMa | Alula Australis | UMa | MS | 11h 18m | +31° 32' | 4.4 | 2" |
| VY UMa | | UMa | Var/CS | 10h 45m | +67° 25' | 5.9-6.5 | N/A |
| M 81 | Bode's Galaxy | UMa | Gx | 09h 56m | +69° 04' | 7.8 | 22' |
| M 82 | Cigar Galaxy | UMa | Gx | 09h 56m | +69° 41' | 9.0 | 9' |
| M 87 | | Vir | Gx | 12h 31m | +12° 23' | 9.6 | 8' |
| M 104 | Sombrero Galaxy | Vir | Gx | 12h 40m | -11° 37' | 9.1 | 9' |
| M 49 | | Vir | Gx | 12h 30m | +08° 00' | 9.3 | 9' |
| M 60 | | Vir | Gx | 12h 44m | +11° 33' | 9.8 | 7' |
| M 86 | | Vir | Gx | 12h 26m | +12° 57' | 9.8 | 10' |
| Gam Vir | Porrima | Vir | MS | 12h 42m | -01° 27' | 2.7 | 2" |
| SS Vir | | Vir | Var/CS | 12h 25m | +00° 48' | 6.0-9.6 | 6.0-9.6 |

# Chart 11

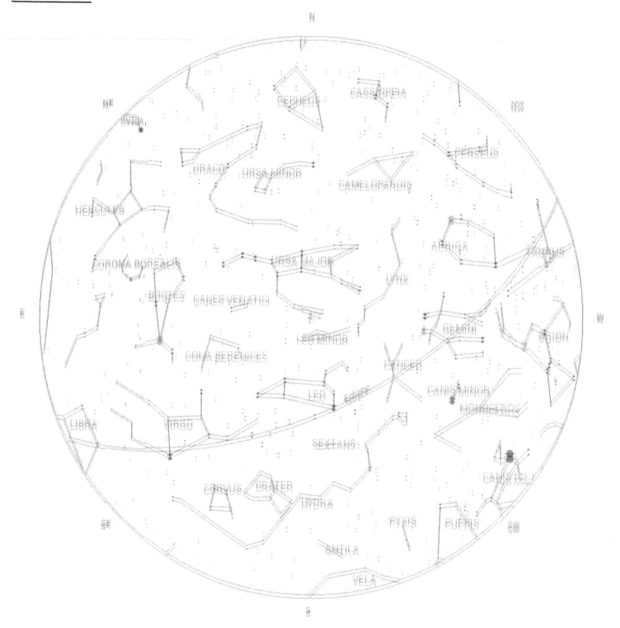

| Designation | Name | Con: | Type | R.A: | Dec: | Mag | Size/Sep |
|---|---|---|---|---|---|---|---|
| Zet Cnc | Tegmen | Cnc | MS | 08h 12m | +17° 39' | 4.7 | 6" |
| Phi2 Cnc | | Cnc | MS | 08h 27m | +26° 56' | 5.6 | 5" |
| Iot Cnc | | Cnc | MS | 08h 47m | +28° 46' | 4.0 | 30" |
| 57 Cnc | | Cnc | MS | 08h 54m | +30° 35' | 5.4 | 56" |
| M 44 | Praesepe | Cnc | OC | 08h 40m | +19° 40' | 3.8 | 70' |
| M 67 | | Cnc | OC | 08h 51m | +11° 48' | 7.4 | 25' |
| X Cnc | | Cnc | Var/CS | 08h 55m | +17° 14' | 5.6-7.5 | N/A |
| M 53 | | Com | GC | 13h 13m | +18° 10' | 7.7 | 13' |
| M 64 | Black Eye Galaxy | Com | Gx | 12h 57m | +21° 41' | 9.3 | 10' |

| Designation | Name | Con: | Type | R.A: | Dec: | Mag | Size/Sep |
|---|---|---|---|---|---|---|---|
| Melotte 111 | Coma Star Cluster | Com | OC | 12h 25m | +26° 06' | 2.9 | 120' |
| NGC 4725 | | Com | Gx | 12h 50m | +25° 30' | 9.9 | 10' |
| 24 Com | | Com | MS | 12h 35m | +18° 23' | 5.0 | 20" |
| M 51 | Whirlpool Galaxy | CVn | Gx | 13h 30m | +47° 12' | 8.7 | 10' |
| M 63 | Sunflower Galaxy | CVn | Gx | 13h 16m | +42° 02' | 9.3 | 12' |
| M 106 | | CVn | Gx | 12h 19m | +47° 18' | 9.1 | 17' |
| M 3 | | CVn | GC | 13h 42m | +28° 23' | 6.3 | 18' |
| Alp CVn | Cor Caroli | CVn | MS | 12h 56m | +38° 19' | 2.9 | 19" |
| M 94 | | CVn | Gx | 12h 51m | +41° 07' | 8.7 | 10' |
| 2 CVn | | CVn | MS | 12h 16m | +40° 40' | 5.7 | 11" |
| NGC 4656 | Hook Galaxy | CVn | Gx | 12h 44m | +32° 10' | 9.7 | 9' |
| NGC 4449 | | CVn | Gx | 12h 28m | +44° 06' | 9.5 | 5' |
| Y CVn | La Superba | CVn | CS | 12h 45m | +45° 26' | 5.2-5.5 | N/A |
| NGC 4490 | Cocoon Galaxy | CVn | Gx | 12h 31m | +41° 39' | 9.8 | 6' |
| NGC 4631 | Whale Galaxy | CVn | Gx | 12h 42m | +32° 33' | 9.5 | 13' |
| RY Dra | | Dra | CS | 12h 56m | +66° 00' | 6.0-8.0 | N/A |
| M 83 | | Hya | Gx | 13h 37m | -29° 52' | 7.8 | 14' |
| M 68 | | Hya | GC | 12h 39m | -26° 45' | 7.3 | 11' |
| NGC 3242 | Ghost of Jupiter | Hya | PN | 10h 25m | -18° 39' | 8.6 | 40" |
| Eps Hya | | Hya | MS | 08h 47m | +06° 25' | 3.4 | 3" |
| M 48 | | Hya | OC | 08h 14m | -05° 45' | 5.5 | 30' |
| U Hya | | Hya | Var/CS | 10h 38m | -13° 23' | 4.8-6.5 | N/A |
| R Leo | Peltier's Variable Star | Leo | Var | 09h 48m | +11° 26' | 4.4-10.5 | N/A |
| Gam Leo | Algieba | Leo | MS | 10h 20m | +19° 50' | 2.0 | 5" |
| 54 Leo | | Leo | MS | 10h 56m | +24° 45' | 4.3 | 6" |
| M 66 | | Leo | Gx | 11h 20m | +13° 00' | 9.7 | 9' |
| NGC 3521 | | Leo | Gx | 11h 06m | -00° 02' | 9.9 | 10' |
| Iot Leo | | Leo | MS | 11h 24m | +10° 32' | 3.9 | 2" |
| 38 Lyn | | Lyn | MS | 09h 19m | +36° 48' | 3.8 | 3" |
| M 40 | Winnecke 4 | UMa | MS | 12h 22m | +58° 05' | 9.6 | |
| M 97 | Owl Nebula | UMa | PN | 11h 15m | +55° 01' | 9.7 | 3' |
| Zet UMa | Mizar & Alcor | UMa | MS | 13h 24m | +54° 56' | 2.1 | 711" |
| Xi UMa | Alula Australis | UMa | MS | 11h 18m | +31° 32' | 4.4 | 2" |
| VY UMa | | UMa | Var/CS | 10h 45m | +67° 25' | 5.9-6.5 | N/A |
| M 81 | Bode's Galaxy | UMa | Gx | 09h 56m | +69° 04' | 7.8 | 22' |
| M 82 | Cigar Galaxy | UMa | Gx | 09h 56m | +69° 41' | 9.0 | 9' |
| M 87 | | Vir | Gx | 12h 31m | +12° 23' | 9.6 | 8' |
| M 104 | Sombrero Galaxy | Vir | Gx | 12h 40m | -11° 37' | 9.1 | 9' |
| M 49 | | Vir | Gx | 12h 30m | +08° 00' | 9.3 | 9' |
| M 60 | | Vir | Gx | 12h 44m | +11° 33' | 9.8 | 7' |
| The Vir | | Vir | MS | 13h 10m | -05° 32' | 4.4 | 70" |
| M 86 | | Vir | Gx | 12h 26m | +12° 57' | 9.8 | 10' |
| Gam Vir | Porrima | Vir | MS | 12h 42m | -01° 27' | 2.7 | 2" |
| SS Vir | | Vir | Var/CS | 12h 25m | +00° 48' | 6.0-9.6 | 6.0-9.6 |

# Chart 12

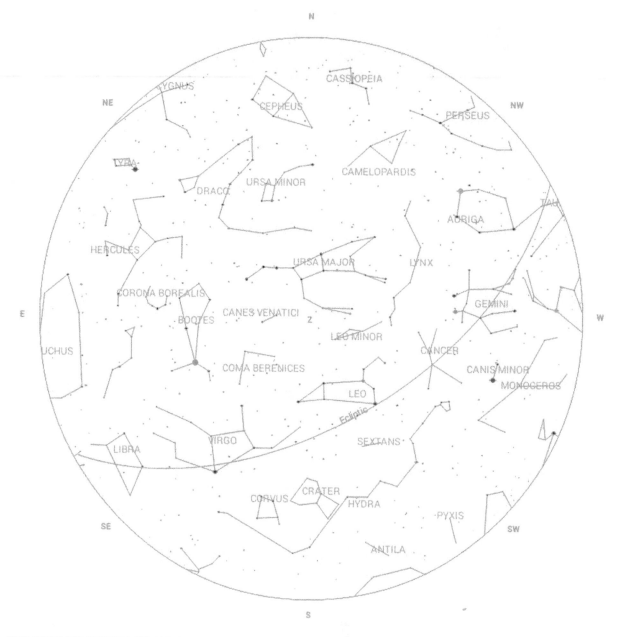

| Designation | Name | Con. | Type | R.A. | Dec. | Mag | Size/Sep |
|---|---|---|---|---|---|---|---|
| Eps Boo | Izar | Boo | MS | 14h 45m | +27° 04' | 2.4 | 3" |
| Xi Boo | | Boo | MS | 14h 51m | +19° 06' | 4.6 | 6" |
| Kap Boo | Asellus Tertius | Boo | MS | 14h 14m | +51° 47' | 4.5 | 13" |
| Pi Boo | | Boo | MS | 14h 41m | +16° 25' | 4.5 | 6" |
| 39 Boo | | Boo | MS | 14h 50m | +48° 43' | 5.7 | 3" |
| Struve 1835 | | Boo | MS | 14h 23m | +08° 27' | 4.9 | 6" |
| M 53 | | Com | GC | 13h 13m | +18° 10' | 7.7 | 13' |
| M 64 | Black Eye Galaxy | Com | Gx | 12h 57m | +21° 41' | 9.3 | 10' |
| Melotte 111 | Coma Star Cluster | Com | OC | 12h 25m | +26° 06' | 2.9 | 120' |

| Designation | Name | Con. | Type | R.A. | Dec. | Mag | Size/Sep |
|---|---|---|---|---|---|---|---|
| 24 Com | | Com | MS | 12h 35m | +18° 23' | 5.0 | 20" |
| Del Crv | Algorab | Crv | MS | 12h 30m | -16° 31' | 5.7 | 24" |
| Struve 1669 | | Crv | MS | 12h 41m | -13° 01' | 5.2 | 5" |
| M 51 | Whirlpool Galaxy | CVn | Gx | 13h 30m | +47° 12' | 8.7 | 10' |
| M 63 | Sunflower Galaxy | CVn | Gx | 13h 16m | +42° 02' | 9.3 | 12' |
| M 106 | | CVn | Gx | 12h 19m | +47° 18' | 9.1 | 17' |
| M 3 | | CVn | GC | 13h 42m | +28° 23' | 6.3 | 18' |
| Alp CVn | Cor Caroli | CVn | MS | 12h 56m | +38° 19' | 2.9 | 19" |
| M 94 | | CVn | Gx | 12h 51m | +41° 07' | 8.7 | 10' |
| 2 CVn | | CVn | MS | 12h 16m | +40° 40' | 5.7 | 11" |
| NGC 4656 | Hook Galaxy | CVn | Gx | 12h 44m | +32° 10' | 9.7 | 9' |
| NGC 4449 | | CVn | Gx | 12h 28m | +44° 06' | 9.5 | 5' |
| Y CVn | La Superba | CVn | CS | 12h 45m | +45° 26' | 5.2-5.5 | N/A |
| NGC 4490 | Cocoon Galaxy | CVn | Gx | 12h 31m | +41° 39' | 9.8 | 6' |
| NGC 4631 | Whale Galaxy | CVn | Gx | 12h 42m | +32° 33' | 9.5 | 13' |
| RY Dra | | Dra | CS | 12h 56m | +66° 00' | 6.0-8.0 | N/A |
| M 83 | | Hya | Gx | 13h 37m | -29° 52' | 7.8 | 14' |
| 54 Hya | | Hya | MS | 14h 46m | -25° 27' | 5.2 | 9" |
| M 68 | | Hya | GC | 12h 39m | -26° 45' | 7.3 | 11' |
| NGC 3242 | Ghost of Jupiter | Hya | PN | 10h 25m | -18° 39' | 8.6 | 40" |
| U Hya | | Hya | Var/CS | 10h 38m | -13° 23' | 4.8-6.5 | N/A |
| R Leo | Peltier's Variable Star | Leo | Var | 09h 48m | +11° 26' | 4.4-10.5 | N/A |
| Gam Leo | Algieba | Leo | MS | 10h 20m | +19° 50' | 2.0 | 5" |
| 54 Leo | | Leo | MS | 10h 56m | +24° 45' | 4.3 | 6" |
| M 66 | | Leo | Gx | 11h 20m | +13° 00' | 9.7 | 9' |
| Iot Leo | | Leo | MS | 11h 24m | +10° 32' | 3.9 | 2" |
| 38 Lyn | | Lyn | MS | 09h 19m | +36° 48' | 3.8 | 3" |
| M 101 | Pinwheel Galaxy | UMa | Gx | 14h 03m | +54° 21' | 8.4 | 22' |
| M 40 | Winnecke 4 | UMa | MS | 12h 22m | +58° 05' | 9.6 | |
| M 97 | Owl Nebula | UMa | PN | 11h 15m | +55° 01' | 9.7 | 3' |
| Zet UMa | Mizar & Alcor | UMa | MS | 13h 24m | +54° 56' | 2.1 | 711" |
| Xi UMa | Alula Australis | UMa | MS | 11h 18m | +31° 32' | 4.4 | 2" |
| VY UMa | | UMa | Var/CS | 10h 45m | +67° 25' | 5.9-6.5 | N/A |
| M 81 | Bode's Galaxy | UMa | Gx | 09h 56m | +69° 04' | 7.8 | 22' |
| M 82 | Cigar Galaxy | UMa | Gx | 09h 56m | +69° 41' | 9.0 | 9' |
| M 87 | | Vir | Gx | 12h 31m | +12° 23' | 9.6 | 8' |
| M 104 | Sombrero Galaxy | Vir | Gx | 12h 40m | -11° 37' | 9.1 | 9' |
| M 49 | | Vir | Gx | 12h 30m | +08° 00' | 9.3 | 9' |
| M 60 | | Vir | Gx | 12h 44m | +11° 33' | 9.8 | 7' |
| The Vir | | Vir | MS | 13h 10m | -05° 32' | 4.4 | 70" |
| M 86 | | Vir | Gx | 12h 26m | +12° 57' | 9.8 | 10' |
| Gam Vir | Porrima | Vir | MS | 12h 42m | -01° 27' | 2.7 | 2" |
| SS Vir | | Vir | Var/CS | 12h 25m | +00° 48' | 6.0-9.6 | 6.0-9.6 |

# Chart 13

| Designation | Name | Con: | Type | R.A. | Dec. | Mag | Size/Sep |
|---|---|---|---|---|---|---|---|
| Eps Boo | Izar | Boo | MS | 14h 45m | +27° 04' | 2.4 | 3" |
| Xi Boo | | Boo | MS | 14h 51m | +19° 06' | 4.6 | 6" |
| Kap Boo | Asellus Tertius | Boo | MS | 14h 14m | +51° 47' | 4.5 | 13" |
| Pi Boo | | Boo | MS | 14h 41m | +16° 25' | 4.5 | 6" |
| 39 Boo | | Boo | MS | 14h 50m | +48° 43' | 5.7 | 3" |
| Struve 1835 | | Boo | MS | 14h 23m | +08° 27' | 4.9 | 6" |
| Mu Boo | Alkalurops | Boo | MS | 15h 24m | +37° 23' | 4.3 | 108" |
| 44 Boo | | Boo | MS | 15h 04m | +47° 39' | 4.8 | 2" |
| M 53 | | Com | GC | 13h 13m | +18° 10' | 7.7 | 13' |

| Designation | Name | Con: | Type | R.A.: | Dec: | Mag | Size/Sep |
|---|---|---|---|---|---|---|---|
| Del Boo | | Boo | MS | 15h 16m | +33° 19' | 3.5 | 105" |
| Nu Boo | | Boo | MS | 15h 31m | +40° 50' | 5.0 | 15' |
| M 64 | Black Eye Galaxy | Com | Gx | 12h 57m | +21° 41' | 8.3 | 10' |
| Melotte 111 | Coma Star Cluster | Com | OC | 12h 25m | +26° 06' | 2.9 | 120' |
| 24 Com | | Com | MS | 12h 35m | +18° 23' | 5.0 | 20" |
| Del Crv | Algorab | Crv | MS | 12h 30m | -16° 31' | 5.7 | 24" |
| Struve 1669 | | Crv | MS | 12h 41m | -13° 01' | 5.2 | 5" |
| M 51 | Whirlpool Galaxy | CVn | Gx | 13h 30m | +47° 12' | 8.7 | 10' |
| M 63 | Sunflower Galaxy | CVn | Gx | 13h 16m | +42° 02' | 8.3 | 12' |
| M 106 | | CVn | Gx | 12h 19m | +47° 18' | 8.1 | 17' |
| M 3 | | CVn | GC | 13h 42m | +28° 23' | 6.3 | 18' |
| Alp CVn | Cor Caroli | CVn | MS | 12h 56m | +38° 19' | 2.9 | 19" |
| M 94 | | CVn | Gx | 12h 51m | +41° 07' | 8.7 | 10' |
| 2 CVn | | CVn | MS | 12h 16m | +40° 40' | 5.7 | 11" |
| NGC 4656 | Hook Galaxy | CVn | Gx | 12h 44m | +32° 10' | 8.7 | 9' |
| NGC 4449 | | CVn | Gx | 12h 28m | +44° 06' | 9.5 | 5' |
| Y CVn | La Superba | CVn | ES | 12h 45m | +45° 26' | 5.3-5.5 | N/A |
| Zet CrB | | CrB | MS | 15h 39m | +36° 38' | 4.6 | 6" |
| R CrB | Fade Out Star | CrB | Var | 15h 49m | +28° 09' | 5.7-14.8 | N/A |
| NGC 4631 | Whale Galaxy | CVn | Gx | 12h 42m | +32° 33' | 9.5 | 13' |
| RY Dra | | Dra | ES | 12h 56m | +66° 00' | 6.0-8.0 | N/A |
| M 83 | | Hya | Gx | 13h 37m | -29° 52' | 7.8 | 14' |
| 54 Hya | | Hya | MS | 14h 46m | -25° 27' | 5.2 | 9" |
| M 68 | | Hya | GC | 12h 39m | -26° 45' | 7.3 | 11' |
| NGC 3242 | Ghost of Jupiter | Hya | PN | 10h 25m | -18° 39' | 8.6 | 40" |
| U Hya | | Hya | Var/ES | 10h 38m | -13° 23' | 4.8-6.5 | N/A |
| Gam Leo | Algieba | Leo | MS | 10h 20m | +19° 50' | 2.0 | 5" |
| 54 Leo | | Leo | MS | 10h 56m | +24° 45' | 4.3 | 6" |
| M 66 | | Leo | Gx | 11h 20m | +13° 00' | 8.7 | 9' |
| Iot Leo | | Leo | MS | 11h 24m | +10° 32' | 3.9 | 2" |
| M 5 | | Ser | GC | 15h 19m | +02° 05' | 5.7 | 23' |
| Del Ser | | Ser | MS | 15h 35m | +10° 32' | 4.2 | 4" |
| M 101 | Pinwheel Galaxy | UMa | Gx | 14h 03m | +54° 21' | 8.4 | 22' |
| M 40 | Winnecke 4 | UMa | MS | 12h 22m | +58° 05' | 9.6 | |
| M 97 | Owl Nebula | UMa | PN | 11h 15m | +55° 01' | 9.7 | 3' |
| Zet UMa | Mizar & Alcor | UMa | MS | 13h 24m | +54° 56' | 2.1 | 711" |
| Xi UMa | Alula Australis | UMa | MS | 11h 18m | +31° 32' | 4.4 | 2" |
| VY UMa | | UMa | Var/ES | 10h 45m | +67° 25' | 5.9-6.5 | N/A |
| M 87 | | Vir | Gx | 12h 31m | +12° 23' | 8.6 | 8' |
| M 104 | Sombrero Galaxy | Vir | Gx | 12h 40m | -11° 37' | 8.1 | 9' |
| M 49 | | Vir | Gx | 12h 30m | +08° 00' | 8.3 | 9' |
| The Vir | | Vir | MS | 13h 10m | -05° 32' | 4.4 | 70" |
| Gam Vir | Porrima | Vir | MS | 12h 42m | -01° 27' | 2.7 | 2" |
| SS Vir | | Vir | Var/ES | 12h 25m | +00° 48' | 6.0-9.6 | 6.0-9.6 |

# Chart 14

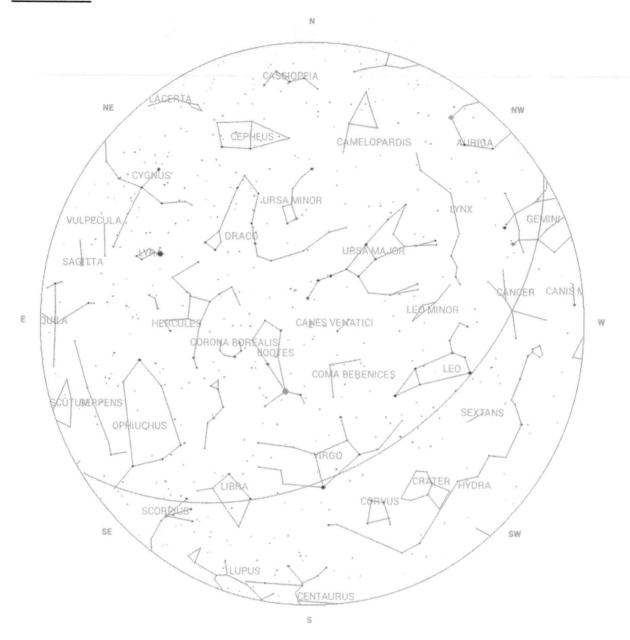

| Designation | Name | Con. | Type | R.A. | Dec. | Mag | Size/Sep |
|---|---|---|---|---|---|---|---|
| Eps Boo | Izar | Boo | MS | 14h 45m | +27° 04' | 2.4 | 3" |
| Xi Boo | | Boo | MS | 14h 51m | +19° 06' | 4.6 | 6" |
| Kap Boo | Asellus Tertius | Boo | MS | 14h 14m | +51° 47' | 4.5 | 13" |
| Pi Boo | | Boo | MS | 14h 41m | +16° 25' | 4.5 | 6" |
| 39 Boo | | Boo | MS | 14h 50m | +48° 43' | 5.7 | 3" |
| Struve 1835 | | Boo | MS | 14h 23m | +08° 27' | 4.9 | 6" |
| Mu Boo | Alkalurops | Boo | MS | 15h 24m | +37° 23' | 4.3 | 108" |
| 44 Boo | | Boo | MS | 15h 04m | +47° 39' | 4.8 | 2" |
| Del Boo | | Boo | MS | 15h 16m | +33° 19' | 3.5 | 105" |

| Designation | Name | Con. | Type | R.A. | Dec. | Mag | Size/Sep |
|---|---|---|---|---|---|---|---|
| Nu Boo | | Boo | MS | 15h 31m | +40° 50' | 5.0 | 15' |
| M 53 | | Com | GC | 13h 13m | +18° 10' | 7.7 | 13' |
| M 64 | Black Eye Galaxy | Com | Gx | 12h 57m | +21° 41' | 9.3 | 10' |
| Melotte 111 | Coma Star Cluster | Com | OC | 12h 25m | +26° 06' | 2.9 | 120' |
| 24 Com | | Com | MS | 12h 35m | +18° 23' | 5.0 | 20" |
| Zet CrB | | CrB | MS | 15h 39m | +36° 38' | 4.6 | 6" |
| Sig CrB | | CrB | MS | 16h 15m | +33° 52' | 5.7 | 7" |
| R CrB | Fade Out Star | CrB | Var | 15h 49m | +28° 09' | 5.7-14.8 | N/A |
| T CrB | Blaze Star | CrB | RN | 16h 00m | +25° 55' | 2.0-10.8 | N/A |
| Del Crv | Algorab | Crv | MS | 12h 30m | -16° 31' | 5.7 | 24" |
| Struve 1669 | | Crv | MS | 12h 41m | -13° 01' | 5.2 | 5" |
| M 51 | Whirlpool Galaxy | CVn | Gx | 13h 30m | +47° 12' | 8.7 | 10' |
| M 63 | Sunflower Galaxy | CVn | Gx | 13h 16m | +42° 02' | 9.3 | 12' |
| M 106 | | CVn | Gx | 12h 19m | +47° 18' | 9.1 | 17' |
| M 3 | | CVn | GC | 13h 42m | +28° 23' | 6.3 | 18' |
| Alp CVn | Cor Caroli | CVn | MS | 12h 56m | +38° 19' | 2.9 | 19" |
| M 94 | | CVn | Gx | 12h 51m | +41° 07' | 8.7 | 10' |
| 2 CVn | | CVn | MS | 12h 16m | +40° 40' | 5.7 | 11" |
| NGC 4449 | | CVn | Gx | 12h 28m | +44° 06' | 9.5 | 5' |
| Y CVn | La Superba | CVn | CS | 12h 45m | +45° 26' | 5.2-5.5 | N/A |
| NGC 4631 | Whale Galaxy | CVn | Gx | 12h 42m | +32° 33' | 9.5 | 13' |
| RY Dra | | Dra | CS | 12h 56m | +66° 00' | 6.0-8.0 | N/A |
| 16/17 Dra | | Dra | MS | 16h 36m | +52° 55' | 5.1 | 90" |
| M 13 | Keystone Cluster | Her | GC | 16h 42m | +36° 27' | 5.8 | 20' |
| Kap Her | Marfik | Her | MS | 16h 41m | +31° 36' | 5.0 | 28" |
| NGC 6229 | | Her | GC | 16h 47m | +47° 32' | 9.4 | 4' |
| Iot Leo | | Leo | MS | 11h 24m | +10° 32' | 3.9 | 2" |
| Alp Lib | Zuben Elgenubi | Lib | MS | 14h 51m | -16° 02' | 2.8 | 230" |
| NGC 5897 | Ghost Globular | Lib | GC | 15h 17m | -21° 01' | 8.4 | 11' |
| Bet Lib | The Emerald Star | Lib | * | 15h 14m | -09° 23' | 2.6 | N/A |
| Struve 1962 | | Lib | MS | 15h 39m | -08° 47' | 5.4 | 12" |
| M 5 | | Ser | GC | 15h 19m | +02° 05' | 5.7 | 23' |
| Del Ser | | Ser | MS | 15h 35m | +10° 32' | 4.2 | 4" |
| M 101 | Pinwheel Galaxy | UMa | Gx | 14h 03m | +54° 21' | 8.4 | 22' |
| M 40 | Winnecke 4 | UMa | MS | 12h 22m | +58° 05' | 9.6 | |
| Zet UMa | Mizar & Alcor | UMa | MS | 13h 24m | +54° 56' | 2.1 | 711" |
| Xi UMa | Alula Australis | UMa | MS | 11h 18m | +31° 32' | 4.4 | 2" |
| M 87 | | Vir | Gx | 12h 31m | +12° 23' | 9.6 | 8' |
| M 104 | Sombrero Galaxy | Vir | Gx | 12h 40m | -11° 37' | 9.1 | 9' |
| M 49 | | Vir | Gx | 12h 30m | +08° 00' | 9.3 | 9' |
| The Vir | | Vir | MS | 13h 10m | -05° 32' | 4.4 | 70" |
| Gam Vir | Porrima | Vir | MS | 12h 42m | -01° 27' | 2.7 | 2" |
| SS Vir | | Vir | Var/CS | 12h 25m | +00° 48' | 6.0-9.6 | 6.0-9.6 |

# Chart 15

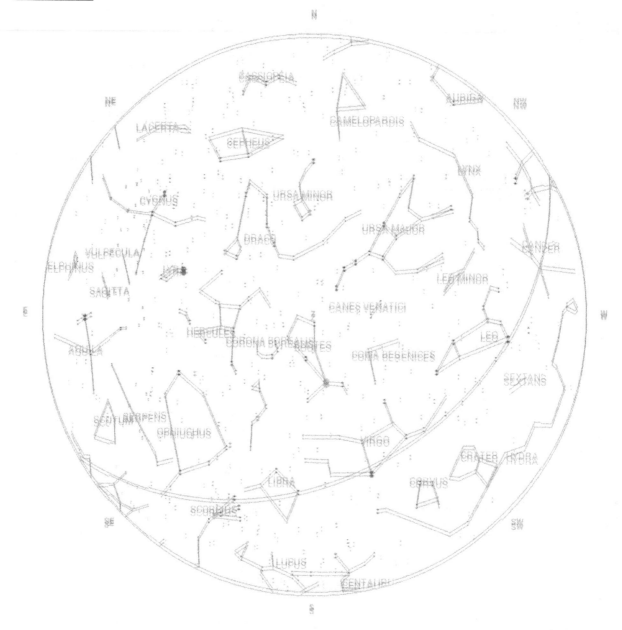

| Designation | Name | Con. | Type | R.A. | Dec. | Mag | Size/Sep |
|---|---|---|---|---|---|---|---|
| Eps Boo | Izar | Boo | MS | 14h 45m | ±27° 04' | 2.4 | 3" |
| Xi Boo | | Boo | MS | 14h 51m | ±19° 06' | 4.6 | 6" |
| Kap Boo | Asellus Tertius | Boo | MS | 14h 14m | ±51° 47' | 4.5 | 13" |
| Pi Boo | | Boo | MS | 14h 41m | ±16° 25' | 4.5 | 6" |
| 39 Boo | | Boo | MS | 14h 50m | ±48° 43' | 5.7 | 3" |
| Struve 1835 | | Boo | MS | 14h 23m | ±08° 27' | 4.8 | 6" |
| Mu Boo | Alkalurops | Boo | MS | 15h 24m | ±37° 23' | 4.3 | 108" |
| 44 Boo | | Boo | MS | 15h 04m | ±47° 39' | 4.8 | 2" |
| Del Boo | | Boo | MS | 15h 16m | ±33° 19' | 3.5 | 105" |

| Designation | Name | Con. | Type | R.A. | Dec. | Mag | Size/Sep |
|---|---|---|---|---|---|---|---|
| Nu Boo | | Boo | MS | 15h 31m | +40° 50' | 5.0 | 15' |
| M 53 | | Com | GC | 13h 13m | +18° 10' | 7.7 | 13' |
| Melotte 111 | Coma Star Cluster | Com | OC | 12h 25m | +26° 06' | 2.9 | 120' |
| 24 Com | | Com | MS | 12h 35m | +18° 23' | 5.0 | 20" |
| Zet CrB | | CrB | MS | 15h 39m | +36° 38' | 4.6 | 6" |
| Sig CrB | | CrB | MS | 16h 15m | +33° 52' | 5.7 | 7" |
| R CrB | Fade Out Star | CrB | Var | 15h 49m | +28° 09' | 5.7-14.8 | N/A |
| T CrB | Blaze Star | CrB | RN | 16h 00m | +25° 55' | 2.0-10.8 | N/A |
| M 3 | | CVn | GC | 13h 42m | +28° 23' | 6.3 | 18' |
| Alp CVn | Cor Caroli | CVn | MS | 12h 56m | +38° 19' | 2.9 | 19" |
| 2 CVn | | CVn | MS | 12h 16m | +40° 40' | 5.7 | 11" |
| Y CVn | La Superba | CVn | CS | 12h 45m | +45° 26' | 5.2-5.5 | N/A |
| RY Dra | | Dra | CS | 12h 56m | +66° 00' | 6.0-8.0 | N/A |
| 16/17 Dra | | Dra | MS | 16h 36m | +52° 55' | 5.1 | 90" |
| Nu Dra | Kuma | Dra | MS | 17h 32m | +55° 10' | 4.9 | 63" |
| Mu Dra | | Dra | MS | 17h 05m | +54° 28' | 5.8 | 2" |
| Psi Dra | | Dra | MS | 17h 42m | +72° 09' | 4.6 | 30" |
| M 13 | Keystone Cluster | Her | GC | 16h 42m | +36° 27' | 5.8 | 20' |
| Kap Her | Marfik | Her | MS | 16h 41m | +31° 36' | 5.0 | 28" |
| M 92 | | Her | GC | 17h 17m | +43° 08' | 6.5 | 14' |
| Alp Her | Rasalgethi | Her | MS | 17h 15m | +14° 23' | 3.1 | 5" |
| Del Her | Sarin | Her | MS | 17h 15m | +24° 50' | 3.1 | 14" |
| Rho Her | | Her | MS | 17h 24m | +37° 09' | 4.2 | 4" |
| Alp Lib | Zuben Elgenubi | Lib | MS | 14h 51m | -16° 02' | 2.8 | 230" |
| Bet Lib | The Emerald Star | Lib | * | 15h 14m | -09° 23' | 2.6 | N/A |
| Struve 1962 | | Lib | MS | 15h 39m | -08° 47' | 5.4 | 12" |
| IC 4665 | Summer Beehive | Oph | OC | 17h 46m | +05° 43' | 5.3 | 70' |
| M 10 | | Oph | GC | 16h 57m | -04° 06' | 6.6 | 20' |
| M 14 | | Oph | GC | 17h 38m | -03° 15' | 7.6 | 11' |
| M 12 | | Oph | GC | 16h 47m | -01° 57' | 6.1 | 16' |
| Rho Oph | | Oph | MS | 16h 26m | -23° 27' | 4.6 | 3" |
| M 19 | | Oph | GC | 17h 03m | -26° 16' | 6.8 | 17' |
| M 62 | | Oph | GC | 17h 01m | -30° 07' | 6.4 | 15' |
| 36 Oph | | Oph | MS | 17h 15m | -26° 36' | 4.3 | 730" |
| Omi Oph | | Oph | MS | 17h 18m | -24° 17' | 5.1 | 10" |
| 61 Oph | | Oph | MS | 17h 45m | +02° 35' | 6.2 | 21" |
| M 5 | | Ser | GC | 15h 19m | +02° 05' | 5.7 | 23' |
| Del Ser | | Ser | MS | 15h 35m | +10° 32' | 4.2 | 4" |
| Zet UMa | Mizar & Alcor | UMa | MS | 13h 24m | +54° 56' | 2.1 | 711" |
| The Vir | | Vir | MS | 13h 10m | -05° 32' | 4.4 | 70" |
| Gam Vir | Porrima | Vir | MS | 12h 42m | -01° 27' | 2.7 | 2" |
| SS Vir | | Vir | Var/CS | 12h 25m | +00° 48' | 6.0-9.6 | 6.0-9.6 |

# Chart 16

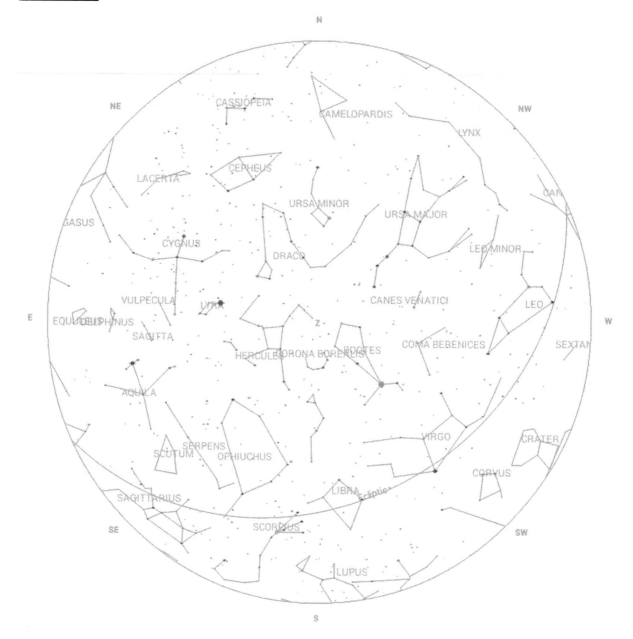

| Designation | Name | Con. | Type | R.A. | Dec. | Mag | Size/Sep |
|---|---|---|---|---|---|---|---|
| Eps Boo | Izar | Boo | MS | 14h 45m | +27° 04' | 2.4 | 3" |
| Xi Boo | | Boo | MS | 14h 51m | +19° 06' | 4.6 | 6" |
| Kap Boo | Asellus Tertius | Boo | MS | 14h 14m | +51° 47' | 4.5 | 13" |
| Pi Boo | | Boo | MS | 14h 41m | +16° 25' | 4.5 | 6" |
| 39 Boo | | Boo | MS | 14h 50m | +48° 43' | 5.7 | 3" |
| Struve 1835 | | Boo | MS | 14h 23m | +08° 27' | 4.9 | 6" |
| Mu Boo | Alkalurops | Boo | MS | 15h 24m | +37° 23' | 4.3 | 108" |
| 44 Boo | | Boo | MS | 15h 04m | +47° 39' | 4.8 | 2" |
| Del Boo | | Boo | MS | 15h 16m | +33° 19' | 3.5 | 105" |

| Designation | Name | Con. | Type | R.A. | Dec. | Mag | Size/Sep |
|---|---|---|---|---|---|---|---|
| Nu Boo | | Boo | MS | 15h 31m | +40° 50' | 5.0 | 15' |
| Zet CrB | | CrB | MS | 15h 39m | +36° 38' | 4.6 | 6" |
| Sig CrB | | CrB | MS | 16h 15m | +33° 52' | 5.7 | 7" |
| R CrB | Fade Out Star | CrB | Var | 15h 49m | +28° 09' | 5.7-14.8 | N/A |
| T CrB | Blaze Star | CrB | RN | 16h 00m | +25° 55' | 2.0-10.8 | N/A |
| M 3 | | CVn | GC | 13h 42m | +28° 23' | 6.3 | 18' |
| 16/17 Dra | | Dra | MS | 16h 36m | +52° 55' | 5.1 | 90" |
| Nu Dra | Kuma | Dra | MS | 17h 32m | +55° 10' | 4.9 | 63" |
| Mu Dra | | Dra | MS | 17h 05m | +54° 28' | 5.8 | 2" |
| Psi Dra | | Dra | MS | 17h 42m | +72° 09' | 4.6 | 30" |
| 39 Dra | | Dra | MS | 18h 24m | +58° 48' | 5.0 | 89" |
| 40/41 Dra | | Dra | MS | 18h 00m | +80° 00' | 5.7 | 222" |
| M 13 | Keystone Cluster | Her | GC | 16h 42m | +36° 27' | 5.8 | 20' |
| Kap Her | Marfik | Her | MS | 16h 41m | +31° 36' | 5.0 | 28" |
| M 92 | | Her | GC | 17h 17m | +43° 08' | 6.5 | 14' |
| Alp Her | Rasalgethi | Her | MS | 17h 15m | +14° 23' | 3.1 | 5" |
| Del Her | Sarin | Her | MS | 17h 15m | +24° 50' | 3.1 | 14" |
| Rho Her | | Her | MS | 17h 24m | +37° 09' | 4.2 | 4" |
| 95 Her | | Her | MS | 18h 02m | +21° 36' | 4.3 | 6" |
| 100 Her | | Her | MS | 18h 08m | +26° 06' | 5.8 | 14" |
| Alp Lib | Zuben Elgenubi | Lib | MS | 14h 51m | -16° 02' | 2.8 | 230" |
| Bet Lib | The Emerald Star | Lib | * | 15h 14m | -09° 23' | 2.6 | N/A |
| Struve 1962 | | Lib | MS | 15h 39m | -08° 47' | 5.4 | 12" |
| Eps Lyr | The Double Double | Lyr | MS | 18h 44m | +39° 40' | 4.7 | 3" |
| Bet Lyr | Sheliak | Lyr | MS | 18h 50m | +33° 22' | 3.2 | 86" |
| Del Lyr | | Lyr | MS | 18h 54m | +36° 58' | 4.2 | 630" |
| Zet Lyr | | Lyr | MS | 18h 45m | +37° 36' | 4.4 | 44" |
| IC 4665 | Summer Beehive | Oph | OC | 17h 46m | +05° 43' | 5.3 | 70' |
| M 10 | | Oph | GC | 16h 57m | -04° 06' | 6.6 | 20' |
| M 12 | | Oph | GC | 16h 47m | -01° 57' | 6.1 | 16' |
| Rho Oph | | Oph | MS | 16h 26m | -23° 27' | 4.6 | 3" |
| M 62 | | Oph | GC | 17h 01m | -30° 07' | 6.4 | 15' |
| 36 Oph | | Oph | MS | 17h 15m | -26° 36' | 4.3 | 730" |
| Omi Oph | | Oph | MS | 17h 18m | -24° 17' | 5.1 | 10" |
| 61 Oph | | Oph | MS | 17h 45m | +02° 35' | 6.2 | 21" |
| NGC 6633 | Tweedledum Cluster | Oph | OC | 18h 27m | +06° 31' | 5.6 | 20' |
| 70 Oph | | Oph | MS | 18h 06m | +02° 30' | 4.2 | 4" |
| IC 4756 | | Ser | OC | 18h 39m | +05° 27' | 5.4 | 39' |
| M 16 | Eagle Nebula | Ser | Neb | 18h 19m | -13° 49' | 6.0 | 9' |
| The Ser | | Ser | MS | 18h 56m | +04° 12' | 4.3 | 22" |
| M 5 | | Ser | GC | 15h 19m | +02° 05' | 5.7 | 23' |
| Del Ser | | Ser | MS | 15h 35m | +10° 32' | 4.2 | 4" |
| Zet UMa | Mizar & Alcor | UMa | MS | 13h 24m | +54° 56' | 2.1 | 711" |
| The Vir | | Vir | MS | 13h 10m | -05° 32' | 4.4 | 70" |

## Chart 17

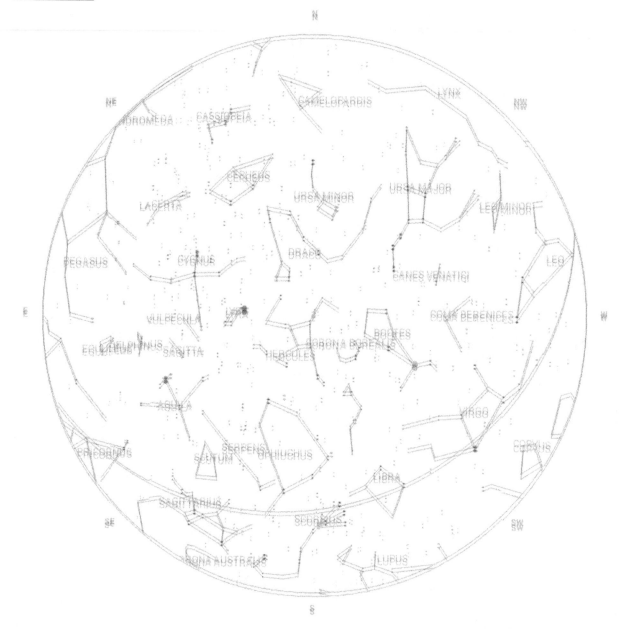

| Designation | Name | Con: | Type | R.A: | Dec: | Mag | Size/Sep |
|---|---|---|---|---|---|---|---|
| 15 Aql | | Aql | MS | 19h 06m | -04° 00' | 5.4 | 39" |
| 57 Aql | | Aql | MS | 19h 55m | -08° 14' | 5.7 | 36" |
| Eps Boo | Izar | Boo | MS | 14h 45m | +27° 04' | 2.4 | 3" |
| Xi Boo | | Boo | MS | 14h 51m | +19° 06' | 4.6 | 6" |
| Kap Boo | Asellus Tertius | Boo | MS | 14h 14m | +51° 47' | 4.5 | 13" |
| Pi Boo | | Boo | MS | 14h 41m | +16° 25' | 4.5 | 6" |
| 39 Boo | | Boo | MS | 14h 50m | +48° 43' | 5.7 | 3" |
| Struve 1835 | | Boo | MS | 14h 23m | +08° 27' | 4.9 | 6" |
| Mu Boo | Alkalurops | Boo | MS | 15h 24m | +37° 23' | 4.3 | 108" |

| Designation | Name | Con: | Type | R.A: | Dec: | Mag | Size/Sep |
|---|---|---|---|---|---|---|---|
| 44 Boo | | Boo | MS | 15h 04m | +47° 39' | 4.8 | 2" |
| Del Boo | | Boo | MS | 15h 16m | +33° 19' | 3.5 | 105" |
| Nu Boo | | Boo | MS | 15h 31m | +40° 50' | 5.0 | 15' |
| Zet CrB | | CrB | MS | 15h 39m | +36° 38' | 4.6 | 6" |
| Sig CrB | | CrB | MS | 16h 15m | +33° 52' | 5.7 | 7" |
| Bet Cyg | Albireo | Cyg | MS | 19h 31m | +27° 58' | 3.1 | 34" |
| Del Cyg | | Cyg | MS | 19h 45m | +45° 08' | 2.9 | 2" |
| 16/17 Dra | | Dra | MS | 16h 36m | +52° 55' | 5.1 | 90" |
| Nu Dra | Kuma | Dra | MS | 17h 32m | +55° 10' | 4.9 | 63" |
| Psi Dra | | Dra | MS | 17h 42m | +72° 09' | 4.6 | 30" |
| 39 Dra | | Dra | MS | 18h 24m | +58° 48' | 5.0 | 89" |
| 40/41 Dra | | Dra | MS | 18h 00m | +80° 00' | 5.7 | 222" |
| Eps Dra | | Dra | MS | 19h 48m | +70° 16' | 3.8 | 3" |
| Kap Her | Marfik | Her | MS | 16h 41m | +31° 36' | 5.0 | 28" |
| Alp Her | Rasalgethi | Her | MS | 17h 15m | +14° 23' | 3.1 | 5" |
| Del Her | Sarin | Her | MS | 17h 15m | +24° 50' | 3.1 | 14" |
| Rho Her | | Her | MS | 17h 24m | +37° 09' | 4.2 | 4" |
| 95 Her | | Her | MS | 18h 02m | +21° 36' | 4.3 | 6" |
| Alp Lib | Zuben Elgenubi | Lib | MS | 14h 51m | -16° 02' | 2.8 | 230" |
| Bet Lib | The Emerald Star | Lib | * | 15h 14m | -09° 23' | 2.6 | N/A |
| Struve 1962 | | Lib | MS | 15h 39m | -08° 47' | 5.4 | 12" |
| Eps Lyr | The Double Double | Lyr | MS | 18h 44m | +39° 40' | 4.7 | 3" |
| Bet Lyr | Sheliak | Lyr | MS | 18h 50m | +33° 22' | 3.3 | 86" |
| Del Lyr | | Lyr | MS | 18h 54m | +36° 58' | 4.2 | 630" |
| Zet Lyr | | Lyr | MS | 18h 45m | +37° 36' | 4.4 | 44" |
| IC 4665 | Summer Beehive | Oph | OC | 17h 46m | +05° 43' | 5.3 | 70' |
| Rho Oph | | Oph | MS | 16h 26m | -23° 27' | 4.6 | 3" |
| 36 Oph | | Oph | MS | 17h 15m | -26° 36' | 4.3 | 730" |
| Omi Oph | | Oph | MS | 17h 18m | -24° 17' | 5.1 | 10" |
| NGC 6633 | Tweedledum Cluster | Oph | OC | 18h 27m | +06° 31' | 5.6 | 20' |
| 70 Oph | | Oph | MS | 18h 06m | +02° 30' | 4.2 | 4" |
| M 6 | Butterfly Cluster | Sco | OC | 17h 40m | -32° 15' | 4.6 | 20' |
| M 7 | | Sco | OC | 17h 54m | -34° 48' | 3.3 | 80' |
| M 4 | | Sco | GC | 16h 24m | -26° 32' | 5.4 | 36' |
| Alp Sco | Antares | Sco | MS | 16h 29m | -26° 26' | 1.0 | 3" |
| Bet Sco | Graffias | Sco | MS | 16h 05m | -19° 48' | 2.6 | 14" |
| Nu Sco | Jabbah | Sco | MS | 16h 12m | -19° 28' | 4.0 | 2" |
| Xi Sco | | Sco | MS | 16h 04m | -11° 22' | 4.2 | 8" |
| Struve 1999 | | Sco | MS | 16h 04m | -11° 22' | 4.2 | 12" |
| IC 4756 | | Ser | OC | 18h 39m | +05° 27' | 5.4 | 39' |
| The Ser | | Ser | MS | 18h 56m | +04° 12' | 4.3 | 22" |
| M 5 | | Ser | GC | 15h 19m | +02° 05' | 5.7 | 23' |
| Del Ser | | Ser | MS | 15h 35m | +10° 32' | 4.2 | 4" |
| Collinder 399 | Coathanger | Vul | Ast | 19h 25m | +20° 11' | 4.8 | 89' |

# Chart 18

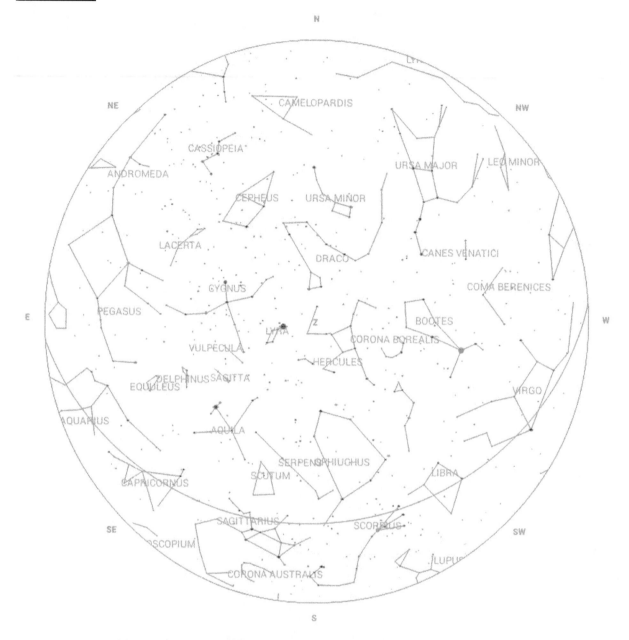

| Designation | Name | Con. | Type | R.A. | Dec. | Mag | Size/Sep |
|---|---|---|---|---|---|---|---|
| 15 Aql | | Aql | MS | 19h 06m | -04° 00' | 5.4 | 39" |
| 57 Aql | | Aql | MS | 19h 55m | -08° 14' | 5.7 | 36" |
| Mu Boo | Alkalurops | Boo | MS | 15h 24m | +37° 23' | 4.3 | 108" |
| 44 Boo | | Boo | MS | 15h 04m | +47° 39' | 4.8 | 2" |
| Del Boo | | Boo | MS | 15h 16m | +33° 19' | 3.5 | 105" |
| Nu Boo | | Boo | MS | 15h 31m | +40° 50' | 5.0 | 15' |
| Zet CrB | | CrB | MS | 15h 39m | +36° 38' | 4.6 | 6" |
| Sig CrB | | CrB | MS | 16h 15m | +33° 52' | 5.7 | 7" |
| Bet Cyg | Albireo | Cyg | MS | 19h 31m | +27° 58' | 3.1 | 34" |

| Designation | Name | Con. | Type | R.A. | Dec. | Mag | Size/Sep |
|---|---|---|---|---|---|---|---|
| Del Cyg | | Cyg | MS | 19h 45m | +45° 08' | 2.9 | 2" |
| NGC 7000 | North American Nebula | Cyg | Neb | 20h 59m | +44° 22' | 4.0 | 120' |
| Gam Del | | Del | MS | 20h 47m | +16° 07' | 3.9 | 9" |
| 16/17 Dra | | Dra | MS | 16h 36m | +52° 55' | 5.1 | 90" |
| Nu Dra | Kuma | Dra | MS | 17h 32m | +55° 10' | 4.9 | 63" |
| Psi Dra | | Dra | MS | 17h 42m | +72° 09' | 4.6 | 30" |
| 39 Dra | | Dra | MS | 18h 24m | +58° 48' | 5.0 | 89" |
| 40/41 Dra | | Dra | MS | 18h 00m | +80° 00' | 5.7 | 222" |
| Eps Dra | | Dra | MS | 19h 48m | +70° 16' | 3.8 | 3" |
| Eps Equ | | Equ | MS | 20h 59m | +04° 18' | 5.2 | 11" |
| Kap Her | Marfik | Her | MS | 16h 41m | +31° 36' | 5.0 | 28" |
| Alp Her | Rasalgethi | Her | MS | 17h 15m | +14° 23' | 3.1 | 5" |
| Del Her | Sarin | Her | MS | 17h 15m | +24° 50' | 3.1 | 14" |
| Rho Her | | Her | MS | 17h 24m | +37° 09' | 4.2 | 4" |
| 95 Her | | Her | MS | 18h 02m | +21° 36' | 4.3 | 6" |
| Eps Lyr | The Double Double | Lyr | MS | 18h 44m | +39° 40' | 4.7 | 3" |
| Bet Lyr | Sheliak | Lyr | MS | 18h 50m | +33° 22' | 3.2 | 86" |
| Del Lyr | | Lyr | MS | 18h 54m | +36° 58' | 4.2 | 630" |
| Zet Lyr | | Lyr | MS | 18h 45m | +37° 36' | 4.4 | 44" |
| IC 4665 | Summer Beehive | Oph | OC | 17h 46m | +05° 43' | 5.3 | 70' |
| Rho Oph | | Oph | MS | 16h 26m | -23° 27' | 4.6 | 3" |
| 36 Oph | | Oph | MS | 17h 15m | -26° 36' | 4.3 | 730" |
| Omi Oph | | Oph | MS | 17h 18m | -24° 17' | 5.1 | 10" |
| NGC 6633 | Tweedledum Cluster | Oph | OC | 18h 27m | +06° 31' | 5.6 | 20' |
| 70 Oph | | Oph | MS | 18h 06m | +02° 30' | 4.2 | 4" |
| M 6 | Butterfly Cluster | Sco | OC | 17h 40m | -32° 15' | 4.6 | 20' |
| M 7 | | Sco | OC | 17h 54m | -34° 48' | 3.3 | 80' |
| M 4 | | Sco | GC | 16h 24m | -26° 32' | 5.4 | 36' |
| Alp Sco | Antares | Sco | MS | 16h 29m | -26° 26' | 1.0 | 3" |
| Bet Sco | Graffias | Sco | MS | 16h 05m | -19° 48' | 2.6 | 14" |
| Nu Sco | Jabbah | Sco | MS | 16h 12m | -19° 28' | 4.0 | 2" |
| Xi Sco | | Sco | MS | 16h 04m | -11° 22' | 4.2 | 8" |
| Struve 1999 | | Sco | MS | 16h 04m | -11° 22' | 4.2 | 12" |
| IC 4756 | | Ser | OC | 18h 39m | +05° 27' | 5.4 | 39' |
| The Ser | | Ser | MS | 18h 56m | +04° 12' | 4.3 | 22" |
| M 5 | | Ser | GC | 15h 19m | +02° 05' | 5.7 | 23' |
| Del Ser | | Ser | MS | 15h 35m | +10° 32' | 4.2 | 4" |
| The Sge | | Sge | MS | 20h 10m | +20° 55' | 4.6 | 84" |
| M 24 | Sagittarius Star Cloud | Sgr | OC | 18h 18m | -18° 24' | 3.1 | 90' |
| M 22 | | Sgr | GC | 18h 36m | -23° 54' | 5.2 | 32' |
| M 8 | Lagoon Nebula | Sgr | Neb | 18h 04m | -24° 23' | 5.0 | 17' |
| Collinder 399 | Coathanger | Vul | Ast | 19h 25m | +20° 11' | 4.8 | 89' |
| NGC 6885 | | Vul | OC | 20h 12m | +26° 29' | 5.7 | 20' |

# Chart 19

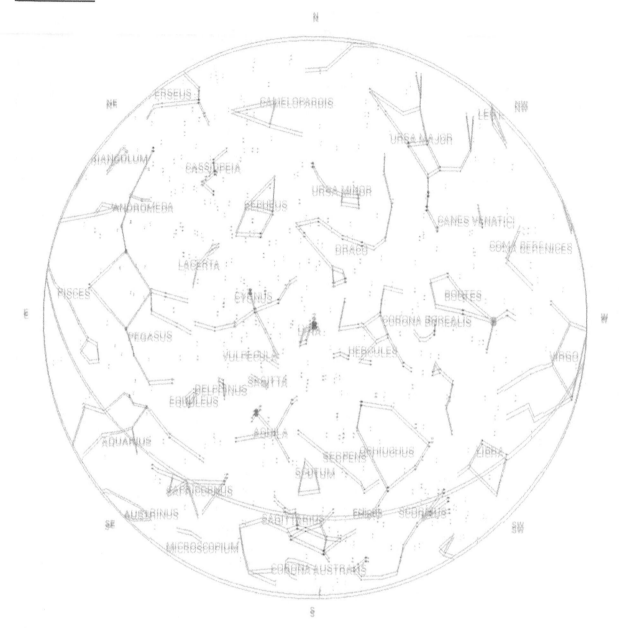

| Designation | Name | Con. | Type | R.A. | Dec. | Mag | Size/Sep |
|---|---|---|---|---|---|---|---|
| 15 Aql | | Aql | MS | 19h 06m | -04° 00' | 5.4 | 39" |
| 57 Aql | | Aql | MS | 19h 55m | -08° 14' | 5.7 | 36" |
| Bet Cyg | Albireo | Cyg | MS | 19h 31m | +27° 58' | 3.1 | 34" |
| Del Cyg | | Cyg | MS | 19h 45m | +45° 08' | 2.9 | 2" |
| 16 Cyg | | Cyg | MS | 19h 42m | +50° 32' | 5.9 | 39" |
| 61 Cyg | | Cyg | MS | 21h 07m | +38° 44' | 5.2 | 28" |
| M 39 | | Cyg | OC | 21h 32m | +48° 26' | 5.3 | 29' |
| Mu Cyg | | Cyg | MS | 21h 44m | +28° 45' | 4.5 | 200" |

| Designation | Name | Con: | Type | R.A.: | Dec: | Mag | Size/Sep |
|---|---|---|---|---|---|---|---|
| V460 | | Cyg | Var/CS | 21h 42m | +35° 31' | 5.6-7.0 | N/A |
| NGC 7000 | North American Nebula | Cyg | Neb | 20h 59m | +44° 22' | 4.0 | 120' |
| | Northern Coalsack | Cyg | DN | 20h 41m | +43° 00' | 6.0 | 60' |
| Gam Del | | Del | MS | 20h 47m | +16° 07' | 3.9 | 9" |
| 16/17 Dra | | Dra | MS | 16h 36m | +52° 55' | 5.1 | 90" |
| Nu Dra | Kuma | Dra | MS | 17h 32m | +55° 10' | 4.9 | 63" |
| Mu Dra | | Dra | MS | 17h 05m | +54° 28' | 5.8 | 2" |
| Psi Dra | | Dra | MS | 17h 42m | +72° 09' | 4.6 | 30" |
| 39 Dra | | Dra | MS | 18h 24m | +58° 48' | 5.0 | 89" |
| 40/41 Dra | | Dra | MS | 18h 00m | +80° 00' | 5.7 | 222" |
| Eps Dra | | Dra | MS | 19h 48m | +70° 16' | 3.8 | 3" |
| Eps Equ | | Equ | MS | 20h 59m | +04° 18' | 5.2 | 11" |
| M 13 | Keystone Cluster | Her | GC | 16h 42m | +36° 27' | 5.8 | 20' |
| Kap Her | Marfik | Her | MS | 16h 41m | +31° 36' | 5.0 | 28" |
| Alp Her | Rasalgethi | Her | MS | 17h 15m | +14° 23' | 3.1 | 5" |
| Del Her | Sarin | Her | MS | 17h 15m | +24° 50' | 3.1 | 14" |
| Rho Her | | Her | MS | 17h 24m | +37° 09' | 4.2 | 4" |
| 95 Her | | Her | MS | 18h 02m | +21° 36' | 4.3 | 6" |
| 100 Her | | Her | MS | 18h 08m | +26° 06' | 5.8 | 14" |
| Eps Lyr | The Double Double | Lyr | MS | 18h 44m | +39° 40' | 4.7 | 3" |
| Bet Lyr | Sheliak | Lyr | MS | 18h 50m | +33° 22' | 3.2 | 86" |
| Del Lyr | | Lyr | MS | 18h 54m | +36° 58' | 4.2 | 630" |
| Zet Lyr | | Lyr | MS | 18h 45m | +37° 36' | 4.4 | 44" |
| IC 4665 | Summer Beehive | Oph | OC | 17h 46m | +05° 43' | 5.3 | 70' |
| M 12 | | Oph | GC | 16h 47m | -01° 57' | 6.1 | 16' |
| Rho Oph | | Oph | MS | 16h 26m | -23° 27' | 4.6 | 3" |
| 36 Oph | | Oph | MS | 17h 15m | -26° 36' | 4.3 | 730" |
| Omi Oph | | Oph | MS | 17h 18m | -24° 17' | 5.1 | 10" |
| 61 Oph | | Oph | MS | 17h 45m | +02° 35' | 6.2 | 21" |
| NGC 6633 | Tweedledum Cluster | Oph | OC | 18h 27m | +06° 31' | 5.6 | 20' |
| 70 Oph | | Oph | MS | 18h 06m | +02° 30' | 4.2 | 4" |
| IC 4756 | | Ser | OC | 18h 39m | +05° 27' | 5.4 | 39' |
| M 16 | Eagle Nebula | Ser | Neb | 18h 19m | -13° 49' | 6.0 | 9' |
| The Ser | | Ser | MS | 18h 56m | +04° 12' | 4.3 | 22" |
| The Sge | | Sge | MS | 20h 10m | +20° 55' | 4.6 | 84" |
| 15 Sge | | Sge | MS | 20h 04m | +17° 04' | 5.8 | 204" |
| M 17 | Swan Nebula | Sgr | Neb | 18h 21m | -16° 11' | 6.0 | 11' |
| M 25 | | Sgr | OC | 18h 32m | -19° 07' | 6.2 | 29' |
| M 24 | Sagittarius Star Cloud | Sgr | OC | 18h 18m | -18° 24' | 3.1 | 90' |
| M 22 | | Sgr | GC | 18h 36m | -23° 54' | 5.2 | 32' |
| M 8 | Lagoon Nebula | Sgr | Neb | 18h 04m | -24° 23' | 5.0 | 17' |
| M 23 | | Sgr | OC | 17h 57m | -18° 59' | 5.9 | 29' |
| Collinder 399 | Coathanger | Vul | Ast | 19h 25m | +20° 11' | 4.0 | 89' |
| NGC 6885 | | Vul | OC | 20h 12m | +26° 29' | 5.7 | 20' |

# Chart 20

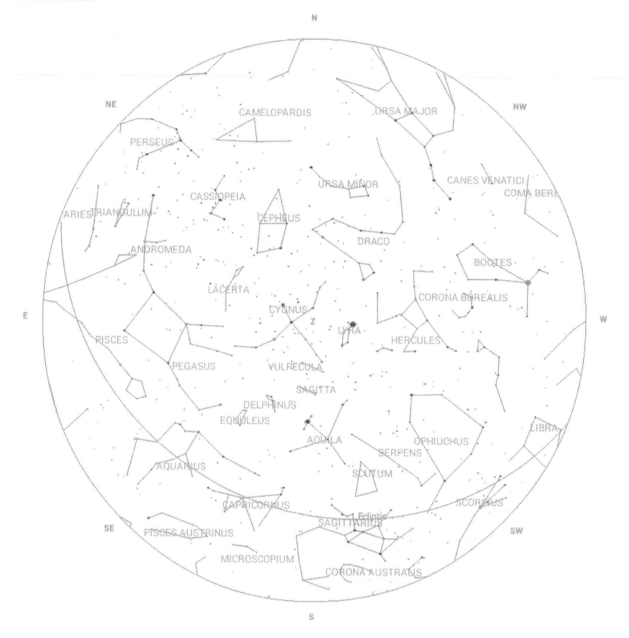

| Designation | Name | Con. | Type | R.A. | Dec. | Mag | Size/Sep |
|---|---|---|---|---|---|---|---|
| 15 Aql | | Aql | MS | 19h 06m | -04° 00' | 5.4 | 39" |
| 57 Aql | | Aql | MS | 19h 55m | -08° 14' | 5.7 | 36" |
| Alp Cap | Al Giedi | Cap | MS | 20h 18m | -12° 32' | 3.6 | 45" |
| Bet Cap | Dabih | Cap | MS | 20h 21m | -14° 47' | 3.1 | 205" |
| Omi Cap | | Cap | MS | 20h 30m | -18° 35' | 5.9 | 22" |
| IC 1396 | Misty Clover Cluster | Cep | OC | 21h 39m | +57° 30' | 5.1 | 89' |
| Bet Cep | Alfirk | Cep | MS | 21h 29m | +70° 34' | 3.2 | 13" |
| Del Cep | | Cep | MS/Var | 22h 29m | +58° 25' | 3.5-4.4 | 41" |
| Struve 2816 | | Cep | MS | 21h 39m | +57° 29' | 5.7 | 20" |

| Designation | Name | Con. | Type | R.A. | Dec. | Mag | Size/Sep |
|---|---|---|---|---|---|---|---|
| Mu Cep | Herschel's Garnet Star | Cep | Var/CS | 21h 44m | +58° 47' | 3.4-5.1 | N/A |
| Struve 2840 | | Cep | MS | 21h 52m | +55° 48' | 5.7 | 18" |
| Xi Cep | Alkurhah | Cep | MS | 22h 04m | +64° 38' | 4.3 | 8" |
| Bet Cyg | Albireo | Cyg | MS | 19h 31m | +27° 58' | 3.1 | 34" |
| Del Cyg | | Cyg | MS | 19h 45m | +45° 08' | 2.9 | 2" |
| 16 Cyg | | Cyg | MS | 19h 42m | +50° 32' | 5.9 | 39" |
| 61 Cyg | | Cyg | MS | 21h 07m | +38° 44' | 5.2 | 29" |
| M 39 | | Cyg | OC | 21h 32m | +48° 26' | 5.3 | 29' |
| Mu Cyg | | Cyg | MS | 21h 44m | +28° 45' | 4.5 | 200" |
| NGC 7000 | North American Nebula | Cyg | Neb | 20h 59m | +44° 22' | 4.0 | 120' |
| Gam Del | | Del | MS | 20h 47m | +16° 07' | 3.9 | 9" |
| Nu Dra | Kuma | Dra | MS | 17h 32m | +55° 10' | 4.9 | 63" |
| Mu Dra | | Dra | MS | 17h 05m | +54° 28' | 5.8 | 2" |
| Psi Dra | | Dra | MS | 17h 42m | +72° 09' | 4.6 | 30" |
| 39 Dra | | Dra | MS | 18h 24m | +58° 48' | 5.0 | 89" |
| 40/41 Dra | | Dra | MS | 18h 00m | +80° 00' | 5.7 | 222" |
| Eps Dra | | Dra | MS | 19h 48m | +70° 16' | 3.8 | 3" |
| Eps Equ | | Equ | MS | 20h 59m | +04° 18' | 5.2 | 11" |
| Alp Her | Rasalgethi | Her | MS | 17h 15m | +14° 23' | 3.1 | 5" |
| Del Her | Sarin | Her | MS | 17h 15m | +24° 50' | 3.1 | 14" |
| Rho Her | | Her | MS | 17h 24m | +37° 09' | 4.2 | 4" |
| 95 Her | | Her | MS | 18h 02m | +21° 36' | 4.3 | 6" |
| 100 Her | | Her | MS | 18h 08m | +26° 06' | 5.8 | 14" |
| 8 Lac | | Lac | MS | 22h 36m | +39° 38' | 5.7 | 82" |
| Eps Lyr | The Double Double | Lyr | MS | 18h 44m | +39° 40' | 4.7 | 3" |
| Bet Lyr | Sheliak | Lyr | MS | 18h 50m | +33° 22' | 3.2 | 86" |
| Del Lyr | | Lyr | MS | 18h 54m | +36° 58' | 4.2 | 630" |
| Zet Lyr | | Lyr | MS | 18h 45m | +37° 36' | 4.4 | 44" |
| IC 4665 | Summer Beehive | Oph | OC | 17h 46m | +05° 43' | 5.3 | 70' |
| 36 Oph | | Oph | MS | 17h 15m | -26° 36' | 4.3 | 730" |
| Omi Oph | | Oph | MS | 17h 18m | -24° 17' | 5.1 | 10" |
| NGC 6633 | Tweedledum Cluster | Oph | OC | 18h 27m | +06° 31' | 5.6 | 20' |
| 70 Oph | | Oph | MS | 18h 06m | +02° 30' | 4.2 | 4" |
| Eps Peg | Enif | Peg | MS | 21h 44m | +09° 52' | 2.1 | 143" |
| IC 4756 | | Ser | OC | 18h 39m | +05° 27' | 5.4 | 39' |
| The Ser | | Ser | MS | 18h 56m | +04° 12' | 4.3 | 22" |
| The Sge | | Sge | MS | 20h 10m | +20° 55' | 4.6 | 84" |
| 15 Sge | | Sge | MS | 20h 04m | +17° 04' | 5.8 | 204" |
| M 24 | Sagittarius Star Cloud | Sgr | OC | 18h 18m | -18° 24' | 3.1 | 90' |
| M 22 | | Sgr | GC | 18h 36m | -23° 54' | 5.2 | 32' |
| M 8 | Lagoon Nebula | Sgr | Neb | 18h 04m | -24° 23' | 5.0 | 17' |
| M 23 | | Sgr | OC | 17h 57m | -18° 59' | 5.9 | 29' |
| Collinder 399 | Coathanger | Vul | Ast | 19h 25m | +20° 11' | 4.8 | 89' |
| NGC 6885 | | Vul | OC | 20h 12m | +26° 29' | 5.7 | 20' |

Chart 21

| Designation | Name | Con: | Type | R.A. | Dec: | Mag | Size/Sep |
|---|---|---|---|---|---|---|---|
| Struve 2404 | | Aql | MS | 18h 51m | +10° 59' | 6.4 | 4" |
| 15 Aql | | Aql | MS | 19h 06m | -04° 00' | 5.4 | 39" |
| 57 Aql | | Aql | MS | 19h 55m | -08° 14' | 5.7 | 36" |
| V Aql | | Aql | Var | 19h 04m | -05° 41' | 6.6-8.4 | N/A |
| 107 Aqr | | Aqr | MS | 23h 47m | -18° 35' | 5.3 | 7" |
| 94 Aqr | | Aqr | MS | 23h 19m | -13° 28' | 5.2 | 13" |
| NGC 7293 | Helix Nebula | Aqr | PN | 22h 30m | -20° 50' | 6.3 | 16' |
| Zet Aqr | | Aqr | MS | 22h 29m | -00° 01' | 3.7 | 2" |
| 41 Aqr | | Aqr | MS | 22h 14m | -21° 04' | 5.3 | 5" |

| Designation | Name | Con: | Type | R.A.: | Dec.: | Mag | Size/Sep |
|---|---|---|---|---|---|---|---|
| 53 Aqr | | Aqr | MS | 22h 27m | -16° 45' | 5.8 | 3" |
| M 2 | | Aqr | GC | 21h 33m | -00° 49' | 6.6 | 16' |
| Alp Cap | Al Giedi | Cap | MS | 20h 18m | -12° 32' | 3.6 | 45" |
| Bet Cap | Dabih | Cap | MS | 20h 21m | -14° 47' | 3.1 | 205" |
| RT Cap | | Cap | Var/ES | 20h 17m | -21° 19' | 6.5-8.1 | N/A |
| Omi Cap | | Cap | MS | 20h 30m | -18° 35' | 5.9 | 22" |
| Sig Cas | | Cas | MS | 23h 59m | +55° 45' | 4.9 | 3" |
| IC 1396 | Misty Clover Cluster | Cep | OC | 21h 39m | +57° 30' | 5.1 | 89' |
| Bet Cep | Alfirk | Cep | MS | 21h 29m | +70° 34' | 3.2 | 13" |
| Del Cep | | Cep | MS/Var | 22h 29m | +58° 25' | 3.5-4.4 | 41" |
| Struve 2816 | | Cep | MS | 21h 39m | +57° 29' | 5.7 | 20" |
| Mu Cep | Herschel's Garnet Star | Cep | Var/ES | 21h 44m | +58° 47' | 3.4-5.1 | N/A |
| Struve 2840 | | Cep | MS | 21h 52m | +55° 48' | 5.7 | 18" |
| Xi Cep | Alkurhah | Cep | MS | 22h 04m | +64° 38' | 4.3 | 8" |
| Omi Cep | | Cep | MS | 23h 19m | +68° 07' | 4.8 | 3" |
| Bet Cyg | Albireo | Cyg | MS | 19h 31m | +27° 58' | 3.1 | 34" |
| Del Cyg | | Cyg | MS | 19h 45m | +45° 08' | 2.9 | 2" |
| 16 Cyg | | Cyg | MS | 19h 42m | +50° 32' | 5.9 | 39" |
| 61 Cyg | | Cyg | MS | 21h 07m | +38° 44' | 5.2 | 29" |
| M 39 | | Cyg | OC | 21h 32m | +48° 26' | 5.3 | 29' |
| Mu Cyg | | Cyg | MS | 21h 44m | +28° 45' | 4.5 | 200" |
| V460 | | Cyg | Var/ES | 21h 42m | +35° 31' | 5.8-7.0 | N/A |
| NGC 7000 | North American Nebula | Cyg | Neb | 20h 59m | +44° 22' | 4.0 | 120' |
| | Northern Coalsack | Cyg | DN | 20h 41m | +43° 00' | 8.0 | 60' |
| Gam Del | | Del | MS | 20h 47m | +16° 07' | 3.9 | 9" |
| 39 Dra | | Dra | MS | 18h 24m | +58° 48' | 5.0 | 89" |
| 40/41 Dra | | Dra | MS | 18h 00m | +80° 00' | 5.7 | 222" |
| Eps Dra | | Dra | MS | 19h 48m | +70° 16' | 3.8 | 3" |
| UX Dra | | Dra | Var/ES | 19h 22m | +76° 34' | 5.9-7.1 | N/A |
| Eps Equ | | Equ | MS | 20h 59m | +04° 18' | 5.2 | 11" |
| 95 Her | | Her | MS | 18h 02m | +21° 36' | 4.3 | 6" |
| 100 Her | | Her | MS | 18h 08m | +26° 06' | 5.8 | 14" |
| 8 Lac | | Lac | MS | 22h 36m | +39° 38' | 5.7 | 82" |
| Eps Lyr | The Double Double | Lyr | MS | 18h 44m | +39° 40' | 4.7 | 3" |
| Bet Lyr | Sheliak | Lyr | MS | 18h 50m | +33° 22' | 3.3 | 86" |
| Del Lyr | | Lyr | MS | 18h 54m | +36° 58' | 4.3 | 630" |
| Zet Lyr | | Lyr | MS | 18h 45m | +37° 36' | 4.4 | 44" |
| M 15 | | Peg | GC | 21h 30m | +12° 10' | 6.3 | 18' |
| Eps Peg | Enif | Peg | MS | 21h 44m | +09° 52' | 2.1 | 143" |
| M 11 | Wild Duck Cluster | Sct | OC | 18h 51m | -06° 16' | 6.1 | 32' |
| The Sge | | Sge | MS | 20h 10m | +20° 55' | 4.6 | 84" |
| 15 Sge | | Sge | MS | 20h 04m | +17° 04' | 5.8 | 204" |
| Collinder 399 | Coathanger | Vul | Ast | 19h 25m | +20° 11' | 4.8 | 89' |
| NGC 6885 | | Vul | OC | 20h 12m | +26° 29' | 5.7 | 20' |

# Chart 22

| Designation | Name | Con. | Type | R.A. | Dec. | Mag | Size/Sep |
|---|---|---|---|---|---|---|---|
| M 31 | Andromeda Galaxy | And | Gx | 00h 43m | +41° 16' | 4.3 | 156' |
| Pi And | | And | MS | 00h 38m | +33° 49' | 4.4 | 36" |
| 15 Aql | | Aql | MS | 19h 06m | -04° 00' | 5.4 | 39" |
| 57 Aql | | Aql | MS | 19h 55m | -08° 14' | 5.7 | 36" |
| 107 Aqr | | Aqr | MS | 23h 47m | -18° 35' | 5.3 | 7" |
| 94 Aqr | | Aqr | MS | 23h 19m | -13° 28' | 5.2 | 13" |
| NGC 7293 | Helix Nebula | Aqr | PN | 22h 30m | -20° 50' | 6.3 | 16' |
| Zet Aqr | | Aqr | MS | 22h 29m | -00° 01' | 3.7 | 2" |

| Designation | Name | Con. | Type | R.A. | Dec. | Mag | Size/Sep |
|---|---|---|---|---|---|---|---|
| 41 Aqr | | Aqr | MS | 22h 14m | -21° 04' | 5.3 | 5" |
| 53 Aqr | | Aqr | MS | 22h 27m | -16° 45' | 5.6 | 3" |
| M 2 | | Aqr | GC | 21h 33m | -00° 49' | 6.6 | 16' |
| M 30 | | Cap | GC | 21h 40m | -23° 11' | 6.9 | 12' |
| Alp Cap | Al Giedi | Cap | MS | 20h 18m | -12° 32' | 3.6 | 45" |
| Bet Cap | Dabih | Cap | MS | 20h 21m | -14° 47' | 3.1 | 205" |
| Omi Cap | | Cap | MS | 20h 30m | -18° 35' | 5.9 | 22" |
| Struve 3053 | | Cas | MS | 00h 03m | +66° 06' | 5.9 | 15" |
| Eta Cas | Achird | Cas | MS | 00h 50m | +57° 54' | 3.6 | 13" |
| Sig Cas | | Cas | MS | 23h 59m | +55° 45' | 4.9 | 3" |
| IC 1396 | Misty Clover Cluster | Cep | OC | 21h 39m | +57° 30' | 5.1 | 89' |
| Bet Cep | Alfirk | Cep | MS | 21h 29m | +70° 34' | 3.2 | 13" |
| Del Cep | | Cep | MS/Var | 22h 29m | +58° 25' | 3.5-4.4 | 41" |
| Struve 2816 | | Cep | MS | 21h 39m | +57° 29' | 5.7 | 20" |
| Mu Cep | Herschel's Garnet Star | Cep | Var/CS | 21h 44m | +58° 47' | 3.4-5.1 | N/A |
| Struve 2840 | | Cep | MS | 21h 52m | +55° 48' | 5.7 | 18" |
| Xi Cep | Alkurhah | Cep | MS | 22h 04m | +64° 38' | 4.3 | 8" |
| Omi Cep | | Cep | MS | 23h 19m | +68° 07' | 4.8 | 3" |
| Bet Cyg | Albireo | Cyg | MS | 19h 31m | +27° 58' | 3.1 | 34" |
| Del Cyg | | Cyg | MS | 19h 45m | +45° 08' | 2.9 | 2" |
| 16 Cyg | | Cyg | MS | 19h 42m | +50° 32' | 5.9 | 39" |
| NGC 6960 | Veil Nebula (West) | Cyg | SNR | 20h 46m | +30° 43' | 7.0 | 63' |
| 61 Cyg | | Cyg | MS | 21h 07m | +38° 44' | 5.2 | 29" |
| M 39 | | Cyg | OC | 21h 32m | +48° 26' | 5.3 | 29' |
| Mu Cyg | | Cyg | MS | 21h 44m | +28° 45' | 4.5 | 200" |
| V460 | | Cyg | Var/CS | 21h 42m | +35° 31' | 5.6-7.0 | N/A |
| NGC 6992 | Veil Nebula (East) | Cyg | SNR | 20h 56m | +31° 43' | 7.0 | 60' |
| NGC 7000 | North American Nebula | Cyg | Neb | 20h 59m | +44° 22' | 4.0 | 120' |
| | Northern Coalsack | Cyg | DN | 20h 41m | +43° 00' | 6.0 | 60' |
| Gam Del | | Del | MS | 20h 47m | +16° 07' | 3.9 | 9" |
| Eps Dra | | Dra | MS | 19h 48m | +70° 16' | 3.8 | 3" |
| Eps Equ | | Equ | MS | 20h 59m | +04° 18' | 5.2 | 11" |
| NGC 7243 | | Lac | OC | 22h 15m | +49° 54' | 6.7 | 29' |
| 8 Lac | | Lac | MS | 22h 36m | +39° 38' | 5.7 | 82" |
| Struve 2470/2474 | The Double Double's Double | Lyr | MS | 19h 09m | +34° 41' | 6.7 | 16" |
| M 15 | | Peg | GC | 21h 30m | +12° 10' | 6.3 | 18' |
| Eps Peg | Enif | Peg | MS | 21h 44m | +09° 52' | 2.1 | 143" |
| TX Psc | | Psc | Var/CS | 23h 46m | +03° 29' | 4.5-5.3 | N/A |
| 55 Psc | | Psc | MS | 00h 40m | +21° 26' | 5.4 | 6" |
| 65 Psc | | Psc | MS | 00h 50m | +27° 43' | 7.0 | 4" |
| The Sge | | Sge | MS | 20h 10m | +20° 55' | 4.6 | 84" |
| 15 Sge | | Sge | MS | 20h 04m | +17° 04' | 5.8 | 204" |
| Collinder 399 | Coathanger | Vul | Ast | 19h 25m | +20° 11' | 4.8 | 89' |
| NGC 6885 | | Vul | OC | 20h 12m | +26° 29' | 5.7 | 20' |

# Chart 23

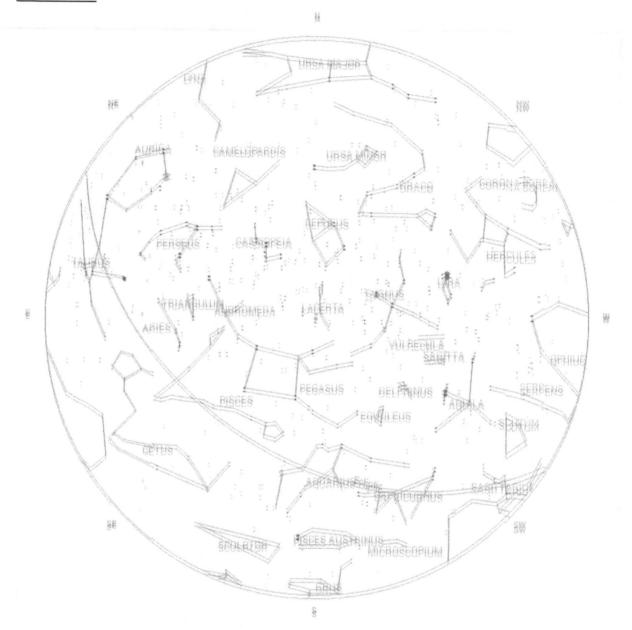

| Designation | Name | Con: | Type | R.A: | Dec: | Mag | Size/Sep |
|---|---|---|---|---|---|---|---|
| M 31 | Andromeda Galaxy | And | Gx | 00h 43m | +41° 16' | 4.3 | 156' |
| NGC 752 | Golf Ball Cluster | And | OC | 01h 58m | +37° 47' | 6.6 | 75' |
| Pi And | | And | MS | 00h 38m | +33° 49' | 4.4 | 36" |
| 107 Aqr | | Aqr | MS | 23h 47m | -18° 35' | 5.3 | 7" |
| 94 Aqr | | Aqr | MS | 23h 19m | -13° 28' | 5.2 | 13" |
| NGC 7293 | Helix Nebula | Aqr | PN | 22h 30m | -20° 50' | 6.3 | 16' |
| Zet Aqr | | Aqr | MS | 22h 29m | -00° 01' | 3.7 | 2" |
| 41 Aqr | | Aqr | MS | 22h 14m | -21° 04' | 5.3 | 5" |

| Designation | Name | Con. | Type | R.A. | Dec. | Mag | Size/Sep |
|---|---|---|---|---|---|---|---|
| 53 Aqr | | Aqr | MS | 22h 27m | -16° 45' | 5.8 | 3" |
| M 2 | | Aqr | GC | 21h 33m | -00° 49' | 6.6 | 16' |
| Lam Ari | | Ari | MS | 01h 59m | +23° 41' | 4.8 | 37" |
| Gam Ari | Mesarthim | Ari | MS | 01h 54m | +19° 22' | 4.6 | 8" |
| M 30 | | Cap | GC | 21h 40m | -23° 11' | 6.9 | 12' |
| Alp Cap | Al Giedi | Cap | MS | 20h 18m | -12° 32' | 3.6 | 45" |
| Bet Cap | Dabih | Cap | MS | 20h 21m | -14° 47' | 3.1 | 205" |
| Omi Cap | | Cap | MS | 20h 30m | -18° 35' | 5.9 | 22" |
| NGC 457 | Owl Cluster | Cas | OC | 01h 20m | +58° 17' | 5.1 | 20' |
| Struve 163 | | Cas | MS | 01h 51m | +64° 51' | 6.5 | 35" |
| M 103 | | Cas | OC | 01h 33m | +60° 39' | 6.9 | 5' |
| Struve 3053 | | Cas | MS | 00h 03m | +66° 06' | 5.9 | 15" |
| Eta Cas | Achird | Cas | MS | 00h 50m | +57° 54' | 3.6 | 13" |
| Sig Cas | | Cas | MS | 23h 59m | +55° 45' | 4.9 | 3" |
| NGC 663 | | Cas | OC | 01h 46m | +61° 14' | 6.4 | 14' |
| IC 1396 | Misty Clover Cluster | Cep | OC | 21h 39m | +57° 30' | 5.1 | 89' |
| Bet Cep | Alfirk | Cep | MS | 21h 29m | +70° 34' | 3.2 | 13" |
| Del Cep | | Cep | MS/Var | 22h 29m | +58° 25' | 3.5-4.4 | 41" |
| Struve 2816 | | Cep | MS | 21h 39m | +57° 29' | 5.7 | 20" |
| Mu Cep | Herschel's Garnet Star | Cep | Var/CS | 21h 44m | +58° 47' | 3.4-5.1 | N/A |
| Struve 2840 | | Cep | MS | 21h 52m | +55° 48' | 5.7 | 18" |
| Xi Cep | Alkurhah | Cep | MS | 22h 04m | +64° 38' | 4.3 | 8" |
| Omi Cep | | Cep | MS | 23h 19m | +68° 07' | 4.8 | 3" |
| NGC 6960 | Veil Nebula (West) | Cyg | SNR | 20h 46m | +30° 43' | 7.0 | 63' |
| 61 Cyg | | Cyg | MS | 21h 07m | +38° 44' | 5.2 | 29" |
| M 39 | | Cyg | OC | 21h 32m | +48° 26' | 5.3 | 29' |
| Mu Cyg | | Cyg | MS | 21h 44m | +28° 45' | 4.5 | 200" |
| NGC 6992 | Veil Nebula (East) | Cyg | SNR | 20h 56m | +31° 43' | 7.0 | 60' |
| NGC 7000 | North American Nebula | Cyg | Neb | 20h 59m | +44° 22' | 4.0 | 120' |
| | Northern Coalsack | Cyg | DN | 20h 41m | +43° 00' | 6.0 | 60' |
| Gam Del | | Del | MS | 20h 47m | +16° 07' | 3.9 | 9" |
| Eps Equ | | Equ | MS | 20h 59m | +04° 18' | 5.2 | 11" |
| NGC 7243 | | Lac | OC | 22h 15m | +49° 54' | 6.7 | 29' |
| 8 Lac | | Lac | MS | 22h 36m | +39° 38' | 5.7 | 82" |
| M 15 | | Peg | GC | 21h 30m | +12° 10' | 6.3 | 18' |
| Eps Peg | Enif | Peg | MS | 21h 44m | +09° 52' | 2.1 | 143" |
| 55 Psc | | Psc | MS | 00h 40m | +21° 26' | 5.4 | 6" |
| 65 Psc | | Psc | MS | 00h 50m | +27° 43' | 7.0 | 4" |
| Psi1 Psc | | Psc | MS | 01h 06m | +21° 28' | 5.3 | 30" |
| Zet Psc | | Psc | MS | 01h 14m | +07° 35' | 5.2 | 23" |
| The Sge | | Sge | MS | 20h 10m | +20° 55' | 4.6 | 84" |
| 15 Sge | | Sge | MS | 20h 04m | +17° 04' | 5.8 | 204" |
| M 33 | Triangulum Galaxy | Tri | Gx | 01h 34m | +30° 40' | 6.4 | 62' |
| NGC 6885 | | Vul | OC | 20h 12m | +26° 29' | 5.7 | 20' |

# Chart 24

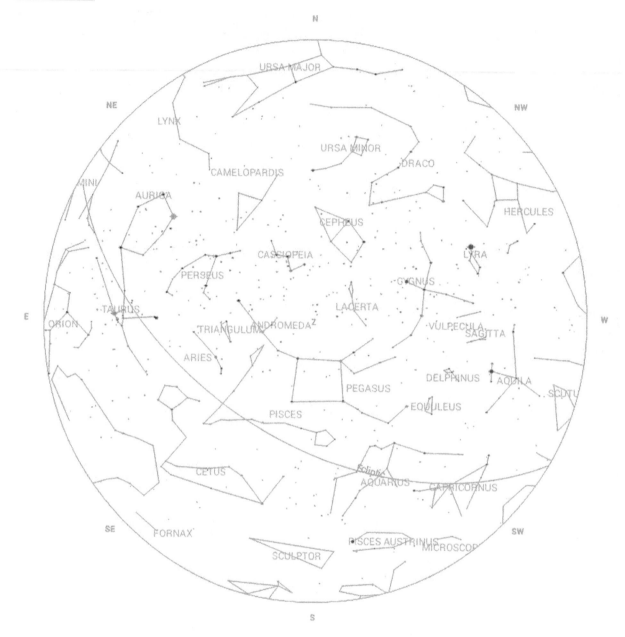

| Designation | Name | Con. | Type | R.A. | Dec. | Mag | Size/Sep |
|---|---|---|---|---|---|---|---|
| Gam And | Almach | And | MS | 02h 04m | +42° 20' | 2.1 | 10" |
| M 31 | Andromeda Galaxy | And | Gx | 00h 43m | +41° 16' | 4.3 | 156' |
| NGC 752 | Golf Ball Cluster | And | OC | 01h 58m | +37° 47' | 6.6 | 75' |
| Pi And | | And | MS | 00h 38m | +33° 49' | 4.4 | 36" |
| 107 Aqr | | Aqr | MS | 23h 47m | -18° 35' | 5.3 | 7" |
| 94 Aqr | | Aqr | MS | 23h 19m | -13° 28' | 5.2 | 13" |
| NGC 7293 | Helix Nebula | Aqr | PN | 22h 30m | -20° 50' | 6.3 | 16' |
| Zet Aqr | | Aqr | MS | 22h 29m | -00° 01' | 3.7 | 2" |

| Designation | Name | Con. | Type | R.A. | Dec. | Mag | Size/Sep |
|---|---|---|---|---|---|---|---|
| 41 Aqr | | Aqr | MS | 22h 14m | -21° 04' | 5.3 | 5" |
| 53 Aqr | | Aqr | MS | 22h 27m | -16° 45' | 5.6 | 3" |
| M 2 | | Aqr | GC | 21h 33m | -00° 49' | 6.6 | 16' |
| Lam Ari | | Ari | MS | 01h 59m | +23° 41' | 4.8 | 37" |
| Gam Ari | Mesarthim | Ari | MS | 01h 54m | +19° 22' | 4.6 | 8" |
| 30 Ari | | Ari | MS | 02h 37m | +24° 39' | 6.5 | 39" |
| M 30 | | Cap | GC | 21h 40m | -23° 11' | 6.9 | 12' |
| NGC 457 | Owl Cluster | Cas | OC | 01h 20m | +58° 17' | 5.1 | 20' |
| Struve 163 | | Cas | MS | 01h 51m | +64° 51' | 6.5 | 35" |
| M 103 | | Cas | OC | 01h 33m | +60° 39' | 6.9 | 5' |
| Struve 3053 | | Cas | MS | 00h 03m | +66° 06' | 5.9 | 15" |
| Eta Cas | Achird | Cas | MS | 00h 50m | +57° 54' | 3.6 | 13" |
| Sig Cas | | Cas | MS | 23h 59m | +55° 45' | 4.9 | 3" |
| Iot Cas | | Cas | MS | 02h 29m | +67° 24' | 4.5 | 7" |
| NGC 659 | Ying Yang Cluster | Cas | OC | 01h 44m | +60° 40' | 7.2 | 5' |
| NGC 663 | | Cas | OC | 01h 46m | +61° 14' | 6.4 | 14' |
| IC 1396 | Misty Clover Cluster | Cep | OC | 21h 39m | +57° 30' | 5.1 | 89' |
| Bet Cep | Alfirk | Cep | MS | 21h 29m | +70° 34' | 3.2 | 13" |
| Del Cep | | Cep | MS/Var | 22h 29m | +58° 25' | 3.5-4.4 | 41" |
| Struve 2816 | | Cep | MS | 21h 39m | +57° 29' | 5.7 | 20" |
| Mu Cep | Herschel's Garnet Star | Cep | Var/CS | 21h 44m | +58° 47' | 3.4-5.1 | N/A |
| Struve 2840 | | Cep | MS | 21h 52m | +55° 48' | 5.7 | 18" |
| Xi Cep | Alkurhah | Cep | MS | 22h 04m | +64° 38' | 4.3 | 8" |
| Omi Cep | | Cep | MS | 23h 19m | +68° 07' | 4.8 | 3" |
| Gam Cet | Kaffajidhma | Cet | MS | 02h 43m | +03° 14' | 3.5 | 3" |
| Omi Cet | Mira | Cet | Var | 02h 19m | -02° 59' | 2.0-10.1 | N/A |
| 61 Cyg | | Cyg | MS | 21h 07m | +38° 44' | 5.2 | 29" |
| M 39 | | Cyg | OC | 21h 32m | +48° 26' | 5.3 | 29' |
| Mu Cyg | | Cyg | MS | 21h 44m | +28° 45' | 4.5 | 200" |
| V460 | | Cyg | Var/CS | 21h 42m | +35° 31' | 5.6-7.0 | N/A |
| NGC 7243 | | Lac | OC | 22h 15m | +49° 54' | 6.7 | 29' |
| 8 Lac | | Lac | MS | 22h 36m | +39° 38' | 5.7 | 82" |
| M 34 | | Per | OC | 02h 42m | +42° 46' | 5.8 | 35' |
| NGC 869/884 | Double Cluster | Per | OC | 02h 21m | +57° 08' | 4.4 | 18' |
| Eta Per | | Per | MS | 02h 51m | +55° 54' | 3.8 | 29" |
| M 15 | | Peg | GC | 21h 30m | +12° 10' | 6.3 | 18' |
| Eps Peg | Enif | Peg | MS | 21h 44m | +09° 52' | 2.1 | 143" |
| Alp Psc | Alrisha | Psc | MS | 02h 02m | +02° 46' | 3.8 | 2" |
| TX Psc | | Psc | Var/CS | 23h 46m | +03° 29' | 4.5-5.3 | N/A |
| 55 Psc | | Psc | MS | 00h 40m | +21° 26' | 5.4 | 6" |
| Psi1 Psc | | Psc | MS | 01h 06m | +21° 28' | 5.3 | 30" |
| Zet Psc | | Psc | MS | 01h 14m | +07° 35' | 5.2 | 23" |
| M 33 | Triangulum Galaxy | Tri | Gx | 01h 34m | +30° 40' | 6.4 | 62' |
| Alp UMi | Polaris | UMi | MS | 02h 51m | +89° 20' | 2.0 | 18" |

# 2020 (July to December)

# Planet Visibility Ratings

| | | Morning Sky | | | | | | | | Evening Sky | | | | | |
|---|---|---|---|---|---|---|---|---|---|---|---|---|---|---|---|
| | | Me | Ve | Ma | Ju | Sa | Ur | Ne | Me | Ve | Ma | Ju | Sa | Ur | Ne |
| Jan | 5th | | | * | | | | | | ** | | | | *** | ** |
| | 15th | | | * | * | | | | | ** | | | | ** | ** |
| | 25th | | | * | * | | | | | ** | | | | ** | ** |
| Feb | 5th | | | * | * | ** | | | ** | ** | | | | *** | ** |
| | 15th | | | * | * | ** | | | | *** | | | | ** | * |
| | 25th | | | * | ** | ** | | | | *** | | | | * | |
| Mar | 5th | *** | | * | ** | ** | | | | *** | | | | * | |
| | 15th | **** | | * | ** | ** | | | | *** | | | | * | |
| | 25th | *** | | * | ** | ** | | | | *** | | | | * | |
| Apr | 5th | *** | | ** | ** | ** | | * | | *** | | | | | |
| | 15th | ** | | ** | *** | *** | | * | | **** | | | | | |
| | 25th | | | ** | *** | *** | | ** | | **** | | | | | |
| May | 5th | | | ** | *** | *** | | ** | | **** | | | | | |
| | 15th | | | ** | **** | *** | * | ** | | *** | | | | | |
| | 25th | | | ** | **** | *** | * | ** | *** | | | | | | |
| Jun | 5th | | | ** | **** | **** | * | *** | **** | | | | | | |
| | 15th | | *** | ** | **** | **** | * | *** | *** | | | | | | |
| | 25th | | *** | *** | **** | **** | * | *** | | | | | | | |
| Jul | 5th | | **** | *** | ***** | **** | ** | **** | | | | | | | |
| | 15th | *** | **** | *** | | ***** | ** | **** | | | | ***** | | | |
| | 25th | *** | **** | *** | | | ** | **** | | | | ***** | ***** | | |
| Aug | 5th | | **** | **** | | | ** | **** | | | | ***** | **** | | |
| | 15th | | *** | **** | | | *** | ***** | | | | **** | **** | | |
| | 25th | | *** | **** | | | *** | ***** | | | | **** | **** | | |
| Sep | 5th | | *** | **** | | | *** | **** | | | | **** | **** | | |
| | 15th | | *** | ***** | | | *** | | ** | | | **** | **** | | ***** |
| | 25th | | ** | ***** | | | *** | | *** | | | *** | *** | | ***** |
| Oct | 5th | | ** | ***** | | | *** | | *** | | | *** | *** | | ***** |
| | 15th | | ** | | | | **** | | *** | | ***** | *** | *** | | **** |
| | 25th | | ** | | | | **** | | | ***** | ***** | *** | *** | | **** |
| Nov | 5th | ** | ** | | | | | | | **** | ** | ** | | **** | **** |
| | 15th | ** | ** | | | | | | | **** | ** | ** | | **** | **** |
| | 25th | | ** | | | | | | | **** | ** | ** | *** | *** | |
| Dec | 5th | | * | | | | | | | *** | ** | ** | *** | *** | |
| | 15th | | * | | | | | | | *** | ** | ** | *** | *** | |
| | 25th | | * | | | | | | | *** | * | ** | *** | ** | |

# Solar and Lunar Eclipses

| Date | Time (UT) | Type | Visible From |
|---|---|---|---|
| Jan 10th | 19:09 | Penumbral Lunar | Africa, Asia, Australia, Europe and north-eastern and northern North America. |
| Jun 5th | 19:24 | Penumbral Lunar | Africa, Asia, Australia, Europe and eastern South America. |
| Jun 21st | 06:41 | Annular Solar | Eastern Africa, Asia and the Indian Ocean. |
| Jul 5th | 04:29 | Penumbral Lunar | Africa, Central America, western Europe, North America and South America. |
| Nov 30th | 09:43 | Penumbral Lunar | Australia, western Europe, North America and South America. |
| Dec 14th | 16:15 | Total Solar | Antarctica, the southern Atlantic, the southern Pacific and South America. |

# Planetary Highlights

| Date | Time (UT) | Elon. | Vis. | Description |
|---|---|---|---|---|
| Jan 16th | 17:45 | 50° W | AM | Mars is 4.8° north of the bright star Antares. (Scorpius) |
| Jan 27th | 19:25 | 39° E | PM | Venus is 0.1° south of Neptune. (Aquarius) |
| Mar 9th | 14:32 | 42° E | PM | Venus is 2.4° north of Uranus. (Aries) |
| Mar 20th | 06:14 | 66° W | AM | Mars is 0.7° south of Jupiter. (Sagittarius) |
| Mar 24th | 02:03 | 25° W | AM | Mercury is at greatest western elongation from the Sun. (Aquarius) |
| Mar 24th | 22:00 | 43° E | PM | Venus is at greatest eastern elongation from the Sun. (Aries) |
| Mar 31st | 10:50 | 67° W | AM | Mars is 0.9° south of Saturn. (Capricornus) |
| Apr 3rd | 11:59 | 44° W | PM | Venus is 0.3° south of the Pleiades star cluster. (Taurus.) |
| May 21st | 07:49 | 19° E | PM | Mercury is 0.9° south of Venus. (Taurus) |
| Jun 4th | 13:04 | 26° E | PM | Mercury is at greatest eastern elongation from the Sun. (Gemini) |
| Jun 12th | 12:20 | 89° W | AM | Mars is 1.7° south of Neptune.(Aquarius) |
| Jul 11th | 17:51 | 42° W | AM | Venus is 1.0° north of the bright star Aldebaran. (Taurus.) |
| Jul 14th | 09:06 | 180° | AN | Jupiter is at opposition. (Sagittarius) |
| Jul 20th | 23:33 | 180° | AN | Saturn is at opposition. (Sagittarius) |
| Aug 12th | 23:49 | 48° W | AM | Venus is at greatest western elongation from the Sun. (Orion) |
| Sep 12th | 09:20 | 180° | AN | Neptune is at opposition. (Aquarius) |
| Oct 1st | 16:00 | 23° E | PM | Mercury is at greatest eastern elongation from the Sun. (Virgo) |
| Oct 2nd | 17:50 | 37° W | AM | Venus is 0.1° south of the bright star Regulus. (Leo.) |
| Oct 14th | 21:17 | 180° | AN | Mars is at opposition. (Pisces) |
| Oct 31st | 17:15 | 180° | AN | Uranus is at opposition. (Aries) |
| Dec 21st | 13:46 | 29° E | PM | Jupiter is 0.1° south of Saturn. (Capricornus) |

# Major Meteor Showers

| Shower Name | Start Date | End Date | Peak | ZHR | Speed | Brightness | Moon |
|---|---|---|---|---|---|---|---|
| Quadrantids | Dec 28th | Jan 12th | Jan 3rd | 120 | *** | ***** | ◑ |
| Lyrids | Apr 18th | Apr 25th | Apr 22nd | 18 | *** | ***** | ● |
| Eta Aquariids | Apr 24th | May 19th | May 7th | 40 | * | **** | ○ |
| June Bootids | Jun 23rd | Jun 25th | Jun 24th | Var | ***** | ***** | ● |
| Alpha Capricornids | Jul 8th | Aug 10th | Jul 27th | 5 | ***** | **** | ◑ |
| Southern Delta Aquariids | Jul 21st | Aug 23rd | Jul 30th | 16 | *** | * | ☽ |
| Perseids | Jul 13th | Aug 26th | Aug 12th | 100 | * | ***** | ● |
| Kappa Cygnids | Aug 6th | Aug 31st | Aug 17th | 3 | ***** | ** | ● |
| Aurigids | Aug 29th | Sep 4th | Sep 1st | 6 | * | **** | ○ |
| September Epsilon Perseids | Sep 5th | Sep 28th | Sep 9th | 5 | * | ** | ◐ |
| Draconids | Oct 6th | Oct 10th | Oct 8th | Var | ***** | *** | ◐ |
| Southern Taurids | Sep 7th | Nov 19th | Oct 10th | 5 | **** | **** | ◐ |
| Orionids | Aug 25th | Nov 19th | Oct 22nd | 15 | * | **** | ◑ |
| Andromedids | Oct 26th | Nov 20th | Nov 8th | Var | ***** | **** | ◐ |
| Northern Taurids | Oct 25th | Dec 4th | Nov 11th | 5 | **** | **** | ● |
| Leonids | Nov 5th | Dec 3rd | Nov 18th | 15 | * | **** | ● |
| Alpha Monocerotids | Nov 21st | Nov 23rd | Nov 21st | Var | * | **** | ◑ |
| Geminids | Nov 30th | Dec 17th | Dec 13th | 120 | **** | *** | ● |
| December Leonis Minorids | Dec 6th | Jan 18th | Dec 20th | 5 | * | ** | ◑ |
| Ursids | Dec 17th | Dec 24th | Dec 22nd | 10 | **** | ** | ◑ |
| Coma Berenicids | Dec 24th | Jan 3rd | Dec 31st | 5 | * | ** | ○ |

# July 1ˢᵗ to 10ᵗʰ, 2020

## The Moon

| 1ˢᵗ | 3ʳᵈ | 5ᵗʰ | 7ᵗʰ | 9ᵗʰ |

| Date | Con | R.A. | Dec | Mag | Diam | Ill. | Elon. | Phase | Close To |
|------|-----|------|-----|-----|------|------|-------|-------|----------|
| 1st | Lib | 15h 22m | -15° 13' | -11.5 | 32' | 83% | 130° E | +G | |
| 2nd | Sco | 16h 19m | -19° 20' | -11.8 | 32' | 91% | 143° E | +G | Antares |
| 3rd | Oph | 17h 18m | -22° 16' | -12.2 | 32' | 96% | 156° E | FM | |
| 4th | Sgr | 18h 18m | -23° 50' | -12.5 | 32' | 99% | 170° E | FM | |
| 5th | Sgr | 19h 17m | -23° 56' | -12.6 | 31' | 100% | 176° W | FM | Jupiter |
| 6th | Cap | 20h 15m | -22° 39' | -12.3 | 31' | 98% | 162° W | FM | Jupiter, Saturn |
| 7th | Cap | 21h 10m | -20° 10' | -12.0 | 31' | 94% | 150° W | -G | |
| 8th | Aqr | 22h 2m | -16° 46' | -11.7 | 30' | 88% | 138° W | -G | |
| 9th | Aqr | 22h 50m | -12° 41' | -11.4 | 30' | 81% | 127° W | -G | Neptune |
| 10th | Aqr | 23h 36m | -8° 11' | -11.0 | 30' | 73% | 116° W | -G | Neptune |

## Mercury and Venus

Mercury
5ᵗʰ

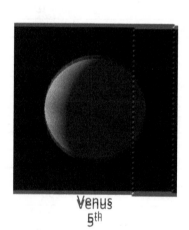

Venus
5ᵗʰ

## Mercury

| Date | Con. | R.A. | Dec. | Mag. | Diam. | Ill. | Elon. | Vis. | Rat. | Close To |
|------|------|------|------|------|-------|------|-------|------|------|----------|
| 1st | Gem | 6h 40m | 18° 36' | 5.0 | 12" | 1% | 1° W | NV | N/A | |
| 3rd | Gem | 6h 35m | 18° 29' | 4.6 | 12" | 1% | 4° W | NV | N/A | |
| 5th | Gem | 6h 30m | 18° 26' | 4.1 | 12" | 3% | 7° W | NV | N/A | |
| 7th | Gem | 6h 27m | 18° 29' | 3.5 | 11" | 5% | 10° W | NV | N/A | |
| 9th | Gem | 6h 24m | 18° 37' | 2.9 | 11" | 7% | 13° W | NV | N/A | |

## Venus

| Date | Con. | R.A. | Dec. | Mag. | Diam. | Ill. | Elon. | Vis. | Rat. | Close To |
|------|------|------|------|------|-------|------|-------|------|------|----------|
| 1st | Tau | 4h 19m | 17° 15' | -4.5 | 43" | 19% | 36° W | AM | **** | Pleiades, Hyades, Aldebaran |
| 3rd | Tau | 4h 21m | 17° 13' | -4.5 | 41" | 21% | 38° W | AM | **** | Pleiades, Hyades, Aldebaran |
| 5th | Tau | 4h 24m | 17° 13' | -4.5 | 40" | 23% | 39° W | AM | **** | Pleiades, Hyades, Aldebaran |
| 7th | Tau | 4h 27m | 17° 16' | -4.5 | 39" | 25% | 40° W | AM | **** | Hyades, Aldebaran |
| 9th | Tau | 4h 31m | 17° 21' | -4.5 | 38" | 26% | 41° W | AM | **** | Hyades, Aldebaran |

## Mars and the Outer Planets

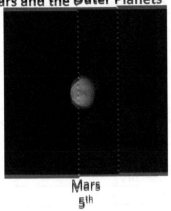

Mars
5th

Jupiter
5th

Saturn
5th

### Mars

| Date | Con. | R.A. | Dec. | Mag. | Diam. | Ill. | Elon. | Vis. | Rat. | Close To |
|------|------|------|------|------|-------|------|-------|------|------|----------|
| 1st | Psc | 0h 12m | -2° 4' | -0.5 | 11" | 84% | 98° W | AM | *** | |
| 5th | Psc | 0h 21m | -1° 13' | -0.6 | 12" | 85% | 100° W | AM | *** | |
| 10th | Cet | 0h 32m | 0° 12' | -0.7 | 12" | 85% | 102° W | AM | *** | |

### The Outer Planets

| Planet | Date | Con. | R.A. | Dec. | Mag. | Diam. | Elon. | Vis. | Rat. | Close To |
|--------|------|------|------|------|------|-------|-------|------|------|----------|
| Jupiter | 5th | Sgr | 19h 41m | -21° 43' | -2.7 | 47" | 170° W | AM | ***** | Moon, Saturn |
| Saturn | 5th | Sgr | 20h 7m | -20° 24' | 0.2 | 18" | 163° W | AM | **** | |
| Uranus | 5th | Ari | 2h 31m | 14° 23' | 5.8 | 3" | 67° W | AM | ** | |
| Neptune | 5th | Aqr | 23h 28m | -4° 36' | 7.9 | 2" | 113° W | AM | **** | |

## Highlights

| Date | Time (UT) | Event |
|------|-----------|-------|
| 1st | 02:47 | Mercury is at inferior conjunction with the Sun. (Not visible.) |
| 2nd | 15:04 | The waxing gibbous Moon is north of the bright star Antares. (Scorpius, evening sky.) |
| 4th | 19:56 | Asteroid Vesta is in conjunction with the Sun. (Not visible.) |
| 5th | 04:29 | Penumbral lunar eclipse: Visible from Africa, Antarctica, the Atlantic, Central America, western Europe, North America, the eastern Pacific and South America. |
| | 04:45 | Full Moon. (Visible all night.) |
| | 20:32 | The full Moon is south of Jupiter. (Visible all night.) |
| | 21:20 | The full Moon is south of dwarf planet Pluto. (Visible all night.) |
| 6th | 09:37 | The just-past full Moon is south of Saturn. (Morning sky.) |
| 10th | 08:48 | The waning gibbous Moon is south of Neptune. (Morning sky.) |

# July 11<sup>th</sup> to 20<sup>th</sup>, 2020

## The Moon

11<sup>th</sup>

13<sup>th</sup>

15<sup>th</sup>

17<sup>th</sup>

19<sup>th</sup>

| Date | Con | R.A. | Dec | Mag | Diam | Ill. | Elon. | Phase | Close To |
|------|-----|------|-----|-----|------|------|-------|-------|----------|
| 11th | Psc | 0h 20m | -3° 26' | -10.7 | 30' | 64% | 106° W | LQ | Mars |
| 12th | Cet | 1h 4m | 1° 24' | -10.3 | 30' | 55% | 96° W | LQ | Mars |
| 13th | Psc | 1h 47m | 6° 11' | -9.9 | 30' | 45% | 86° W | LQ | Uranus |
| 14th | Ari | 2h 32m | 10° 45' | -9.4 | 30' | 36% | 76° W | -Cr | Uranus |
| 15th | Ari | 3h 18m | 14° 57' | -8.9 | 30' | 27% | 66° W | -Cr | Uranus, Pleiades |
| 16th | Tau | 4h 7m | 18° 37' | -8.3 | 30' | 19% | 54° W | -Cr | Venus, Pleiades, Hyades, Aldebar |
| 17th | Tau | 4h 59m | 21° 31' | -7.5 | 31' | 11% | 42° W | NM | Venus, Hyades, Aldebaran |
| 18th | Tau | 5h 54m | 23° 24' | -6.6 | 31' | 6% | 30° W | NM | Mercury |
| 19th | Gem | 6h 51m | 24° 4' | -5.5 | 31' | 2% | 16° W | NM | Mercury |
| 20th | Gem | 7h 50m | 23° 21' | -4.3 | 32' | 0% | 3° W | NM | |

## Mercury and Venus

Mercury
15<sup>th</sup>

Venus
15<sup>th</sup>

**Mercury**

| Date | Con. | R.A. | Dec. | Mag. | Diam. | Ill. | Elon. | Vis. | Rat. | Close To |
|------|------|------|------|------|-------|------|-------|------|------|----------|
| 11th | Gem | 6h 23m | 18° 50' | 2.4 | 10" | 11% | 15° W | AM | ** | |
| 13th | Gem | 6h 23m | 19° 6' | 1.9 | 10" | 14% | 17° W | AM | *** | |
| 15th | Gem | 6h 25m | 19° 25' | 1.5 | 9" | 19% | 19° W | AM | *** | |
| 17th | Gem | 6h 28m | 19° 47' | 1.1 | 9" | 24% | 20° W | AM | *** | |
| 19th | Gem | 6h 33m | 20° 9' | 0.8 | 8" | 29% | 21° W | AM | *** | Moon |

**Venus**

| Date | Con. | R.A. | Dec. | Mag. | Diam. | Ill. | Elon. | Vis. | Rat. | Close To |
|------|------|------|------|------|-------|------|-------|------|------|----------|
| 11th | Tau | 4h 35m | 17° 28' | -4.5 | 36" | 28% | 42° W | AM | **** | Hyades, Aldebaran |
| 13th | Tau | 4h 39m | 17° 36' | -4.5 | 35" | 30% | 43° W | AM | **** | Hyades, Aldebaran |
| 15th | Tau | 4h 44m | 17° 45' | -4.5 | 34" | 31% | 44° W | AM | **** | Hyades, Aldebaran |
| 17th | Tau | 4h 49m | 17° 55' | -4.5 | 33" | 33% | 45° W | AM | **** | Moon, Hyades, Aldebaran |
| 19th | Tau | 4h 55m | 18° 6' | -4.4 | 32" | 34% | 46° W | AM | **** | Hyades, Aldebaran |

## Mars and the Outer Planets

Mars
15ᵗʰ

Jupiter
15ᵗʰ

Saturn
15ᵗʰ

**Mars**

| Date | Con. | R.A. | Dec. | Mag. | Diam. | Ill. | Elon. | Vis. | Rat. | Close To |
|------|------|------|------|------|-------|------|-------|------|------|----------|
| 11th | Cet | 0h 34m | 0° 0' | -0.7 | 12" | 85% | 103° W | AM | *** | Moon |
| 15th | Cet | 0h 42m | 0° 46' | -0.8 | 13" | 85% | 105° W | AM | *** | |
| 20th | Cet | 0h 52m | 1° 42' | -0.9 | 13" | 85% | 107° W | AM | *** | |

**The Outer Planets**

| Planet | Date | Con. | R.A. | Dec. | Mag. | Diam. | Elon. | Vis. | Rat. | Close To |
|--------|------|------|------|------|------|-------|-------|------|------|----------|
| Jupiter | 15th | Sgr | 19h 36m | -21° 56' | -2.8 | 48" | 179° E | AN | ***** | Saturn |
| Saturn | 15th | Sgr | 20h 4m | -20° 34' | 0.1 | 18" | 174° W | AN | ***** | |
| Uranus | 15th | Ari | 2h 32m | 14° 29' | 5.8 | 4" | 77° W | AM | ** | Moon |
| Neptune | 15th | Aqr | 23h 28m | -4° 38' | 7.9 | 2" | 123° W | AM | **** | |

## Highlights

| Date | Time (UT) | Event |
|------|-----------|-------|
| 11ᵗʰ | 17:51 | Venus is 1.0° north of the bright star Aldebaran. (Taurus, morning sky.) |
| | 19:00 | The nearly last quarter Moon is south of Mars. (Morning sky.) |
| 12ᵗʰ | 06:48 | Mercury is stationary prior to resuming prograde motion. (Morning sky.) |
| | 23:30 | Last Quarter Moon. (Morning sky.) |
| 14ᵗʰ | 09:06 | Jupiter is at opposition. (Visible all night.) |
| | 13:13 | The waning crescent Moon is south of Uranus. (Morning sky.) |
| 16ᵗʰ | 00:49 | The waning crescent Moon is south of the Pleiades star cluster. (Taurus, morning sky.) |
| 17ᵗʰ | 00:18 | The waning crescent Moon is north of the bright star Aldebaran. (Taurus, morning sky.) |
| | 06:35 | The waning crescent Moon is north of Venus. (Morning sky.) |
| | N/A | Good opportunity to see Earthshine on the waning crescent Moon. (Morning sky.) |
| 19ᵗʰ | 02:48 | The nearly new Moon is north of Mercury. (Morning sky.) |
| 20ᵗʰ | 17:33 | New Moon. (Not visible.) |
| | 23:33 | Saturn is at opposition. (Visible all night.) |

# July 21st to 31st, 2020

## The Moon

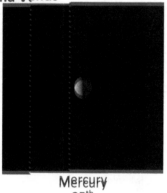

| 21st | 23rd | 25th | 27th | 29th | 31st |

| Date | Con | R.A. | Dec | Mag | Diam | Ill. | Elon. | Phase | Close To |
|------|-----|------|------|-----|------|------|-------|-------|----------|
| 21st | Cnc | 8h 49m | 21° 14' | -5.0 | 32' | 1% | 11° E | NM | Praesepe |
| 22nd | Leo | 9h 47m | 17° 48' | -6.2 | 32' | 4% | 25° E | NM | Regulus |
| 23rd | Leo | 10h 43m | 13° 18' | -7.3 | 32' | 10% | 38° E | NM | Regulus |
| 24th | Leo | 11h 37m | 8° 2' | -8.2 | 32' | 18% | 50° E | +Cr | |
| 25th | Vir | 12h 30m | 2° 19' | -8.9 | 32' | 27% | 62° E | +Cr | |
| 26th | Vir | 13h 22m | -3° 30' | -9.6 | 32' | 38% | 74° E | FQ | Spica |
| 27th | Vir | 14h 15m | -9° 6' | -10.1 | 32' | 50% | 87° E | FQ | |
| 28th | Lib | 15h 8m | -14° 11' | -10.6 | 32' | 61% | 99° E | FQ | |
| 29th | Sco | 16h 4m | -18° 27' | -11.0 | 32' | 72% | 112° E | +G | Antares |
| 30th | Oph | 17h 1m | -21° 39' | -11.4 | 32' | 81% | 125° E | +G | Antares |
| 31st | Sgr | 18h 0m | -23° 34' | -11.7 | 32' | 89% | 139° E | +G | |

## Mercury and Venus

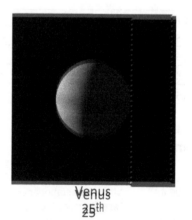

Mercury
25th

Venus
25th

### Mercury

| Date | Con. | R.A. | Dec. | Mag. | Diam. | Ill. | Elon. | Vis. | Rat. | Close To |
|------|------|------|------|------|-------|------|-------|------|------|----------|
| 21st | Gem | 6h 39m | 20° 32' | 0.5 | 8" | 34% | 21° W | AM | *** | |
| 23rd | Gem | 6h 47m | 20° 52' | 0.2 | 8" | 41% | 21° W | AM | *** | |
| 25th | Gem | 6h 56m | 21° 10' | -0.1 | 7" | 47% | 21° W | AM | *** | |
| 27th | Gem | 7h 7m | 21° 22' | -0.3 | 7" | 54% | 20° W | AM | *** | |
| 29th | Gem | 7h 20m | 21° 29' | -0.6 | 7" | 61% | 19° W | AM | *** | |
| 31st | Gem | 7h 33m | 21° 28' | -0.8 | 6" | 68% | 18° W | AM | ** | |

## Venus

| Date | Con. | R.A. | Dec. | Mag. | Diam. | Ill. | Elon. | Vis. | Rat. | Close To |
|------|------|------|------|------|-------|------|-------|------|------|----------|
| 21st | Tau | 5h 0m | 18° 18' | -4.4 | 31" | 36% | 46° W | AM | **** | Hyades, Aldebaran |
| 23rd | Tau | 5h 6m | 18° 29' | -4.4 | 30" | 37% | 47° W | AM | **** | Hyades, Aldebaran |
| 25th | Tau | 5h 13m | 18° 41' | -4.4 | 30" | 39% | 47° W | AM | **** | Aldebaran |
| 27th | Tau | 5h 19m | 18° 52' | -4.4 | 29" | 40% | 47° W | AM | **** | |
| 29th | Tau | 5h 26m | 19° 4' | -4.4 | 28" | 41% | 48° W | AM | **** | |
| 31st | Tau | 5h 33m | 19° 15' | -4.4 | 27" | 43% | 48° W | AM | **** | |

## Mars and the Outer Planets

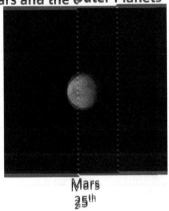

Mars
25th

Jupiter
25th

Saturn
25th

### Mars

| Date | Con. | R.A. | Dec. | Mag. | Diam. | Ill. | Elon. | Vis. | Rat. | Close To |
|------|------|------|------|------|-------|------|-------|------|------|----------|
| 21st | Cet | 0h 54m | 1° 52' | -0.9 | 13" | 85% | 108° W | AM | *** | |
| 25th | Cet | 1h 1m | 2° 34' | -1.0 | 14" | 86% | 110° W | AM | *** | |
| 31st | Psc | 1h 12m | 3° 31' | -1.1 | 15" | 86% | 113° W | AM | *** | |

### The Outer Planets

| Planet | Date | Con. | R.A. | Dec. | Mag. | Diam. | Elon. | Vis. | Rat. | Close To |
|--------|------|------|------|------|------|-------|-------|------|------|----------|
| Jupiter | 25th | Sgr | 19h 30m | -22° 9' | -2.7 | 47" | 167° E | PM | ***** | Saturn |
| Saturn | 25th | Sgr | 20h 1m | -20° 44' | 0.1 | 18" | 175° E | AN | ***** | |
| Uranus | 25th | Ari | 2h 33m | 14° 32' | 5.8 | 4" | 87° W | AM | ** | |
| Neptune | 25th | Aqr | 23h 27m | -4° 42' | 7.8 | 2" | 133° W | AM | **** | |

## Highlights

| Date | Time (UT) | Event |
|------|-----------|-------|
| 22nd | 15:04 | Mercury is at greatest western elongation from the Sun. (Morning sky.) |
| | 21:55 | The waxing crescent Moon is north of the bright star Regulus. (Leo, evening sky.) |
| 23rd | N/A | Good opportunity to see Earthshine on the waxing crescent Moon. (Evening sky.) |
| 26th | 11:56 | The nearly first quarter Moon is north of the bright star Spica. (Virgo, evening sky.) |
| 27th | 12:33 | First Quarter Moon. (Evening sky.) |
| 28th | N/A | The Piscis Austrinid meteor shower is at its maximum. (ZHR: 5) |
| 29th | 23:43 | The waxing gibbous Moon is north of the bright star Antares. (Scorpius, evening sky.) |
| 30th | N/A | The Delta Aquariid meteor shower is at its maximum. (ZHR: 16) |
| | N/A | The Alpha Capricornid meteor shower is at its maximum. (ZHR: 5) |

# August 1st to 10th, 2020

## The Moon

| 1st | 3rd | 5th | 7th | 9th |

| Date | Con | R.A. | Dec | Mag | Diam | Ill. | Elon. | Phase | Close To |
|------|-----|------|-----|-----|------|------|-------|-------|----------|
| 1st | Sgr | 18h 58m | -24° 5' | -12.0 | 31' | 95% | 153° E | +G | Jupiter |
| 2nd | Sgr | 19h 56m | -23° 13' | -12.4 | 31' | 98% | 166° E | FM | Jupiter, Saturn |
| 3rd | Cap | 20h 52m | -21° 6' | -12.6 | 31' | 100% | 179° E | FM | |
| 4th | Cap | 21h 44m | -17° 58' | -12.5 | 30' | 99% | 169° W | FM | |
| 5th | Aqr | 22h 34m | -14° 4' | -12.2 | 30' | 96% | 157° W | FM | |
| 6th | Aqr | 23h 21m | -9° 39' | -11.9 | 30' | 92% | 147° W | -G | Neptune |
| 7th | Psc | 0h 6m | -4° 55' | -11.6 | 30' | 86% | 136° W | -G | Neptune |
| 8th | Cet | 0h 49m | 0° 3' | -11.3 | 30' | 79% | 126° W | -G | Mars |
| 9th | Psc | 1h 33m | 4° 46' | -10.9 | 30' | 70% | 116° W | -G | Mars |
| 10th | Cet | 2h 17m | 9° 25' | -10.6 | 30' | 61% | 106° W | LQ | Uranus |

## Mercury and Venus

Mercury
5th

Venus
5th

### Mercury

| Date | Con. | R.A. | Dec. | Mag. | Diam. | Ill. | Elon. | Vis. | Rat. | Close To |
|------|------|------|------|------|-------|------|-------|------|------|----------|
| 1st | Gem | 7h 41m | 21° 24' | -0.9 | 6" | 71% | 17° W | AM | ** | |
| 3rd | Gem | 7h 56m | 21° 10' | -1.0 | 6" | 78% | 15° W | NV | N/A | |
| 5th | Cnc | 8h 12m | 20° 45' | -1.2 | 6" | 84% | 13° W | NV | N/A | Praesepe |
| 7th | Cnc | 8h 28m | 20° 9' | -1.3 | 5" | 89% | 11° W | NV | N/A | Praesepe |
| 9th | Cnc | 8h 45m | 19° 23' | -1.5 | 5" | 93% | 8° W | NV | N/A | Praesepe |

**Venus**

| Date | Con. | R.A. | Dec. | Mag. | Diam. | Ill. | Elon. | Vis. | Rat. | Close To |
|------|------|------|------|------|-------|------|-------|------|------|----------|
| **1st** | Tau | 5h 37m | 19° 20' | -4.4 | 27" | 43% | 48° W | AM | **** | |
| **3rd** | Tau | 5h 44m | 19° 29' | -4.4 | 26" | 45% | 48° W | AM | **** | |
| **5th** | Ori | 5h 51m | 19° 38' | -4.4 | 26" | 46% | 48° W | AM | **** | |
| **7th** | Ori | 5h 59m | 19° 46' | -4.3 | 25" | 47% | 48° W | AM | *** | |
| **9th** | Ori | 6h 7m | 19° 53' | -4.3 | 25" | 48% | 48° W | AM | *** | |

## Mars and the Outer Planets

Mars
5th

Jupiter
5th

Saturn
5th

**Mars**

| Date | Con. | R.A. | Dec. | Mag. | Diam. | Ill. | Elon. | Vis. | Rat. | Close To |
|------|------|------|------|------|-------|------|-------|------|------|----------|
| **1st** | Psc | 1h 13m | 3° 40' | -1.1 | 15" | 86% | 114° W | AM | *** | |
| **5th** | Psc | 1h 20m | 4° 14' | -1.2 | 15" | 87% | 116° W | AM | *** | |
| **10th** | Psc | 1h 27m | 4° 52' | -1.3 | 16" | 88% | 119° W | AM | **** | |

**The Outer Planets**

| Planet | Date | Con. | R.A. | Dec. | Mag. | Diam. | Elon. | Vis. | Rat. | Close To |
|--------|------|------|------|------|------|-------|-------|------|------|----------|
| Jupiter | 5th | Sgr | 19h 25m | -22° 22' | -2.7 | 47" | 155° E | PM | ***** | Saturn |
| Saturn | 5th | Sgr | 19h 58m | -20° 54' | 0.2 | 18" | 164° E | PM | **** | |
| Uranus | 5th | Ari | 2h 33m | 14° 35' | 5.8 | 4" | 98° W | AM | ** | |
| Neptune | 5th | Aqr | 23h 26m | -4° 47' | 7.8 | 2" | 144° W | AM | **** | |

## Highlights

| Date | Time (UT) | Event |
|------|-----------|-------|
| 1st | 23:55 | The waxing gibbous Moon is south of Jupiter. (Evening sky.) |
| 2nd | 06:13 | The nearly full Moon is south of dwarf planet Pluto. (Evening sky.) |
| | 12:41 | The nearly full Moon is south of Saturn. (Evening sky.) |
| 3rd | 15:59 | Full Moon (Visible all night.) |
| 6th | 14:44 | The waning gibbous Moon is south of Neptune. (Morning sky.) |
| 9th | 09:26 | The waning gibbous Moon is south of Mars. (Morning sky.) |
| 10th | 19:54 | The nearly last quarter Moon is south of Uranus. (Morning sky.) |

# August 11<sup>th</sup> to 20<sup>th</sup>, 2020

## The Moon

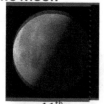    

11<sup>th</sup>  13<sup>th</sup>  15<sup>th</sup>  17<sup>th</sup>  19<sup>th</sup>

| Date | Con | R.A. | Dec | Mag | Diam | Ill. | Elon. | Phase | Close To |
|------|-----|------|-----|-----|------|------|-------|-------|----------|
| 11th | Ari | 3h 2m | 13° 44' | -10.2 | 30' | 52% | 96° W | LQ | Uranus, Pleiades |
| 12th | Tau | 3h 49m | 17° 33' | -9.8 | 30' | 43% | 85° W | LQ | Pleiades, Hyades, Aldebaran |
| 13th | Tau | 4h 39m | 20° 42' | -9.3 | 30' | 33% | 74° W | =Cr | Hyades, Aldebaran |
| 14th | Tau | 5h 33m | 22° 57' | -8.7 | 31' | 24% | 61° W | =Cr | |
| 15th | Gem | 6h 29m | 24° 4' | -8.0 | 31' | 16% | 48° W | =Cr | Venus |
| 16th | Gem | 7h 27m | 23° 53' | -7.1 | 32' | 9% | 34° W | NM | |
| 17th | Cnc | 8h 26m | 22° 15' | -6.1 | 32' | 4% | 21° W | NM | Praesepe |
| 18th | Leo | 9h 25m | 19° 13' | -4.9 | 32' | 1% | 7° W | NM | Mercury, Praesepe, Regulus |
| 19th | Leo | 10h 23m | 14° 58' | -4.6 | 33' | 0% | 7° E | NM | Mercury, Regulus |
| 20th | Leo | 11h 19m | 9° 45' | -5.9 | 33' | 3% | 20° E | NM | |

## Mercury and Venus

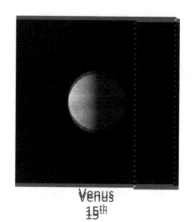

Mercury
15<sup>th</sup>

Venus
15<sup>th</sup>

### Mercury

| Date | Con. | R.A. | Dec. | Mag. | Diam. | Ill. | Elon. | Vis. | Rat. | Close To |
|------|------|------|------|------|-------|------|-------|------|------|----------|
| 11th | Cnc | 9h 2m | 18° 27' | -1.6 | 5" | 96% | 6° W | NV | N/A | Praesepe |
| 13th | Cnc | 9h 18m | 17° 23' | -1.7 | 5" | 98% | 4° W | NV | N/A | Praesepe |
| 15th | Leo | 9h 35m | 16° 11' | -1.9 | 5" | 99% | 2° W | NV | N/A | Regulus |
| 17th | Leo | 9h 51m | 14° 53' | -1.9 | 5" | 100% | 0° E | NV | N/A | Regulus |
| 19th | Leo | 10h 6m | 13° 30' | -1.7 | 5" | 100% | 2° E | NV | N/A | Moon, Regulus |

## Venus

| Date | Con. | R.A. | Dec. | Mag. | Diam. | Ill. | Elon. | Vis. | Rat. | Close To |
|------|------|------|------|------|-------|------|-------|------|------|----------|
| 11th | Ori | 6h 15m | 19° 59' | -4.3 | 24" | 49% | 48° W | AM | *** | |
| 13th | Gem | 6h 23m | 20° 3' | -4.3 | 23" | 50% | 48° W | AM | *** | |
| 15th | Gem | 6h 31m | 20° 5' | -4.3 | 23" | 51% | 48° W | AM | *** | Moon |
| 17th | Gem | 6h 39m | 20° 7' | -4.3 | 22" | 52% | 47° W | AM | *** | |
| 19th | Gem | 6h 48m | 20° 6' | -4.3 | 22" | 54% | 47° W | AM | *** | |

## Mars and the Outer Planets

Mars
15th

Jupiter
15th

Saturn
15th

### Mars

| Date | Con. | R.A. | Dec. | Mag. | Diam. | Ill. | Elon. | Vis. | Rat. | Close To |
|------|------|------|------|------|-------|------|-------|------|------|----------|
| 11th | Psc | 1h 29m | 4° 59' | -1.3 | 16" | 88% | 119° W | AM | **** | |
| 15th | Psc | 1h 34m | 5° 26' | -1.4 | 16" | 88% | 122° W | AM | **** | |
| 20th | Psc | 1h 39m | 5° 54' | -1.5 | 17" | 89% | 125° W | AM | **** | |

### The Outer Planets

| Planet | Date | Con. | R.A. | Dec. | Mag. | Diam. | Elon. | Vis. | Rat. | Close To |
|--------|------|------|------|------|------|-------|-------|------|------|----------|
| Jupiter | 15th | Sgr | 19h 21m | -22° 31' | -2.7 | 46" | 145° E | PM | **** | Saturn |
| Saturn | 15th | Sgr | 19h 55m | -21° 3' | 0.2 | 18" | 153° E | PM | **** | |
| Uranus | 15th | Ari | 2h 33m | 14° 35' | 5.7 | 4" | 107° W | AM | *** | |
| Neptune | 15th | Aqr | 23h 25m | -4° 52' | 7.8 | 2" | 154° W | AM | ***** | |

## Highlights

| Date | Time (UT) | Event |
|------|-----------|-------|
| 11th | 16:45 | Last Quarter Moon. (Morning sky.) |
| 12th | 11:53 | The just-past last quarter Moon is south of the Pleiades star cluster. (Taurus, morning sky.) |
| | 23:49 | Venus is at greatest western elongation from the Sun. (Morning sky.) |
| 13th | 11:11 | The waning crescent Moon is north of the bright star Aldebaran. (Taurus, morning sky.) |
| | N/A | The Perseid meteor shower is at its maximum. (ZHR: 100) |
| 15th | 13:27 | Uranus is stationary prior to beginning retrograde motion. (Morning sky.) |
| | 14:18 | The waning crescent Moon is north of Venus. (Morning sky.) |
| | N/A | Good opportunity to see Earthshine on the waning crescent Moon. (Morning sky.) |
| 17th | 14:54 | Mercury is at superior conjunction with the Sun. (Not visible.) |
| 18th | N/A | The Kappa Cygnid meteor shower is at its maximum. (ZHR: 3) |
| 19th | 02:42 | New Moon. (Not visible.) |

# August 21ˢᵗ to 31ˢᵗ, 2020

## The Moon

| 21ˢᵗ | 23ʳᵈ | 25ᵗʰ | 27ᵗʰ | 29ᵗʰ | 31ˢᵗ |

| Date | Con | R.A. | Dec | Mag | Diam | Ill. | Elon. | Phase | Close To |
|------|-----|------|-----|-----|------|------|-------|-------|----------|
| 21st | Vir | 12h 14m | 3° 57' | -7.1 | 33' | 8% | 33° E | NM | |
| 22nd | Vir | 13h 7m | -2° 3' | -8.0 | 33' | 16% | 45° E | +Cr | Spica |
| 23rd | Vir | 14h 1m | -7° 54' | -8.8 | 33' | 25% | 58° E | +Cr | Spica |
| 24th | Lib | 14h 55m | -13° 15' | -9.5 | 32' | 36% | 70° E | +Cr | |
| 25th | Lib | 15h 51m | -17° 46' | -10.0 | 32' | 47% | 83° E | FQ | Antares |
| 26th | Oph | 16h 48m | -21° 13' | -10.5 | 32' | 58% | 96° E | FQ | Antares |
| 27th | Sgr | 17h 46m | -23° 24' | -10.9 | 31' | 69% | 110° E | +G | |
| 28th | Sgr | 18h 44m | -24° 13' | -11.2 | 31' | 78% | 124° E | +G | Jupiter |
| 29th | Sgr | 19h 41m | -23° 40' | -11.6 | 31' | 86% | 137° E | +G | Jupiter, Saturn |
| 30th | Cap | 20h 37m | -21° 51' | -11.9 | 31' | 93% | 150° E | +G | Saturn |
| 31st | Cap | 21h 29m | -18° 59' | -12.2 | 30' | 97% | 162° E | FM | |

## Mercury and Venus

Mercury
25ᵗʰ

Venus
25ᵗʰ

**Mercury**

| Date | Con. | R.A. | Dec. | Mag. | Diam. | Ill. | Elon. | Vis. | Rat. | Close To |
|------|------|------|------|------|-------|------|-------|------|------|----------|
| 21st | Leo | 10h 21m | 12° 3' | -1.5 | 5" | 99% | 4° E | NV | N/A | Regulus |
| 23rd | Leo | 10h 35m | 10° 34' | -1.3 | 5" | 98% | 6° E | NV | N/A | Regulus |
| 25th | Leo | 10h 49m | 9° 2' | -1.1 | 5" | 97% | 8° E | NV | N/A | |
| 27th | Leo | 11h 3m | 7° 30' | -0.9 | 5" | 95% | 9° E | NV | N/A | |
| 29th | Leo | 11h 16m | 5° 57' | -0.8 | 5" | 94% | 11° E | NV | N/A | |
| 31st | Leo | 11h 28m | 4° 24' | -0.7 | 5" | 93% | 12° E | NV | N/A | |

**Venus**

| Date | Con. | R.A. | Dec. | Mag. | Diam. | Ill. | Elon. | Vis. | Rat. | Close To |
|------|------|------|------|------|-------|------|-------|------|------|----------|
| 21st | Gem | 6h 57m | 20° 4' | -4.3 | 22" | 55% | 47° W | AM | *** | |
| 23rd | Gem | 7h 5m | 20° 0' | -4.2 | 21" | 56% | 46° W | AM | *** | |
| 25th | Gem | 7h 14m | 19° 54' | -4.2 | 21" | 56% | 46° W | AM | *** | |
| 27th | Gem | 7h 23m | 19° 47' | -4.2 | 20" | 57% | 46° W | AM | *** | |
| 29th | Gem | 7h 32m | 19° 37' | -4.2 | 20" | 58% | 45° W | AM | *** | |
| 31st | Gem | 7h 41m | 19° 26' | -4.2 | 20" | 59% | 45° W | AM | *** | |

## Mars and the Outer Planets

Mars
25th

Jupiter
25th

Saturn
25th

**Mars**

| Date | Con. | R.A. | Dec. | Mag. | Diam. | Ill. | Elon. | Vis. | Rat. | Close To |
|------|------|------|------|------|-------|------|-------|------|------|----------|
| 21st | Psc | 1h 40m | 5° 59' | -1.5 | 17" | 90% | 126° W | AM | **** | |
| 25th | Psc | 1h 44m | 6° 17' | -1.6 | 18" | 90% | 129° W | AM | **** | |
| 31st | Psc | 1h 48m | 6° 37' | -1.8 | 19" | 92% | 133° W | AM | **** | |

**The Outer Planets**

| Planet | Date | Con. | R.A. | Dec. | Mag. | Diam. | Elon. | Vis. | Rat. | Close To |
|--------|------|------|------|------|------|-------|-------|------|------|----------|
| Jupiter | 25th | Sgr | 19h 18m | -22° 38' | -2.6 | 45" | 135° E | PM | **** | Saturn |
| Saturn | 25th | Sgr | 19h 53m | -21° 10' | 0.3 | 18" | 144° E | PM | **** | |
| Uranus | 25th | Ari | 2h 33m | 14° 35' | 5.7 | 4" | 116° W | AM | *** | |
| Neptune | 25th | Aqr | 23h 25m | -4° 59' | 7.8 | 2" | 163° W | AM | ***** | |

## Highlights

| Date | Time (UT) | Event |
|------|-----------|-------|
| 22nd | 21:01 | The waxing crescent Moon is north of the bright star Spica. (Virgo, evening sky.) |
| | N/A | Good opportunity to see Earthshine on the waxing crescent Moon. (Evening sky.) |
| 25th | 17:58 | First Quarter Moon. (Evening sky.) |
| 26th | 04:33 | The just-past first quarter Moon is north of Antares. (Scorpius, evening sky.) |
| 29th | 02:50 | The waxing gibbous Moon is south of Jupiter. (Evening sky.) |
| | 10:01 | The waxing gibbous Moon is south of the dwarf planet Pluto. (Evening sky.) |
| | 15:18 | The waxing gibbous Moon is south of Saturn. (Evening sky.) |

# September 1st to 10th, 2020

## The Moon

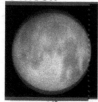    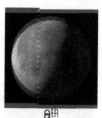

| | 1st | | 3rd | | 5th | | 7th | | 9th |

| Date | Con. | R.A. | Dec. | Mag | Diam | Ill. | Elon. | Phase | Close To |
|------|------|------|------|-----|------|------|-------|-------|----------|
| 1st | Aqr | 22h 19m | -15° 16' | -12.5 | 30' | 99% | 174° E | FM | |
| 2nd | Aqr | 23h 7m | -10° 58' | -12.6 | 30' | 100% | 175° W | FM | Neptune |
| 3rd | Aqr | 23h 52m | -6° 16' | -12.3 | 30' | 98% | 165° W | FM | Neptune |
| 4th | Cet | 0h 36m | -1° 24' | -12.1 | 30' | 95% | 155° W | -G | |
| 5th | Psc | 1h 19m | 3° 29' | -11.8 | 29' | 90% | 145° W | -G | Mars |
| 6th | Psc | 2h 3m | 8° 14' | -11.5 | 29' | 84% | 135° W | -G | Mars, Uranus |
| 7th | Ari | 2h 48m | 12° 40' | -11.2 | 30' | 76% | 124° W | -G | Uranus |
| 8th | Tau | 3h 34m | 16° 39' | -10.8 | 30' | 68% | 114° W | -G | Pleiades |
| 9th | Tau | 4h 22m | 19° 59' | -10.5 | 30' | 59% | 103° W | LQ | Pleiades, Hyades, Aldebaran |
| 10th | Tau | 5h 14m | 22° 31' | -10.1 | 30' | 49% | 91° W | LQ | Hyades, Aldebaran |

## Mercury and Venus

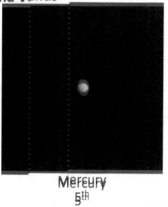 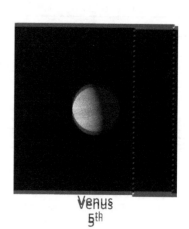

Mercury
5th

Venus
5th

### Mercury

| Date | Con. | R.A. | Dec. | Mag. | Diam. | Ill. | Elon. | Vis. | Rat. | Close To |
|------|------|------|------|------|-------|------|-------|------|------|----------|
| 1st | Leo | 11h 35m | 3° 38' | -0.6 | 5" | 92% | 13° E | NV | N/A | |
| 3rd | Vir | 11h 47m | 2° 6' | -0.5 | 5" | 90% | 14° E | NV | N/A | |
| 5th | Vir | 11h 58m | 0° 35' | -0.4 | 5" | 89% | 15° E | NV | N/A | |
| 7th | Vir | 12h 10m | 0° 55' | -0.3 | 5" | 87% | 16° E | PM | ** | |
| 9th | Vir | 12h 21m | -2° 23' | -0.3 | 5" | 85% | 17° E | PM | ** | |

## Venus

| Date | Con. | R.A. | Dec. | Mag. | Diam. | Ill. | Elon. | Vis. | Rat. | Close To |
|------|------|------|------|------|-------|------|-------|------|------|----------|
| 1st | Gem | 7h 45m | 19° 19' | -4.2 | 19" | 60% | 45° W | AM | *** | |
| 3rd | Gem | 7h 54m | 19° 5' | -4.2 | 19" | 61% | 44° W | AM | *** | |
| 5th | Cnc | 8h 3m | 18° 48' | -4.2 | 19" | 62% | 44° W | AM | *** | Praesepe |
| 7th | Cnc | 8h 12m | 18° 30' | -4.2 | 18" | 62% | 43° W | AM | *** | Praesepe |
| 9th | Cnc | 8h 21m | 18° 9' | -4.2 | 18" | 63% | 43° W | AM | *** | Praesepe |

## Mars and the Outer Planets

Mars
5th

Jupiter
5th

Saturn
5th

## Mars

| Date | Con. | R.A. | Dec. | Mag. | Diam. | Ill. | Elon. | Vis. | Rat. | Close To |
|------|------|------|------|------|-------|------|-------|------|------|----------|
| 1st | Psc | 1h 48m | 6° 39' | -1.8 | 19" | 92% | 134° W | AM | **** | |
| 5th | Psc | 1h 50m | 6° 47' | -1.9 | 20" | 93% | 137° W | AM | **** | Moon |
| 10th | Psc | 1h 50m | 6° 51' | -2.0 | 20" | 94% | 143° W | AM | **** | |

## The Outer Planets

| Planet | Date | Con. | R.A. | Dec. | Mag. | Diam. | Elon. | Vis. | Rat. | Close To |
|--------|------|------|------|------|------|-------|-------|------|------|----------|
| Jupiter | 5th | Sgr | 19h 16m | -22° 42' | -2.5 | 44" | 124° E | PM | **** | Saturn |
| Saturn | 5th | Sgr | 19h 51m | -21° 16' | 0.3 | 18" | 133° E | PM | **** | |
| Uranus | 5th | Ari | 2h 32m | 14° 32' | 5.7 | 4" | 126° W | AM | *** | |
| Neptune | 5th | Aqr | 23h 23m | -5° 6' | 7.8 | 2" | 174° W | AN | ***** | |

## Highlights

| Date | Time (UT) | Event |
|------|-----------|-------|
| 1st | N/A | The Alpha Aurigid meteor shower is at its maximum. (ZHR: 6) |
| 2nd | 05:23 | Full Moon. (Visible all night.) |
| | 18:14 | Dwarf planet Ceres is at opposition. (Visible all night.) |
| | 19:34 | The full Moon is south of Neptune. (Visible all night.) |
| 6th | 05:45 | The waning gibbous Moon is south of Mars. (Morning sky.) |
| 7th | 04:07 | The waning gibbous Moon is south of Uranus. (Morning sky.) |
| 8th | 17:52 | The waning gibbous Moon is south of the Pleaides star cluster. (Taurus, morning sky.) |
| 9th | 17:56 | The nearly last quarter Moon is north of Aldebaran. (Taurus, morning sky.) |
| | 18:04 | Mars is stationary prior to beginning retrograde motion. (Morning sky.) |
| 10th | 09:26 | Last Quarter Moon. (Morning sky.) |
| | N/A | The Epsilon Perseid meteor shower is at its maximum. (ZHR: 5) |

# September 11<sup>th</sup> to 20<sup>th</sup>, 2020

## The Moon

11<sup>th</sup>

13<sup>th</sup>

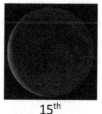
15<sup>th</sup>

17<sup>th</sup>

19<sup>th</sup>

| Date | Con | R.A. | Dec | Mag | Diam | Ill. | Elon. | Phase | Close To |
|------|-----|------|-----|-----|------|------|-------|-------|----------|
| 11th | Gem | 6h 8m | 24° 1' | -9.6 | 31' | 39% | 78° W | LQ | |
| 12th | Gem | 7h 4m | 24° 18' | -9.1 | 31' | 29% | 65° W | -Cr | |
| 13th | Cnc | 8h 2m | 23° 14' | -8.4 | 32' | 20% | 51° W | -Cr | Venus, Praesepe |
| 14th | Cnc | 9h 1m | 20° 46' | -7.6 | 32' | 12% | 37° W | NM | Venus, Praesepe |
| 15th | Leo | 9h 59m | 16° 59' | -6.6 | 33' | 6% | 24° W | NM | Regulus |
| 16th | Leo | 10h 56m | 12° 4' | -5.4 | 33' | 2% | 11° W | NM | |
| 17th | Vir | 11h 51m | 6° 19' | -4.4 | 33' | 0% | 3° E | NM | |
| 18th | Vir | 12h 47m | 0° 9' | -5.5 | 33' | 2% | 15° E | NM | Mercury, Spica |
| 19th | Vir | 13h 42m | -6° 2' | -6.7 | 33' | 6% | 28° E | NM | Mercury, Spica |
| 20th | Lib | 14h 38m | -11° 48' | -7.8 | 33' | 13% | 41° E | +Cr | |

## Mercury and Venus

Mercury
15<sup>th</sup>

Venus
15<sup>th</sup>

**Mercury**

| Date | Con. | R.A. | Dec. | Mag. | Diam. | Ill. | Elon. | Vis. | Rat. | Close To |
|------|------|------|------|------|-------|------|-------|------|------|----------|
| 11th | Vir | 12h 31m | -3° 50' | -0.2 | 5" | 84% | 18° E | PM | ** | |
| 13th | Vir | 12h 42m | -5° 14' | -0.2 | 5" | 82% | 19° E | PM | ** | |
| 15th | Vir | 12h 52m | -6° 36' | -0.1 | 6" | 80% | 20° E | PM | ** | Spica |
| 17th | Vir | 13h 2m | -7° 56' | -0.1 | 6" | 78% | 20° E | PM | *** | Spica |
| 19th | Vir | 13h 12m | -9° 13' | -0.1 | 6" | 76% | 21° E | PM | *** | Moon, Spica |

**Venus**

| Date | Con. | R.A. | Dec. | Mag. | Diam. | Ill. | Elon. | Vis. | Rat. | Close To |
|------|------|------|------|------|-------|------|-------|------|------|----------|
| 11th | Cnc | 8h 31m | 17° 47' | -4.2 | 18" | 64% | 42° W | AM | *** | Praesepe |
| 13th | Cnc | 8h 40m | 17° 22' | -4.1 | 18" | 65% | 42° W | AM | *** | Moon, Praesepe |
| 15th | Cnc | 8h 49m | 16° 56' | -4.1 | 17" | 66% | 41° W | AM | *** | Praesepe |
| 17th | Cnc | 8h 58m | 16° 27' | -4.1 | 17" | 67% | 41° W | AM | *** | Praesepe |
| 19th | Cnc | 9h 7m | 15° 57' | -4.1 | 17" | 67% | 40° W | AM | *** | Praesepe |

## Mars and the Outer Planets

Mars
15th

Jupiter
15th

Saturn
15th

**Mars**

| Date | Con. | R.A. | Dec. | Mag. | Diam. | Ill. | Elon. | Vis. | Rat. | Close To |
|------|------|------|------|------|-------|------|-------|------|------|----------|
| 11th | Psc | 1h 50m | 6° 52' | -2.1 | 20" | 94% | 142° W | AM | **** | |
| 15th | Psc | 1h 49m | 6° 50' | -2.2 | 21" | 96% | 146° W | AM | ***** | |
| 20th | Psc | 1h 47m | 6° 43' | -2.3 | 22" | 97% | 151° W | AM | ***** | |

**The Outer Planets**

| Planet | Date | Con. | R.A. | Dec. | Mag. | Diam. | Elon. | Vis. | Rat. | Close To |
|--------|------|------|------|------|------|-------|-------|------|------|----------|
| Jupiter | 15th | Sgr | 19h 15m | -22° 43' | -2.5 | 43" | 115° E | PM | **** | Saturn |
| Saturn | 15th | Sgr | 19h 50m | -21° 20' | 0.4 | 18" | 124° E | PM | **** | |
| Uranus | 15th | Ari | 2h 32m | 14° 28' | 5.7 | 4" | 136° W | AM | *** | |
| Neptune | 15th | Aqr | 23h 22m | -5° 12' | 7.8 | 2" | 177° E | AN | ***** | |

## Highlights

| Date | Time (UT) | Event |
|------|-----------|-------|
| 12th | 09:20 | Neptune is at opposition. (Visible all night.) |
| | 23:20 | Jupiter is stationary prior to resuming prograde motion. (Evening sky.) |
| 13th | 10:21 | Venus is 2.3° south of the Praesepe star cluster. (Cancer, morning sky.) |
| 14th | 02:05 | The waning crescent Moon is north of the Praesepe star cluster. (Cancer, morning sky.) |
| | 03:24 | The waning crescent Moon is north of Venus. (Morning sky.) |
| | N/A | Good opportunity to see Earthshine on the waning crescent Moon. (Morning sky.) |
| 15th | 17:03 | The waning crescent Moon is north of the bright star Regulus. (Leo, morning sky.) |
| 17th | 11:01 | New Moon. (Not visible.) |
| 18th | 22:48 | The waxing crescent Moon is north of Mercury. (Evening sky.) |
| 19th | 03:55 | The waxing crescent Moon is north of the bright star Spica. (Evening sky.) |
| 20th | N/A | Good opportunity to see Earthshine on the waxing crescent Moon. (Evening sky.) |

# September 21ˢᵗ to 30ᵗʰ, 2020

## The Moon

| 21ˢᵗ | 23ʳᵈ | 25ᵗʰ | 27ᵗʰ | 29ᵗʰ |

| Date | Con | R.A. | Dec | Mag | Diam | Ill. | Elon. | Phase | Close To |
|------|-----|------|-----|-----|------|------|-------|-------|----------|
| 21st | Lib | 15h 34m | -16° 48' | -8.6 | 33' | 22% | 55° E | +Cr | |
| 22nd | Oph | 16h 33m | -20° 41' | -9.3 | 32' | 33% | 68° E | +Cr | Antares |
| 23rd | Oph | 17h 32m | -23° 15' | -9.8 | 32' | 44% | 82° E | FQ | |
| 24th | Sgr | 18h 31m | -24° 24' | -10.3 | 31' | 55% | 96° E | FQ | Jupiter |
| 25th | Sgr | 19h 29m | -24° 7' | -10.7 | 31' | 65% | 110° E | FQ | Jupiter, Saturn |
| 26th | Cap | 20h 24m | -22° 34' | -11.1 | 31' | 74% | 123° E | +G | Saturn |
| 27th | Cap | 21h 17m | -19° 54' | -11.4 | 30' | 83% | 135° E | +G | |
| 28th | Aqr | 22h 7m | -16° 21' | -11.7 | 30' | 89% | 147° E | +G | |
| 29th | Aqr | 22h 55m | -12° 10' | -12.0 | 30' | 95% | 158° E | +G | Neptune |
| 30th | Aqr | 23h 40m | -7° 33' | -12.3 | 30' | 98% | 168° E | FM | Neptune |

## Mercury and Venus

Mercury
25ᵗʰ

Venus
25ᵗʰ

### Mercury

| Date | Con. | R.A. | Dec. | Mag. | Diam. | Ill. | Elon. | Vis. | Rat. | Close To |
|------|------|------|------|------|-------|------|-------|------|------|----------|
| 21st | Vir | 13h 22m | -10° 37' | 0.0 | 6" | 74% | 22° E | PM | *** | Spica |
| 23rd | Vir | 13h 31m | -11° 37' | 0.0 | 6" | 72% | 22° E | PM | *** | Spica |
| 25th | Vir | 13h 40m | -12° 44' | 0.0 | 6" | 69% | 22° E | PM | *** | Spica |
| 27th | Vir | 13h 48m | -13° 47' | 0.0 | 6" | 67% | 23° E | PM | *** | Spica |
| 29th | Vir | 13h 57m | -14° 46' | 0.0 | 7" | 64% | 23° E | PM | *** | Spica |

## Venus

| Date | Con. | R.A. | Dec. | Mag. | Diam. | Ill. | Elon. | Vis. | Rat. | Close To |
|------|------|------|------|------|-------|------|-------|------|------|----------|
| 21st | Cnc | 9h 16m | 15° 25' | -4.1 | 17" | 68% | 40° W | AM | ** | Praesepe |
| 23rd | Leo | 9h 26m | 14° 51' | -4.1 | 16" | 69% | 39° W | AM | ** | |
| 25th | Leo | 9h 35m | 14° 15' | -4.1 | 16" | 70% | 39° W | AM | ** | Regulus |
| 27th | Leo | 9h 44m | 13° 38' | -4.1 | 16" | 70% | 38° W | AM | ** | Regulus |
| 29th | Leo | 9h 53m | 12° 59' | -4.1 | 16" | 71% | 38° W | AM | ** | Regulus |

## Mars and the Outer Planets

Mars
25th

Jupiter
25th

Saturn
25th

## Mars

| Date | Con. | R.A. | Dec. | Mag. | Diam. | Ill. | Elon. | Vis. | Rat. | Close To |
|------|------|------|------|------|-------|------|-------|------|------|----------|
| 21st | Psc | 1h 46m | 6° 41' | -2.3 | 22" | 97% | 153° W | AM | ***** | |
| 25th | Psc | 1h 43m | 6° 31' | -2.4 | 22" | 98% | 157° W | AM | ***** | |
| 30th | Psc | 1h 38m | 6° 16' | -2.5 | 22" | 99% | 163° W | AM | ***** | |

## The Outer Planets

| Planet | Date | Con. | R.A. | Dec. | Mag. | Diam. | Elon. | Vis. | Rat. | Close To |
|--------|------|------|------|------|------|-------|-------|------|------|----------|
| Jupiter | 25th | Sgr | 19h 16m | -22° 42' | -2.4 | 41" | 107° E | PM | *** | Moon, Saturn |
| Saturn | 25th | Sgr | 19h 49m | -21° 22' | 0.4 | 17" | 115° E | PM | *** | Moon |
| Uranus | 25th | Ari | 2h 31m | 14° 23' | 5.7 | 4" | 145° W | AM | *** | |
| Neptune | 25th | Aqr | 23h 21m | -5° 19' | 7.8 | 2" | 168° E | PM | ***** | |

## Highlights

| Date | Time (UT) | Event |
|------|-----------|-------|
| 22nd | 03:11 | Mercury is 0.3° north of the bright star Spica. (Virgo, evening sky.) |
| | 09:17 | The waxing crescent Moon is north of the bright star Antares. (Scorpius, evening sky.) |
| | 13:32 | Autumn Equinox. |
| 24th | 01:56 | First Quarter Moon. (Evening sky.) |
| 25th | 06:49 | The just-past first quarter Moon is south of Jupiter. (Evening sky.) |
| | 13:57 | The just-past first quarter Moon is south of dwarf planet Pluto. (Evening sky.) |
| | 21:08 | The just-past first quarter Moon is south of Saturn. (Evening sky.) |
| 29th | 01:06 | Saturn is stationary prior to resuming prograde motion. (Evening sky.) |
| 30th | 03:00 | The nearly full Moon is south of Neptune. (Evening sky.) |

# October 1st to 10th, 2020

## The Moon

1st

3rd

5th

7th

9th

| Date | Con | R.A. | Dec | Mag | Diam | Ill. | Elon. | Phase | Close To |
|---|---|---|---|---|---|---|---|---|---|
| 1st | Psc | 0h 24m | -2° 40' | -12.6 | 30' | 100% | 178° E | FM | |
| 2nd | Cet | 1h 8m | 2° 16' | -12.5 | 29' | 99% | 172° W | FM | Mars |
| 3rd | Psc | 1h 51m | 7° 6' | -12.3 | 29' | 98% | 162° W | FM | Mars, Uranus |
| 4th | Ari | 2h 35m | 11° 40' | -12.0 | 29' | 94% | 152° W | -G | Uranus |
| 5th | Ari | 3h 21m | 15° 49' | -11.7 | 30' | 89% | 141° W | -G | Pleiades |
| 6th | Tau | 4h 9m | 19° 22' | -11.4 | 30' | 82% | 130° W | -G | Pleiades, Hyades, Aldebaran |
| 7th | Tau | 4h 59m | 22° 8' | -11.1 | 30' | 74% | 119° W | -G | Hyades, Aldebaran |
| 8th | Tau | 5h 51m | 23° 57' | -10.7 | 30' | 65% | 107° W | LQ | |
| 9th | Gem | 6h 46m | 24° 37' | -10.4 | 31' | 55% | 94° W | LQ | |
| 10th | Gem | 7h 42m | 24° 1' | -9.9 | 31' | 45% | 81° W | LQ | |

## Mercury and Venus

Mercury
5th

Venus
5th

**Mercury**

| Date | Con. | R.A. | Dec. | Mag. | Diam. | Ill. | Elon. | Vis. | Rat. | Close To |
|---|---|---|---|---|---|---|---|---|---|---|
| 1st | Vir | 14h 4m | -15° 40' | 0.1 | 7" | 60% | 23° E | PM | *** | Spica |
| 3rd | Vir | 14h 11m | -16° 28' | 0.1 | 7" | 57% | 23° E | PM | *** | |
| 5th | Vir | 14h 18m | -17° 10' | 0.1 | 7" | 53% | 23° E | PM | *** | |
| 7th | Lib | 14h 23m | -17° 44' | 0.2 | 8" | 48% | 22° E | PM | *** | |
| 9th | Lib | 14h 27m | -18° 10' | 0.3 | 8" | 43% | 22° E | PM | *** | |

**Venus**

| Date | Con. | R.A. | Dec. | Mag. | Diam. | Ill. | Elon. | Vis. | Rat. | Close To |
|------|------|------|------|------|-------|------|-------|------|------|----------|
| **1st** | Leo | 10h 2m | 12° 18' | -4.1 | 16" | 72% | 37° W | AM | ** | Regulus |
| **3rd** | Leo | 10h 11m | 11° 36' | -4.1 | 15" | 72% | 37° W | AM | ** | Regulus |
| **5th** | Leo | 10h 20m | 10° 52' | -4.1 | 15" | 73% | 36° W | AM | ** | Regulus |
| **7th** | Leo | 10h 29m | 10° 7' | -4.1 | 15" | 74% | 36° W | AM | ** | Regulus |
| **9th** | Leo | 10h 38m | 9° 21' | -4.1 | 15" | 75% | 36° W | AM | ** | Regulus |

## Mars and the Outer Planets

Mars
5th

Jupiter
5th

Saturn
5th

**Mars**

| Date | Con. | R.A. | Dec. | Mag. | Diam. | Ill. | Elon. | Vis. | Rat. | Close To |
|------|------|------|------|------|-------|------|-------|------|------|----------|
| **1st** | Psc | 1h 37m | 6° 13' | -2.5 | 22" | 99% | 164° W | AM | ***** | |
| **5th** | Psc | 1h 33m | 5° 58' | -2.6 | 23" | 99% | 168° W | AM | ***** | |
| **10th** | Psc | 1h 26m | 5° 39' | -2.6 | 23" | 100% | 175° W | AN | ***** | |

**The Outer Planets**

| Planet | Date | Con. | R.A. | Dec. | Mag. | Diam. | Elon. | Vis. | Rat. | Close To |
|--------|------|------|------|------|------|-------|-------|------|------|----------|
| Jupiter | 5th | Sgr | 19h 19m | -22° 38' | -2.3 | 40" | 98° E | PM | *** | Saturn |
| Saturn | 5th | Sgr | 19h 49m | -21° 22' | 0.5 | 17" | 106° E | PM | *** | |
| Uranus | 5th | Ari | 2h 29m | 14° 16' | 5.7 | 4" | 154° W | AM | *** | |
| Neptune | 5th | Aqr | 23h 21m | -5° 25' | 7.8 | 2" | 159° E | PM | ***** | |

## Highlights

| Date | Time (UT) | Event |
|------|-----------|-------|
| 1st | 16:00 | Mercury is at greatest eastern elongation from the Sun. (Evening sky.) |
| | 21:06 | Full Moon. (Visible all night.) |
| 2nd | 17:50 | Venus is 0.1° south of the bright star Regulus. (Leo, morning sky.) |
| 3rd | 04:36 | The waning gibbous Moon is south of Mars. (Morning sky.) |
| | 19:02 | Dwarf planet Pluto is stationary prior to resuming prograde motion. (Evening sky.) |
| 4th | 10:02 | The waning gibbous Moon is south of Uranus. (Morning sky.) |
| 6th | 00:09 | The waning gibbous Moon is south of the Pleiades star cluster. (Taurus, morning sky.) |
| | 23:56 | The waning gibbous Moon is north of the bright star Aldebaran. (Taurus, morning sky.) |
| 8th | N/A | The Draconid meteor shower is at its maximum. (ZHR: Variable.) |
| 10th | 00:40 | Last Quarter Moon. (Morning sky.) |

# October 11th to 20th, 2020

## The Moon

| 11th | 13th | 15th | 17th | 19th |

| Date | Con | R.A. | Dec | Mag | Diam | Ill. | Elon. | Phase | Close To |
|------|-----|------|-----|-----|------|------|-------|-------|----------|
| 11th | Cnc | 8h 39m | 22° 6' | -9.4 | 31' | 35% | 67° W | -Cr | Praesepe |
| 12th | Leo | 9h 35m | 18° 52' | -8.8 | 32' | 25% | 54° W | -Cr | Regulus |
| 13th | Leo | 10h 31m | 14° 28' | -8.0 | 32' | 16% | 41° W | -Cr | Venus, Regulus |
| 14th | Leo | 11h 26m | 9° 4' | -7.0 | 33' | 8% | 28° W | NM | Venus |
| 15th | Vir | 12h 21m | 3° 2' | -5.9 | 33' | 3% | 15° W | NM | |
| 16th | Vir | 13h 17m | -3° 18' | -4.6 | 33' | 0% | 2° W | NM | Spica |
| 17th | Vir | 14h 13m | -9° 28' | -5.0 | 33' | 1% | 11° E | NM | Mercury |
| 18th | Lib | 15h 11m | -15° 3' | -6.3 | 33' | 4% | 24° E | NM | Mercury |
| 19th | Sco | 16h 11m | -19° 35' | -7.4 | 33' | 11% | 38° E | NM | Antares |
| 20th | Oph | 17h 12m | -22° 46' | -8.3 | 32' | 19% | 52° E | +Cr | Antares |

## Mercury and Venus

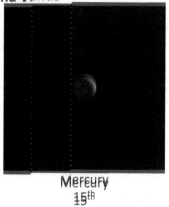

Mercury
15th

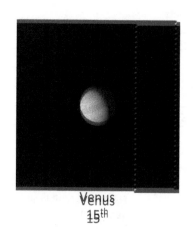

Venus
15th

## Mercury

| Date | Con. | R.A. | Dec. | Mag. | Diam. | Ill. | Elon. | Vis. | Rat. | Close To |
|------|------|------|------|------|-------|------|-------|------|------|----------|
| 11th | Lib | 14h 31m | -18° 26' | 0.4 | 8" | 38% | 21° E | PM | *** | |
| 13th | Lib | 14h 32m | -18° 31' | 0.6 | 9" | 31% | 19° E | PM | *** | |
| 15th | Lib | 14h 32m | -18° 21' | 0.9 | 9" | 25% | 17° E | PM | *** | |
| 17th | Lib | 14h 29m | -17° 57' | 1.4 | 9" | 18% | 15° E | NV | N/A | Moon |
| 19th | Lib | 14h 25m | -17° 14' | 2.0 | 10" | 12% | 12° E | NV | N/A | |

## Venus

| Date | Con. | R.A. | Dec. | Mag. | Diam. | Ill. | Elon. | Vis. | Rat. | Close To |
|------|------|------|------|------|-------|------|-------|------|------|----------|
| 11th | Leo | 10h 47m | 8° 33' | -4.1 | 15" | 75% | 35° W | AM | ** | Regulus |
| 13th | Leo | 10h 56m | 7° 44' | -4.1 | 14" | 76% | 35° W | AM | ** | Moon |
| 15th | Leo | 11h 5m | 6° 55' | -4.0 | 14" | 76% | 34° W | AM | ** | |
| 17th | Leo | 11h 14m | 6° 4' | -4.0 | 14" | 77% | 34° W | AM | ** | |
| 19th | Leo | 11h 23m | 5° 12' | -4.0 | 14" | 78% | 34° W | AM | ** | |

## Mars and the Outer Planets

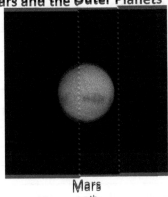

Mars
15th

Jupiter
15th

Saturn
15th

## Mars

| Date | Con. | R.A. | Dec. | Mag. | Diam. | Ill. | Elon. | Vis. | Rat. | Close To |
|------|------|------|------|------|-------|------|-------|------|------|----------|
| 11th | Psc | 1h 25m | 5° 36' | -2.6 | 22" | 100% | 176° W | AN | ***** | |
| 15th | Psc | 1h 20m | 5° 21' | -2.6 | 22" | 100% | 179° E | AN | ***** | |
| 20th | Psc | 1h 14m | 5° 6' | -2.5 | 22" | 100% | 173° E | AN | ***** | |

## The Outer Planets

| Planet | Date | Con. | R.A. | Dec. | Mag. | Diam. | Elon. | Vis. | Rat. | Close To |
|--------|------|------|------|------|------|-------|-------|------|------|----------|
| Jupiter | 15th | Sgr | 19h 22m | -22° 31' | -2.3 | 39" | 90° E | PM | *** | Saturn |
| Saturn | 15th | Sgr | 19h 50m | -21° 21' | 0.5 | 17" | 97° E | PM | *** | |
| Uranus | 15th | Ari | 2h 28m | 14° 9' | 5.7 | 4" | 164° W | AM | **** | |
| Neptune | 15th | Aqr | 23h 20m | -5° 30' | 7.8 | 2" | 149° E | PM | **** | |

## Highlights

| Date | Time (UT) | Event |
|------|-----------|-------|
| 11th | 13:39 | The just-past last quarter Moon is north of the Praesepe star cluster. (Cancer, morning sky.) |
| 13th | 00:52 | The waning crescent Moon is north of the bright star Regulus. (Leo, morning sky.) |
| | 23:18 | The waning crescent Moon is north of Venus. (Morning sky.) |
| | N/A | Good opportunity to see Earthshine on the waning crescent Moon. (Morning sky.) |
| 14th | 04:29 | Mercury is stationary prior to beginning retrograde motion. (Evening sky.) |
| | 21:17 | Mars is at opposition. (Visible all night.) |
| 16th | 19:32 | New Moon. (Not visible.) |
| 18th | N/A | The Epsilon Geminid meteor shower is at its maximum. (ZHR: 3) |
| 19th | 20:28 | The waxing crescent Moon is north of the bright star Antares. (Scorpius, evening sky.) |
| 20th | N/A | Good opportunity to see Earthshine on the waxing crescent Moon. (Evening sky.) |

# October 21st to 31st, 2020

## The Moon

| 21st | 23rd | 25th | 27th | 29th | 31st |

| Date | Con | R.A. | Dec | Mag | Diam | Ill. | Elon. | Phase | Close To |
|------|-----|------|-----|-----|------|------|-------|-------|----------|
| 21st | Sgr | 18h 13m | -24° 27' | -9.0 | 32' | 28% | 67° E | +Cr | |
| 22nd | Sgr | 19h 13m | -24° 35' | -9.6 | 31' | 39% | 81° E | FQ | Jupiter, Saturn |
| 23rd | Cap | 20h 11m | -23° 19' | -10.1 | 31' | 49% | 94° E | FQ | Jupiter, Saturn |
| 24th | Cap | 21h 5m | -20° 51' | -10.5 | 31' | 60% | 107° E | FQ | |
| 25th | Cap | 21h 56m | -17° 27' | -10.9 | 30' | 69% | 119° E | +G | |
| 26th | Aqr | 22h 44m | -13° 22' | -11.2 | 30' | 78% | 130° E | +G | Neptune |
| 27th | Aqr | 23h 29m | -8° 49' | -11.5 | 30' | 85% | 140° E | +G | Neptune |
| 28th | Psc | 0h 13m | -3° 59' | -11.8 | 30' | 91% | 150° E | +G | |
| 29th | Cet | 0h 57m | 0° 59' | -12.1 | 29' | 96% | 160° E | FM | Mars |
| 30th | Psc | 1h 40m | 5° 53' | -12.4 | 29' | 99% | 170° E | FM | Mars, Uranus |
| 31st | Ari | 2h 24m | 10° 35' | -12.6 | 29' | 100% | 180° E | FM | Uranus |

## Mercury and Venus

Mercury
25th

Venus
25th

### Mercury

| Date | Con. | R.A. | Dec. | Mag. | Diam. | Ill. | Elon. | Vis. | Rat. | Close To |
|------|------|------|------|------|-------|------|-------|------|------|----------|
| 21st | Vir | 14h 19m | -16° 14' | 2.9 | 10" | 6% | 8° E | NV | N/A | |
| 23rd | Vir | 14h 11m | -14° 58' | 4.0 | 10" | 2% | 4° E | NV | N/A | |
| 25th | Vir | 14h 2m | -13° 31' | 5.5 | 10" | 0% | 0° E | NV | N/A | Spica |
| 27th | Vir | 13h 53m | -12° 0' | 4.4 | 10" | 1% | 4° W | NV | N/A | Spica |
| 29th | Vir | 13h 46m | -10° 37' | 2.9 | 10" | 5% | 8° W | NV | N/A | Spica |
| 31st | Vir | 13h 41m | -9° 29' | 1.8 | 9" | 12% | 11° W | NV | N/A | Spica |

**Venus**

| Date | Con. | R.A. | Dec. | Mag. | Diam. | Ill. | Elon. | Vis. | Rat. | Close To |
|------|------|------|------|------|-------|------|-------|------|------|----------|
| **21st** | Leo | 11h 32m | 4° 20' | -4.0 | 14" | 78% | 33° W | AM | ** | |
| **23rd** | Vir | 11h 41m | 3° 27' | -4.0 | 14" | 79% | 33° W | AM | ** | |
| **25th** | Vir | 11h 50m | 2° 33' | -4.0 | 14" | 80% | 33° W | AM | ** | |
| **27th** | Vir | 11h 59m | 1° 38' | -4.0 | 13" | 80% | 32° W | AM | ** | |
| **29th** | Vir | 12h 8m | 0° 44' | -4.0 | 13" | 81% | 32° W | AM | ** | |
| **31st** | Vir | 12h 17m | 0° 11' | -4.0 | 13" | 81% | 32° W | AM | ** | |

## Mars and the Outer Planets

Mars
25th

Jupiter
25th

Saturn
25th

**Mars**

| Date | Con. | R.A. | Dec. | Mag. | Diam. | Ill. | Elon. | Vis. | Rat. | Close To |
|------|------|------|------|------|-------|------|-------|------|------|----------|
| **21st** | Psc | 1h 13m | 5° 4' | -2.5 | 22" | 100% | 172° E | AN | ***** | |
| **25th** | Psc | 1h 8m | 4° 55' | -2.3 | 21" | 99% | 167° E | PM | ***** | |
| **31st** | Psc | 1h 3m | 4° 49' | -2.2 | 20" | 98% | 160° E | PM | ***** | |

**The Outer Planets**

| Planet | Date | Con. | R.A. | Dec. | Mag. | Diam. | Elon. | Vis. | Rat. | Close To |
|--------|------|------|------|------|------|-------|-------|------|------|----------|
| Jupiter | 25th | Sgr | 19h 27m | -22° 22' | -2.2 | 38" | 82° E | PM | *** | Saturn |
| Saturn | 25th | Sgr | 19h 51m | -21° 17' | 0.6 | 17" | 88° E | PM | *** | |
| Uranus | 25th | Ari | 2h 26m | 14° 2' | 5.7 | 4" | 174° W | AN | **** | |
| Neptune | 25th | Aqr | 23h 19m | -5° 35' | 7.8 | 2" | 140° E | PM | **** | |

## Highlights

| Date | Time (UT) | Event |
|------|-----------|-------|
| 21st | N/A | The Orionid meteor shower is at its maximum. (ZHR: 25) |
| 22nd | 17:04 | The nearly first quarter Moon is south of Jupiter. (Evening sky.) |
| | 22:51 | The nearly first quarter Moon is south of the dwarf planet Pluto. (Evening sky.) |
| 23rd | 04:08 | The almost first quarter Moon is south of Saturn. (Evening sky.) |
| | 13:24 | First Quarter Moon. (Evening sky.) |
| 24th | N/A | The Leo Minorid meteor shower is at its maximum. (ZHR: 2) |
| 25th | 18:17 | Mercury is at inferior conjunction with the Sun. (Not visible.) |
| 27th | 06:58 | The waxing gibbous Moon is south of Neptune. (Evening sky.) |
| 29th | 15:02 | The waxing gibbous Moon is south of Mars. (Evening sky.) |
| 31st | 12:32 | The almost full Moon is south of Uranus. (Visible all night.) |
| | 14:50 | Full Moon. (Visible all night.) |
| | 17:15 | Uranus is at opposition. (Visible all night.) |

# November 1st to 10th, 2020

## The Moon

    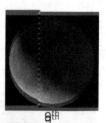

1st       3rd       5th       7th       9th

| Date | Con | R.A. | Dec | Mag | Diam | Ill. | Elon. | Phase | | Close To |
|------|-----|------|-----|-----|------|------|-------|-------|--|----------|
| 1st | Ari | 3h 9m | 14° 55' | -12.5 | 29' | 99% | 170° W | FM | | Uranus, Pleiades |
| 2nd | Tau | 3h 56m | 18° 40' | -12.2 | 30' | 97% | 159° W | FM | | Pleiades, Hyades, Aldebaran |
| 3rd | Tau | 4h 46m | 21° 41' | -11.9 | 30' | 93% | 147° W | -G | | Hyades, Aldebaran |
| 4th | Tau | 5h 38m | 23° 46' | -11.6 | 30' | 87% | 135° W | -G | | |
| 5th | Gem | 6h 32m | 24° 45' | -11.3 | 30' | 80% | 123° W | -G | | |
| 6th | Gem | 7h 27m | 24° 31' | -11.0 | 31' | 71% | 110° W | -G | | |
| 7th | Cnc | 8h 22m | 23° 0' | -10.6 | 31' | 61% | 97° W | LQ | | Praesepe |
| 8th | Cnc | 9h 17m | 20° 14' | -10.2 | 31' | 51% | 85° W | LQ | | Praesepe |
| 9th | Leo | 10h 12m | 16° 19' | -9.7 | 32' | 40% | 72° W | LQ | | Regulus |
| 10th | Leo | 11h 5m | 11° 25' | -9.1 | 32' | 29% | 60° W | -Cr | | |

## Mercury and Venus

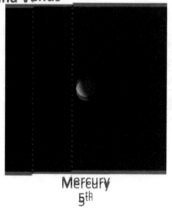 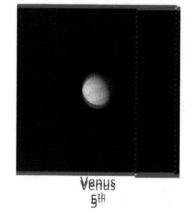

Mercury                 Venus
5th                     5th

## Mercury

| Date | Con. | R.A. | Dec. | Mag. | Diam. | Ill. | Elon. | Vis. | Rat. | Close To |
|------|------|------|------|------|-------|------|-------|------|------|----------|
| 1st | Vir | 13h 39m | -9° 3' | 1.3 | 9" | 16% | 13° W | NV | N/A | Spica |
| 3rd | Vir | 13h 38m | -8° 28' | 0.6 | 8" | 26% | 14° W | NV | N/A | Spica |
| 5th | Vir | 13h 39m | -8° 17' | 0.1 | 8" | 35% | 16° W | AM | ** | Spica |
| 7th | Vir | 13h 43m | -8° 26' | -0.2 | 7" | 45% | 17° W | AM | *** | Spica |
| 9th | Vir | 13h 49m | -8° 53' | -0.4 | 7" | 54% | 18° W | AM | *** | Spica |

## Venus

| Date | Con. | R.A. | Dec. | Mag. | Diam. | Ill. | Elon. | Vis. | Rat. | Close To |
|------|------|------|------|------|-------|------|-------|------|------|----------|
| 1st | Vir | 12h 22m | 0° 39' | -4.0 | 13" | 82% | 32° W | AM | ** | |
| 3rd | Vir | 12h 31m | -1° 35' | -4.0 | 13" | 82% | 31° W | AM | ** | |
| 5th | Vir | 12h 40m | -2° 30' | -4.0 | 13" | 83% | 31° W | AM | ** | |
| 7th | Vir | 12h 49m | -3° 26' | -4.0 | 13" | 83% | 31° W | AM | ** | Spica |
| 9th | Vir | 12h 58m | -4° 21' | -4.0 | 13" | 84% | 30° W | AM | ** | Spica |

## Mars and the Outer Planets

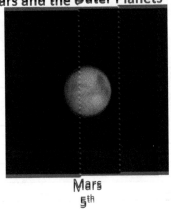

Mars
5th

Jupiter
5th

Saturn
5th

## Mars

| Date | Con. | R.A. | Dec. | Mag. | Diam. | Ill. | Elon. | Vis. | Rat. | Close To |
|------|------|------|------|------|-------|------|-------|------|------|----------|
| 1st | Psc | 1h 2m | 4° 49' | -2.1 | 20" | 98% | 159° E | PM | ***** | |
| 5th | Psc | 1h 0m | 4° 51' | -2.0 | 19" | 97% | 154° E | PM | **** | |
| 10th | Psc | 0h 58m | 4° 60' | -1.8 | 18" | 96% | 148° E | PM | **** | |

## The Outer Planets

| Planet | Date | Con. | R.A. | Dec. | Mag. | Diam. | Elon. | Vis. | Rat. | Close To |
|--------|------|------|------|------|------|-------|-------|------|------|----------|
| Jupiter | 5th | Sgr | 19h 34m | -22° 8' | -2.1 | 37" | 72° E | PM | ** | Saturn |
| Saturn | 5th | Sgr | 19h 54m | -21° 13' | 0.6 | 16" | 77° E | PM | ** | |
| Uranus | 5th | Ari | 2h 25m | 13° 53' | 5.7 | 4" | 175° E | AN | **** | |
| Neptune | 5th | Aqr | 23h 18m | -5° 38' | 7.8 | 2" | 129° E | PM | **** | |

## Highlights

| Date | Time (UT) | Event |
|------|-----------|-------|
| 2nd | 08:17 | The waning gibbous Moon is south of the Pleiades star cluster. (Taurus, morning sky.) |
| 3rd | 08:15 | The waning gibbous Moon is north of the bright star Aldebaran. (Taurus, morning sky.) |
| | 08:16 | Mercury is stationary prior to resuming prograde motion. (Morning sky.) |
| 7th | 19:04 | The nearly last quarter Moon is north of the Praesepe star cluster. (Morning sky.) |
| 8th | 13:47 | Last Quarter Moon. (Morning sky.) |
| 9th | 11:34 | The just-past last quarter Moon is north of the bright star Regulus. (Morning sky.) |
| 10th | 16:57 | Mercury is at greatest western elongation from the Sun. (Morning sky.) |

# November 11<sup>th</sup> to 20<sup>th</sup>, 2020

## The Moon

 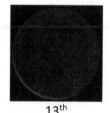  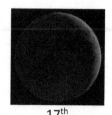 

| | 11<sup>th</sup> | | 13<sup>th</sup> | | 15<sup>th</sup> | | 17<sup>th</sup> | | 19<sup>th</sup> |

| Date | Con | R.A. | Dec | Mag | Diam | Ill. | Elon. | Phase | Close To |
|------|-----|------|-----|-----|------|------|-------|-------|----------|
| 11th | Vir | 11h 58m | 5° 45' | -8.3 | 33' | 19% | 47° W | -Cr | |
| 12th | Vir | 12h 52m | 0° 23' | -7.5 | 33' | 11% | 35° W | NM | Venus, Spica |
| 13th | Vir | 13h 47m | -6° 37' | -6.4 | 33' | 5% | 22° W | NM | Mercury, Venus, Spica |
| 14th | Lib | 14h 44m | -12° 32' | -5.0 | 33' | 1% | 9° W | NM | Mercury |
| 15th | Lib | 15h 43m | -17° 41' | -4.4 | 33' | 0% | 5° E | NM | Antares |
| 16th | Oph | 16h 44m | -21° 38' | -5.8 | 33' | 2% | 19° E | NM | Antares |
| 17th | Sgr | 17h 47m | -24° 4' | -6.9 | 33' | 7% | 34° E | NM | |
| 18th | Sgr | 18h 50m | -24° 51' | -7.9 | 32' | 15% | 48° E | +Cr | |
| 19th | Sgr | 19h 51m | -24° 4' | -8.7 | 32' | 23% | 62° E | +Cr | Jupiter, Saturn |
| 20th | Cap | 20h 48m | -21° 55' | -9.3 | 31' | 33% | 76° E | +Cr | |

## Mercury and Venus

Mercury
15<sup>th</sup>

Venus
15<sup>th</sup>

**Mercury**

| Date | Con. | R.A. | Dec. | Mag. | Diam. | Ill. | Elon. | Vis. | Rat. | Close To |
|------|------|------|------|------|-------|------|-------|------|------|----------|
| 11th | Vir | 13h 57m | -9° 33' | -0.6 | 7" | 62% | 18° W | AM | *** | Spica |
| 13th | Vir | 14h 5m | -10° 23' | -0.7 | 6" | 69% | 18° W | AM | ** | Moon |
| 15th | Vir | 14h 15m | -11° 20' | -0.7 | 6" | 74% | 17° W | AM | ** | |
| 17th | Lib | 14h 26m | -12° 21' | -0.7 | 6" | 79% | 17° W | AM | ** | |
| 19th | Lib | 14h 37m | -13° 25' | -0.7 | 6" | 83% | 16° W | AM | ** | |

**Venus**

| Date | Con. | R.A. | Dec. | Mag. | Diam. | Ill. | Elon. | Vis. | Rat. | Close To |
|------|------|------|------|------|-------|------|-------|------|------|----------|
| 11th | Vir | 13h 7m | -5° 17' | -4.0 | 13" | 84% | 30° W | AM | ** | Spica |
| 13th | Vir | 13h 16m | -6° 12' | -4.0 | 12" | 85% | 30° W | AM | ** | Moon, Spica |
| 15th | Vir | 13h 25m | -7° 7' | -4.0 | 12" | 85% | 30° W | AM | ** | Spica |
| 17th | Vir | 13h 35m | -8° 1' | -4.0 | 12" | 86% | 29° W | AM | ** | Spica |
| 19th | Vir | 13h 44m | -8° 55' | -4.0 | 12" | 86% | 29° W | AM | ** | Spica |

## Mars and the Outer Planets

Mars
15th

Jupiter
15th

Saturn
15th

**Mars**

| Date | Con. | R.A. | Dec. | Mag. | Diam. | Ill. | Elon. | Vis. | Rat. | Close To |
|------|------|------|------|------|-------|------|-------|------|------|----------|
| 11th | Psc | 0h 57m | 5° 2' | -1.8 | 18" | 96% | 147° E | PM | **** | |
| 15th | Psc | 0h 57m | 5° 15' | -1.7 | 17" | 95% | 143° E | PM | **** | |
| 20th | Psc | 0h 57m | 5° 36' | -1.5 | 16" | 94% | 138° E | PM | **** | |

**The Outer Planets**

| Planet | Date | Con. | R.A. | Dec. | Mag. | Diam. | Elon. | Vis. | Rat. | Close To |
|--------|------|------|------|------|------|-------|-------|------|------|----------|
| Jupiter | 15th | Sgr | 19h 40m | -21° 52' | -2.1 | 36" | 64° E | PM | ** | Saturn |
| Saturn | 15th | Sgr | 19h 57m | -21° 4' | 0.6 | 16" | 68° E | PM | ** | |
| Uranus | 15th | Ari | 2h 23m | 13° 45' | 5.7 | 4" | 165° E | PM | **** | |
| Neptune | 15th | Aqr | 23h 18m | -5° 40' | 7.9 | 2" | 118° E | PM | **** | |

## Highlights

| Date | Time (UT) | Event |
|------|-----------|-------|
| 12th | 21:34 | The waning crescent Moon is north of Venus. (Morning sky.) |
| | N/A | Good opportunity to see Earthshine on the waning crescent Moon. (Morning sky.) |
| | N/A | The Northern Taurid meteor shower is at its maximum. (ZHR: 5) |
| 13th | 01:24 | The waning crescent Moon is north of the bright star Spica. (Virgo, morning sky.) |
| | 21:14 | The waning crescent Moon is north of Mercury. (Morning sky.) |
| 15th | 05:08 | New Moon. (Not visible.) |
| | 07:34 | Venus is 4.1° north of the bright star Spica. (Virgo, morning sky.) |
| | 19:42 | Mars is stationary prior to resuming prograde motion. (Evening sky.) |
| 18th | N/A | Good opportunity to see Earthshine on the waxing crescent Moon. (Evening sky.) |
| | N/A | The Leonid meteor shower is at its maximum. (ZHR: 20) |
| 19th | 06:14 | The waxing crescent Moon is south of dwarf planet Pluto. (Evening sky.) |
| | 07:52 | The waxing crescent Moon is south of Jupiter. (Evening sky.) |
| | 14:19 | The waxing crescent Moon is south of Saturn. (Evening sky.) |

# November 21ˢᵗ to 30ᵗʰ, 2020

## The Moon

| | 21ˢᵗ | | 23ʳᵈ | | 25ᵗʰ | | 27ᵗʰ | | 29ᵗʰ |

| Date | Con | R.A. | Dec. | Mag. | Diam | Ill. | Elon. | Phase | Close To |
|------|-----|------|------|------|------|------|-------|-------|----------|
| 21st | Cap | 21h 41m | -18° 42' | -9.8 | 31' | 43% | 88° E | FQ | |
| 22nd | Aqr | 22h 31m | -14° 43' | -10.3 | 30' | 53% | 99° E | FQ | Neptune |
| 23rd | Aqr | 23h 18m | -10° 12' | -10.6 | 30' | 63% | 110° E | FQ | Neptune |
| 24th | Psc | 0h 2m | -5° 23' | -11.0 | 30' | 72% | 120° E | +G | Neptune |
| 25th | Cet | 0h 45m | 0° 26' | -11.3 | 30' | 80% | 130° E | +G | Mars |
| 26th | Psc | 1h 28m | 4° 31' | -11.6 | 29' | 87% | 140° E | +G | Mars |
| 27th | Cet | 2h 12m | 9° 19' | -11.9 | 29' | 93% | 149° E | +G | Uranus |
| 28th | Ari | 2h 57m | 13° 47' | -12.2 | 30' | 97% | 159° E | FM | Uranus |
| 29th | Tau | 3h 44m | 17° 45' | -12.5 | 30' | 99% | 170° E | FM | Pleiades, Hyades |
| 30th | Tau | 4h 33m | 21° 1' | -12.7 | 30' | 100% | 179° W | FM | Pleiades, Hyades, Aldebaran |

## Mercury and Venus

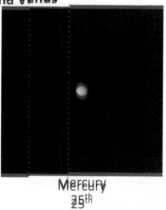

Mercury
25ᵗʰ

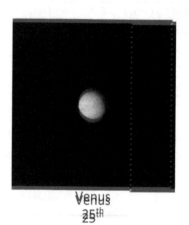

Venus
25ᵗʰ

## Mercury

| Date | Con. | R.A. | Dec. | Mag. | Diam. | Ill. | Elon. | Vis. | Rat. | Close To |
|------|------|------|------|------|-------|------|-------|------|------|----------|
| 21st | Lib | 14h 48m | -14° 30' | -0.7 | 5" | 86% | 15° W | AM | ** | |
| 23rd | Lib | 15h 0m | -15° 35' | -0.7 | 5" | 89% | 14° W | NV | N/A | |
| 25th | Lib | 15h 12m | -16° 38' | -0.7 | 5" | 91% | 14° W | NV | N/A | |
| 27th | Lib | 15h 24m | -17° 39' | -0.7 | 5" | 93% | 13° W | NV | N/A | |
| 29th | Lib | 15h 36m | -18° 38' | -0.7 | 5" | 95% | 13° W | NV | N/A | |

## Venus

| Date | Con. | R.A. | Dec. | Mag. | Diam. | Ill. | Elon. | Vis. | Rat. | Close To |
|------|------|------|------|------|-------|------|-------|------|------|----------|
| 21st | Vir | 13h 53m | -9° 48' | -4.0 | 12" | 87% | 39° W | AM | ** | Spica |
| 23rd | Vir | 14h 3m | -10° 40' | -4.0 | 12" | 87% | 39° W | AM | ** | Spica |
| 25th | Vir | 14h 12m | -11° 31' | -4.0 | 12" | 87% | 38° W | AM | ** | |
| 27th | Vir | 14h 22m | -12° 21' | -4.0 | 12" | 88% | 38° W | AM | ** | |
| 29th | Lib | 14h 31m | -13° 11' | -4.0 | 12" | 88% | 28° W | AM | * | |

## Mars and the Outer Planets

Mars
25th

Jupiter
25th

Saturn
25th

### Mars

| Date | Con. | R.A. | Dec. | Mag. | Diam. | Ill. | Elon. | Vis. | Rat. | Close To |
|------|------|------|------|------|-------|------|-------|------|------|----------|
| 21st | Psc | 0h 58m | 5° 41' | -1.5 | 16" | 94% | 137° E | PM | **** | |
| 25th | Psc | 0h 59m | 6° 3' | -1.3 | 16" | 93% | 133° E | PM | **** | Moon |
| 30th | Psc | 1h 2m | 6° 36' | -1.2 | 15" | 92% | 129° E | PM | **** | |

### The Outer Planets

| Planet | Date | Con. | R.A. | Dec. | Mag. | Diam. | Elon. | Vis. | Rat. | Close To |
|--------|------|------|------|------|------|-------|-------|------|------|----------|
| Jupiter | 25th | Sgr | 19h 48m | -21° 34' | -2.0 | 35" | 55° E | PM | ** | Saturn |
| Saturn | 25th | Sgr | 20h 0m | -20° 55' | 0.6 | 16" | 58° E | PM | ** | |
| Uranus | 25th | Ari | 2h 22m | 13° 38' | 5.7 | 4" | 154° E | PM | *** | |
| Neptune | 25th | Aqr | 23h 18m | -5° 41' | 7.9 | 2" | 108° E | PM | *** | |

## Highlights

| Date | Time (UT) | Event |
|------|-----------|-------|
| 22nd | 04:46 | First Quarter Moon. (Evening sky.) |
| | N/A | The Alpha Monocerotid meteor shower is at its maximum. (ZHR: Variable.) |
| 23rd | 10:52 | The just-past first quarter Moon is south of Neptune. (Evening sky.) |
| 25th | 19:21 | The waxing gibbous Moon is south of Mars. (Evening sky.) |
| 27th | 15:36 | The waxing gibbous Moon is south of Uranus. (Evening sky.) |
| 29th | 05:27 | Neptune is stationary prior to resuming prograde motion. (Evening sky.) |
| | 13:46 | The nearly full Moon is south of the Pleiades star cluster. (Evening sky.) |
| 30th | 09:30 | Full Moon. (Visible all night.) |
| | 09:43 | Penumbral lunar eclipse. Visible from Asia, the northern Atlantic, Australia, western Europe, North America, the Pacific and South America. |
| | 13:47 | The full Moon is north of the bright star Aldebaran. (Taurus, visible all night.) |

# December 1ˢᵗ to 10ᵗʰ, 2020

## The Moon

1ˢᵗ     3ʳᵈ     5ᵗʰ     7ᵗʰ     9ᵗʰ

| Date | Con | R.A. | Dec | Mag | Diam | Ill. | Elon. | Phase | Close To |
|------|-----|------|-----|-----|------|------|-------|-------|----------|
| 1st | Tau | 5h 25m | 23° 24' | -12.4 | 30' | 99% | 167° W | FM | |
| 2nd | Gem | 6h 19m | 24° 41' | -12.1 | 30' | 96% | 154° W | FM | |
| 3rd | Gem | 7h 14m | 24° 45' | -11.8 | 30' | 91% | 142° W | -G | |
| 4th | Cnc | 8h 9m | 23° 33' | -11.5 | 31' | 84% | 129° W | -G | Praesepe |
| 5th | Cnc | 9h 4m | 21° 6' | -11.2 | 31' | 76% | 116° W | -G | Praesepe |
| 6th | Leo | 9h 58m | 17° 30' | -10.8 | 31' | 66% | 104° W | -G | Regulus |
| 7th | Leo | 10h 50m | 12° 57' | -10.4 | 32' | 56% | 92° W | LQ | Regulus |
| 8th | Vir | 11h 42m | 7° 39' | -9.9 | 32' | 45% | 80° W | LQ | |
| 9th | Vir | 12h 34m | 1° 50' | -9.3 | 32' | 34% | 68° W | -Cr | |
| 10th | Vir | 13h 26m | -4° 11' | -8.6 | 33' | 23% | 56° W | -Cr | Spica |

## Mercury and Venus

Mercury
5ᵗʰ

Venus
5ᵗʰ

### Mercury

| Date | Con. | R.A. | Dec. | Mag. | Diam. | Ill. | Elon. | Vis. | Rat. | Close To |
|------|------|------|------|------|-------|------|-------|------|------|----------|
| 1st | Lib | 15h 49m | -19° 33' | -0.8 | 5" | 96% | 11° W | NV | N/A | Antares |
| 3rd | Sco | 16h 2m | -20° 25' | -0.8 | 5" | 97% | 10° W | NV | N/A | Antares |
| 5th | Sco | 16h 15m | -21° 14' | -0.8 | 5" | 98% | 9° W | NV | N/A | Antares |
| 7th | Oph | 16h 28m | -21° 58' | -0.8 | 5" | 98% | 8° W | NV | N/A | Antares |
| 9th | Oph | 16h 41m | -22° 39' | -0.9 | 5" | 99% | 6° W | NV | N/A | Antares |

**Venus**

| Date | Con. | R.A. | Dec. | Mag. | Diam. | Ill. | Elon. | Vis. | Rat. | Close To |
|------|------|------|------|------|-------|------|-------|------|------|----------|
| 1st | Lib | 14h 41m | -13° 59' | -4.0 | 12" | 89% | 28° W | AM | * | |
| 3rd | Lib | 14h 51m | -14° 45' | -4.0 | 12" | 89% | 27° W | AM | * | |
| 5th | Lib | 15h 1m | -15° 30' | -4.0 | 11" | 90% | 27° W | AM | * | |
| 7th | Lib | 15h 11m | -16° 14' | -4.0 | 11" | 90% | 27° W | AM | * | |
| 9th | Lib | 15h 21m | -16° 56' | -4.0 | 11" | 90% | 27° W | AM | * | |

## Mars and the Outer Planets

Mars
5th

Jupiter
5th

Saturn
5th

**Mars**

| Date | Con. | R.A. | Dec. | Mag. | Diam. | Ill. | Elon. | Vis. | Rat. | Close To |
|------|------|------|------|------|-------|------|-------|------|------|----------|
| 1st | Psc | 1h 3m | 6° 43' | -1.1 | 15" | 92% | 128° E | PM | *** | |
| 5th | Psc | 1h 6m | 7° 13' | -1.0 | 14" | 92% | 124° E | PM | *** | |
| 10th | Psc | 1h 11m | 7° 53' | -0.8 | 13" | 91% | 120° E | PM | *** | |

**The Outer Planets**

| Planet | Date | Con. | R.A. | Dec. | Mag. | Diam. | Elon. | Vis. | Rat. | Close To |
|--------|------|------|------|------|------|-------|-------|------|------|----------|
| Jupiter | 5th | Sgr | 19h 56m | -21° 12' | -2.0 | 34" | 47° E | PM | ** | Saturn |
| Saturn | 5th | Sgr | 20h 4m | -20° 45' | 0.6 | 16" | 49° E | PM | ** | |
| Uranus | 5th | Ari | 2h 20m | 13° 32' | 5.7 | 4" | 143° E | PM | *** | |
| Neptune | 5th | Aqr | 23h 18m | -5° 41' | 7.9 | 2" | 97° E | PM | *** | |

## Highlights

| Date | Time (UT) | Event |
|------|-----------|-------|
| 5th | 00:06 | The waning gibbous Moon is north of the Praesepe star cluster. (Cancer, morning sky.) |
| 6th | 16:38 | The waning gibbous Moon is north of the bright star Regulus. (Leo, morning sky.) |
| 8th | 00:37 | Last Quarter Moon. (Morning sky.) |
| 9th | N/A | The Monocerotid meteor shower is at its maximum. (ZHR: 2) |
| 10th | 12:46 | The waning crescent Moon is north of the bright star Spica. (Virgo, morning sky.) |

# December 11<sup>th</sup> to 20<sup>th</sup>, 2020

## The Moon

| 11<sup>th</sup> | 13<sup>th</sup> | 15<sup>th</sup> | 17<sup>th</sup> | 19<sup>th</sup> |

| Date | Con | R.A. | Dec | Mag | Diam | Ill. | Elon. | Phase | Close To |
|------|-----|------|-----|-----|------|------|-------|-------|----------|
| 11th | Vir | 14h 20m | -10° 6' | -7.8 | 33' | 14% | 44° W | -Cr | |
| 12th | Lib | 15h 16m | -15° 30' | -6.8 | 33' | 7% | 31° W | NM | Venus |
| 13th | Sco | 16h 16m | -19° 59' | -5.6 | 33' | 2% | 17° W | NM | Venus, Antares |
| 14th | Oph | 17h 18m | -23° 9' | -4.1 | 33' | 0% | 3° W | NM | Mercury |
| 15th | Sgr | 18h 22m | -24° 43' | -5.1 | 33' | 1% | 12° E | NM | |
| 16th | Sgr | 19h 25m | -24° 37' | -6.3 | 32' | 4% | 27° E | NM | Jupiter, Saturn |
| 17th | Cap | 20h 25m | -22° 58' | -7.4 | 32' | 10% | 41° E | NM | Jupiter, Saturn |
| 18th | Cap | 21h 21m | -20° 4' | -8.2 | 31' | 18% | 54° E | +Cr | |
| 19th | Aqr | 22h 14m | -16° 14' | -8.9 | 31' | 26% | 66° E | +Cr | |
| 20th | Aqr | 23h 2m | -11° 47' | -9.4 | 30' | 36% | 77° E | +Cr | Neptune |

## Mercury and Venus

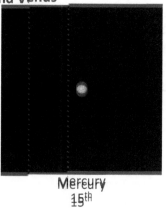

Mercury
15<sup>th</sup>

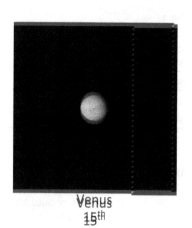

Venus
15<sup>th</sup>

### Mercury

| Date | Con. | R.A. | Dec. | Mag. | Diam. | Ill. | Elon. | Vis. | Rat. | Close To |
|------|------|------|------|------|-------|------|-------|------|------|----------|
| 11th | Oph | 16h 55m | -23° 14' | -0.9 | 5" | 99% | 5° W | NV | N/A | Antares |
| 13th | Oph | 17h 8m | -23° 45' | -1.0 | 5" | 100% | 4° W | NV | N/A | Antares |
| 15th | Oph | 17h 22m | -24° 12' | -1.1 | 5" | 100% | 3° W | NV | N/A | |
| 17th | Oph | 17h 35m | -24° 33' | -1.1 | 5" | 100% | 2° W | NV | N/A | |
| 19th | Sgr | 17h 49m | -24° 49' | -1.2 | 5" | 100% | 0° W | NV | N/A | |

## Venus

| Date | Con. | R.A. | Dec. | Mag. | Diam. | Ill. | Elon. | Vis. | Rat. | Close To |
|------|------|------|------|------|-------|------|-------|------|------|----------|
| 11th | Lib | 15h 31m | -17° 36' | -4.0 | 11" | 91% | 26° W | AM | * | |
| 13th | Lib | 15h 41m | -18° 15' | -4.0 | 11" | 91% | 26° W | AM | * | Moon |
| 15th | Lib | 15h 51m | -18° 51' | -4.0 | 11" | 91% | 26° W | AM | * | Antares |
| 17th | Lib | 16h 1m | -19° 26' | -3.9 | 11" | 92% | 25° W | AM | * | Antares |
| 19th | Sco | 16h 12m | -19° 58' | -3.9 | 11" | 92% | 25° W | AM | * | Antares |

## Mars and the Outer Planets

Mars
15th

Jupiter
15th

Saturn
15th

## Mars

| Date | Con. | R.A. | Dec. | Mag. | Diam. | Ill. | Elon. | Vis. | Rat. | Close To |
|------|------|------|------|------|-------|------|-------|------|------|----------|
| 11th | Psc | 1h 12m | 8° 2' | -0.8 | 13" | 91% | 119° E | PM | *** | |
| 15th | Psc | 1h 16m | 8° 38' | -0.7 | 12" | 90% | 116° E | PM | *** | |
| 20th | Psc | 1h 23m | 9° 25' | -0.5 | 12" | 90% | 112° E | PM | *** | |

## The Outer Planets

| Planet | Date | Con. | R.A. | Dec. | Mag. | Diam. | Elon. | Vis. | Rat. | Close To |
|--------|------|------|------|------|------|-------|-------|------|------|----------|
| Jupiter | 15th | Sgr | 20h 5m | -20° 48' | -2.0 | 34" | 38° E | PM | ** | Saturn |
| Saturn | 15th | Sgr | 20h 8m | -20° 33' | 0.6 | 15" | 39° E | PM | ** | |
| Uranus | 15th | Ari | 2h 19m | 13° 27' | 5.7 | 4" | 131° E | PM | *** | |
| Neptune | 15th | Aqr | 23h 18m | -5° 39' | 7.9 | 2" | 86° E | PM | *** | |

## Highlights

| Date | Time (UT) | Event |
|------|-----------|-------|
| 11th | N/A | Good opportunity to see Earthshine on the waning crescent Moon. (Morning sky.) |
| 13th | 21:05 | The waning crescent Moon is north of Venus. (Morning sky.) |
| 14th | 16:15 | Total solar eclipse: Visible from Antarctica, the southern Atlantic, the southern Pacific and South America. |
| | 16:17 | New Moon. (Not visible.) |
| | N/A | The Geminid meteor shower is at its maximum. (ZHR: 120) |
| 17th | 04:00 | The waxing crescent Moon is south of Jupiter. (Evening sky.) |
| | 04:39 | The waxing crescent Moon is south of Saturn. (Evening sky.) |
| 18th | N/A | Good opportunity to see Earthshine on the waxing crescent Moon. (Evening sky.) |
| 20th | 03:09 | Mercury is at superior conjunction with the Sun. (Not visible.) |
| | 21:18 | The nearly first quarter Moon is south of Neptune. (Evening sky.) |
| | N/A | The Leonis Minorid meteor shower is at its maximum. (ZHR: 5) |

# December 21st to 31st, 2020

## The Moon

| 21st | 23rd | 25th | 27th | 29th | 31st |

| Date | Con | R.A. | Dec | Mag | Diam | Ill. | Elon. | Phase | Close To |
|------|-----|------|-----|-----|------|------|-------|-------|----------|
| 21st | Aqr | 23h 48m | -6° 57' | -9.9 | 30' | 45% | 87° E | FQ | Neptune |
| 22nd | Cet | 0h 32m | -1° 58' | -10.3 | 30' | 55% | 97° E | FQ | |
| 23rd | Psc | 1h 16m | 3° 2' | -10.7 | 30' | 64% | 107° E | FQ | Mars |
| 24th | Psc | 1h 59m | 7° 54' | -11.0 | 30' | 73% | 116° E | +G | Mars, Uranus |
| 25th | Ari | 2h 43m | 12° 28' | -11.4 | 30' | 81% | 126° E | +G | Uranus |
| 26th | Tau | 3h 29m | 16° 36' | -11.7 | 30' | 88% | 137° E | +G | Pleiades |
| 27th | Tau | 4h 18m | 20° 7' | -11.9 | 30' | 93% | 148° E | +G | Pleiades, Hyades, Aldebaran |
| 28th | Tau | 5h 9m | 22° 47' | -12.2 | 30' | 97% | 160° E | FM | Hyades, Aldebaran |
| 29th | Gem | 6h 3m | 24° 26' | -12.5 | 30' | 100% | 172° E | FM | |
| 30th | Gem | 6h 59m | 24° 51' | -12.6 | 31' | 100% | 175° W | FM | |
| 31st | Gem | 7h 55m | 23° 58' | -12.3 | 31' | 98% | 162° W | FM | Praesepe |

## Mercury and Venus

Mercury
25th

Venus
25th

**Mercury**

| Date | Con. | R.A. | Dec. | Mag. | Diam. | Ill. | Elon. | Vis. | Rat. | Close To |
|------|------|------|------|------|-------|------|-------|------|------|----------|
| 21st | Sgr | 18h 3m | -24° 60' | -1.2 | 5" | 100% | 1° E | NV | N/A | |
| 23rd | Sgr | 18h 17m | -25° 5' | -1.1 | 5" | 100% | 2° E | NV | N/A | |
| 25th | Sgr | 18h 31m | -25° 4' | -1.1 | 5" | 100% | 3° E | NV | N/A | |
| 27th | Sgr | 18h 46m | -24° 58' | -1.0 | 5" | 99% | 5° E | NV | N/A | |
| 29th | Sgr | 19h 0m | -24° 45' | -1.0 | 5" | 99% | 6° E | NV | N/A | |
| 31st | Sgr | 19h 14m | -24° 27' | -1.0 | 5" | 98% | 7° E | NV | N/A | |

**Venus**

| Date | Con. | R.A. | Dec. | Mag. | Diam. | Ill. | Elon. | Vis. | Rat. | Close To |
|------|------|------|------|------|-------|------|-------|------|------|----------|
| 21st | Sco | 16h 22m | -20° 28' | -3.9 | 11" | 92% | 24° W | AM | * | Antares |
| 23rd | Oph | 16h 33m | -20° 56' | -3.9 | 11" | 93% | 24° W | AM | * | Antares |
| 25th | Oph | 16h 43m | -21° 21' | -3.9 | 11" | 93% | 24° W | AM | * | Antares |
| 27th | Oph | 16h 54m | -21° 44' | -3.9 | 11" | 93% | 23° W | AM | * | Antares |
| 29th | Oph | 17h 5m | -22° 4' | -3.9 | 11" | 94% | 23° W | AM | * | Antares |
| 31st | Oph | 17h 15m | -22° 22' | -3.9 | 11" | 94% | 22° W | AM | * | |

## Mars and the Outer Planets

Mars
25th

Jupiter
25th

Saturn
25th

**Mars**

| Date | Con. | R.A. | Dec. | Mag. | Diam. | Ill. | Elon. | Vis. | Rat. | Close To |
|------|------|------|------|------|-------|------|-------|------|------|----------|
| 21st | Psc | 1h 24m | 9° 34' | -0.5 | 12" | 90% | 111° E | PM | *** | |
| 25th | Psc | 1h 30m | 10° 14' | -0.4 | 11" | 89% | 108° E | PM | *** | |
| 31st | Psc | 1h 39m | 11° 15' | -0.3 | 10" | 89% | 104° E | PM | *** | |

**The Outer Planets**

| Planet | Date | Con. | R.A. | Dec. | Mag. | Diam. | Elon. | Vis. | Rat. | Close To |
|--------|------|------|------|------|------|-------|-------|------|------|----------|
| Jupiter | 25th | Cap | 20h 14m | -20° 20' | -2.0 | 33" | 29° E | PM | * | Saturn |
| Saturn | 25th | Cap | 20h 12m | -20° 20' | 0.6 | 15" | 29° E | PM | ** | |
| Uranus | 25th | Ari | 2h 18m | 13° 23' | 5.7 | 4" | 120° E | PM | *** | Moon |
| Neptune | 25th | Aqr | 23h 18m | -5° 36' | 7.9 | 2" | 75° E | PM | ** | |

## Highlights

| Date | Time (UT) | Event |
|------|-----------|-------|
| 21st | 10:04 | Winter Solstice. |
| | 13:46 | Jupiter is 0.1° south of Saturn. (Evening sky.) |
| | 23:42 | First Quarter Moon. (Evening sky.) |
| 22nd | 18:51 | Venus is 5.7° north of the bright star Antares. (Scorpius, morning sky.) |
| 23rd | 18:09 | The waxing gibbous Moon is south of Mars. (Evening sky.) |
| | N/A | The Ursid meteor shower is at its maximum. (ZHR: 10) |
| 24th | 23:51 | The waxing gibbous Moon is south of Uranus. (Evening sky.) |
| 26th | 20:11 | The waxing gibbous Moon is south of the Pleiades star cluster. (Taurus, evening sky.) |
| 27th | 19:28 | The waxing gibbous Moon is north of the bright star Aldebaran. (Taurus, evening sky.) |
| 30th | 03:29 | Full Moon. (Visible all night.) |

# Planet Visibility Ratings

## Morning Sky

| | | Me | Ve | Ma | Ju | Sa | Ur | Ne |
|---|---|---|---|---|---|---|---|---|
| Jan | 5th | | * | | | | | |
| | 15th | | * | | | | | |
| | 25th | | * | | | | | |
| Feb | 5th | | | | | | | |
| | 15th | | | | | ** | | |
| | 25th | **** | | | * | ** | | |
| Mar | 5th | **** | | | * | ** | | |
| | 15th | *** | | | ** | ** | | |
| | 25th | ** | | | ** | ** | | |
| Apr | 5th | | | | ** | ** | | * |
| | 15th | | | | ** | ** | | * |
| | 25th | | | | ** | ** | | ** |
| May | 5th | | | | *** | *** | | ** |
| | 15th | | | | *** | *** | | ** |
| | 25th | | | | *** | *** | * | ** |
| Jun | 5th | | | | *** | *** | * | *** |
| | 15th | | | | **** | **** | * | *** |
| | 25th | *** | | | **** | **** | * | *** |
| Jul | 5th | *** | | | **** | **** | ** | *** |
| | 15th | *** | | | ***** | **** | ** | **** |
| | 25th | | | | ***** | ***** | ** | **** |
| Aug | 5th | | | | ***** | | ** | **** |
| | 15th | | | | ***** | | ** | **** |
| | 25th | | | | | | *** | ***** |
| Sep | 5th | | | | | | *** | ***** |
| | 15th | | | | | | *** | |
| | 25th | | | | | | *** | |
| Oct | 5th | | | | | | *** | |
| | 15th | | | | | | **** | |
| | 25th | ** | | | | | **** | |
| Nov | 5th | | | | | | | |
| | 15th | | | | | | | |
| | 25th | | | * | | | | |
| Dec | 5th | | | * | | | | |
| | 15th | | | * | | | | |
| | 25th | | | * | | | | |

## Evening Sky

| | | Me | Ve | Ma | Ju | Sa | Ur | Ne |
|---|---|---|---|---|---|---|---|---|
| Jan | 5th | | | ** | * | ** | *** | ** |
| | 15th | ** | | ** | | | ** | ** |
| | 25th | *** | | ** | | | ** | ** |
| Feb | 5th | | | ** | | | ** | * |
| | 15th | | | ** | | | ** | * |
| | 25th | | | ** | | | ** | |
| Mar | 5th | | | ** | | | * | |
| | 15th | | | * | | | * | |
| | 25th | | | * | | | * | |
| Apr | 5th | | | * | | | * | |
| | 15th | | | * | | | | |
| | 25th | | | * | | | | |
| May | 5th | ** | | * | | | | |
| | 15th | *** | | * | | | | |
| | 25th | *** | * | * | | | | |
| Jun | 5th | | * | * | | | | |
| | 15th | | * | * | | | | |
| | 25th | | * | * | | | | |
| Jul | 5th | | * | * | | | | |
| | 15th | | ** | * | | | | |
| | 25th | | ** | * | | | | |
| Aug | 5th | | ** | * | | ***** | | |
| | 15th | | ** | * | | **** | | |
| | 25th | ** | ** | | ***** | **** | | |
| Sep | 5th | *** | ** | | ***** | **** | | |
| | 15th | *** | ** | | ***** | **** | | ***** |
| | 25th | *** | *** | | ***** | **** | | ***** |
| Oct | 5th | | *** | | **** | *** | | ***** |
| | 15th | | *** | | **** | *** | | **** |
| | 25th | | *** | | **** | *** | | **** |
| Nov | 5th | | **** | | *** | *** | **** | **** |
| | 15th | | **** | | *** | ** | **** | **** |
| | 25th | | **** | | *** | ** | **** | *** |
| Dec | 5th | | **** | | *** | ** | *** | *** |
| | 15th | | **** | | ** | ** | *** | *** |
| | 25th | ** | *** | | ** | ** | *** | ** |

# Solar and Lunar Eclipses

| Date | Time (UT) | Type | Visible From |
|---|---|---|---|
| May 26th | 11:18 | Total Lunar | Eastern Asia, Australia, North America, South America and the Pacific. |
| Jun 10th | 10:43 | Annular Solar | Northern Asia, northern Europe, northern and north-eastern North America. |
| Nov 19th | 09:03 | Partial Lunar | Asia, Australia, far western Europe, North America and South America. |
| Dec 4th | 07:35 | Total Solar | Antarctica and the southern Atlantic. |

# Planetary Highlights

| Date | Time (UT) | Elon. | Vis. | Description |
|---|---|---|---|---|
| Jan 21st | 23:21 | 90° E | PM | Mars is 1.7° north of Uranus. (Aries) |
| Jan 24th | 01:50 | 19° E | PM | Mercury is at greatest eastern elongation from the Sun. (Capricornus) |
| Mar 2nd | 21:49 | 73° E | PM | Mars is 2.7° south of the Pleiades star cluster. (Taurus) |
| Mar 5th | 06:44 | 26° W | AM | Mercury is 0.3° north of Jupiter. (Capricornus_ |
| Mar 6th | 11:13 | 26° W | AM | Mercury is at greatest western elongation from the Sun. (Capricornus) |
| May 17th | 05:43 | 23° E | PM | Mercury is at greatest eastern elongation from the Sun. (Taurus) |
| May 29th | 05:19 | 18° E | PM | Mercury is 0.4° south of Venus. (Taurus) |
| Jun 23rd | 07:10 | 38° E | PM | Mars is 0.0° north of the Praesepe star cluster. (Cancer) |
| Jul 3rd | 01:53 | 28° E | PM | Venus is north of the Praesepe star cluster. (Cancer) |
| Jul 4th | 19:41 | 23° W | AM | Mercury is at greatest western elongation from the Sun. (Taurus) |
| Jul 13th | 07:04 | 30° E | PM | Venus is 0.5° north of Mars. (Leo) |
| Jul 21st | 12:42 | 31° E | PM | Venus is 1.2° north of the bright star Regulus. (Leo) |
| Jul 29th | 04:19 | 24° E | PM | Mars is 0.7° north of the bright star Regulus. (Leo) |
| Aug 2nd | 11:16 | 180° | AN | Saturn is at opposition. (Capricornus) |
| Aug 20th | 09:38 | 180° | AN | Jupiter is at opposition. (Aquarius) |
| Sep 4th | 23:11 | 37° E | PM | Venus is 1.8° north of the bright star Spica. (Virgo) |
| Sep 14th | 04:06 | 24° E | AN | Mercury is at greatest eastern elongation from the Sun. (Virgo) |
| Sep 14th | 19:26 | 180° | AN | Neptune is at opposition. (Aquarius) |
| Oct 16th | 06:35 | 46° E | PM | Venus is 1.5° north of the bright star Antares. (Scorpius) |
| Oct 29th | 20:38 | 48° E | PM | Venus is at greatest eastern elongation from the Sun. (Ophiuchus) |
| Nov 5th | 04:31 | 180° | AN | Uranus is at opposition (Aries) |

# Major Meteor Showers

| Shower Name | Start Date | End Date | Peak | ZHR | Speed | Brightness | Moon |
|---|---|---|---|---|---|---|---|
| Quadrantids | Dec 28th | Jan 12th | Jan 3rd | 120 | *** | ***** | ◐ |
| Lyrids | Apr 18th | Apr 25th | Apr 22nd | 18 | *** | ***** | ◑ |
| Eta Aquariids | Apr 24th | May 19th | May 7th | 40 | * | **** | ◕ |
| June Bootids | Jun 23rd | Jun 25th | Jun 24th | Var | ***** | ***** | ○ |
| Alpha Capricornids | Jul 8th | Aug 10th | Jul 27th | 5 | ***** | **** | ◔ |
| Southern Delta Aquariids | Jul 21st | Aug 23rd | Jul 30th | 16 | *** | * | ◑ |
| Perseids | Jul 13th | Aug 26th | Aug 12th | 100 | * | ***** | ● |
| Kappa Cygnids | Aug 6th | Aug 31st | Aug 17th | 3 | ***** | ** | ◑ |
| Aurigids | Aug 29th | Sep 4th | Sep 1st | 6 | * | **** | ● |
| September Epsilon Perseids | Sep 5th | Sep 28th | Sep 9th | 5 | * | ** | ◑ |
| Draconids | Oct 6th | Oct 10th | Oct 8th | Var | ***** | *** | ● |
| Southern Taurids | Sep 7th | Nov 19th | Oct 10th | 5 | **** | **** | ● |
| Orionids | Aug 25th | Nov 19th | Oct 22nd | 15 | * | **** | ◔ |
| Andromedids | Oct 26th | Nov 20th | Nov 8th | Var | ***** | **** | ● |
| Northern Taurids | Oct 25th | Dec 4th | Nov 11th | 5 | **** | **** | ◑ |
| Leonids | Nov 5th | Dec 3rd | Nov 18th | 15 | * | **** | ○ |
| Alpha Monocerotids | Nov 21st | Nov 23rd | Nov 21st | Var | * | **** | ◔ |
| Geminids | Nov 30th | Dec 17th | Dec 13th | 120 | **** | *** | ◑ |
| December Leonis Minorids | Dec 6th | Jan 18th | Dec 20th | 5 | * | ** | ○ |
| Ursids | Dec 17th | Dec 24th | Dec 22nd | 10 | **** | ** | ◕ |
| Coma Berenicids | Dec 24th | Jan 3rd | Dec 31st | 5 | * | ** | ● |

# January 1st to 10th, 2021

## The Moon

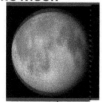    

| 1st | 3rd | 5th | 7th | 9th |

| Date | Con | R.A. | Dec | Mag | Diam | Ill. | Elon. | Phase | Close To |
|------|-----|------|-----|-----|------|------|-------|-------|----------|
| 1st | Cnc | 8h 51m | 21° 46' | -12.0 | 31' | 94% | 149° W | -G | Praesepe |
| 2nd | Leo | 9h 46m | 18° 23' | -11.7 | 31' | 88% | 137° W | -G | Regulus |
| 3rd | Leo | 10h 39m | 14° 0' | -11.3 | 32' | 80% | 125° W | -G | Regulus |
| 4th | Leo | 11h 30m | 8° 51' | -11.0 | 32' | 71% | 113° W | -G | |
| 5th | Vir | 12h 21m | 3° 13' | -10.5 | 32' | 60% | 101° W | LQ | |
| 6th | Vir | 13h 12m | -2° 40' | -10.1 | 32' | 49% | 90° W | LQ | Spica |
| 7th | Vir | 14h 4m | -8° 28' | -9.5 | 32' | 38% | 78° W | LQ | Spica |
| 8th | Lib | 14h 58m | -13° 53' | -8.9 | 32' | 27% | 65° W | -Cr | |
| 9th | Lib | 15h 54m | -18° 33' | -8.1 | 33' | 17% | 52° W | -Cr | Antares |
| 10th | Oph | 16h 54m | -22° 7' | -7.2 | 32' | 9% | 39° W | NM | Antares |

## Mercury and Venus

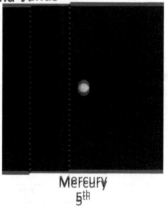 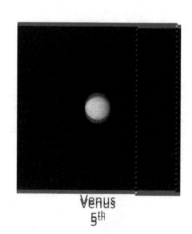

| Mercury | Venus |
| 5th | 5th |

### Mercury

| Date | Con. | R.A. | Dec. | Mag. | Diam. | Ill. | Elon. | Vis. | Rat. | Close To |
|------|------|------|------|------|-------|------|-------|------|------|----------|
| 1st | Sgr | 19h 21m | -24° 15' | -1.0 | 5" | 98% | 8° E | NV | N/A | |
| 3rd | Sgr | 19h 35m | -23° 47' | -0.9 | 5" | 97% | 9° E | NV | N/A | |
| 5th | Sgr | 19h 49m | -23° 13' | -0.9 | 5" | 96% | 11° E | NV | N/A | Jupiter, Saturn |
| 7th | Sgr | 20h 3m | -22° 33' | -0.9 | 5" | 94% | 12° E | NV | N/A | Jupiter, Saturn |
| 9th | Cap | 20h 17m | -21° 47' | -0.9 | 5" | 93% | 13° E | NV | N/A | Jupiter, Saturn |

## Venus

| Date | Con. | R.A. | Dec. | Mag. | Diam. | Ill. | Elon. | Vis. | Rat. | Close To |
|------|------|------|------|------|-------|------|-------|------|------|----------|
| 1st | Oph | 17h 21m | -22° 30' | -3.9 | 11" | 94% | 22° W | AM | * | |
| 3rd | Oph | 17h 31m | -22° 44' | -3.9 | 11" | 94% | 21° W | AM | * | |
| 5th | Oph | 17h 42m | -22° 55' | -3.9 | 11" | 95% | 21° W | AM | * | |
| 7th | Sgr | 17h 53m | -23° 3' | -3.9 | 11" | 95% | 20° W | AM | * | |
| 9th | Sgr | 18h 4m | -23° 8' | -3.9 | 11" | 95% | 20° W | AM | * | |

## Mars and the Outer Planets

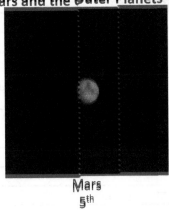

Mars
5th

Jupiter
5th

Saturn
5th

## Mars

| Date | Con. | R.A. | Dec. | Mag. | Diam. | Ill. | Elon. | Vis. | Rat. | Close To |
|------|------|------|------|------|-------|------|-------|------|------|----------|
| 1st | Psc | 1h 41m | 11° 26' | -0.2 | 10" | 89% | 103° E | PM | *** | |
| 5th | Ari | 1h 47m | 12° 8' | -0.1 | 10" | 89% | 100° E | PM | ** | |
| 10th | Ari | 1h 56m | 13° 1' | 0.0 | 10" | 89% | 97° E | PM | ** | |

## The Outer Planets

| Planet | Date | Con. | R.A. | Dec. | Mag. | Diam. | Elon. | Vis. | Rat. | Close To |
|--------|------|------|------|------|------|-------|-------|------|------|----------|
| Jupiter | 5th | Cap | 20h 35m | -19° 47' | -2.0 | 33" | 20° E | PM | * | Mercury, Saturn |
| Saturn | 5th | Cap | 20h 17m | -20° 4' | 0.6 | 15" | 18° E | PM | ** | Mercury |
| Uranus | 5th | Ari | 2h 18m | 13° 21' | 5.7 | 4" | 108° E | PM | *** | Mars |
| Neptune | 5th | Aqr | 23h 19m | -5° 31' | 7.9 | 2" | 63° E | PM | ** | |

## Highlights

| Date | Time (UT) | Event |
|------|-----------|-------|
| 1st | 08:16 | The waning gibbous Moon is north of the Praesepe star cluster. (Cancer, morning sky.) |
| 2nd | 20:40 | The waning gibbous Moon is north of the bright star Regulus. (Leo, morning sky.) |
| 4th | N/A | The Quadrantid meteor shower is at its maximum. (ZHR: 120) |
| 6th | 09:38 | Last Quarter Moon. (Morning sky.) |
| | 18:08 | The last quarter Moon is north of the bright star Spica. (Virgo, morning sky.) |
| 10th | 01:09 | The waning crescent Moon is north of the bright star Antares. (Scorpius, morning sky.) |
| | N/A | Good opportunity to see Earthshine on the waning crescent Moon. (Morning sky.) |

# January 11th to 20th, 2021

## The Moon

| | 11th | 13th | 15th | 17th | 19th |

| Date | Con | R.A. | Dec | Mag | Diam | Ill. | Elon. | Phase | | Close To |
|------|-----|------|-----|-----|------|------|-------|-------|---|----------|
| 11th | Sgr | 17h 56m | -24° 17' | -6.2 | 32' | 4% | 24° W | NM | | Venus |
| 12th | Sgr | 18h 58m | -24° 51' | -4.9 | 32' | 1% | 10° W | NM | | Venus |
| 13th | Sgr | 20h 0m | -23° 49' | -4.4 | 32' | 0% | 5° E | NM | | Mercury, Jupiter, Saturn |
| 14th | Cap | 20h 58m | -21° 22' | -5.7 | 31' | 2% | 18° E | NM | | Mercury, Jupiter, Saturn |
| 15th | Cap | 21h 53m | -17° 49' | -6.7 | 31' | 6% | 31° E | NM | | |
| 16th | Aqr | 22h 44m | -13° 29' | -7.6 | 31' | 12% | 42° E | NM | | Neptune |
| 17th | Aqr | 23h 31m | -8° 40' | -8.4 | 30' | 20% | 53° E | +Cr | | Neptune |
| 18th | Psc | 0h 17m | -3° 38' | -9.0 | 30' | 28% | 64° E | +Cr | | |
| 19th | Cet | 1h 1m | 1° 27' | -9.5 | 30' | 37% | 73° E | +Cr | | |
| 20th | Psc | 1h 44m | 6° 25' | -10.0 | 30' | 46% | 83° E | FQ | | Mars, Uranus |

## Mercury and Venus

Mercury
15th

Venus
15th

**Mercury**

| Date | Con. | R.A. | Dec. | Mag. | Diam. | Ill. | Elon. | Vis. | Rat. | | Close To |
|------|------|------|------|------|-------|------|-------|------|------|---|----------|
| 11th | Cap | 20h 30m | -20° 55' | -0.9 | 5" | 90% | 15° E | NV | N/A | | Jupiter, Saturn |
| 13th | Cap | 20h 44m | -19° 57' | -0.9 | 6" | 87% | 16° E | PM | ** | | Moon, Jupiter, Saturn |
| 15th | Cap | 20h 56m | -18° 54' | -0.9 | 6" | 83% | 17° E | PM | ** | | Jupiter, Saturn |
| 17th | Cap | 21h 8m | -17° 48' | -0.8 | 6" | 78% | 18° E | PM | ** | | Jupiter |
| 19th | Cap | 21h 20m | -16° 38' | -0.8 | 6" | 73% | 18° E | PM | *** | | |

## Venus

| Date | Con. | R.A. | Dec. | Mag. | Diam. | Ill. | Elon. | Vis. | Rat. | Close To |
|------|------|------|------|------|-------|------|-------|------|------|----------|
| 11th | Sgr | 18h 15m | -23° 11' | -3.9 | 10" | 95% | 19° W | AM | * | Moon |
| 13th | Sgr | 18h 26m | -23° 10' | -3.9 | 10" | 96% | 19° W | AM | * | |
| 15th | Sgr | 18h 37m | -23° 7' | -3.9 | 10" | 96% | 18° W | AM | * | |
| 17th | Sgr | 18h 48m | -23° 1' | -3.9 | 10" | 96% | 18° W | AM | * | |
| 19th | Sgr | 18h 59m | -22° 52' | -3.9 | 10" | 96% | 17° W | AM | * | |

## Mars and the Outer Planets

Mars
15th

Jupiter
15th

Saturn
15th

### Mars

| Date | Con. | R.A. | Dec. | Mag. | Diam. | Ill. | Elon. | Vis. | Rat. | Close To |
|------|------|------|------|------|-------|------|-------|------|------|----------|
| 11th | Ari | 1h 58m | 13° 12' | 0.0 | 9" | 89% | 96° E | PM | ** | Uranus |
| 15th | Ari | 2h 5m | 13° 55' | 0.1 | 9" | 89% | 94° E | PM | ** | Uranus |
| 20th | Ari | 2h 15m | 14° 49' | 0.2 | 9" | 89% | 91° E | PM | ** | Moon, Uranus |

### The Outer Planets

| Planet | Date | Con. | R.A. | Dec. | Mag. | Diam. | Elon. | Vis. | Rat. | Close To |
|--------|------|------|------|------|------|-------|-------|------|------|----------|
| Jupiter | 15th | Cap | 20h 34m | -19° 13' | -1.9 | 33" | 11° E | NV | N/A | Mercury, Saturn |
| Saturn | 15th | Cap | 20h 22m | -19° 48' | 0.6 | 15" | 8° E | NV | N/A | Mercury |
| Uranus | 15th | Ari | 2h 18m | 13° 20' | 5.7 | 4" | 97° E | PM | ** | Mars |
| Neptune | 15th | Aqr | 23h 20m | -5° 25' | 7.9 | 2" | 53° E | PM | ** | |

## Highlights

| Date | Time (UT) | Event |
|------|-----------|-------|
| 11th | 11:01 | Mercury is 1.5° south of Jupiter. (Evening sky.) |
| | 20:52 | The waning crescent Moon is south of Venus. (Morning sky.) |
| 13th | 05:01 | New Moon. (Not visible.) |
| 14th | 06:51 | The waxing crescent Moon is south of Mercury. (Evening sky.) |
| | 11:19 | Uranus is stationary prior to resuming prograde motion. (Evening sky.) |
| | 12:35 | Dwarf planet Pluto is in conjunction with the Sun. (Not visible.) |
| 16th | N/A | Good opportunity to see Earthshine on the waxing crescent Moon. (Evening sky.) |
| 17th | 05:34 | The waxing crescent Moon is south of Neptune. (Evening sky.) |
| 20th | 21:02 | First Quarter Moon. (Evening sky.) |

# January 21st to 31st, 2021

## The Moon

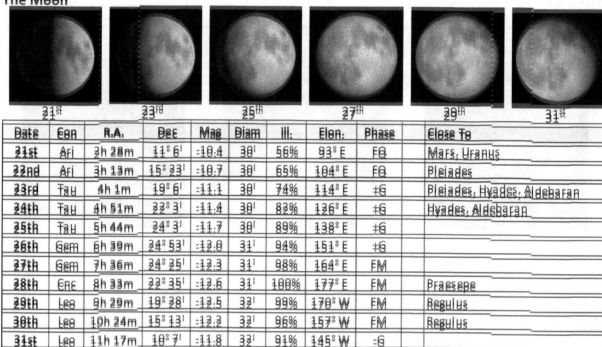

| 21st | 23rd | 25th | 27th | 29th | 31st |

| Date | Con | R.A. | Dec | Mag | Diam | Ill. | Elon. | Phase | Close To |
|------|-----|------|-----|-----|------|------|-------|-------|----------|
| 21st | Ari | 2h 28m | 11° 6' | -10.4 | 30' | 56% | 93° E | FQ | Mars, Uranus |
| 22nd | Ari | 3h 13m | 15° 23' | -10.7 | 30' | 65% | 104° E | FQ | Pleiades |
| 23rd | Tau | 4h 1m | 19° 6' | -11.1 | 30' | 74% | 114° E | +G | Pleiades, Hyades, Aldebaran |
| 24th | Tau | 4h 51m | 22° 3' | -11.4 | 30' | 82% | 126° E | +G | Hyades, Aldebaran |
| 25th | Tau | 5h 44m | 24° 3' | -11.7 | 30' | 89% | 138° E | +G | |
| 26th | Gem | 6h 39m | 24° 53' | -12.0 | 31' | 94% | 151° E | +G | |
| 27th | Gem | 7h 36m | 24° 25' | -12.3 | 31' | 98% | 164° E | FM | |
| 28th | Cnc | 8h 33m | 22° 35' | -12.6 | 31' | 100% | 177° E | FM | Praesepe |
| 29th | Leo | 9h 29m | 19° 28' | -12.5 | 32' | 99% | 170° W | FM | Regulus |
| 30th | Leo | 10h 24m | 15° 13' | -12.2 | 32' | 96% | 157° W | FM | Regulus |
| 31st | Leo | 11h 17m | 10° 7' | -11.8 | 32' | 91% | 145° W | -G | |

## Mercury and Venus

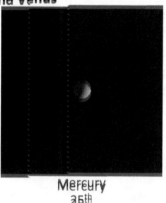

Mercury
25th

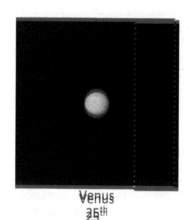

Venus
25th

### Mercury

| Date | Con. | R.A. | Dec. | Mag. | Diam. | Ill. | Elon. | Vis. | Rat. | Close To |
|------|------|------|------|------|-------|------|-------|------|------|----------|
| 21st | Cap | 21h 30m | -15° 28' | -0.7 | 6" | 66% | 19° E | PM | *** | |
| 23rd | Cap | 21h 38m | -14° 18' | -0.6 | 7" | 58% | 19° E | PM | *** | |
| 25th | Cap | 21h 45m | -13° 12' | -0.4 | 7" | 49% | 18° E | PM | *** | |
| 27th | Cap | 21h 50m | -12° 14' | -0.1 | 8" | 40% | 17° E | PM | *** | |
| 29th | Cap | 21h 52m | -11° 26' | 0.3 | 8" | 30% | 16° E | PM | ** | |
| 31st | Cap | 21h 51m | -10° 53' | 0.9 | 9" | 21% | 14° E | NV | N/A | |

## Venus

| Date | Con. | R.A. | Dec. | Mag. | Diam. | Ill. | Elon. | Vis. | Rat. | Close To |
|------|------|------|------|------|-------|------|-------|------|------|----------|
| 21st | Sgr | 19h 9m | -22° 40' | -3.9 | 10" | 97% | 16° W | AM | * | |
| 23rd | Sgr | 19h 20m | -22° 26' | -3.9 | 10" | 97% | 16° W | AM | * | |
| 25th | Sgr | 19h 31m | -22° 9' | -3.9 | 10" | 97% | 15° W | AM | * | |
| 27th | Sgr | 19h 42m | -21° 49' | -3.9 | 10" | 97% | 15° W | NV | N/A | |
| 29th | Sgr | 19h 52m | -21° 26' | -3.9 | 10" | 97% | 14° W | NV | N/A | Saturn |
| 31st | Sgr | 20h 3m | -21° 1' | -3.9 | 10" | 98% | 13° W | NV | N/A | Saturn |

## Mars and the Outer Planets

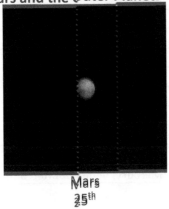

Mars
25th

Jupiter
25th

Saturn
25th

## Mars

| Date | Con. | R.A. | Dec. | Mag. | Diam. | Ill. | Elon. | Vis. | Rat. | Close To |
|------|------|------|------|------|-------|------|-------|------|------|----------|
| 21st | Ari | 2h 17m | 14° 59' | 0.2 | 9" | 89% | 90° E | PM | ** | Moon, Uranus |
| 25th | Ari | 2h 25m | 15° 43' | 0.3 | 8" | 89% | 88° E | PM | ** | Uranus |
| 31st | Ari | 2h 37m | 16° 44' | 0.4 | 8" | 89% | 85° E | PM | ** | Uranus |

## The Outer Planets

| Planet | Date | Con. | R.A. | Dec. | Mag. | Diam. | Elon. | Vis. | Rat. | Close To |
|--------|------|------|------|------|------|-------|-------|------|------|----------|
| Jupiter | 25th | Cap | 20h 44m | -18° 38' | -1.9 | 32" | 3° E | NV | N/A | Saturn |
| Saturn | 25th | Cap | 20h 27m | -19° 32' | 0.6 | 15" | 1° W | NV | N/A | |
| Uranus | 25th | Ari | 2h 18m | 13° 22' | 5.8 | 4" | 86° E | PM | ** | Mars |
| Neptune | 25th | Aqr | 23h 21m | -5° 19' | 7.9 | 2" | 42° E | PM | ** | |

## Highlights

| Date | Time (UT) | Event |
|------|-----------|-------|
| 21st | 05:48 | The just-past first quarter Moon is south of Mars. (Evening sky.) |
| | 06:22 | The just-past first quarter Moon is south of Uranus. (Evening sky.0 |
| | 23:21 | Mars is 1.7° north of Uranus. (Evening sky.) |
| 23rd | 05:30 | The waxing gibbous Moon is south of the Pleiades star cluster. (Taurus, evening sky.) |
| 24th | 01:50 | Mercury is at greatest eastern elongation from the Sun. (Evening sky.) |
| | 05:31 | The waxing gibbous Moon is north of the bright star Aldebaran. (Taurus, evening sky.) |
| | 05:43 | Saturn is in conjunction with the Sun. (Not visible.) |
| 28th | 14:10 | The almost full Moon is north of the Praesepe star cluster. (Cancer, visible all night.) |
| | 19:17 | Full Moon. (Visible all night.) |
| 29th | 06:04 | Jupiter is in conjunction with the Sun. (Not visible.) |
| 30th | 02:13 | Mercury is stationary prior to beginning retrograde motion. (Evening sky.) |
| | 06:06 | The waning gibbous Moon is north of the bright star Regulus. (Leo, morning sky.) |

# February 1st to 10th, 2021

## The Moon

1st

3rd

5th

7th

9th

| Date | Con | R.A. | Dec | Mag | Diam | Ill. | Elon. | Phase | Close To |
|------|-----|------|-----|-----|------|------|-------|-------|----------|
| 1st | Vir | 12h 9m | 4° 26' | -11.5 | 32' | 84% | 133° W | -G | |
| 2nd | Vir | 13h 0m | -1° 30' | -11.1 | 32' | 74% | 121° W | -G | Spica |
| 3rd | Vir | 13h 52m | -7° 22' | -10.7 | 32' | 64% | 109° W | LQ | Spica |
| 4th | Lib | 14h 45m | -12° 52' | -10.2 | 32' | 53% | 97° W | LQ | |
| 5th | Lib | 15h 40m | -17° 40' | -9.7 | 32' | 41% | 84° W | LQ | |
| 6th | Oph | 16h 37m | -21° 27' | -9.1 | 32' | 30% | 71° W | -Cr | Antares |
| 7th | Oph | 17h 37m | -23° 56' | -8.4 | 32' | 21% | 57° W | -Cr | |
| 8th | Sgr | 18h 38m | -24° 56' | -7.6 | 32' | 12% | 43° W | NM | |
| 9th | Sgr | 19h 38m | -24° 24' | -6.7 | 32' | 6% | 29° W | NM | |
| 10th | Cap | 20h 37m | -22° 25' | -5.6 | 31' | 2% | 15° W | NM | Mercury, Venus, Jupiter, Saturn |

## Mercury and Venus

Mercury
5th

Venus
5th

### Mercury

| Date | Con. | R.A. | Dec. | Mag. | Diam. | Ill. | Elon. | Vis. | Rat. | Close To |
|------|------|------|------|------|-------|------|-------|------|------|----------|
| 1st | Cap | 21h 50m | -10° 41' | 1.3 | 9" | 16% | 12° E | NV | N/A | |
| 3rd | Cap | 21h 45m | -10° 32' | 2.2 | 9" | 9% | 9° E | NV | N/A | |
| 5th | Cap | 21h 38m | -10° 40' | 3.3 | 10" | 4% | 5° E | NV | N/A | |
| 7th | Cap | 21h 30m | -11° 4' | 4.4 | 10" | 1% | 1° E | NV | N/A | Jupiter |
| 9th | Aqr | 21h 20m | -11° 39' | 4.5 | 10" | 1% | 3° W | NV | N/A | Venus, Jupiter |

## Venus

| Date | Con. | R.A. | Dec. | Mag. | Diam. | Ill. | Elon. | Vis. | Rat. | Close To |
|------|------|------|------|------|-------|------|-------|------|------|----------|
| 1st | Cap | 20h 8m | -20° 48' | -3.9 | 10" | 98% | 13° W | NV | N/A | Saturn |
| 3rd | Cap | 20h 19m | -20° 19' | -3.9 | 10" | 98% | 12° W | NV | N/A | Jupiter, Saturn |
| 5th | Cap | 20h 29m | -19° 48' | -3.9 | 10" | 98% | 12° W | NV | N/A | Jupiter, Saturn |
| 7th | Cap | 20h 40m | -19° 14' | -3.9 | 10" | 98% | 11° W | NV | N/A | Jupiter, Saturn |
| 9th | Cap | 20h 50m | -18° 38' | -3.9 | 10" | 98% | 11° W | NV | N/A | Mercury, Jupiter, Saturn |

## Mars and the Outer Planets

Mars
5th

Jupiter
5th

Saturn
5th

### Mars

| Date | Con. | R.A. | Dec. | Mag. | Diam. | Ill. | Elon. | Vis. | Rat. | Close To |
|------|------|------|------|------|-------|------|-------|------|------|----------|
| 1st | Ari | 2h 39m | 16° 55' | 0.5 | 8" | 89% | 85° E | PM | ** | |
| 5th | Ari | 2h 48m | 17° 35' | 0.5 | 8" | 89% | 83° E | PM | ** | |
| 10th | Ari | 2h 59m | 18° 25' | 0.6 | 7" | 89% | 81° E | PM | ** | |

### The Outer Planets

| Planet | Date | Con. | R.A. | Dec. | Mag. | Diam. | Elon. | Vis. | Rat. | Close To |
|--------|------|------|------|------|------|-------|-------|------|------|----------|
| Jupiter | 5th | Cap | 20h 54m | -17° 57' | -1.9 | 33" | 6° W | NV | N/A | Venus, Saturn |
| Saturn | 5th | Cap | 20h 33m | -19° 14' | 0.6 | 15" | 11° W | NV | N/A | Venus |
| Uranus | 5th | Ari | 2h 18m | 13° 25' | 5.8 | 4" | 75° E | PM | ** | Mars |
| Neptune | 5th | Aqr | 23h 22m | -5° 10' | 7.9 | 2" | 31° E | PM | * | |

## Highlights

| Date | Time (UT) | Event |
|------|-----------|-------|
| 2nd | 22:13 | The waning gibbous Moon is north of the bright star Spica. (Virgo, morning sky.) |
| 4th | 17:38 | Last Quarter Moon. (Morning sky.) |
| 6th | 09:13 | The waning crescent Moon is north of the bright star Antares. (Scorpius, morning sky.) |
| 8th | 13:42 | Mercury is at inferior conjunction with the Sun. (Not visible.) |
| | N/A | Good opportunity to see Earthshine on the waning crescent Moon. (Morning sky.) |

# February 11th to 20th, 2021

## The Moon

| 11th | 13th | 15th | 17th | 19th |

| Date | Con | R.A. | Dec | Mag | Diam | Ill. | Elon. | Phase | Close To |
|------|-----|------|-----|-----|------|------|-------|-------|----------|
| 11th | Cap | 21h 33m | -19° 13' | -4.5 | 31' | 0% | 2° W | NM | Mercury, Venus, Jupiter |
| 12th | Aqr | 22h 25m | -15° 7' | -4.9 | 31' | 1% | 10° E | NM | |
| 13th | Aqr | 23h 14m | -10° 24' | -6.0 | 30' | 3% | 21° E | NM | Neptune |
| 14th | Psc | 0h 0m | -5° 21' | -7.0 | 30' | 8% | 32° E | NM | Neptune |
| 15th | Cet | 0h 45m | 0° 11' | -7.8 | 30' | 14% | 42° E | +Cr | |
| 16th | Psc | 1h 29m | 4° 55' | -8.5 | 30' | 21% | 52° E | +Cr | |
| 17th | Cet | 2h 13m | 9° 45' | -9.1 | 30' | 29% | 62° E | +Cr | Uranus |
| 18th | Ari | 2h 57m | 14° 13' | -9.6 | 30' | 38% | 72° E | FQ | Mars, Uranus |
| 19th | Tau | 3h 44m | 18° 7' | -10.0 | 30' | 47% | 83° E | FQ | Mars, Pleiades, Hyades |
| 20th | Tau | 4h 33m | 21° 20' | -10.4 | 30' | 57% | 94° E | FQ | Pleiades, Hyades, Aldebaran |

## Mercury and Venus

Mercury
15th

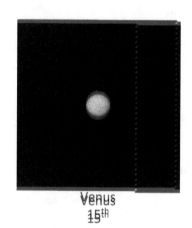

Venus
15th

### Mercury

| Date | Con. | R.A. | Dec. | Mag. | Diam. | Ill. | Elon. | Vis. | Rat. | Close To |
|------|------|------|------|------|-------|------|-------|------|------|----------|
| 11th | Aqr | 21h 11m | -12° 21' | 3.6 | 10" | 3% | 7° W | NV | N/A | Moon, Venus, Jupiter, Saturn |
| 13th | Aqr | 21h 3m | -13° 6' | 2.7 | 10" | 7% | 11° W | NV | N/A | Venus, Jupiter, Saturn |
| 15th | Aqr | 20h 57m | -13° 49' | 2.0 | 10" | 12% | 15° W | NV | N/A | Venus, Jupiter, Saturn |
| 17th | Cap | 20h 53m | -14° 28' | 1.5 | 10" | 18% | 18° W | AM | *** | Venus, Jupiter, Saturn |
| 19th | Cap | 20h 51m | -15° 1' | 1.1 | 9" | 23% | 20° W | AM | *** | Jupiter, Saturn |

## Venus

| Date | Con. | R.A. | Dec. | Mag. | Diam. | Ill. | Elon. | Vis. | Rat. | Close To |
|------|------|------|------|------|-------|------|-------|------|------|----------|
| 11th | Cap | 21h 0m | -18° 0' | -3.9 | 10" | 98% | 10° W | NV | N/A | Moon, Mercury, Jupiter, Saturn |
| 13th | Cap | 21h 10m | -17° 20' | -3.9 | 10" | 99% | 10° W | NV | N/A | Mercury, Jupiter, Saturn |
| 15th | Cap | 21h 20m | -16° 38' | -3.9 | 10" | 99% | 9° W | NV | N/A | Mercury, Jupiter |
| 17th | Cap | 21h 30m | -15° 55' | -3.9 | 10" | 99% | 8° W | NV | N/A | Mercury, Jupiter |
| 19th | Cap | 21h 40m | -15° 9' | -3.9 | 10" | 99% | 8° W | NV | N/A | Jupiter |

## Mars and the Outer Planets

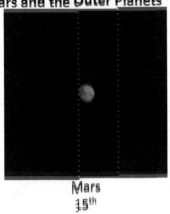

Mars
15th

Jupiter
15th

Saturn
15th

### Mars

| Date | Con. | R.A. | Dec. | Mag. | Diam. | Ill. | Elon. | Vis. | Rat. | Close To |
|------|------|------|------|------|-------|------|-------|------|------|----------|
| 11th | Ari | 3h 1m | 18° 34' | 0.6 | 7" | 89% | 80° E | PM | ** | |
| 15th | Ari | 3h 10m | 19° 13' | 0.7 | 7" | 89% | 78° E | PM | ** | Pleiades |
| 20th | Ari | 3h 22m | 19° 58' | 0.8 | 7" | 89% | 76° E | PM | ** | Pleiades |

### The Outer Planets

| Planet | Date | Con. | R.A. | Dec. | Mag. | Diam. | Elon. | Vis. | Rat. | Close To |
|--------|------|------|------|------|------|-------|-------|------|------|----------|
| Jupiter | 15th | Cap | 21h 4m | -17° 18' | -2.0 | 33" | 13° W | NV | N/A | Mercury, Venus, Saturn |
| Saturn | 15th | Cap | 20h 37m | -18° 57' | 0.7 | 15" | 20° W | AM | ** | Mercury |
| Uranus | 15th | Ari | 2h 19m | 13° 30' | 5.8 | 4" | 66° E | PM | ** | |
| Neptune | 15th | Aqr | 23h 24m | -5° 3' | 8.0 | 2" | 22° E | PM | * | |

## Highlights

| Date | Time (UT) | Event |
|------|-----------|-------|
| 11th | 19:06 | New Moon. (Not visible.) |
| 13th | 18:13 | The waxing crescent Moon is south of Neptune. (Evening sky.) |
| 15th | N/A | Good opportunity to see Earthshine on the waxing crescent Moon. (Evening sky.) |
| 17th | 15:26 | The waxing crescent Moon is south of Uranus. (Evening sky.) |
| 19th | 00:11 | The almost first quarter Moon is south of Mars. (Evening sky.) |
| | 11:52 | The almost first quarter Moon is north of the Pleiades star cluster. (Taurus, evening sky.) |
| | 18:48 | First Quarter Moon. (Evening sky.) |
| 20th | 11:55 | The just-past first quarter Moon is north of the bright star Aldebaran. (Taurus, evening sky.) |
| | 13:25 | Mercury is stationary prior to resuming prograde motion. (Morning sky.) |

# February 21st to 28th, 2021

## The Moon

21st

23rd

25th

27th

| Date | Con | R.A. | Dec | Mag | Diam | Ill. | Elon. | Phase | Close To |
|------|-----|------|-----|-----|------|------|-------|-------|----------|
| 21st | Tau | 5h 24m | 23° 39' | -10.8 | 30' | 66% | 106° E | +G | |
| 22nd | Gem | 6h 18m | 24° 53' | -11.1 | 30' | 75% | 119° E | +G | |
| 23rd | Gem | 7h 14m | 24° 54' | -11.5 | 31' | 83% | 132° E | +G | |
| 24th | Cnc | 8h 11m | 23° 34' | -11.8 | 31' | 90% | 145° E | +G | Praesepe |
| 25th | Cnc | 9h 7m | 20° 53' | -12.1 | 32' | 96% | 158° E | FM | Praesepe |
| 26th | Leo | 10h 3m | 16° 57' | -12.4 | 32' | 99% | 171° E | FM | Regulus |
| 27th | Leo | 10h 58m | 11° 59' | -12.6 | 32' | 100% | 176° W | FM | |
| 28th | Vir | 11h 51m | 6° 17' | -12.3 | 33' | 98% | 164° W | FM | |

## Mercury and Venus

Mercury
24th

Venus
24th

**Mercury**

| Date | Con. | R.A. | Dec. | Mag. | Diam. | Ill. | Elon. | Vis. | Rat. | Close To |
|------|------|------|------|------|-------|------|-------|------|------|----------|
| 21st | Cap | 20h 51m | -15° 27' | 0.8 | 9" | 29% | 22° W | AM | *** | Jupiter, Saturn |
| 23rd | Cap | 20h 53m | -15° 47' | 0.6 | 9" | 34% | 24° W | AM | *** | Jupiter, Saturn |
| 25th | Cap | 20h 56m | -15° 60' | 0.4 | 8" | 39% | 25° W | AM | **** | Jupiter, Saturn |
| 27th | Cap | 21h 1m | -16° 6' | 0.3 | 8" | 44% | 25° W | AM | **** | Jupiter, Saturn |

**Venus**

| Date | Con. | R.A. | Dec. | Mag. | Diam. | Ill. | Elon. | Vis. | Rat. | Close To |
|------|------|------|------|------|-------|------|-------|------|------|----------|
| **21st** | Cap | 21h 50m | -14° 22' | -3.9 | 10" | 99% | 7° W | NV | N/A | |
| **23rd** | Aqr | 22h 0m | -13° 33' | -3.9 | 10" | 99% | 7° W | NV | N/A | |
| **25th** | Aqr | 22h 9m | -12° 43' | -3.9 | 10" | 99% | 6° W | NV | N/A | |
| **27th** | Aqr | 22h 19m | -11° 51' | -3.9 | 10" | 99% | 6° W | NV | N/A | |

## Mars and the Outer Planets

Mars
24th

Jupiter
24th

Saturn
24th

**Mars**

| Date | Con. | R.A. | Dec. | Mag. | Diam. | Ill. | Elon. | Vis. | Rat. | Close To |
|------|------|------|------|------|-------|------|-------|------|------|----------|
| **21st** | Ari | 3h 24m | 20° 7' | 0.8 | 7" | 89% | 76° E | PM | ** | Pleiades |
| **24th** | Tau | 3h 31m | 20° 32' | 0.8 | 7" | 89% | 75° E | PM | ** | Pleiades |
| **28th** | Tau | 3h 41m | 21° 5' | 0.9 | 6" | 90% | 74° E | PM | ** | Pleiades |

**The Outer Planets**

| Planet | Date | Con. | R.A. | Dec. | Mag. | Diam. | Elon. | Vis. | Rat. | Close To |
|--------|------|------|------|------|------|-------|-------|------|------|----------|
| Jupiter | 24th | Cap | 21h 12m | -16° 42' | -2.0 | 33" | 20° W | AM | * | Mercury, Saturn |
| Saturn | 24th | Cap | 20h 42m | -18° 43' | 0.7 | 15" | 27° W | AM | ** | Mercury |
| Uranus | 24th | Ari | 2h 20m | 13° 36' | 5.8 | 3" | 57° E | PM | ** | |
| Neptune | 24th | Aqr | 23h 25m | -4° 55' | 8.0 | 2" | 13° E | NV | N/A | |

## Highlights

| Date | Time (UT) | Event |
|------|-----------|-------|
| 25th | 01:04 | The waxing gibbous Moon is north of the Praesepe star cluster. (Cancer, evening sky.) |
| 26th | 13:30 | The nearly full Moon is north of the bright star Regulus. (Leo, evening sky.) |
| 27th | 08:18 | Full Moon. (Visible all night.) |

# March 1<sup>st</sup> to 10<sup>th</sup>, 2021

## The Moon

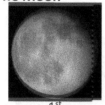
1<sup>st</sup>

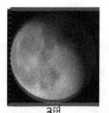
3<sup>rd</sup>

5<sup>th</sup>

7<sup>th</sup>

9<sup>th</sup>

| Date | Con | R.A. | Dec | Mag | Diam | Ill. | Elon. | Phase | Close To |
|------|-----|------|-----|-----|------|------|-------|-------|----------|
| 1st | Vir | 12h 44m | 0° 10' | -12.0 | 33' | 94% | 151° W | =G | Spica |
| 2nd | Vir | 13h 37m | -5° 58' | -11.6 | 33' | 87% | 139° W | =G | Spica |
| 3rd | Lib | 14h 31m | -11° 46' | -11.2 | 33' | 78% | 127° W | =G | |
| 4th | Lib | 15h 27m | -16° 52' | -10.8 | 32' | 67% | 114° W | =G | |
| 5th | Oph | 16h 24m | -20° 57' | -10.4 | 32' | 56% | 100° W | LQ | Antares |
| 6th | Oph | 17h 24m | -23° 45' | -9.9 | 32' | 45% | 86° W | LQ | |
| 7th | Sgr | 18h 24m | -25° 4' | -9.4 | 32' | 34% | 72° W | =Cr | |
| 8th | Sgr | 19h 24m | -24° 51' | -8.7 | 31' | 24% | 58° W | =Cr | |
| 9th | Cap | 20h 22m | -23° 13' | -8.0 | 31' | 16% | 44° W | =Cr | Saturn |
| 10th | Cap | 21h 17m | -20° 20' | -7.2 | 31' | 9% | 32° W | NM | Mercury, Jupiter, Saturn |

## Mercury and Venus

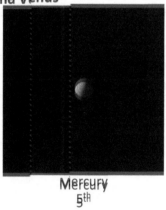

Mercury
5<sup>th</sup>

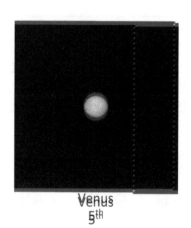

Venus
5<sup>th</sup>

### Mercury

| Date | Con. | R.A. | Dec. | Mag. | Diam. | Ill. | Elon. | Vis. | Rat. | Close To |
|------|------|------|------|------|-------|------|-------|------|------|----------|
| 1st | Cap | 21h 7m | -16° 6' | 0.3 | 8" | 48% | 26° W | AM | **** | Jupiter, Saturn |
| 3rd | Cap | 21h 14m | -15° 59' | 0.2 | 7" | 52% | 26° W | AM | **** | Jupiter, Saturn |
| 5th | Cap | 21h 21m | -15° 46' | 0.2 | 7" | 55% | 26° W | AM | **** | Jupiter, Saturn |
| 7th | Cap | 21h 30m | -15° 27' | 0.1 | 7" | 58% | 26° W | AM | *** | Jupiter |
| 9th | Cap | 21h 38m | -15° 2' | 0.1 | 7" | 61% | 25° W | AM | *** | Jupiter |

## Venus

| Date | Con. | R.A. | Dec. | Mag. | Diam. | Ill. | Elon. | Vis. | Rat. | Close To |
|------|------|------|------|------|-------|------|-------|------|------|----------|
| 1st | Aqr | 22h 29m | -10° 58' | -3.9 | 10" | 99% | 5° W | NV | N/A | |
| 3rd | Aqr | 22h 38m | -10° 4' | -3.9 | 10" | 100% | 5° W | NV | N/A | |
| 5th | Aqr | 22h 48m | -9° 9' | -3.9 | 10" | 100% | 4° W | NV | N/A | |
| 7th | Aqr | 22h 57m | -8° 13' | -3.9 | 10" | 100% | 4° W | NV | N/A | |
| 9th | Aqr | 23h 6m | -7° 15' | -3.9 | 10" | 100% | 3° W | NV | N/A | |

## Mars and the Outer Planets

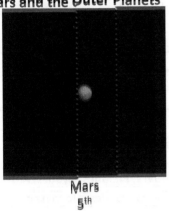

Mars
5th

Jupiter
5th

Saturn
5th

## Mars

| Date | Con. | R.A. | Dec. | Mag. | Diam. | Ill. | Elon. | Vis. | Rat. | Close To |
|------|------|------|------|------|-------|------|-------|------|------|----------|
| 1st | Tau | 3h 43m | 21° 13' | 0.9 | 6" | 90% | 73° E | PM | ** | Pleiades |
| 5th | Tau | 3h 53m | 21° 44' | 1.0 | 6" | 90% | 72° E | PM | ** | Pleiades, Hyades |
| 10th | Tau | 4h 5m | 22° 20' | 1.0 | 6" | 90% | 70° E | PM | * | Pleiades, Hyades, Aldebaran |

## The Outer Planets

| Planet | Date | Con. | R.A. | Dec. | Mag. | Diam. | Ill. | Elon. | Vis. | Rat. | Close To |
|--------|------|------|------|------|------|-------|------|-------|------|------|----------|
| Jupiter | 5th | Cap | 21h 21m | -16° 6' | -2.0 | 33" | | 26° W | AM | * | Mercury, Saturn |
| Saturn | 5th | Cap | 20h 46m | -18° 28' | 0.7 | 15" | | 35° W | AM | ** | Mercury |
| Uranus | 5th | Ari | 2h 22m | 13° 42' | 5.8 | 3" | | 49° E | PM | * | |
| Neptune | 5th | Aqr | 23h 26m | -4° 47' | 8.0 | 2" | | 5° E | NV | N/A | Venus |

## Highlights

| Date | Time (UT) | Event |
|------|-----------|-------|
| 2nd | 07:42 | The waning gibbous Moon is north of the bright star Spica. (Virgo, morning sky.) |
| | 21:49 | Mars is 2.7° south of the Pleiades star cluster. (Taurus, evening sky.) |
| 5th | 06:44 | Mercury is 0.3° north of Jupiter. (Morning sky.) |
| | 14:48 | The nearly last quarter Moon is north of the bright star Antares. (Scorpius, morning sky.) |
| 6th | 01:31 | Last Quarter Moon. (Morning sky.) |
| | 11:13 | Mercury is at greatest western elongation from the Sun. (Morning sky.) |
| 8th | 07:10 | Asteroid Vesta is at opposition. (Visible all night.) |
| | 23:03 | The waning crescent Moon is south of dwarf planet Pluto. (Morning sky.) |
| 9th | 22:40 | The waning crescent Moon is south of Saturn. (Morning sky.) |
| 10th | 16:56 | The waning crescent Moon is south of Jupiter. (Morning sky.) |
| | N/A | Good opportunity to see Earthshine on the waning crescent Moon. (Morning sky.) |

# March 11<sup>th</sup> to 20<sup>th</sup>, 2021

## The Moon

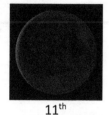
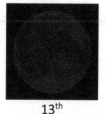
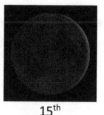

| 11<sup>th</sup> | 13<sup>th</sup> | 15<sup>th</sup> | 17<sup>th</sup> | 19<sup>th</sup> |

| Date | Con | R.A. | Dec | Mag | Diam | Ill. | Elon. | Phase | | Close To |
|------|-----|------|-----|-----|------|------|-------|-------|---|----------|
| 11th | Aqr | 22h 9m | -16° 29' | -6.2 | 31' | 4% | 19° W | NM | | Mercury, Jupiter |
| 12th | Aqr | 22h 58m | -11° 56' | -5.2 | 30' | 1% | 8° W | NM | | Venus, Neptune |
| 13th | Aqr | 23h 45m | -6° 57' | -4.4 | 30' | 0% | 3° E | NM | | Venus, Neptune |
| 14th | Cet | 0h 30m | -1° 45' | -5.3 | 30' | 1% | 13° E | NM | | |
| 15th | Psc | 1h 14m | 3° 25' | -6.3 | 30' | 4% | 23° E | NM | | |
| 16th | Psc | 1h 58m | 8° 25' | -7.2 | 30' | 9% | 33° E | NM | | Uranus |
| 17th | Ari | 2h 43m | 13° 4' | -7.9 | 30' | 15% | 43° E | +Cr | | Uranus |
| 18th | Tau | 3h 29m | 17° 11' | -8.6 | 29' | 22% | 54° E | +Cr | | Pleiades |
| 19th | Tau | 4h 16m | 20° 38' | -9.1 | 30' | 30% | 65° E | +Cr | | Mars, Pleiades, Hyades, Aldebara |
| 20th | Tau | 5h 7m | 23° 15' | -9.6 | 30' | 39% | 77° E | FQ | | Mars, Hyades, Aldebaran |

## Mercury and Venus

Mercury
15<sup>th</sup>

Venus
15<sup>th</sup>

**Mercury**

| Date | Con. | R.A. | Dec. | Mag. | Diam. | Ill. | Elon. | Vis. | Rat. | | Close To |
|------|------|------|------|------|-------|------|-------|------|------|---|----------|
| 11th | Cap | 21h 48m | -14° 31' | 0.1 | 7" | 64% | 25° W | AM | *** | | Moon, Jupiter |
| 13th | Cap | 21h 58m | -13° 55' | 0.1 | 6" | 66% | 24° W | AM | *** | | Jupiter |
| 15th | Aqr | 22h 8m | -13° 14' | 0.0 | 6" | 69% | 23° W | AM | *** | | Jupiter |
| 17th | Aqr | 22h 19m | -12° 27' | 0.0 | 6" | 71% | 23° W | AM | *** | | |
| 19th | Aqr | 22h 29m | -11° 35' | -0.1 | 6" | 73% | 22° W | AM | *** | | |

**Venus**

| Date | Con. | R.A. | Dec. | Mag. | Diam. | Ill. | Elon. | Vis. | Rat. | Close To |
|------|------|------|------|------|-------|------|-------|------|------|----------|
| 11th | Aqr | 23h 15m | -6° 17' | -3.9 | 10" | 100% | 3° W | NV | N/A | Neptune |
| 13th | Aqr | 23h 25m | -5° 19' | -3.9 | 10" | 100% | 2° W | NV | N/A | Moon, Neptune |
| 15th | Aqr | 23h 34m | -4° 20' | -3.9 | 10" | 100% | 2° W | NV | N/A | Neptune |
| 17th | Aqr | 23h 43m | -3° 20' | -3.9 | 10" | 100% | 1° W | NV | N/A | Neptune |
| 19th | Psc | 23h 52m | -2° 20' | -3.9 | 10" | 100% | 1° W | NV | N/A | |

## Mars and the Outer Planets

Mars
15th

Jupiter
15th

Saturn
15th

**Mars**

| Date | Con. | R.A. | Dec. | Mag. | Diam. | Ill. | Elon. | Vis. | Rat. | Close To |
|------|------|------|------|------|-------|------|-------|------|------|----------|
| 11th | Tau | 4h 8m | 22° 26' | 1.1 | 6" | 90% | 70° E | PM | * | Pleiades, Hyades, Aldebaran |
| 15th | Tau | 4h 18m | 22° 52' | 1.1 | 6" | 90% | 69° E | PM | * | Pleiades, Hyades, Aldebaran |
| 20th | Tau | 4h 30m | 23° 21' | 1.2 | 6" | 91% | 68° E | PM | * | Moon, Hyades, Aldebaran |

**The Outer Planets**

| Planet | Date | Con. | R.A. | Dec. | Mag. | Diam. | Elon. | Vis. | Rat. | Close To |
|--------|------|------|------|------|------|-------|-------|------|------|----------|
| Jupiter | 15th | Cap | 21h 29m | -15° 27' | -2.0 | 34" | 33° W | AM | ** | Mercury, Saturn |
| Saturn | 15th | Cap | 20h 50m | -18° 13' | 0.7 | 16" | 43° W | AM | ** | |
| Uranus | 15th | Ari | 2h 23m | 13° 51' | 5.8 | 3" | 40° E | PM | * | |
| Neptune | 15th | Aqr | 23h 27m | -4° 38' | 8.0 | 2" | 4° W | NV | N/A | Venus |

## Highlights

| Date | Time (UT) | Event |
|------|-----------|-------|
| 11th | 00:24 | The waning crescent Moon is south of Mercury. (Morning sky.) |
| | 11:29 | Neptune is in conjunction with the Sun. (Not visible.) |
| 13th | 10:22 | New Moon. (Not visible.) |
| 16th | N/A | Good opportunity to see Earthshine on the waxing crescent Moon. (Evening sky.) |
| 17th | 02:06 | The waxing crescent Moon is south of Uranus. (Evening sky.) |
| 18th | 22:25 | The waxing crescent Moon is south of the Pleiades star cluster. (Taurus, evening sky.) |
| 19th | 18:25 | The waxing crescent Moon is south of Mars. (Evening sky.) |
| | 22:31 | The waxing crescent Moon is north of the bright star Aldebaran. (Taurus, evening sky.) |
| 20th | 09:38 | Spring equinox. |

# March 21ˢᵗ to 31ˢᵗ, 2021

## The Moon

 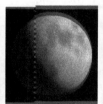    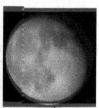

| 21ˢᵗ | 23ʳᵈ | 25ᵗʰ | 27ᵗʰ | 29ᵗʰ | 31ˢᵗ |

| Date | Con | R.A. | Dec | Mag | Diam | Ill. | Elon. | Phase | Close To |
|------|-----|------|-----|-----|------|------|-------|-------|----------|
| 21st | Tau | 5h 59m | 24° 51' | -10.1 | 30' | 49% | 89° E | FQ | |
| 22nd | Gem | 6h 53m | 25° 18' | -10.5 | 30' | 59% | 101° E | FQ | |
| 23rd | Gem | 7h 49m | 24° 28' | -10.9 | 31' | 68% | 114° E | +G | |
| 24th | Cnc | 8h 44m | 22° 19' | -11.2 | 31' | 78% | 127° E | +G | Praesepe |
| 25th | Leo | 9h 40m | 18° 53' | -11.6 | 32' | 86% | 140° E | +G | Regulus |
| 26th | Leo | 10h 34m | 14° 19' | -11.9 | 32' | 93% | 153° E | +G | Regulus |
| 27th | Leo | 11h 28m | 8° 49' | -12.3 | 33' | 97% | 166° E | FM | |
| 28th | Vir | 12h 22m | 2° 42' | -12.6 | 33' | 100% | 178° E | FM | |
| 29th | Vir | 13h 16m | -3° 40' | -12.4 | 33' | 99% | 169° W | FM | Spica |
| 30th | Vir | 14h 11m | -9° 52' | -12.1 | 33' | 96% | 156° W | FM | Spica |
| 31st | Lib | 15h 8m | -15° 29' | -11.7 | 33' | 89% | 143° W | -G | |

## Mercury and Venus

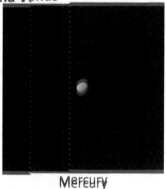

Mercury
25ᵗʰ

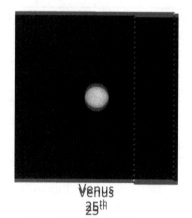

Venus
25ᵗʰ

### Mercury

| Date | Con. | R.A. | Dec. | Mag. | Diam. | Ill. | Elon. | Vis. | Rat. | Close To |
|------|------|------|------|------|-------|------|-------|------|------|----------|
| 21st | Aqr | 22h 41m | -10° 38' | -0.1 | 6" | 75% | 21° W | AM | *** | |
| 23rd | Aqr | 22h 52m | -9° 36' | -0.1 | 6" | 77% | 20° W | AM | *** | |
| 25th | Aqr | 23h 4m | -8° 29' | -0.2 | 6" | 79% | 19° W | AM | ** | |
| 27th | Aqr | 23h 15m | -7° 17' | -0.3 | 6" | 81% | 17° W | AM | ** | Neptune |
| 29th | Aqr | 23h 28m | -6° 1' | -0.4 | 5" | 83% | 16° W | AM | ** | Neptune |
| 31st | Aqr | 23h 40m | -4° 40' | -0.4 | 5" | 85% | 15° W | AM | ** | Neptune |

## Venus

| Date | Con. | R.A. | Dec. | Mag. | Diam. | Ill. | Elon. | Vis. | Rat. | Close To |
|------|------|------|------|------|-------|------|-------|------|------|----------|
| 21st | Psc | 0h 1m | -1° 20' | -3.9 | 10" | 100% | 1° W | NV | N/A | |
| 23rd | Psc | 0h 10m | 0° 20' | -3.9 | 10" | 100% | 0° W | NV | N/A | |
| 25th | Psc | 0h 19m | 0° 41' | -3.9 | 10" | 100% | 0° E | NV | N/A | |
| 27th | Cet | 0h 29m | 1° 41' | -3.9 | 10" | 100% | 1° E | NV | N/A | |
| 29th | Cet | 0h 38m | 2° 42' | -3.9 | 10" | 100% | 1° E | NV | N/A | |
| 31st | Psc | 0h 47m | 3° 42' | -3.9 | 10" | 100% | 2° E | NV | N/A | |

## Mars and the Outer Planets

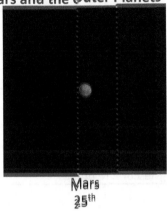

Mars
25th

Jupiter
25th

Saturn
25th

## Mars

| Date | Con. | R.A. | Dec. | Mag. | Diam. | Ill. | Elon. | Vis. | Rat. | Close To |
|------|------|------|------|------|-------|------|-------|------|------|----------|
| 21st | Tau | 4h 33m | 23° 26' | 1.2 | 6" | 91% | 67° E | PM | * | Hyades, Aldebaran |
| 25th | Tau | 4h 43m | 23° 46' | 1.2 | 6" | 91% | 66° E | PM | * | Hyades, Aldebaran |
| 31st | Tau | 4h 59m | 24° 11' | 1.3 | 5" | 91% | 65° E | PM | * | Hyades, Aldebaran |

## The Outer Planets

| Planet | Date | Con. | R.A. | Dec. | Mag. | Diam. | Elon. | Vis. | Rat. | Close To |
|--------|------|------|------|------|------|-------|-------|------|------|----------|
| Jupiter | 25th | Cap | 21h 38m | -14° 48' | -2.0 | 34" | 40° W | AM | ** | |
| Saturn | 25th | Cap | 20h 53m | -17° 60' | 0.8 | 16" | 51° W | AM | ** | |
| Uranus | 25th | Ari | 2h 25m | 14° 0' | 5.9 | 3" | 33° E | PM | * | |
| Neptune | 25th | Aqr | 23h 29m | -4° 39' | 8.0 | 3" | 13° W | NV | N/A | Mercury |

## Highlights

| Date | Time (UT) | Event |
|------|-----------|-------|
| 21st | 14:41 | First Quarter Moon. (Evening sky.) |
| 22nd | 13:00 | Mars is 7.0° north of the bright star Aldebaran. (Taurus, evening sky.) |
| 24th | 09:36 | The waxing gibbous Moon is north of the Praesepe star cluster. (Cancer, evening sky.) |
| 26th | 01:16 | The waxing gibbous Moon is north of the bright star Regulus. (Leo, evening sky.) |
| | 06:21 | Venus is at superior conjunction with the Sun. (Not visible.) |
| 28th | 18:49 | Full Moon. (Visible all night.) |
| 29th | 15:11 | The just-past full Moon is north of the bright star Spica. (Virgo, visible all night.) |

# April 1st to 10th, 2021

## The Moon

| | 1st | | 3rd | | 5th | | 7th | | 9th |

| Date | Con | R.A. | Dec | Mag | Diam | Ill. | Elon. | Phase | Close To |
|------|-----|------|-----|-----|------|------|-------|-------|----------|
| 1st | Sco | 16h 7m | -20° 7' | -11.4 | 33' | 81% | 129° W | -G | Antares |
| 2nd | Oph | 17h 8m | -23° 24' | -11.0 | 32' | 71% | 115° W | -G | Antares |
| 3rd | Sgr | 18h 9m | -25° 8' | -10.5 | 32' | 60% | 100° W | LQ | |
| 4th | Sgr | 19h 10m | -25° 17' | -10.1 | 32' | 49% | 86° W | LQ | |
| 5th | Cap | 20h 9m | -23° 55' | -9.6 | 31' | 38% | 72° W | LQ | Saturn |
| 6th | Cap | 21h 5m | -21° 16' | -9.0 | 31' | 28% | 59° W | -Cr | Jupiter, Saturn |
| 7th | Cap | 21h 57m | -17° 37' | -8.4 | 31' | 20% | 47° W | -Cr | Jupiter |
| 8th | Aqr | 22h 47m | -13° 13' | -7.6 | 30' | 12% | 36° W | NM | Neptune |
| 9th | Aqr | 23h 33m | -8° 20' | -6.8 | 30' | 7% | 25° W | NM | Neptune |
| 10th | Psc | 0h 18m | -3° 12' | -5.8 | 30' | 3% | 15° W | NM | Mercury, Neptune |

## Mercury and Venus

Mercury
5th

Venus
5th

### Mercury

| Date | Con. | R.A. | Dec. | Mag. | Diam. | Ill. | Elon. | Vis. | Rat. | Close To |
|------|------|------|------|------|-------|------|-------|------|------|----------|
| 1st | Aqr | 23h 46m | -3° 58' | -0.5 | 5" | 86% | 14° W | NV | N/A | Neptune |
| 3rd | Psc | 23h 59m | -2° 31' | -0.6 | 5" | 88% | 13° W | NV | N/A | |
| 5th | Psc | 0h 12m | -1° 0' | -0.7 | 5" | 90% | 12° W | NV | N/A | |
| 7th | Psc | 0h 25m | 0° 35' | -0.9 | 5" | 92% | 10° W | NV | N/A | |
| 9th | Cet | 0h 39m | 2° 14' | -1.0 | 5" | 94% | 9° W | NV | N/A | |

**Venus**

| Date | Con. | R.A. | Dec. | Mag. | Diam. | Ill. | Elon. | Vis. | Rat. | Close To |
|------|------|------|------|------|-------|------|-------|------|------|----------|
| 1st | Psc | 0h 51m | 4° 12' | -3.9 | 10" | 100% | 2° E | NV | N/A | |
| 3rd | Psc | 1h 0m | 5° 11' | -3.9 | 10" | 100% | 2° E | NV | N/A | |
| 5th | Psc | 1h 10m | 6° 11' | -3.9 | 10" | 100% | 3° E | NV | N/A | |
| 7th | Psc | 1h 19m | 7° 9' | -3.9 | 10" | 100% | 3° E | NV | N/A | |
| 9th | Psc | 1h 28m | 8° 7' | -3.9 | 10" | 100% | 4° E | NV | N/A | |

## Mars and the Outer Planets

Mars
5th

Jupiter
5th

Saturn
5th

**Mars**

| Date | Con. | R.A. | Dec. | Mag. | Diam. | Ill. | Elon. | Vis. | Rat. | Close To |
|------|------|------|------|------|-------|------|-------|------|------|----------|
| 1st | Tau | 5h 1m | 24° 15' | 1.3 | 5" | 91% | 64° E | PM | * | Hyades, Aldebaran |
| 5th | Tau | 5h 12m | 24° 28' | 1.3 | 5" | 92% | 63° E | PM | * | Aldebaran |
| 10th | Tau | 5h 25m | 24° 41' | 1.4 | 5" | 92% | 62° E | PM | * | |

**The Outer Planets**

| Planet | Date | Con. | R.A. | Dec. | Mag. | Diam. | Elon. | Vis. | Rat. | Close To |
|--------|------|------|------|------|------|-------|-------|------|------|----------|
| Jupiter | 5th | Cap | 21h 46m | -14° 7' | -2.1 | 35" | 48° W | AM | ** | |
| Saturn | 5th | Cap | 20h 57m | -17° 47' | 0.8 | 16" | 60° W | AM | ** | Moon |
| Uranus | 5th | Ari | 2h 27m | 14° 11' | 5.9 | 3" | 22° E | PM | * | |
| Neptune | 5th | Aqr | 23h 30m | -4° 20' | 8.0 | 2" | 22° W | AM | * | |

## Highlights

| Date | Time (UT) | Event |
|------|-----------|-------|
| 1st | 19:47 | The waning gibbous Moon is north of the bright star Antares. (Scorpius, morning sky.) |
| 4th | 10:03 | Last Quarter Moon. (Morning sky.) |
| 5th | 05:16 | The just-past last quarter Moon is south of the dwarf planet Pluto. (Morning sky.) |
| 6th | 08:47 | The waning crescent Moon is south of Saturn. (Morning sky.) |
| 7th | 06:36 | The waning crescent Moon Is south of Jupiter. (Morning sky.) |
| 8th | N/A | Good opportunity to see Earthshine on the waning crescent Moon. (Morning sky.) |
| 9th | 11:01 | The waning crescent Moon is south of Neptune. (Morning sky.) |

# April 11<sup>th</sup> to 20<sup>th</sup>, 2021

## The Moon

| | 11<sup>th</sup> | | 13<sup>th</sup> | | 15<sup>th</sup> | | 17<sup>th</sup> | | 19<sup>th</sup> |

| Date | Con | R.A. | Dec | Mag | Diam | Ill. | Elon. | Phase | Close To |
|------|-----|------|-----|-----|------|------|-------|-------|----------|
| 11th | Cet | 1h 2m | 2° 1' | -4.8 | 30' | 0% | 5° W | NM | Mercury, Venus |
| 12th | Psc | 1h 45m | 7° 6' | -4.5 | 30' | 0% | 5° E | NM | Mercury, Venus, Uranus |
| 13th | Ari | 2h 30m | 11° 54' | -5.5 | 29' | 2% | 16° E | NM | Venus, Uranus |
| 14th | Ari | 3h 15m | 16° 13' | -6.5 | 29' | 5% | 26° E | NM | Uranus, Pleiades |
| 15th | Tau | 4h 3m | 19° 55' | -7.3 | 29' | 10% | 37° E | NM | Pleiades, Hyades, Aldebaran |
| 16th | Tau | 4h 52m | 22° 48' | -8.1 | 30' | 16% | 48° E | +Cr | Hyades, Aldebaran |
| 17th | Tau | 5h 43m | 24° 43' | -8.7 | 30' | 24% | 60° E | +Cr | Mars |
| 18th | Gem | 6h 36m | 25° 31' | -9.3 | 30' | 33% | 73° E | +Cr | |
| 19th | Gem | 7h 31m | 25° 6' | -9.8 | 30' | 42% | 85° E | FQ | |
| 20th | Cnc | 8h 25m | 23° 26' | -10.2 | 31' | 52% | 98° E | FQ | Praesepe |

## Mercury and Venus

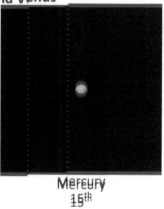

Mercury
15<sup>th</sup>

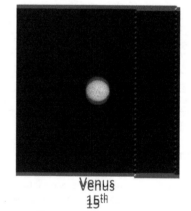

Venus
15<sup>th</sup>

### Mercury

| Date | Con. | R.A. | Dec. | Mag. | Diam. | Ill. | Elon. | Vis. | Rat. | Close To |
|------|------|------|------|------|-------|------|-------|------|------|----------|
| 11th | Psc | 0h 53m | 3° 56' | -1.2 | 5" | 96% | 7° W | NV | N/A | Moon |
| 13th | Psc | 1h 7m | 5° 40' | -1.4 | 5" | 98% | 5° W | NV | N/A | Venus |
| 15th | Psc | 1h 22m | 7° 27' | -1.7 | 5" | 99% | 3° W | NV | N/A | Venus |
| 17th | Psc | 1h 37m | 9° 15' | -2.0 | 5" | 100% | 1° W | NV | N/A | Venus |
| 19th | Ari | 1h 52m | 11° 3' | -2.2 | 5" | 100% | 1° E | NV | N/A | Venus |

## Venus

| Date | Con. | R.A. | Dec. | Mag. | Diam. | Ill. | Elon. | Vis. | Rat. | Close To |
|------|------|------|------|------|-------|------|-------|------|------|----------|
| 11th | Psc | 1h 37m | 9° 5' | -3.9 | 10" | 100% | 4° E | NV | N/A | Moon |
| 13th | Psc | 1h 47m | 10° 1' | -3.9 | 10" | 100% | 5° E | NV | N/A | Moon, Mercury |
| 15th | Ari | 1h 56m | 10° 57' | -3.9 | 10" | 100% | 5° E | NV | N/A | Mercury |
| 17th | Ari | 2h 5m | 11° 51' | -3.9 | 10" | 100% | 6° E | NV | N/A | Mercury |
| 19th | Ari | 2h 15m | 12° 45' | -3.9 | 10" | 99% | 6° E | NV | N/A | Mercury, Uranus |

## Mars and the Outer Planets

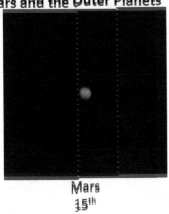

Mars
15th

Jupiter
15th

Saturn
15th

### Mars

| Date | Con. | R.A. | Dec. | Mag. | Diam. | Ill. | Elon. | Vis. | Rat. | Close To |
|------|------|------|------|------|-------|------|-------|------|------|----------|
| 11th | Tau | 5h 28m | 24° 43' | 1.4 | 5" | 92% | 62° E | PM | * | |
| 15th | Tau | 5h 38m | 24° 49' | 1.4 | 5" | 92% | 61° E | PM | * | |
| 20th | Tau | 5h 52m | 24° 53' | 1.5 | 5" | 93% | 59° E | PM | * | |

### The Outer Planets

| Planet | Date | Con. | R.A. | Dec. | Mag. | Diam. | Elon. | Vis. | Rat. | Close To |
|--------|------|------|------|------|------|-------|-------|------|------|----------|
| Jupiter | 15th | Cap | 21h 53m | -13° 32' | -2.1 | 36" | 55° W | AM | ** | |
| Saturn | 15th | Cap | 20h 59m | -17° 37' | 0.7 | 16" | 69° W | AM | ** | |
| Uranus | 15th | Ari | 2h 30m | 14° 22' | 5.9 | 3" | 14° E | NV | N/A | Venus |
| Neptune | 15th | Aqr | 23h 32m | -4° 12' | 7.9 | 2" | 31° W | AM | * | |

## Highlights

| Date | Time (UT) | Event |
|------|-----------|-------|
| 11th | 20:51 | Dwarf planet Ceres is in conjunction with the Sun. (Not visible.) |
| 13th | 02:31 | New Moon. (Not visible.) |
| 15th | 03:30 | The waxing crescent Moon is south of the Pleiades star cluster. (Taurus, evening sky.) |
| | N/A | Good opportunity to see Earthshine on the waxing crescent Moon. (Evening sky.) |
| 16th | 03:48 | The waxing crescent Moon is north of the bright star Aldebaran. (Taurus, evening sky.) |
| 17th | 10:49 | The waxing crescent Moon is south of Mars. (Evening sky.) |
| 19th | 01:37 | Mercury is at superior conjunction with the Sun. (Not visible.) |
| 20th | 07:00 | First Quarter Moon. (Evening sky.) |
| | 18:16 | The first quarter Moon is north of the Praesepe star cluster. (Evening sky.) |

# April 21st to 30th, 2021

## The Moon

| 21st | 23rd | 25th | 27th | 29th |

| Date | Con | R.A. | Dec | Mag | Diam | Ill. | Elon. | Phase | Close To |
|------|-----|------|-----|-----|------|------|-------|-------|----------|
| 21st | Cnc | 9h 19m | 20° 31' | -10.6 | 31' | 62% | 110° E | FQ | Praesepe |
| 22nd | Leo | 10h 13m | 16° 27' | -11.0 | 32' | 72% | 123° E | +G | Regulus |
| 23rd | Leo | 11h 6m | 11° 23' | -11.4 | 32' | 82% | 135° E | +G | |
| 24th | Vir | 11h 58m | 5° 34' | -11.8 | 33' | 90% | 147° E | +G | |
| 25th | Vir | 12h 51m | 0° 44' | -12.1 | 33' | 96% | 160° E | FM | Spica |
| 26th | Vir | 13h 46m | -7° 8' | -12.5 | 33' | 99% | 172° E | FM | Spica |
| 27th | Lib | 14h 43m | -13° 12' | -12.6 | 33' | 100% | 174° W | FM | |
| 28th | Lib | 15h 42m | -18° 28' | -12.2 | 33' | 97% | 160° W | FM | Antares |
| 29th | Oph | 16h 44m | -22° 28' | -11.9 | 33' | 92% | 146° W | -G | Antares |
| 30th | Sgr | 17h 48m | -24° 53' | -11.5 | 33' | 84% | 131° W | -G | |

## Mercury and Venus

Mercury
25th

Venus
25th

**Mercury**

| Date | Con. | R.A. | Dec. | Mag. | Diam. | Ill. | Elon. | Vis. | Rat. | Close To |
|------|------|------|------|------|-------|------|-------|------|------|----------|
| 21st | Ari | 2h 8m | 12° 50' | -2.0 | 5" | 99% | 3° E | NV | N/A | Venus |
| 23rd | Ari | 2h 24m | 14° 34' | -1.8 | 5" | 98% | 5° E | NV | N/A | Venus, Uranus |
| 25th | Ari | 2h 41m | 16° 14' | -1.6 | 5" | 95% | 7° E | NV | N/A | Venus, Uranus |
| 27th | Ari | 2h 57m | 17° 48' | -1.5 | 5" | 92% | 9° E | NV | N/A | Venus |
| 29th | Ari | 3h 13m | 19° 15' | -1.3 | 6" | 87% | 11° E | NV | N/A | Venus, Pleiades |

## Venus

| Date | Con. | R.A. | Dec. | Mag. | Diam. | Ill. | Elon. | Vis. | Rat. | Close To |
|------|------|------|------|------|-------|------|-------|------|------|----------|
| **21st** | Ari | 2h 24m | 13° 37' | -3.9 | 10" | 99% | 7° E | NV | N/A | Mercury, Uranus |
| **23rd** | Ari | 2h 34m | 14° 28' | -3.9 | 10" | 99% | 7° E | NV | N/A | Mercury, Uranus |
| **25th** | Ari | 2h 43m | 15° 17' | -3.9 | 10" | 99% | 8° E | NV | N/A | Mercury, Uranus |
| **27th** | Ari | 2h 53m | 16° 5' | -3.9 | 10" | 99% | 8° E | NV | N/A | Mercury |
| **29th** | Ari | 3h 3m | 16° 52' | -3.9 | 10" | 99% | 9° E | NV | N/A | Mercury |

## Mars and the Outer Planets

Mars
25th

Jupiter
25th

Saturn
25th

### Mars

| Date | Con. | R.A. | Dec. | Mag. | Diam. | Ill. | Elon. | Vis. | Rat. | Close To |
|------|------|------|------|------|-------|------|-------|------|------|----------|
| **21st** | Tau | 5h 54m | 24° 54' | 1.5 | 5" | 93% | 59° E | PM | * | |
| **25th** | Gem | 6h 5m | 24° 53' | 1.5 | 5" | 93% | 58° E | PM | * | |
| **30th** | Gem | 6h 18m | 24° 49' | 1.6 | 5" | 93% | 57° E | PM | * | |

### The Outer Planets

| Planet | Date | Con. | R.A. | Dec. | Mag. | Diam. | Elon. | Vis. | Rat. | Close To |
|--------|------|------|------|------|------|-------|-------|------|------|----------|
| Jupiter | 25th | Aqr | 21h 59m | -13° 1' | -2.2 | 37" | 63° W | AM | ** | |
| Saturn | 25th | Cap | 21h 2m | -17° 29' | 0.7 | 17" | 78° W | AM | ** | |
| Uranus | 25th | Ari | 2h 32m | 14° 33' | 5.9 | 3" | 5° E | NV | N/A | Mercury, Venus |
| Neptune | 25th | Aqr | 23h 33m | -4° 5' | 7.9 | 2" | 40° W | AM | ** | |

## Highlights

| Date | Time (UT) | Event |
|------|-----------|-------|
| 22nd | 05:29 | The asteroid Vesta is stationary prior to resuming prograde motion. (Evening sky.) |
| | 09:23 | The waxing gibbous Moon is north of the bright star Regulus. (Leo, evening sky.) |
| | N/A | The Lyrid meteor shower is at its maximum. (ZHR: 18) |
| 26th | 03:53 | The nearly full Moon is north of the bright star Spica. (Virgo, evening sky.) |
| 27th | 03:32 | Full Moon. (Visible all night.) |
| 28th | 02:39 | Dwarf planet Pluto is stationary prior to beginning retrograde motion. (Morning sky.) |
| 29th | 07:19 | The waning gibbous Moon is north of the bright star Antares. (Scorpius, morning sky.) |
| 30th | 23:26 | Uranus is in conjunction with the Sun. (Not visible.) |

# May 1st to 10th, 2021

## The Moon

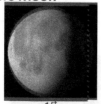    

| 1st | 3rd | 5th | 7th | 9th |

| Date | Con | R.A. | Dec | Mag | Diam | Ill. | Elon. | Phase | Close To |
|------|-----|------|-----|-----|------|------|-------|-------|----------|
| 1st | Sgr | 18h 52m | -25° 35' | -11.1 | 32' | 75% | 116° W | -G | |
| 2nd | Sgr | 19h 53m | -24° 37' | -10.7 | 32' | 64% | 101° W | LQ | |
| 3rd | Cap | 20h 51m | -22° 14' | -10.3 | 31' | 53% | 88° W | LQ | Saturn |
| 4th | Cap | 21h 45m | -18° 44' | -9.8 | 31' | 43% | 75° W | LQ | Jupiter, Saturn |
| 5th | Aqr | 22h 36m | -14° 26' | -9.3 | 31' | 33% | 64° W | -Cr | Jupiter |
| 6th | Aqr | 23h 23m | -9° 37' | -8.7 | 30' | 24% | 53° W | -Cr | Neptune |
| 7th | Psc | 0h 8m | -4° 31' | -8.0 | 30' | 16% | 43° W | -Cr | Neptune |
| 8th | Cet | 0h 51m | 0° 41' | -7.3 | 30' | 10% | 33° W | NM | |
| 9th | Psc | 1h 35m | 5° 49' | -6.4 | 30' | 5% | 23° W | NM | |
| 10th | Ari | 2h 18m | 10° 43' | -5.4 | 29' | 2% | 13° W | NM | Uranus |

## Mercury and Venus

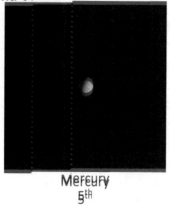

Mercury
5th

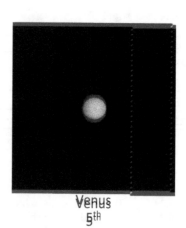

Venus
5th

### Mercury

| Date | Con. | R.A. | Dec. | Mag. | Diam. | Ill. | Elon. | Vis. | Rat. | Close To |
|------|------|------|------|------|-------|------|-------|------|------|----------|
| 1st | Ari | 3h 29m | 20° 33' | -1.1 | 6" | 82% | 13° E | NV | N/A | Venus, Pleiades |
| 3rd | Tau | 3h 44m | 21° 42' | -0.9 | 6" | 76% | 15° E | PM | ** | Venus, Pleiades |
| 5th | Tau | 3h 59m | 22° 40' | -0.7 | 6" | 70% | 17° E | PM | ** | Venus, Pleiades, Hyades, Al |
| 7th | Tau | 4h 13m | 23° 29' | -0.5 | 6" | 64% | 19° E | PM | *** | Venus, Pleiades, Hyades, Al |
| 9th | Tau | 4h 27m | 24° 8' | -0.4 | 7" | 58% | 20° E | PM | *** | Venus, Hyades, Aldebaran |

## Venus

| Date | Con. | R.A. | Dec. | Mag. | Diam. | Ill. | Elon. | Vis. | Rat. | Close To |
|------|------|------|------|------|-------|------|-------|------|------|----------|
| 1st | Ari | 3h 13m | 17° 36' | -3.9 | 10" | 99% | 9° E | NV | N/A | Mercury, Pleiades |
| 3rd | Ari | 3h 23m | 18° 19' | -3.9 | 10" | 99% | 10° E | NV | N/A | Mercury, Pleiades |
| 5th | Tau | 3h 33m | 18° 60' | -3.9 | 10" | 98% | 11° E | NV | N/A | Mercury, Pleiades |
| 7th | Tau | 3h 43m | 19° 38' | -3.9 | 10" | 98% | 11° E | NV | N/A | Mercury, Pleiades |
| 9th | Tau | 3h 53m | 20° 15' | -3.9 | 10" | 98% | 12° E | NV | N/A | Mercury, Pleiades, Hyades |

## Mars and the Outer Planets

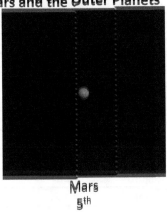

Mars
5th

Jupiter
5th

Saturn
5th

## Mars

| Date | Con. | R.A. | Dec. | Mag. | Diam. | Ill. | Elon. | Vis. | Rat. | Close To |
|------|------|------|------|------|-------|------|-------|------|------|----------|
| 1st | Gem | 6h 21m | 24° 48' | 1.6 | 5" | 93% | 56° E | PM | * | |
| 5th | Gem | 6h 32m | 24° 40' | 1.6 | 5" | 94% | 55° E | PM | * | |
| 10th | Gem | 6h 45m | 24° 27' | 1.6 | 4" | 94% | 54° E | PM | * | |

## The Outer Planets

| Planet | Date | Con. | R.A. | Dec. | Mag. | Diam. | Elon. | Vis. | Rat. | Close To |
|--------|------|------|------|------|------|-------|-------|------|------|----------|
| Jupiter | 5th | Aqr | 22h 5m | -12° 33' | -2.2 | 38" | 71° W | AM | *** | Moon |
| Saturn | 5th | Cap | 21h 3m | -17° 24' | 0.7 | 17" | 87° W | AM | *** | |
| Uranus | 5th | Ari | 2h 34m | 14° 44' | 5.9 | 3" | 4° W | NV | N/A | |
| Neptune | 5th | Aqr | 23h 34m | -3° 58' | 7.9 | 2" | 49° W | AM | ** | |

## Highlights

| Date | Time (UT) | Event |
|------|-----------|-------|
| 2nd | 13:34 | The nearly last quarter Moon is south of dwarf planet Pluto. (Morning sky.) |
| 3rd | 17:17 | The almost last quarter Moon is south of Saturn. (Morning sky.) |
| | 18:32 | Mercury is 2.3° south of the Pleiades star cluster. (Taurus, evening sky.) |
| | 19:51 | Last Quarter Moon. (Morning sky.) |
| 4th | 20:33 | The just-past last quarter Moon is south of Jupiter. (Morning sky.) |
| 6th | 18:33 | The waning crescent Moon is south of Neptune. (Morning sky.) |
| | N/A | The Eta Aquariid meteor shower is at its maximum. (ZHR: 70) |
| 8th | N/A | Good opportunity to see Earthshine on the waning crescent Moon. (Morning sky.) |
| 9th | N/A | The Eta Lyrid meteor shower is at its maximum. (ZHR: 3) |
| 10th | 21:39 | Mercury is 8.0° north of the bright star Aldebaran. (Taurus, evening sky.) |

# May 11th to 20th, 2021

## The Moon

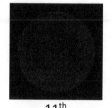
11th

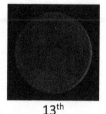
13th

15th

17th

19th

| Date | Con | R.A. | Dec | Mag | Diam | Ill. | Elon. | Phase | Close To |
|------|-----|------|-----|-----|------|------|-------|-------|----------|
| 11th | Ari | 3h 3m | 15° 11' | -4.3 | 29' | 0% | 3° W | NM | Uranus, Pleiades |
| 12th | Tau | 3h 50m | 19° 5' | -4.7 | 29' | 0% | 8° E | NM | Venus, Pleiades, Hyades, Aldebar |
| 13th | Tau | 4h 39m | 22° 13' | -5.8 | 29' | 3% | 19° E | NM | Mercury, Venus, Hyades, Aldebar |
| 14th | Tau | 5h 30m | 24° 25' | -6.8 | 30' | 6% | 31° E | NM | Mercury |
| 15th | Gem | 6h 23m | 25° 32' | -7.6 | 30' | 12% | 43° E | NM | Mars |
| 16th | Gem | 7h 16m | 25° 27' | -8.3 | 30' | 19% | 56° E | +Cr | Mars |
| 17th | Cnc | 8h 10m | 24° 7' | -8.9 | 30' | 27% | 68° E | +Cr | Praesepe |
| 18th | Cnc | 9h 4m | 21° 36' | -9.5 | 31' | 37% | 81° E | +Cr | Praesepe |
| 19th | Leo | 9h 56m | 17° 57' | -10.0 | 31' | 47% | 93° E | FQ | Regulus |
| 20th | Leo | 10h 48m | 13° 20' | -10.4 | 31' | 57% | 105° E | FQ | Regulus |

## Mercury and Venus

Mercury
15th

Venus
15th

**Mercury**

| Date | Con. | R.A. | Dec. | Mag. | Diam. | Ill. | Elon. | Vis. | Rat. | Close To |
|------|------|------|------|------|-------|------|-------|------|------|----------|
| 11th | Tau | 4h 39m | 24° 37' | -0.1 | 7" | 52% | 21° E | PM | *** | Venus, Hyades, Aldebaran |
| 13th | Tau | 4h 50m | 24° 58' | 0.1 | 7" | 46% | 22° E | PM | *** | Moon, Venus, Hyades, Aldeb |
| 15th | Tau | 5h 0m | 25° 10' | 0.3 | 8" | 41% | 23° E | PM | *** | Venus, Hyades, Aldebaran |
| 17th | Tau | 5h 10m | 25° 15' | 0.5 | 8" | 36% | 23° E | PM | *** | Venus, Aldebaran |
| 19th | Tau | 5h 17m | 25° 13' | 0.7 | 9" | 31% | 23° E | PM | *** | Venus |

**Venus**

| Date | Con. | R.A. | Dec. | Mag. | Diam. | Ill. | Elon. | Vis. | Rat. | Close To |
|------|------|------|------|------|-------|------|-------|------|------|----------|
| 11th | Tau | 4h 3m | 20° 50' | -3.9 | 10" | 98% | 12° E | NV | N/A | Mercury, Pleiades, Hyades, Aldeb |
| 13th | Tau | 4h 13m | 21° 22' | -3.9 | 10" | 98% | 13° E | NV | N/A | Moon, Mercury, Pleiades, Hyades |
| 15th | Tau | 4h 24m | 21° 52' | -3.9 | 10" | 97% | 14° E | NV | N/A | Mercury, Pleiades, Hyades, Aldeb |
| 17th | Tau | 4h 34m | 22° 19' | -3.9 | 10" | 97% | 14° E | NV | N/A | Mercury, Hyades, Aldebaran |
| 19th | Tau | 4h 45m | 22° 44' | -3.9 | 10" | 97% | 15° E | NV | N/A | Mercury, Hyades, Aldebaran |

## Mars and the Outer Planets

Mars
15th

Jupiter
15th

Saturn
15th

**Mars**

| Date | Con. | R.A. | Dec. | Mag. | Diam. | Ill. | Elon. | Vis. | Rat. | Close To |
|------|------|------|------|------|-------|------|-------|------|------|----------|
| 11th | Gem | 6h 48m | 24° 24' | 1.6 | 4" | 94% | 53° E | PM | * | |
| 15th | Gem | 6h 58m | 24° 10' | 1.7 | 4" | 94% | 52° E | PM | * | Moon |
| 20th | Gem | 7h 12m | 23° 49' | 1.7 | 4" | 95% | 51° E | PM | * | |

**The Outer Planets**

| Planet | Date | Con. | R.A. | Dec. | Mag. | Diam. | Elon. | Vis. | Rat. | Close To |
|--------|------|------|------|------|------|-------|-------|------|------|----------|
| Jupiter | 15th | Aqr | 22h 10m | -12° 10' | -2.3 | 39" | 80° W | AM | *** | |
| Saturn | 15th | Cap | 21h 4m | -17° 22' | 0.7 | 17" | 96° W | AM | *** | |
| Uranus | 15th | Ari | 2h 36m | 14° 54' | 5.9 | 3" | 13° W | NV | N/A | |
| Neptune | 15th | Aqr | 23h 35m | -3° 53' | 7.9 | 2" | 59° W | AM | ** | |

## Highlights

| Date | Time (UT) | Event |
|------|-----------|-------|
| 11th | 19:01 | New Moon. (Not visible.) |
| 13th | 09:04 | The waxing crescent Moon is north of the bright star Aldebaran. (Taurus, evening sky.) |
| | 19:29 | The waxing crescent Moon is south of Mercury. (Evening sky.) |
| 14th | N/A | Good opportunity to see Earthshine on the waxing crescent Moon. (Evening sky.) |
| 16th | 04:28 | The waxing crescent Moon is north of Mars. (Evening sky.) |
| 17th | 05:43 | Mercury is at greatest eastern elongation from the Sun. (Evening sky.) |
| 18th | 02:02 | The nearly first quarter Moon is north of the Praesepe star cluster. (Cancer, evening sky.) |
| 19th | 17:08 | The almost first quarter Moon is north of the bright star Regulus. (Leo, evening sky.) |
| | 19:13 | First Quarter Moon. (Evening sky.) |

# May 21st to 31st, 2021

## The Moon

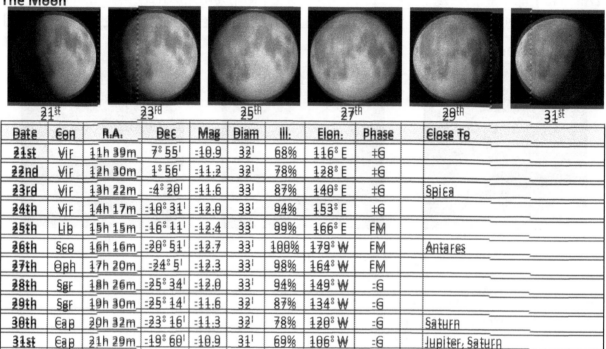

| 21st | 23rd | 25th | 27th | 29th | 31st |

| Date | Con | R.A. | Dec | Mag | Diam | Ill. | Elon. | Phase | Close To |
|------|-----|------|-----|-----|------|------|-------|-------|----------|
| 21st | Vir | 11h 39m | 7° 55' | -10.9 | 32' | 68% | 116° E | +G | |
| 22nd | Vir | 12h 30m | 1° 56' | -11.2 | 32' | 78% | 128° E | +G | |
| 23rd | Vir | 13h 22m | -4° 20' | -11.6 | 33' | 87% | 140° E | +G | Spica |
| 24th | Vir | 14h 17m | -10° 31' | -12.0 | 33' | 94% | 153° E | +G | |
| 25th | Lib | 15h 15m | -16° 11' | -12.4 | 33' | 99% | 166° E | FM | |
| 26th | Sco | 16h 16m | -20° 51' | -12.7 | 33' | 100% | 179° W | FM | Antares |
| 27th | Oph | 17h 20m | -24° 5' | -12.3 | 33' | 98% | 164° W | FM | |
| 28th | Sgr | 18h 26m | -25° 34' | -12.0 | 33' | 94% | 149° W | -G | |
| 29th | Sgr | 19h 30m | -25° 14' | -11.6 | 32' | 87% | 134° W | -G | |
| 30th | Cap | 20h 32m | -23° 16' | -11.3 | 32' | 78% | 120° W | -G | Saturn |
| 31st | Cap | 21h 29m | -19° 60' | -10.9 | 31' | 69% | 106° W | -G | Jupiter, Saturn |

## Mercury and Venus

Mercury

25th

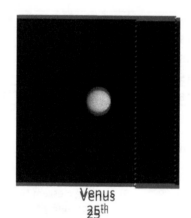

Venus
25th

### Mercury

| Date | Con. | R.A. | Dec. | Mag. | Diam. | Ill. | Elon. | Vis. | Rat. | Close To |
|------|------|------|------|------|-------|------|-------|------|------|----------|
| 21st | Tau | 5h 24m | 25° 4' | 1.0 | 9" | 27% | 23° E | PM | *** | Venus |
| 23rd | Tau | 5h 29m | 24° 50' | 1.3 | 9" | 23% | 22° E | PM | *** | Venus |
| 25th | Tau | 5h 33m | 24° 30' | 1.6 | 10" | 19% | 21° E | PM | *** | Venus |
| 27th | Tau | 5h 35m | 24° 6' | 1.9 | 10" | 15% | 19° E | PM | *** | Venus |
| 29th | Tau | 5h 36m | 23° 38' | 2.3 | 11" | 11% | 18° E | PM | *** | Venus |
| 31st | Tau | 5h 36m | 23° 6' | 2.8 | 11" | 8% | 16° E | PM | ** | Venus |

## Venus

| Date | Con. | R.A. | Dec. | Mag. | Diam. | Ill. | Elon. | Vis. | Rat. | Close To |
|------|------|------|------|------|-------|------|-------|------|------|----------|
| 21st | Tau | 4h 55m | 23° 6' | -3.9 | 10" | 97% | 15° E | PM | * | Mercury, Hyades, Aldebaran |
| 23rd | Tau | 5h 6m | 23° 26' | -3.9 | 10" | 96% | 16° E | PM | * | Mercury, Hyades, Aldebaran |
| 25th | Tau | 5h 16m | 23° 43' | -3.9 | 10" | 96% | 17° E | PM | * | Mercury |
| 27th | Tau | 5h 27m | 23° 57' | -3.9 | 10" | 96% | 17° E | PM | * | Mercury |
| 29th | Tau | 5h 38m | 24° 8' | -3.9 | 10" | 96% | 18° E | PM | * | Mercury |
| 31st | Tau | 5h 48m | 24° 17' | -3.9 | 10" | 95% | 19° E | PM | * | Mercury |

## Mars and the Outer Planets

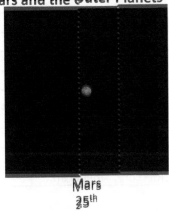

Mars
25th

Jupiter
25th

Saturn
25th

## Mars

| Date | Con. | R.A. | Dec. | Mag. | Diam. | Ill. | Elon. | Vis. | Rat. | Close To |
|------|------|------|------|------|-------|------|-------|------|------|----------|
| 21st | Gem | 7h 14m | 23° 44' | 1.7 | 4" | 95% | 50° E | PM | * | |
| 25th | Gem | 7h 25m | 23° 23' | 1.7 | 4" | 95% | 49° E | PM | * | |
| 31st | Gem | 7h 41m | 22° 47' | 1.7 | 4" | 95% | 47° E | PM | * | |

## The Outer Planets

| Planet | Date | Con. | R.A. | Dec. | Mag. | Diam. | Elon. | Vis. | Rat. | Close To |
|--------|------|------|------|------|------|-------|-------|------|------|----------|
| Jupiter | 25th | Aqr | 22h 13m | -11° 52' | -2.4 | 40" | 89° W | AM | *** | |
| Saturn | 25th | Cap | 21h 4m | -17° 23' | 0.6 | 17" | 106° W | AM | *** | |
| Uranus | 25th | Ari | 2h 39m | 15° 4' | 5.9 | 3" | 23° W | AM | * | |
| Neptune | 25th | Aqr | 23h 35m | -3° 49' | 7.9 | 2" | 69° W | AM | ** | |

## Highlights

| Date | Time (UT) | Event |
|------|-----------|-------|
| 23rd | 12:12 | The waxing gibbous Moon is north of the bright star Spica. (Virgo, evening sky.) |
| | 18:50 | Saturn is stationary prior to beginning retrograde motion. (Morning sky.) |
| 24th | N/A | The May Camelopardalid meteor shower is at its maximum. (ZHR: Variable.) |
| 26th | 11:15 | Full Moon. (Visible all night.) |
| | 11:18 | Total lunar eclipse. Visible from eastern Asia, Australia, North America, South America and the Pacific. |
| | 15:56 | The full Moon is north of the bright star Antares. (Scorpius, visible all night.) |
| 29th | 05:19 | Mercury is 0.4° south of Venus. (Evening sky.) |
| | 19:54 | The waning gibbous Moon is south of dwarf planet Pluto. (Morning sky.) |
| 30th | 01:41 | Mercury is stationary prior to beginning retrograde motion. (Evening sky.) |
| 31st | 00:13 | The waning gibbous Moon is south of Saturn. (Morning sky.) |

# June 1st to 10th, 2021

## The Moon

| 1st | 3rd | 5th | 7th | 9th |

| Date | Con | R.A. | Dec | Mag | Diam | Ill. | Elon. | Phase | Close To |
|------|-----|------|-----|-----|------|------|-------|-------|----------|
| 1st | Aqr | 22h 22m | -15° 48' | -10.5 | 31' | 58% | 94° W | LQ | Jupiter |
| 2nd | Aqr | 23h 10m | -11° 0' | -10.0 | 31' | 48% | 83° W | LQ | Neptune |
| 3rd | Aqr | 23h 56m | -5° 53' | -9.6 | 30' | 38% | 72° W | LQ | Neptune |
| 4th | Cet | 0h 41m | 0° 39' | -9.0 | 30' | 29% | 62° W | -Cr | |
| 5th | Psc | 1h 24m | 4° 31' | -8.4 | 30' | 21% | 53° W | -Cr | |
| 6th | Cet | 2h 7m | 9° 29' | -7.8 | 29' | 13% | 43° W | -Cr | Uranus |
| 7th | Ari | 2h 52m | 14° 5' | -7.0 | 29' | 8% | 33° W | NM | Uranus |
| 8th | Tau | 3h 38m | 18° 9' | -6.1 | 29' | 3% | 22° W | NM | Pleiades |
| 9th | Tau | 4h 27m | 21° 30' | -5.0 | 29' | 1% | 11° W | NM | Pleiades, Hyades, Aldebaran |
| 10th | Tau | 5h 17m | 23° 58' | -4.0 | 30' | 0% | 0° E | NM | Mercury, Aldebaran |

## Mercury and Venus

Mercury
5th

Venus
5th

**Mercury**

| Date | Con. | R.A. | Dec. | Mag. | Diam. | Ill. | Elon. | Vis. | Rat. | Close To |
|------|------|------|------|------|-------|------|-------|------|------|----------|
| 1st | Tau | 5h 36m | 22° 50' | 3.0 | 11" | 7% | 14° E | NV | N/A | Venus |
| 3rd | Tau | 5h 33m | 22° 14' | 3.5 | 12" | 5% | 12° E | NV | N/A | Venus |
| 5th | Tau | 5h 30m | 21° 37' | 4.1 | 12" | 3% | 9° E | NV | N/A | |
| 7th | Tau | 5h 26m | 21° 0' | 4.7 | 12" | 1% | 6° E | NV | N/A | |
| 9th | Tau | 5h 22m | 20° 24' | 5.3 | 12" | 0% | 3° E | NV | N/A | |

**Venus**

| Date | Con. | R.A. | Dec. | Mag. | Diam. | Ill. | Elon. | Vis. | Rat. | Close To |
|------|------|------|------|------|-------|------|-------|------|------|----------|
| 1st | Tau | 5h 54m | 24° 20' | -3.9 | 10" | 95% | 19° E | PM | * | Mercury |
| 3rd | Gem | 6h 5m | 24° 25' | -3.9 | 10" | 95% | 20° E | PM | * | Mercury |
| 5th | Gem | 6h 15m | 24° 26' | -3.9 | 10" | 95% | 20° E | PM | * | |
| 7th | Gem | 6h 26m | 24° 25' | -3.9 | 10" | 94% | 21° E | PM | * | |
| 9th | Gem | 6h 37m | 24° 20' | -3.9 | 11" | 94% | 21° E | PM | * | |

## Mars and the Outer Planets

Mars
5th

Jupiter
5th

Saturn
5th

**Mars**

| Date | Con. | R.A. | Dec. | Mag. | Diam. | Ill. | Elon. | Vis. | Rat. | Close To |
|------|------|------|------|------|-------|------|-------|------|------|----------|
| 1st | Gem | 7h 43m | 22° 41' | 1.7 | 4" | 96% | 46° E | PM | * | |
| 5th | Gem | 7h 54m | 22° 13' | 1.8 | 4" | 96% | 45° E | PM | * | |
| 10th | Cnc | 8h 7m | 21° 35' | 1.8 | 4" | 96% | 43° E | PM | * | Praesepe |

**The Outer Planets**

| Planet | Date | Con. | R.A. | Dec. | Mag. | Diam. | Elon. | Vis. | Rat. | Close To |
|--------|------|------|------|------|------|-------|-------|------|------|----------|
| Jupiter | 5th | Aqr | 22h 16m | -11° 40' | -2.5 | 42" | 100° W | AM | *** | |
| Saturn | 5th | Cap | 21h 4m | -17° 26' | 0.5 | 18" | 118° W | AM | *** | |
| Uranus | 5th | Ari | 2h 41m | 15° 15' | 5.9 | 3" | 33° W | AM | * | |
| Neptune | 5th | Aqr | 23h 36m | -3° 46' | 7.9 | 2" | 80° W | AM | *** | |

## Highlights

| Date | Time (UT) | Event |
|------|-----------|-------|
| 1st | 10:15 | The nearly last quarter Moon is south of Jupiter. (Morning sky.) |
| 2nd | 07:25 | Last Quarter Moon. (Morning sky.) |
| | 23:46 | The last quarter Moon is south of Neptune. (Morning sky.) |
| 7th | 04:56 | The waning crescent Moon is south of Uranus. (Morning sky.) |
| | N/A | Good opportunity to see Earthshine on the waning crescent Moon. (Morning sky.) |
| 8th | 17:30 | The waning crescent Moon is south of the Pleiades star cluster. (Taurus, morning sky.) |
| 10th | 10:43 | Annular solar eclipse. Visible from the Arctic, northern Asia, northern Europe, northern and north-eastern North America |
| | 10:53 | New Moon. |

# June 11th to 20th, 2021

## The Moon

| | 11th | | 13th | | 15th | | 17th | | 19th | |
|---|---|---|---|---|---|---|---|---|---|---|

| Date | Con | R.A. | Dec | Mag | Diam | Ill. | Elon. | Phase | Close To |
|---|---|---|---|---|---|---|---|---|---|
| 11th | Gem | 6h 10m | 25° 22' | -5.1 | 30' | 1% | 13° E | NM | Venus |
| 12th | Gem | 7h 4m | 25° 34' | -6.2 | 30' | 4% | 25° E | NM | Venus |
| 13th | Gem | 7h 58m | 24° 32' | -7.1 | 30' | 9% | 37° E | NM | Mars, Praesepe |
| 14th | Cnc | 8h 51m | 22° 17' | -7.9 | 30' | 15% | 50° E | +Cr | Mars, Praesepe |
| 15th | Leo | 9h 43m | 18° 56' | -8.6 | 31' | 23% | 62° E | +Cr | Regulus |
| 16th | Leo | 10h 34m | 14° 36' | -9.3 | 31' | 32% | 74° E | +Cr | Regulus |
| 17th | Leo | 11h 24m | 9° 29' | -9.8 | 31' | 43% | 85° E | FQ | |
| 18th | Vir | 12h 14m | 3° 48' | -10.3 | 32' | 54% | 96° E | FQ | |
| 19th | Vir | 13h 4m | -2° 12' | -10.7 | 32' | 65% | 108° E | FQ | Spica |
| 20th | Vir | 13h 56m | -8° 15' | -11.1 | 33' | 75% | 120° E | +G | Spica |

## Mercury and Venus

Mercury
15th

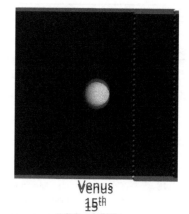

Venus
15th

### Mercury

| Date | Con. | R.A. | Dec. | Mag. | Diam. | Ill. | Elon. | Vis. | Rat. | Close To |
|---|---|---|---|---|---|---|---|---|---|---|
| 11th | Tau | 5h 17m | 19° 50' | 5.4 | 12" | 0% | 0° W | NV | N/A | |
| 13th | Tau | 5h 13m | 19° 20' | 4.9 | 12" | 1% | 4° W | NV | N/A | Aldebaran |
| 15th | Tau | 5h 9m | 18° 55' | 4.2 | 12" | 2% | 7° W | NV | N/A | Aldebaran |
| 17th | Tau | 5h 5m | 18° 35' | 3.7 | 12" | 4% | 10° W | NV | N/A | Hyades, Aldebaran |
| 19th | Tau | 5h 3m | 18° 22' | 3.1 | 11" | 6% | 12° W | NV | N/A | Hyades, Aldebaran |

## Venus

| Date | Con. | R.A. | Dec. | Mag. | Diam. | Ill. | Elon. | Vis. | Rat. | Close To |
|------|------|------|------|------|-------|------|-------|------|------|----------|
| 11th | Gem | 6h 47m | 24° 13' | -3.9 | 11" | 94% | 22° E | PM | * | Moon |
| 13th | Gem | 6h 58m | 24° 3' | -3.9 | 11" | 93% | 23° E | PM | * | |
| 15th | Gem | 7h 9m | 23° 50' | -3.9 | 11" | 93% | 23° E | PM | * | |
| 17th | Gem | 7h 19m | 23° 35' | -3.9 | 11" | 93% | 24° E | PM | * | |
| 19th | Gem | 7h 30m | 23° 16' | -3.9 | 11" | 92% | 24° E | PM | * | |

## Mars and the Outer Planets

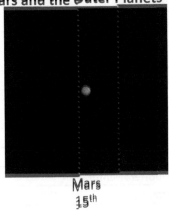

Mars
15th

Jupiter
15th

Saturn
15th

## Mars

| Date | Con. | R.A. | Dec. | Mag. | Diam. | Ill. | Elon. | Vis. | Rat. | Close To |
|------|------|------|------|------|-------|------|-------|------|------|----------|
| 11th | Cnc | 8h 9m | 21° 27' | 1.8 | 4" | 96% | 42° E | PM | * | Praesepe |
| 15th | Cnc | 8h 20m | 20° 53' | 1.8 | 4" | 96% | 41° E | PM | * | Praesepe |
| 20th | Cnc | 8h 32m | 20° 8' | 1.8 | 4" | 97% | 39° E | PM | * | Praesepe |

## The Outer Planets

| Planet | Date | Con. | R.A. | Dec. | Mag. | Diam. | Elon. | Vis. | Rat. | Close To |
|--------|------|------|------|------|------|-------|-------|------|------|----------|
| Jupiter | 15th | Aqr | 22h 17m | -11° 35' | -2.5 | 43" | 110° W | AM | **** | |
| Saturn | 15th | Cap | 21h 3m | -17° 33' | 0.5 | 18" | 128° W | AM | **** | |
| Uranus | 15th | Ari | 2h 43m | 15° 24' | 5.8 | 3" | 43° W | AM | * | |
| Neptune | 15th | Aqr | 23h 36m | -3° 44' | 7.9 | 2" | 90° W | AM | *** | |

s

## Highlights

| Date | Time (UT) | Event |
|------|-----------|-------|
| 11th | 01:07 | Mercury is at inferior conjunction with the Sun. (Not visible.) |
| 12th | 05:32 | The waxing crescent Moon is north of Venus. (Evening sky.) |
| 13th | 21:16 | The waxing crescent Moon is north of Mars. (Evening sky.) |
| | N/A | Good opportunity to see Earthshine on the waxing crescent Moon. (Evening sky.) |
| 14th | 05:56 | The waxing crescent Moon is north of the Praesepe star cluster. (Evening sky.) |
| 16th | 00:38 | The waxing crescent Moon is north of the bright star Regulus. (Leo, evening sky.) |
| 18th | 03:55 | First Quarter Moon (Evening sky.) |
| 19th | 22:32 | The just-past first quarter Moon is north of the bright star Spica. (Virgo, evening sky.) |

# June 21st to 30th, 2021

## The Moon

| 21st | 23rd | 25th | 27th | 29th |

| Date | Con | R.A. | Dec | Mag | Diam | Ill. | Elon. | Phase | Close To |
|------|-----|------|-----|-----|------|------|-------|-------|----------|
| 21st | Lib | 14h 51m | -14° 0' | -11.5 | 33' | 85% | 132° E | +G | |
| 22nd | Lib | 15h 49m | -19° 1' | -11.9 | 33' | 92% | 146° E | +G | Antares |
| 23rd | Oph | 16h 52m | -22° 51' | -12.3 | 33' | 98% | 161° E | FM | Antares |
| 24th | Sgr | 17h 57m | -25° 7' | -12.6 | 33' | 100% | 176° E | FM | |
| 25th | Sgr | 19h 2m | -25° 34' | -12.5 | 33' | 99% | 169° W | FM | |
| 26th | Sgr | 20h 6m | -24° 14' | -12.1 | 32' | 96% | 154° W | FM | |
| 27th | Cap | 21h 7m | -21° 23' | -11.8 | 32' | 90% | 140° W | -G | Saturn |
| 28th | Aqr | 22h 2m | -17° 23' | -11.4 | 32' | 82% | 127° W | -G | Jupiter |
| 29th | Aqr | 22h 54m | -12° 38' | -11.1 | 31' | 73% | 115° W | -G | Jupiter, Neptune |
| 30th | Aqr | 23h 42m | -7° 29' | -10.7 | 31' | 64% | 104° W | LQ | Neptune |

## Mercury and Venus

Mercury
25th

Venus
25th

**Mercury**

| Date | Con. | R.A. | Dec. | Mag. | Diam. | Ill. | Elon. | Vis. | Rat. | Close To |
|------|------|------|------|------|-------|------|-------|------|------|----------|
| 21st | Tau | 5h 2m | 18° 16' | 2.6 | 11" | 9% | 15° W | NV | N/A | Hyades, Aldebaran |
| 23rd | Tau | 5h 1m | 18° 17' | 2.2 | 10" | 12% | 17° W | AM | *** | Hyades, Aldebaran |
| 25th | Tau | 5h 2m | 18° 25' | 1.8 | 10" | 16% | 19° W | AM | *** | Hyades, Aldebaran |
| 27th | Tau | 5h 5m | 18° 38' | 1.5 | 10" | 20% | 20° W | AM | *** | Hyades, Aldebaran |
| 29th | Tau | 5h 8m | 18° 57' | 1.2 | 9" | 24% | 22° W | AM | *** | Aldebaran |

**Venus**

| Date | Con. | R.A. | Dec. | Mag. | Diam. | Ill. | Elon. | Vis. | Rat. | Close To |
|------|------|------|------|------|-------|------|-------|------|------|----------|
| 21st | Gem | 7h 40m | 22° 55' | -3.9 | 11" | 92% | 25° E | PM | * | |
| 23rd | Gem | 7h 51m | 22° 32' | -3.9 | 11" | 91% | 25° E | PM | * | |
| 25th | Gem | 8h 1m | 22° 6' | -3.9 | 11" | 91% | 26° E | PM | * | Praesepe |
| 27th | Cnc | 8h 11m | 21° 37' | -3.9 | 11" | 91% | 26° E | PM | * | Mars, Praesepe |
| 29th | Cnc | 8h 21m | 21° 6' | -3.9 | 11" | 90% | 27° E | PM | * | Mars, Praesepe |

## Mars and the Outer Planets

Mars
25th

Jupiter
25th

Saturn
25th

**Mars**

| Date | Con. | R.A. | Dec. | Mag. | Diam. | Ill. | Elon. | Vis. | Rat. | Close To |
|------|------|------|------|------|-------|------|-------|------|------|----------|
| 21st | Cnc | 8h 35m | 19° 59' | 1.8 | 4" | 97% | 38° E | PM | * | Praesepe |
| 25th | Cnc | 8h 45m | 19° 20' | 1.8 | 4" | 97% | 37° E | PM | * | Praesepe |
| 30th | Cnc | 8h 58m | 18° 28' | 1.8 | 4" | 97% | 35° E | PM | * | Venus, Praesepe |

**The Outer Planets**

| Planet | Date | Con. | R.A. | Dec. | Mag. | Diam. | Elon. | Vis. | Rat. | Close To |
|--------|------|------|------|------|------|-------|-------|------|------|----------|
| Jupiter | 25th | Aqr | 22h 17m | -11° 38' | -2.6 | 45" | 120° W | AM | **** | |
| Saturn | 25th | Cap | 21h 1m | -17° 41' | 0.4 | 18" | 139° W | AM | **** | |
| Uranus | 25th | Ari | 2h 45m | 15° 32' | 5.8 | 3" | 53° W | AM | * | |
| Neptune | 25th | Aqr | 23h 36m | -3° 44' | 7.9 | 2" | 100° W | AM | *** | |

## Highlights

| Date | Time (UT) | Event |
|------|-----------|-------|
| 21st | 03:33 | Summer Solstice. |
| | 03:49 | Jupiter is stationary prior to beginning retrograde motion. (Morning sky.) |
| 22nd | 22:33 | Mercury is stationary prior to beginning retrograde motion. (Morning sky.) |
| 23rd | 04:34 | The nearly full Moon is north of the bright star Antares. (Scorpius, visible all night.) |
| | 07:10 | Mars is 0.0° north of the Praesepe star cluster. (Cancer, evening sky.) |
| 24th | 18:40 | Full Moon. (Visible all night.) |
| 26th | 03:50 | Neptune is stationary prior to beginning retrograde motion. (Morning sky.) |
| | 07:33 | The waning gibbous Moon is south of the dwarf planet Pluto. (Morning sky.) |
| 27th | 10:31 | The waning gibbous Moon is south of Saturn. (Morning sky.) |
| | N/A | The Bootid meteor shower is at its maximum. (ZHR: Variable.) |
| 28th | 23:10 | The waning gibbous Moon is south of Jupiter. (Morning sky.) |
| 30th | 10:33 | The nearly last quarter Moon is south of Neptune. (Morning sky.) |

# July 1<sup>st</sup> to 10<sup>th</sup>, 2021

## The Moon

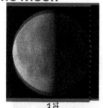
1<sup>st</sup>

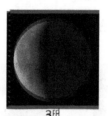
3<sup>rd</sup>

5<sup>th</sup>

7<sup>th</sup>

9<sup>th</sup>

| Date | Con | R.A. | Dec | Mag | Diam | Ill. | Elon. | Phase | Close To |
|------|-----|------|-----|-----|------|------|-------|-------|----------|
| 1st | Cet | 0h 27m | -2° 10' | -10.3 | 30' | 54% | 94° W | LQ | |
| 2nd | Psc | 1h 11m | 3° 7' | -9.9 | 30' | 44% | 84° W | LQ | |
| 3rd | Psc | 1h 55m | 8° 12' | -9.4 | 30' | 35% | 74° W | =Cr | |
| 4th | Ari | 2h 39m | 12° 55' | -8.8 | 30' | 26% | 64° W | =Cr | Uranus |
| 5th | Tau | 3h 25m | 17° 8' | -8.2 | 29' | 18% | 53° W | =Cr | Uranus, Pleiades |
| 6th | Tau | 4h 13m | 20° 42' | -7.5 | 30' | 11% | 43° W | NM | Pleiades, Hyades, Aldebaran |
| 7th | Tau | 5h 3m | 23° 24' | -6.7 | 30' | 6% | 31° W | NM | Mercury, Hyades, Aldebaran |
| 8th | Tau | 5h 55m | 25° 6' | -5.7 | 30' | 2% | 19° W | NM | Mercury |
| 9th | Gem | 6h 49m | 25° 37' | -4.6 | 30' | 0% | 7° W | NM | |
| 10th | Gem | 7h 44m | 24° 53' | -4.6 | 30' | 0% | 6° E | NM | |

## Mercury and Venus

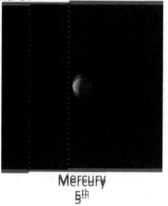
Mercury
5<sup>th</sup>

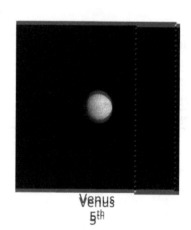
Venus
5<sup>th</sup>

### Mercury

| Date | Con. | R.A. | Dec. | Mag. | Diam. | Ill. | Elon. | Vis. | Rat. | Close To |
|------|------|------|------|------|-------|------|-------|------|------|----------|
| 1st | Tau | 5h 13m | 19° 20' | 0.9 | 9" | 29% | 22° W | AM | *** | Aldebaran |
| 3rd | Tau | 5h 19m | 19° 47' | 0.6 | 8" | 33% | 23° W | AM | *** | |
| 5th | Tau | 5h 27m | 20° 16' | 0.4 | 8" | 38% | 23° W | AM | *** | |
| 7th | Tau | 5h 36m | 20° 46' | 0.2 | 7" | 44% | 23° W | AM | *** | Moon |
| 9th | Tau | 5h 46m | 21° 16' | -0.1 | 7" | 49% | 22° W | AM | *** | |

## Venus

| Date | Con. | R.A. | Dec. | Mag. | Diam. | Ill. | Elon. | Vis. | Rat. | Close To |
|------|------|------|------|------|-------|------|-------|------|------|----------|
| 1st | Cnc | 8h 32m | 20° 33' | -3.9 | 11" | 90% | 27° E | PM | * | Mars, Praesepe |
| 3rd | Cnc | 8h 42m | 19° 57' | -3.9 | 11" | 89% | 28° E | PM | * | Mars, Praesepe |
| 5th | Cnc | 8h 51m | 19° 19' | -3.9 | 11" | 89% | 28° E | PM | * | Mars, Praesepe |
| 7th | Cnc | 9h 1m | 18° 39' | -3.9 | 11" | 88% | 29° E | PM | * | Mars, Praesepe |
| 9th | Cnc | 9h 11m | 17° 57' | -3.9 | 12" | 88% | 29° E | PM | * | Mars, Praesepe |

## Mars and the Outer Planets

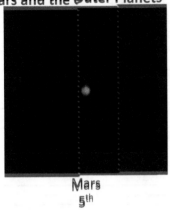

Mars
5th

Jupiter
5th

Saturn
5th

## Mars

| Date | Con. | R.A. | Dec. | Mag. | Diam. | Ill. | Elon. | Vis. | Rat. | Close To |
|------|------|------|------|------|-------|------|-------|------|------|----------|
| 1st | Cnc | 9h 0m | 18° 18' | 1.8 | 4" | 97% | 34° E | PM | * | Venus, Praesepe |
| 5th | Cnc | 9h 10m | 17° 34' | 1.8 | 4" | 97% | 33° E | PM | * | Venus, Praesepe |
| 10th | Cnc | 9h 22m | 16° 36' | 1.8 | 4" | 98% | 31° E | PM | * | Venus |

## The Outer Planets

| Planet | Date | Con. | R.A. | Dec. | Mag. | Diam. | Elon. | Vis. | Rat. | Close To |
|--------|------|------|------|------|------|-------|-------|------|------|----------|
| Jupiter | 5th | Aqr | 22h 16m | -11° 47' | -2.7 | 46" | 131° W | AM | **** | |
| Saturn | 5th | Cap | 20h 59m | -17° 52' | 0.4 | 18" | 150° W | AM | **** | |
| Uranus | 5th | Ari | 3h 46m | 15° 38' | 5.8 | 3" | 63° W | AM | ** | Moon |
| Neptune | 5th | Aqr | 23h 36m | -3° 45' | 7.9 | 2" | 111° W | AM | *** | |

## Highlights

| Date | Time (UT) | Event |
|------|-----------|-------|
| 1st | 21:11 | Last Quarter Moon. (Morning sky.) |
| 3rd | 01:53 | Venus is north of the Praesepe star cluster. (Cancer, evening sky.) |
| 4th | 16:32 | The waning crescent Moon is south of Uranus. (Morning sky.) |
| | 19:41 | Mercury is at greatest western elongation from the Sun. (Morning sky.) |
| 5th | 22:18 | The waning crescent Moon is south of the Pleiades star cluster. (Taurus, morning sky.) |
| 6th | 22:32 | The waning crescent Moon is north of the bright star Aldebaran. (Taurus, morning sky.) |
| | N/A | Good opportunity to see Earthshine on the waning crescent Moon. (Morning sky.) |
| 8th | 03:18 | The waning crescent Moon is north of Mercury. (Morning sky.) |
| 10th | 01:17 | New Moon. (Not visible.) |

# July 11th to 20th, 2021

## The Moon

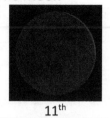
11th

13th

15th

17th

19th

| Date | Con | R.A. | Dec | Mag | Diam | Ill. | Elon. | Phase | Close To |
|------|-----|------|-----|-----|------|------|-------|-------|----------|
| 11th | Cnc | 8h 38m | 22° 53' | -5.7 | 30' | 2% | 19° E | NM | Venus, Mars, Praesepe |
| 12th | Leo | 9h 31m | 19° 43' | -6.7 | 31' | 6% | 31° E | NM | Venus, Mars, Regulus |
| 13th | Leo | 10h 23m | 15° 34' | -7.6 | 31' | 12% | 43° E | NM | Regulus |
| 14th | Leo | 11h 13m | 10° 35' | -8.4 | 31' | 20% | 54° E | +Cr | |
| 15th | Vir | 12h 2m | 5° 3' | -9.1 | 32' | 29% | 66° E | +Cr | |
| 16th | Vir | 12h 51m | 0° 49' | -9.6 | 32' | 40% | 77° E | FQ | Spica |
| 17th | Vir | 13h 42m | -6° 46' | -10.2 | 32' | 51% | 88° E | FQ | Spica |
| 18th | Lib | 14h 34m | -12° 28' | -10.6 | 32' | 62% | 101° E | FQ | |
| 19th | Lib | 15h 30m | -17° 35' | -11.0 | 33' | 73% | 113° E | +G | |
| 20th | Oph | 16h 29m | -21° 43' | -11.4 | 33' | 83% | 127° E | +G | Antares |

## Mercury and Venus

Mercury
15th

Venus
15th

**Mercury**

| Date | Con. | R.A. | Dec. | Mag. | Diam. | Ill. | Elon. | Vis. | Rat. | Close To |
|------|------|------|------|------|-------|------|-------|------|------|----------|
| 11th | Ori | 5h 57m | 21° 45' | -0.3 | 7" | 55% | 22° W | AM | *** | |
| 13th | Gem | 6h 10m | 22° 10' | -0.5 | 6" | 61% | 20° W | AM | *** | |
| 15th | Gem | 6h 24m | 22° 30' | -0.7 | 6" | 68% | 19° W | AM | *** | |
| 17th | Gem | 6h 39m | 22° 45' | -0.9 | 6" | 74% | 17° W | AM | ** | |
| 19th | Gem | 6h 55m | 22° 51' | -1.0 | 6" | 80% | 15° W | AM | ** | |

**Venus**

| Date | Con. | R.A. | Dec. | Mag. | Diam. | Ill. | Elon. | Vis. | Rat. | Close To |
|------|------|------|------|------|-------|------|-------|------|------|----------|
| 11th | Cnc | 9h 21m | 17° 13' | -3.9 | 12" | 87% | 29° E | PM | * | Moon, Mars |
| 13th | Leo | 9h 30m | 16° 28' | -3.9 | 12" | 87% | 30° E | PM | ** | Mars, Regulus |
| 15th | Leo | 9h 40m | 15° 41' | -3.9 | 12" | 86% | 30° E | PM | ** | Mars, Regulus |
| 17th | Leo | 9h 49m | 14° 52' | -3.9 | 12" | 86% | 30° E | PM | ** | Mars, Regulus |
| 19th | Leo | 9h 58m | 14° 1' | -3.9 | 12" | 86% | 31° E | PM | ** | Mars, Regulus |

## Mars and the Outer Planets

Mars
15th

Jupiter
15th

Saturn
15th

**Mars**

| Date | Con. | R.A. | Dec. | Mag. | Diam. | Ill. | Elon. | Vis. | Rat. | Close To |
|------|------|------|------|------|-------|------|-------|------|------|----------|
| 11th | Leo | 9h 25m | 16° 24' | 1.8 | 4" | 98% | 30° E | PM | * | Moon, Venus |
| 15th | Leo | 9h 35m | 15° 36' | 1.8 | 4" | 98% | 29° E | PM | * | Venus, Regulus |
| 20th | Leo | 9h 47m | 14° 33' | 1.8 | 4" | 98% | 27° E | PM | * | Venus, Regulus |

**The Outer Planets**

| Planet | Date | Con. | R.A. | Dec. | Mag. | Diam. | Elon. | Vis. | Rat. | Close To |
|--------|------|------|------|------|------|-------|-------|------|------|----------|
| Jupiter | 15th | Aqr | 22h 14m | -12° 3' | -2.7 | 47" | 141° W | AM | ***** | |
| Saturn | 15th | Cap | 20h 56m | -18° 4' | 0.3 | 19" | 161° W | AM | **** | |
| Uranus | 15th | Ari | 2h 47m | 15° 44' | 5.8 | 4" | 73° W | AM | ** | |
| Neptune | 15th | Aqr | 23h 36m | -3° 47' | 7.9 | 2" | 121° W | AM | **** | |

## Highlights

| Date | Time (UT) | Event |
|------|-----------|-------|
| 12th | 07:45 | The waxing crescent Moon is north of Venus. (Evening sky.) |
| | 08:52 | The waxing crescent Moon is north of Mars. (Evening sky.) |
| | N/A | Good opportunity to see Earthshine on the waxing crescent Moon. (Evening sky.) |
| 13th | 04:28 | The waxing crescent Moon is north of the bright star Regulus. (Leo, evening sky.) |
| | 07:04 | Venus is 0.5° north of Mars. (Evening sky.) |
| 17th | 04:27 | The almost first quarter Moon is north of the bright star Spica. (Virgo, evening sky.) |
| | 10:11 | First Quarter Moon. (Evening sky.) |
| | 20:29 | Dwarf planet Pluto is at opposition. (Visible all night.) |
| 20th | 11:19 | The waxing gibbous Moon is north of the bright star Antares. (Scorpius, evening sky.) |

# July 21st to 31st, 2021

## The Moon

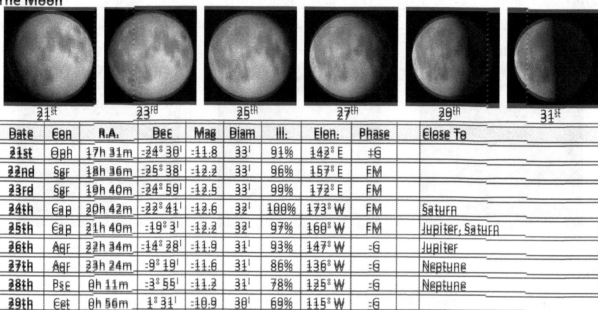

| 21st | 23rd | 25th | 27th | 29th | 31st |

| Date | Con | R.A. | Dec | Mag | Diam | Ill. | Elon. | Phase | Close To |
|---|---|---|---|---|---|---|---|---|---|
| 21st | Oph | 17h 31m | -24° 30' | -11.8 | 33' | 91% | 142° E | +G | |
| 22nd | Sgr | 18h 36m | -25° 38' | -12.2 | 33' | 96% | 157° E | FM | |
| 23rd | Sgr | 19h 40m | -24° 59' | -12.5 | 33' | 99% | 172° E | FM | |
| 24th | Cap | 20h 42m | -22° 41' | -12.6 | 32' | 100% | 173° W | FM | Saturn |
| 25th | Cap | 21h 40m | -19° 3' | -12.2 | 32' | 97% | 160° W | FM | Jupiter, Saturn |
| 26th | Aqr | 22h 34m | -14° 28' | -11.9 | 31' | 93% | 147° W | -G | Jupiter |
| 27th | Aqr | 23h 24m | -9° 19' | -11.6 | 31' | 86% | 136° W | -G | Neptune |
| 28th | Psc | 0h 11m | -3° 55' | -11.2 | 31' | 78% | 125° W | -G | Neptune |
| 29th | Cet | 0h 56m | 1° 31' | -10.9 | 30' | 69% | 115° W | -G | |
| 30th | Psc | 1h 41m | 6° 45' | -10.5 | 30' | 60% | 105° W | LQ | |
| 31st | Ari | 2h 25m | 11° 39' | -10.1 | 30' | 51% | 94° W | LQ | Uranus |

## Mercury and Venus

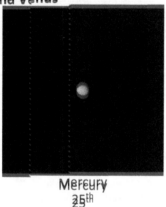

Mercury
25th

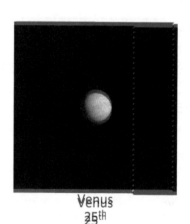

Venus
25th

### Mercury

| Date | Con. | R.A. | Dec. | Mag. | Diam. | Ill. | Elon. | Vis. | Rat. | Close To |
|---|---|---|---|---|---|---|---|---|---|---|
| 21st | Gem | 7h 11m | 22° 49' | -1.2 | 6" | 85% | 13° W | NV | N/A | |
| 23rd | Gem | 7h 29m | 22° 37' | -1.4 | 5" | 90% | 11° W | NV | N/A | |
| 25th | Gem | 7h 47m | 22° 13' | -1.6 | 5" | 94% | 8° W | NV | N/A | |
| 27th | Cnc | 8h 5m | 21° 40' | -1.7 | 5" | 97% | 6° W | NV | N/A | Praesepe |
| 29th | Cnc | 8h 22m | 20° 56' | -1.9 | 5" | 99% | 3° W | NV | N/A | Praesepe |
| 31st | Cnc | 8h 40m | 20° 2' | -2.0 | 5" | 100% | 1° W | NV | N/A | Praesepe |

## Venus

| Date | Con. | R.A. | Dec. | Mag. | Diam. | Ill. | Elon. | Vis. | Rat. | Close To |
|---|---|---|---|---|---|---|---|---|---|---|
| 21st | Leo | 10h 8m | 13° 9' | -3.9 | 13" | 85% | 31° E | PM | ** | Mars, Regulus |
| 23rd | Leo | 10h 17m | 12° 16' | -3.9 | 13" | 84% | 31° E | PM | ** | Mars, Regulus |
| 25th | Leo | 10h 26m | 11° 22' | -3.9 | 13" | 84% | 32° E | PM | ** | Mars, Regulus |
| 27th | Leo | 10h 35m | 10° 26' | -3.9 | 13" | 83% | 32° E | PM | ** | Mars, Regulus |
| 29th | Leo | 10h 44m | 9° 30' | -3.9 | 13" | 83% | 32° E | PM | ** | Mars, Regulus |
| 31st | Leo | 10h 53m | 8° 32' | -3.9 | 13" | 83% | 32° E | PM | ** | Mars |

## Mars and the Outer Planets

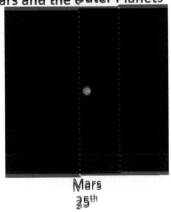

Mars
25th

Jupiter
25th

Saturn
25th

## Mars

| Date | Con. | R.A. | Dec. | Mag. | Diam. | Ill. | Elon. | Vis. | Rat. | Close To |
|---|---|---|---|---|---|---|---|---|---|---|
| 21st | Leo | 9h 49m | 14° 21' | 1.8 | 4" | 98% | 26° E | PM | * | Venus, Regulus |
| 25th | Leo | 9h 59m | 13° 29' | 1.8 | 4" | 98% | 25° E | PM | * | Venus, Regulus |
| 31st | Leo | 10h 13m | 12° 8' | 1.8 | 4" | 99% | 23° E | PM | * | Venus, Regulus |

## The Outer Planets

| Planet | Date | Con. | R.A. | Dec. | Mag. | Diam. | Elon. | Vis. | Rat. | Close To |
|---|---|---|---|---|---|---|---|---|---|---|
| Jupiter | 25th | Aqr | 22h 11m | -12° 24' | -2.8 | 48" | 153° W | AM | ***** | Moon |
| Saturn | 25th | Cap | 20h 53m | -18° 16' | 0.2 | 19" | 172° W | AN | ***** | Moon |
| Uranus | 25th | Ari | 2h 48m | 15° 48' | 5.8 | 4" | 83° W | AM | ** | |
| Neptune | 25th | Aqr | 23h 35m | -3° 51' | 7.8 | 2" | 131° W | AM | **** | |

## Highlights

| Date | Time (UT) | Event |
|---|---|---|
| 21st | 12:42 | Venus is 1.2° north of the bright star Regulus. (Leo, evening sky.) |
| 23rd | 14:34 | The nearly full Moon is south of dwarf planet Pluto. (Evening sky.) |
| 24th | 02:38 | Full Moon. (Visible all night.) |
| | 15:44 | The full Moon is south of Saturn. (Visible all night.) |
| 26th | 01:01 | The waning gibbous Moon is south of Jupiter. (Morning sky.) |
| 27th | 17:05 | The waning gibbous Moon is south of Neptune. (Morning sky.) |
| 28th | N/A | The Piscis Austrinid meteor shower is at its maximum. (ZHR: 5) |
| 29th | 04:19 | Mars is 0.7° north of the bright star Regulus. (Leo, evening sky.) |
| 30th | N/A | The Delta Aquariid meteor shower is at its maximum. (ZHR: 16) |
| | N/A | The Alpha Capricornid meteor shower is at its maximum. (ZHR: 5) |
| 31st | 13:17 | Last Quarter Moon. (Morning sky.) |
| | 23:11 | The last quarter Moon is south of Uranus. (Morning sky.) |

# August 1<sup>st</sup> to 10<sup>th</sup>, 2021

## The Moon

1<sup>st</sup>

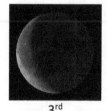
3<sup>rd</sup>

5<sup>th</sup>

7<sup>th</sup>

9<sup>th</sup>

| Date | Con | R.A. | Dec | Mag | Diam | Ill. | Elon. | Phase | Close To |
|------|-----|------|-----|-----|------|------|-------|-------|----------|
| 1st | Ari | 3h 11m | 16° 4' | -9.7 | 30' | 41% | 84° W | LQ | Uranus, Pleiades |
| 2nd | Tau | 3h 58m | 19° 50' | -9.2 | 30' | 32% | 73° W | -Cr | Pleiades, Hyades, Aldebaran |
| 3rd | Tau | 4h 47m | 22° 48' | -8.7 | 30' | 23% | 62° W | -Cr | Hyades, Aldebaran |
| 4th | Tau | 5h 39m | 24° 48' | -8.0 | 30' | 16% | 50° W | -Cr | |
| 5th | Gem | 6h 33m | 25° 40' | -7.3 | 30' | 10% | 37° W | NM | |
| 6th | Gem | 7h 27m | 25° 16' | -6.4 | 30' | 5% | 25° W | NM | |
| 7th | Cnc | 8h 22m | 23° 36' | -5.3 | 30' | 1% | 12° W | NM | Praesepe |
| 8th | Cnc | 9h 16m | 20° 41' | -4.4 | 31' | 0% | 1° E | NM | Mercury, Praesepe |
| 9th | Leo | 10h 9m | 16° 41' | -5.2 | 31' | 1% | 13° E | NM | Mercury, Mars, Regulus |
| 10th | Leo | 11h 0m | 11° 48' | -6.3 | 31' | 4% | 25° E | NM | Venus, Mars |

## Mercury and Venus

Mercury
5<sup>th</sup>

Venus
5<sup>th</sup>

**Mercury**

| Date | Con. | R.A. | Dec. | Mag. | Diam. | Ill. | Elon. | Vis. | Rat. | Close To |
|------|------|------|------|------|-------|------|-------|------|------|----------|
| 1st | Cnc | 8h 49m | 19° 32' | -2.0 | 5" | 100% | 0° E | NV | N/A | Praesepe |
| 3rd | Cnc | 9h 5m | 18° 27' | -1.8 | 5" | 99% | 3° E | NV | N/A | Praesepe |
| 5th | Cnc | 9h 22m | 17° 15' | -1.6 | 5" | 99% | 5° E | NV | N/A | |
| 7th | Leo | 9h 38m | 15° 58' | -1.4 | 5" | 97% | 7° E | NV | N/A | Regulus |
| 9th | Leo | 9h 53m | 14° 37' | -1.2 | 5" | 96% | 9° E | NV | N/A | Moon, Regulus |

**Venus**

| Date | Con. | R.A. | Dec. | Mag. | Diam. | Ill. | Elon. | Vis. | Rat. | Close To |
|------|------|------|------|------|-------|------|-------|------|------|----------|
| 1st | Leo | 10h 57m | 8° 3' | -4.0 | 13" | 82% | 33° E | PM | ** | |
| 3rd | Leo | 11h 6m | 7° 5' | -4.0 | 13" | 82% | 33° E | PM | ** | |
| 5th | Leo | 11h 15m | 6° 5' | -4.0 | 13" | 81% | 33° E | PM | ** | |
| 7th | Leo | 11h 23m | 5° 5' | -4.0 | 13" | 80% | 33° E | PM | ** | |
| 9th | Leo | 11h 32m | 4° 5' | -4.0 | 13" | 80% | 34° E | PM | ** | |

## Mars and the Outer Planets

Mars
5th

Jupiter
5th

Saturn
5th

**Mars**

| Date | Con. | R.A. | Dec. | Mag. | Diam. | Ill. | Elon. | Vis. | Rat. | Close To |
|------|------|------|------|------|-------|------|-------|------|------|----------|
| 1st | Leo | 10h 16m | 11° 54' | 1.8 | 4" | 99% | 22° E | PM | * | Regulus |
| 5th | Leo | 10h 25m | 10° 58' | 1.8 | 4" | 99% | 21° E | PM | * | Regulus |
| 10th | Leo | 10h 37m | 9° 47' | 1.8 | 4" | 99% | 19° E | PM | * | Moon, Mercury, Regulus |

**The Outer Planets**

| Planet | Date | Con. | R.A. | Dec. | Mag. | Diam. | Elon. | Vis. | Rat. | Close To |
|--------|------|------|------|------|------|-------|-------|------|------|----------|
| Jupiter | 5th | Aqr | 22h 6m | -12° 52' | -2.8 | 49" | 164° W | AM | ***** | |
| Saturn | 5th | Cap | 20h 50m | -18° 31' | 0.2 | 19" | 177° E | AN | ***** | |
| Uranus | 5th | Ari | 2h 49m | 15° 51' | 5.8 | 4" | 93° W | AM | ·** | |
| Neptune | 5th | Aqr | 23h 35m | -3° 56' | 7.8 | 2" | 142° W | AM | **** | |

## Highlights

| Date | Time (UT) | Event |
|------|-----------|-------|
| 1st | 13:55 | Mercury is at superior conjunction with the Sun. (Not visible.) |
| 2nd | 05:51 | The waning crescent Moon is south of the Pleiades star cluster. (Taurus, morning sky.) |
| | 11:16 | Saturn is at opposition. (Visible all night.) |
| 3rd | 05:33 | The waning crescent Moon is north of the bright star Aldebaran. (Taurus, morning sky.) |
| 5th | N/A | Good opportunity to see Earthshine on the waning crescent Moon. (Morning sky.) |
| 8th | 13:51 | New Moon. (Not visible.) |
| 10th | 00:54 | The waxing crescent Moon is north of Mars. (Evening sky.) |

# August 11th to 20th, 2021

## The Moon

| | 11th | 13th | 15th | 17th | 19th |

| Date | Con | R.A. | Dec | Mag | Diam | Ill. | Elon. | Phase | Close To |
|------|-----|------|-----|-----|------|------|-------|-------|----------|
| 11th | Vir | 11h 50m | 6° 15' | -7.3 | 32' | 10% | 36° E | NM | Venus |
| 12th | Vir | 12h 40m | 0° 21' | -8.2 | 32' | 17% | 48° E | +Cr | Spica |
| 13th | Vir | 13h 30m | -5° 38' | -8.9 | 32' | 27% | 59° E | +Cr | Spica |
| 14th | Vir | 14h 22m | -11° 25' | -9.5 | 32' | 37% | 71° E | +Cr | |
| 15th | Lib | 15h 16m | -16° 38' | -10.1 | 32' | 48% | 84° E | FQ | |
| 16th | Sco | 16h 13m | -20° 57' | -10.5 | 32' | 60% | 97° E | FQ | Antares |
| 17th | Oph | 17h 13m | -24° 3' | -11.0 | 32' | 71% | 111° E | +G | Antares |
| 18th | Sgr | 18h 15m | -25° 36' | -11.3 | 32' | 81% | 126° E | +G | |
| 19th | Sgr | 19h 18m | -25° 29' | -11.7 | 32' | 89% | 141° E | +G | |
| 20th | Cap | 20h 20m | -23° 44' | -12.0 | 32' | 95% | 155° E | +G | Saturn |

## Mercury and Venus

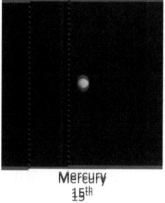 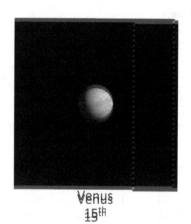

Mercury
15th

Venus
15th

### Mercury

| Date | Con. | R.A. | Dec. | Mag. | Diam. | Ill. | Elon. | Vis. | Rat. | Close To |
|------|------|------|------|------|-------|------|-------|------|------|----------|
| 11th | Leo | 10h 7m | 13° 13' | -1.0 | 5" | 94% | 10° E | NV | N/A | Mars, Regulus |
| 13th | Leo | 10h 21m | 11° 46' | -0.8 | 5" | 92% | 12° E | NV | N/A | Mars, Regulus |
| 15th | Leo | 10h 34m | 10° 18' | -0.7 | 5" | 90% | 14° E | NV | N/A | Mars, Regulus |
| 17th | Leo | 10h 47m | 8° 49' | -0.6 | 5" | 88% | 15° E | NV | N/A | Mars, Regulus |
| 19th | Leo | 11h 0m | 7° 20' | -0.4 | 5" | 86% | 16° E | PM | ** | Mars |

## Venus

| Date | Con. | R.A. | Dec. | Mag. | Diam. | Ill. | Elon. | Vis. | Rat. | Close To |
|------|------|------|------|------|-------|------|-------|------|------|----------|
| 11th | Vir | 11h 41m | 3° 4' | -4.0 | 13" | 79% | 34° E | PM | ** | Moon |
| 13th | Vir | 11h 49m | 2° 2' | -4.0 | 14" | 79% | 34° E | PM | ** | |
| 15th | Vir | 11h 58m | 1° 0' | -4.0 | 14" | 78% | 34° E | PM | ** | |
| 17th | Vir | 12h 6m | 0° 1' | -4.0 | 14" | 77% | 35° E | PM | ** | |
| 19th | Vir | 12h 15m | -1° 3' | -4.0 | 14" | 77% | 35° E | PM | ** | |

## Mars and the Outer Planets

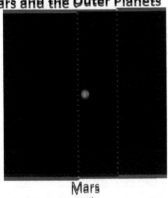

Mars
15th

Jupiter
15th

Saturn
15th

## Mars

| Date | Con. | R.A. | Dec. | Mag. | Diam. | Ill. | Elon. | Vis. | Rat. | Close To |
|------|------|------|------|------|-------|------|-------|------|------|----------|
| 11th | Leo | 10h 40m | 9° 32' | 1.8 | 4" | 99% | 19° E | PM | * | Mercury, Regulus |
| 15th | Leo | 10h 49m | 8° 34' | 1.8 | 4" | 99% | 17° E | PM | * | Mercury |
| 20th | Leo | 11h 1m | 7° 20' | 1.8 | 4" | 99% | 15° E | PM | * | Mercury |

## The Outer Planets

| Planet | Date | Con. | R.A. | Dec. | Mag. | Diam. | Elon. | Vis. | Rat. | Close To |
|--------|------|------|------|------|------|-------|-------|------|------|----------|
| Jupiter | 15th | Aqr | 22h 1m | -13° 20' | -2.9 | 49" | 175° W | AN | ***** | |
| Saturn | 15th | Cap | 20h 47m | -18° 43' | 0.2 | 19" | 167° E | PM | **** | |
| Uranus | 15th | Ari | 2h 49m | 15° 52' | 5.7 | 4" | 103° W | AM | ** | |
| Neptune | 15th | Aqr | 23h 34m | -4° 1' | 7.8 | 2" | 152° W | AM | **** | |

## Highlights

| Date | Time (UT) | Event |
|------|-----------|-------|
| 11th | 05:51 | The waxing crescent Moon is north of Venus. (Evening sky.) |
| | N/A | Good opportunity to see Earthshine on the waxing crescent Moon. (Evening sky.) |
| 13th | 08:16 | The waxing crescent Moon is north of the bright star Spica. (Virgo, evening sky.) |
| | N/A | The Perseid meteor shower is at its maximum. (ZHR: 100) |
| 15th | 15:20 | First Quarter Moon. (Evening sky.) |
| 16th | 18:29 | The just-past first quarter Moon is north of the bright star Antares. (Scorpius, evening sky.) |
| 18th | N/A | The Kappa Cygnid meteor shower is at its maximum. (ZHR: 3) |
| 19th | 04:08 | Mercury is 0.1° south of Mars. (Evening sky.) |
| | 11:14 | The waxing gibbous Moon is south of the dwarf planet Pluto. (Evening sky.) |
| 20th | 00:12 | Uranus is stationary prior to beginning retrograde motion. (Morning sky.) |
| | 09:38 | Jupiter is at opposition. (Visible all night.) |
| | 21:58 | The waxing gibbous Moon is south of Saturn. (Evening sky.) |

# August 21st to 31st, 2021

## The Moon

| 21st | 23rd | 25th | 27th | 29th | 31st |

| Date | Con | R.A. | Dec | Mag | Diam | Ill. | Elon. | Phase | Close To |
|------|-----|------|-----|-----|------|------|-------|-------|----------|
| 21st | Cap | 21h 19m | -20° 33' | -12.4 | 32' | 99% | 169° E | FM | Jupiter, Saturn |
| 22nd | Aqr | 22h 14m | -16° 16' | -12.6 | 32' | 100% | 178° W | FM | Jupiter |
| 23rd | Aqr | 23h 5m | -11° 15' | -12.4 | 31' | 99% | 166° W | FM | Neptune |
| 24th | Aqr | 23h 53m | -5° 49' | -12.1 | 31' | 95% | 155° W | FM | Neptune |
| 25th | Cet | 0h 40m | 0° 17' | -11.8 | 30' | 90% | 144° W | -G | |
| 26th | Psc | 1h 25m | 5° 8' | -11.5 | 30' | 83% | 134° W | -G | |
| 27th | Cet | 2h 10m | 10° 15' | -11.1 | 30' | 75% | 124° W | -G | Uranus |
| 28th | Ari | 2h 55m | 14° 53' | -10.8 | 30' | 67% | 113° W | -G | Uranus |
| 29th | Tau | 3h 42m | 18° 55' | -10.4 | 30' | 58% | 103° W | LQ | Pleiades, Hyades |
| 30th | Tau | 4h 31m | 22° 10' | -10.0 | 30' | 48% | 91° W | LQ | Pleiades, Hyades, Aldebaran |
| 31st | Tau | 5h 21m | 24° 29' | -9.6 | 30' | 39% | 79° W | LQ | Aldebaran |

## Mercury and Venus

Mercury
25th

Venus
25th

**Mercury**

| Date | Con. | R.A. | Dec. | Mag. | Diam. | Ill. | Elon. | Vis. | Rat. | Close To |
|------|------|------|------|------|-------|------|-------|------|------|----------|
| 21st | Leo | 11h 12m | 5° 51' | -0.4 | 5" | 85% | 17° E | PM | ** | Mars |
| 23rd | Leo | 11h 23m | 4° 22' | -0.3 | 5" | 83% | 18° E | PM | ** | Mars |
| 25th | Leo | 11h 34m | 2° 55' | -0.2 | 6" | 81% | 19° E | PM | ** | Mars |
| 27th | Vir | 11h 45m | 1° 28' | -0.1 | 6" | 79% | 20° E | PM | *** | Mars |
| 29th | Vir | 11h 55m | 0° 3' | -0.1 | 6" | 77% | 21° E | PM | *** | Mars |
| 31st | Vir | 12h 5m | -1° 20' | 0.0 | 6" | 75% | 21° E | PM | *** | Mars |

### Venus

| Date | Con. | R.A. | Dec. | Mag. | Diam. | Ill. | Elon. | Vis. | Rat. | Close To |
|------|------|------|------|------|-------|------|-------|------|------|----------|
| 21st | Vir | 12h 23m | -2° 5' | -4.0 | 14" | 76% | 35° E | PM | ** | |
| 23rd | Vir | 12h 32m | -3° 7' | -4.0 | 14" | 76% | 35° E | PM | ** | |
| 25th | Vir | 12h 40m | -4° 8' | -4.0 | 14" | 75% | 36° E | PM | ** | |
| 27th | Vir | 12h 49m | -5° 10' | -4.0 | 15" | 74% | 36° E | PM | ** | Spica |
| 29th | Vir | 12h 57m | -6° 11' | -4.0 | 15" | 74% | 36° E | PM | ** | Spica |
| 31st | Vir | 13h 6m | -7° 11' | -4.0 | 15" | 73% | 37° E | PM | ** | Spica |

## Mars and the Outer Planets

Mars
25th

Jupiter
25th

Saturn
25th

### Mars

| Date | Con. | R.A. | Dec. | Mag. | Diam. | Ill. | Elon. | Vis. | Rat. | Close To |
|------|------|------|------|------|-------|------|-------|------|------|----------|
| 21st | Leo | 11h 3m | 7° 5' | 1.8 | 4" | 99% | 15° E | PM | * | Mercury |
| 25th | Leo | 11h 13m | 6° 4' | 1.8 | 4" | 99% | 14° E | NV | N/A | Mercury |
| 31st | Leo | 11h 27m | 4° 32' | 1.8 | 4" | 100% | 12° E | NV | N/A | Mercury |

### The Outer Planets

| Planet | Date | Con. | R.A. | Dec. | Mag. | Diam. | Elon. | Vis. | Rat. | Close To |
|--------|------|------|------|------|------|-------|-------|------|------|----------|
| Jupiter | 25th | Cap | 21h 56m | -13° 48' | -2.9 | 49" | 175° E | AN | ***** | |
| Saturn | 25th | Cap | 20h 44m | -18° 54' | 0.3 | 18" | 157° E | PM | **** | |
| Uranus | 25th | Ari | 2h 49m | 15° 52' | 5.7 | 4" | 112° W | AM | *** | |
| Neptune | 25th | Aqr | 23h 33m | -4° 7' | 7.8 | 2" | 161° W | AM | **** | |

## Highlights

| Date | Time (UT) | Event |
|------|-----------|-------|
| 22nd | 06:08 | The almost full Moon is south of Jupiter. (Visible all night.) |
| | 12:03 | Full Moon. (Visible all night.) |
| 24th | 02:07 | The waning gibbous Moon is south of Neptune. (Morning sky.) |
| 28th | 10:22 | The waning gibbous Moon is south of Uranus. (Morning sky.) |
| 29th | 15:04 | The almost last quarter Moon is south of the Pleiades star cluster. (Taurus, morning sky.) |
| 30th | 07:14 | Last Quarter Moon. (Morning sky.) |
| | 15:12 | The last quarter Moon is south of the bright star Aldebaran. (Taurus, evening sky.) |

# September 1ˢᵗ to 10ᵗʰ, 2021

## The Moon

| 1ˢᵗ | 3ʳᵈ | 5ᵗʰ | 7ᵗʰ | 9ᵗʰ |

| Date | Con | R.A. | Dec | Mag | Diam | Ill. | Elon. | Phase | Close To |
|------|-----|------|-----|-----|------|------|-------|-------|----------|
| 1st | Gem | 6h 14m | 25° 42' | -9.1 | 30' | 30% | 67° W | -Cr | |
| 2nd | Gem | 7h 8m | 25° 43' | -8.5 | 30' | 21% | 55° W | -Cr | |
| 3rd | Cnc | 8h 3m | 24° 28' | -7.8 | 30' | 14% | 42° W | -Cr | Praesepe |
| 4th | Cnc | 8h 57m | 21° 55' | -6.9 | 31' | 7% | 29° W | NM | Praesepe |
| 5th | Leo | 9h 51m | 18° 13' | -5.9 | 31' | 3% | 17° W | NM | Regulus |
| 6th | Leo | 10h 43m | 13° 29' | -4.8 | 32' | 1% | 4° W | NM | Regulus |
| 7th | Leo | 11h 34m | 7° 59' | -4.7 | 32' | 0% | 7° E | NM | Mars |
| 8th | Vir | 12h 25m | 1° 59' | -5.9 | 32' | 3% | 19° E | NM | Mercury, Mars |
| 9th | Vir | 13h 16m | -4° 13' | -7.0 | 32' | 8% | 31° E | NM | Mercury, Venus, Spica |
| 10th | Vir | 14h 8m | -10° 13' | -7.9 | 32' | 15% | 43° E | +Cr | Venus, Spica |

## Mercury and Venus

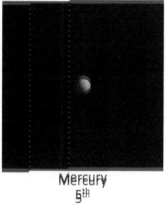 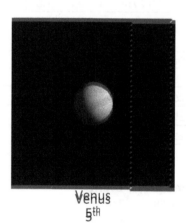

Mercury
5ᵗʰ

Venus
5ᵗʰ

### Mercury

| Date | Con. | R.A. | Dec. | Mag. | Diam. | Ill. | Elon. | Vis. | Rat. | Close To |
|------|------|------|------|------|-------|------|-------|------|------|----------|
| 1st | Vir | 12h 10m | -2° 1' | 0.0 | 6" | 73% | 22° E | PM | *** | |
| 3rd | Vir | 12h 19m | -3° 21' | 0.0 | 6" | 71% | 22° E | PM | *** | |
| 5th | Vir | 12h 28m | -4° 38' | 0.1 | 6" | 69% | 23° E | PM | *** | |
| 7th | Vir | 12h 37m | -5° 53' | 0.1 | 6" | 67% | 23° E | PM | *** | |
| 9th | Vir | 12h 45m | -7° 4' | 0.1 | 7" | 64% | 23° E | PM | *** | Moon, Spica |

## Venus

| Date | Con. | R.A. | Dec. | Mag. | Diam. | Ill. | Elon. | Vis. | Rat. | Close To |
|---|---|---|---|---|---|---|---|---|---|---|
| 1st | Vir | 13h 10m | -7° 41' | -4.0 | 15" | 73% | 37° E | PM | ** | Spica |
| 3rd | Vir | 13h 18m | -8° 41' | -4.0 | 15" | 72% | 37° E | PM | ** | Spica |
| 5th | Vir | 13h 27m | -9° 40' | -4.1 | 16" | 71% | 37° E | PM | ** | Spica |
| 7th | Vir | 13h 36m | -10° 38' | -4.1 | 16" | 71% | 38° E | PM | ** | Spica |
| 9th | Vir | 13h 44m | -11° 36' | -4.1 | 16" | 70% | 38° E | PM | ** | Moon, Spica |

## Mars and the Outer Planets

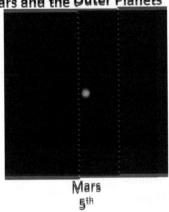

Mars
5th

Jupiter
5th

Saturn
5th

## Mars

| Date | Con. | R.A. | Dec. | Mag. | Diam. | Ill. | Elon. | Vis. | Rat. | Close To |
|---|---|---|---|---|---|---|---|---|---|---|
| 1st | Leo | 11h 29m | 4° 17' | 1.8 | 4" | 100% | 12° E | NV | N/A | |
| 5th | Vir | 11h 38m | 3° 14' | 1.8 | 4" | 100% | 10° E | NV | N/A | |
| 10th | Vir | 11h 50m | 1° 56' | 1.8 | 4" | 100% | 9° E | NV | N/A | |

## The Outer Planets

| Planet | Date | Con. | R.A. | Dec. | Mag. | Diam. | Elon. | Vis. | Rat. | Close To |
|---|---|---|---|---|---|---|---|---|---|---|
| Jupiter | 5th | Cap | 21h 51m | -14° 16' | -2.8 | 49" | 163° E | PM | ***** | |
| Saturn | 5th | Cap | 20h 42m | -19° 5' | 0.3 | 18" | 146° E | PM | **** | |
| Uranus | 5th | Ari | 2h 49m | 15° 50' | 5.7 | 4" | 122° W | AM | *** | |
| Neptune | 5th | Aqr | 23h 32m | -4° 14' | 7.8 | 2" | 171° W | AN | ***** | |

## Highlights

| Date | Time (UT) | Event |
|---|---|---|
| 1st | N/A | The Alpha Aurigid meteor shower is at its maximum. (ZHR: 6) |
| 3rd | N/A | Good opportunity to see Earthshine on the waning crescent Moon. (Morning sky.) |
| 4th | 02:51 | The waning crescent Moon is north of the Praesepe star cluster. (Cancer, morning sky.) |
| | 23:11 | Venus is 1.8° north of the bright star Spica. (Virgo, evening sky.) |
| 5th | 20:36 | The waning crescent Moon is north of the bright star Regulus. (Leo, morning sky.) |
| 7th | 00:52 | New Moon. (Not visible.) |
| 8th | 21:33 | The waxing crescent Moon is north of Mercury. (Evening sky.) |
| 9th | 16:49 | The waxing crescent Moon is north of the bright star Spica. (Virgo, evening sky.) |
| 10th | 02:13 | The waxing crescent Moon is north of Venus. (Evening sky.) |
| | N/A | The Epsilon Perseid meteor shower is at maximum. (ZHR: 5) |
| | N/A | Good opportunity to see Earthshine on the waxing crescent Moon. (Evening sky.) |

# September 11<sup>th</sup> to 20<sup>th</sup>, 2021

## The Moon

11<sup>th</sup>

13<sup>th</sup>

15<sup>th</sup>

17<sup>th</sup>

19<sup>th</sup>

| Date | Con | R.A. | Dec | Mag | Diam | Ill. | Elon. | Phase | Close To |
|------|-----|------|-----|-----|------|------|-------|-------|----------|
| 11th | Lib | 15h 3m | -15° 43' | -8.7 | 32' | 24% | 56° E | +Cr | |
| 12th | Lib | 15h 59m | -20° 20' | -9.4 | 32' | 35% | 69° E | +Cr | Antares |
| 13th | Oph | 16h 59m | -23° 43' | -9.9 | 32' | 46% | 83° E | FQ | Antares |
| 14th | Sgr | 18h 1m | -25° 37' | -10.4 | 32' | 57% | 98° E | FQ | |
| 15th | Sgr | 19h 3m | -25° 53' | -10.9 | 32' | 68% | 112° E | +G | |
| 16th | Sgr | 20h 4m | -24° 31' | -11.2 | 32' | 78% | 127° E | +G | Saturn |
| 17th | Cap | 21h 2m | -21° 43' | -11.6 | 32' | 86% | 140° E | +G | Jupiter, Saturn |
| 18th | Cap | 21h 57m | -17° 46' | -11.9 | 31' | 93% | 153° E | +G | Jupiter |
| 19th | Aqr | 22h 48m | -12° 58' | -12.2 | 31' | 97% | 165° E | FM | Neptune |
| 20th | Aqr | 23h 37m | -7° 40' | -12.5 | 31' | 100% | 176° E | FM | Neptune |

## Mercury and Venus

Mercury
15<sup>th</sup>

Venus
15<sup>th</sup>

**Mercury**

| Date | Con. | R.A. | Dec. | Mag. | Diam. | Ill. | Elon. | Vis. | Rat. | Close To |
|------|------|------|------|------|-------|------|-------|------|------|----------|
| 11th | Vir | 12h 53m | -8° 11' | 0.1 | 7" | 61% | 24° E | PM | *** | Spica |
| 13th | Vir | 13h 0m | -9° 14' | 0.2 | 7" | 58% | 24° E | PM | *** | Spica |
| 15th | Vir | 13h 7m | -10° 12' | 0.2 | 7" | 55% | 23° E | PM | *** | Spica |
| 17th | Vir | 13h 13m | -11° 4' | 0.3 | 7" | 51% | 23° E | PM | *** | Spica |
| 19th | Vir | 13h 18m | -11° 49' | 0.3 | 8" | 47% | 23° E | PM | *** | Spica |

**Venus**

| Date | Con. | R.A. | Dec. | Mag. | Diam. | Ill. | Elon. | Vis. | Rat. | Close To |
|------|------|------|------|------|-------|------|-------|------|------|----------|
| 11th | Vir | 13h 53m | -12° 32' | -4.1 | 16" | 69% | 38° E | PM | ** | Spica |
| 13th | Vir | 14h 1m | -13° 28' | -4.1 | 16" | 69% | 39° E | PM | ** | Spica |
| 15th | Vir | 14h 10m | -14° 22' | -4.1 | 17" | 68% | 39° E | PM | ** | |
| 17th | Vir | 14h 19m | -15° 15' | -4.1 | 17" | 67% | 40° E | PM | ** | |
| 19th | Lib | 14h 28m | -16° 7' | -4.1 | 17" | 67% | 40° E | PM | *** | |

## Mars and the Outer Planets

Mars
15<sup>th</sup>

Jupiter
15<sup>th</sup>

Saturn
15<sup>th</sup>

**Mars**

| Date | Con. | R.A. | Dec. | Mag. | Diam. | Ill. | Elon. | Vis. | Rat. | Close To |
|------|------|------|------|------|-------|------|-------|------|------|----------|
| 11th | Vir | 11h 53m | 1° 40' | 1.8 | 4" | 100% | 8° E | NV | N/A | |
| 15th | Vir | 12h 2m | 0° 37' | 1.7 | 4" | 100% | 7° E | NV | N/A | |
| 20th | Vir | 12h 14m | 0° 42' | 1.7 | 4" | 100% | 6° E | NV | N/A | |

**The Outer Planets**

| Planet | Date | Con. | R.A. | Dec. | Mag. | Diam. | Elon. | Vis. | Rat. | Close To |
|--------|------|------|------|------|------|-------|-------|------|------|----------|
| Jupiter | 15th | Cap | 21h 47m | -14° 38' | -2.8 | 48" | 153° E | PM | ***** | |
| Saturn | 15th | Cap | 20h 40m | -19° 12' | 0.4 | 18" | 137° E | PM | **** | |
| Uranus | 15th | Ari | 2h 48m | 15° 47' | 5.7 | 4" | 131° W | AM | *** | |
| Neptune | 15th | Aqr | 23h 31m | -4° 21' | 7.8 | 2" | 179° E | AN | ***** | |

## Highlights

| Date | Time (UT) | Event |
|------|-----------|-------|
| 13<sup>th</sup> | 00:59 | The almost first quarter Moon is north of the bright star Antares. (Scorpius, evening sky.) |
| | 20:40 | First Quarter Moon. (Evening sky.) |
| 14<sup>th</sup> | 04:16 | Mercury is at greatest eastern elongation from the Sun. (Evening sky.) |
| | 19:26 | Neptune is at opposition. (Visible all night.) |
| 16<sup>th</sup> | 05:15 | The waxing gibbous Moon is south of dwarf planet Pluto. (Evening sky.) |
| 17<sup>th</sup> | 03:46 | The waxing gibbous Moon is south of Saturn. (Evening sky.) |
| 18<sup>th</sup> | 07:37 | The waxing gibbous Moon is south of Jupiter. (Evening sky.) |
| 20<sup>th</sup> | 09:25 | The almost full Moon is south of Neptune. (Visible all night.) |
| | 23:55 | Full Moon. (Visible all night.) |

# September 21st to 30th, 2021

## The Moon

    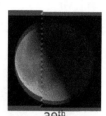

21st  23rd  25th  27th  29th

| Date | Con | R.A. | Dec | Mag | Diam | Ill. | Elon. | Phase | Close To |
|------|-----|------|-----|-----|------|------|-------|-------|----------|
| 21st | Psc | 0h 24m | -2° 7' | -12.5 | 30' | 100% | 173° W | FM | |
| 22nd | Psc | 1h 9m | 3° 24' | -12.3 | 30' | 98% | 162° W | FM | |
| 23rd | Psc | 1h 54m | 8° 42' | -12.0 | 30' | 94% | 152° W | -G | |
| 24th | Ari | 2h 39m | 13° 35' | -11.7 | 30' | 88% | 143° W | -G | Uranus |
| 25th | Tau | 3h 26m | 17° 53' | -11.4 | 30' | 81% | 131° W | -G | Uranus, Pleiades |
| 26th | Tau | 4h 14m | 21° 26' | -11.1 | 30' | 73% | 120° W | -G | Pleiades, Hyades, Aldebaran |
| 27th | Tau | 5h 4m | 24° 4' | -10.7 | 30' | 65% | 108° W | LQ | Hyades, Aldebaran |
| 28th | Tau | 5h 56m | 25° 40' | -10.4 | 30' | 56% | 96° W | LQ | |
| 29th | Gem | 6h 49m | 26° 6' | -9.9 | 30' | 46% | 84° W | LQ | |
| 30th | Gem | 7h 43m | 25° 17' | -9.5 | 30' | 36% | 71° W | -Cr | |

## Mercury and Venus

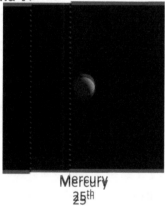 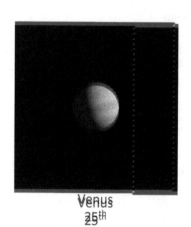

Mercury
25th

Venus
25th

### Mercury

| Date | Con. | R.A. | Dec. | Mag. | Diam. | Ill. | Elon. | Vis. | Rat. | Close To |
|------|------|------|------|------|-------|------|-------|------|------|----------|
| 21st | Vir | 13h 23m | -12° 27' | 0.4 | 8" | 43% | 22° E | PM | *** | Spica |
| 23rd | Vir | 13h 26m | -12° 56' | 0.5 | 8" | 38% | 21° E | PM | *** | Spica |
| 25th | Vir | 13h 28m | -13° 15' | 0.7 | 9" | 33% | 20° E | PM | *** | Spica |
| 27th | Vir | 13h 28m | -13° 21' | 0.9 | 9" | 27% | 18° E | PM | *** | Spica |
| 29th | Vir | 13h 27m | -13° 13' | 1.2 | 9" | 22% | 16° E | PM | ** | Spica |

## Venus

| Date | Con. | R.A. | Dec. | Mag. | Diam. | Ill. | Elon. | Vis. | Rat. | Close To |
|------|------|------|------|------|-------|------|-------|------|------|----------|
| 21st | Lib | 14h 36m | -16° 58' | -4.1 | 17" | 66% | 40° E | PM | *** | |
| 23rd | Lib | 14h 45m | -17° 47' | -4.1 | 18" | 65% | 41° E | PM | *** | |
| 25th | Lib | 14h 54m | -18° 34' | -4.2 | 18" | 64% | 41° E | PM | *** | |
| 27th | Lib | 15h 3m | -19° 20' | -4.2 | 18" | 64% | 42° E | PM | *** | |
| 29th | Lib | 15h 12m | -20° 4' | -4.2 | 19" | 63% | 42° E | PM | *** | |

## Mars and the Outer Planets

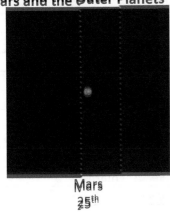

Mars
25th

Jupiter
25th

Saturn
25th

## Mars

| Date | Con. | R.A. | Dec. | Mag. | Diam. | Ill. | Elon. | Vis. | Rat. | Close To |
|------|------|------|------|------|-------|------|-------|------|------|----------|
| 21st | Vir | 12h 16m | 0° 58' | 1.7 | 4" | 100% | 5° E | NV | N/A | |
| 25th | Vir | 12h 26m | -2° 1' | 1.7 | 4" | 100% | 4° E | NV | N/A | |
| 30th | Vir | 12h 38m | -3° 20' | 1.7 | 4" | 100% | 3° E | NV | N/A | |

## The Outer Planets

| Planet | Date | Con. | R.A. | Dec. | Mag. | Diam. | Ill. | Elon. | Vis. | Rat. | Close To |
|--------|------|------|------|------|------|-------|------|-------|------|------|----------|
| Jupiter | 25th | Cap | 21h 43m | -14° 55' | -2.7 | 47" | | 143° E | PM | ***** | |
| Saturn | 25th | Cap | 20h 38m | -19° 18' | 0.4 | 18" | | 127° E | PM | **** | |
| Uranus | 25th | Ari | 2h 47m | 15° 43' | 5.7 | 4" | | 140° W | AM | *** | Moon |
| Neptune | 25th | Aqr | 23h 30m | -4° 27' | 7.8 | 2" | | 170° E | AN | ***** | |

## Highlights

| Date | Time (UT) | Event |
|------|-----------|-------|
| 22nd | 17:00 | Mercury is 1.6° south of the bright star Spica. (Virgo, evening sky.) |
| | 19:22 | Autumn Equinox. |
| 24th | 15:44 | The waning gibbous Moon is south of Uranus. (Morning sky.) |
| 25th | 21:00 | The waning gibbous Moon is south of the Pleiades star cluster. (Taurus, morning sky.) |
| 26th | 21:03 | The waning gibbous Moon is north of the bright star Aldebaran. (Taurus, morning sky.) |
| 27th | 04:07 | Mercury is stationary prior to beginning retrograde motion. (Evening sky.) |
| 29th | 01:58 | Last Quarter Moon. (Morning sky.) |

# October 1ˢᵗ to 10ᵗʰ, 2021

## The Moon

  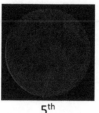  

| | 1ˢᵗ | | 3ʳᵈ | | 5ᵗʰ | | 7ᵗʰ | | 9ᵗʰ |

| Date | Con | R.A. | Dec | Mag | Diam | Ill. | Elon. | Phase | Close To |
|------|-----|------|-----|-----|------|------|-------|-------|----------|
| 1st | Cnc | 8h 37m | 23° 13' | -8.9 | 30' | 27% | 58° W | -Cr | Praesepe |
| 2nd | Leo | 9h 30m | 19° 56' | -8.3 | 31' | 19% | 46° W | -Cr | Regulus |
| 3rd | Leo | 10h 23m | 15° 34' | -7.5 | 31' | 11% | 34° W | NM | Regulus |
| 4th | Leo | 11h 14m | 10° 18' | -6.5 | 32' | 5% | 22° W | NM | |
| 5th | Vir | 12h 5m | 4° 21' | -5.3 | 32' | 1% | 10° W | NM | Mars |
| 6th | Vir | 12h 57m | -1° 57' | -4.3 | 33' | 0% | 2° E | NM | Mercury, Mars, Spica |
| 7th | Vir | 13h 50m | -8° 16' | -5.4 | 33' | 2% | 14° E | NM | Mercury, Spica |
| 8th | Lib | 14h 45m | -14° 11' | -6.6 | 33' | 6% | 27° E | NM | |
| 9th | Lib | 15h 43m | -19° 17' | -7.7 | 33' | 13% | 41° E | +Cr | Venus, Antares |
| 10th | Oph | 16h 43m | -23° 11' | -8.5 | 33' | 21% | 55° E | +Cr | Venus, Antares |

## Mercury and Venus

Mercury
5ᵗʰ

Venus
5ᵗʰ

**Mercury**

| Date | Con. | R.A. | Dec. | Mag. | Diam. | Ill. | Elon. | Vis. | Rat. | Close To |
|------|------|------|------|------|-------|------|-------|------|------|----------|
| 1st | Vir | 13h 24m | -12° 48' | 1.7 | 10" | 16% | 13° E | NV | N/A | Spica |
| 3rd | Vir | 13h 20m | -12° 6' | 2.3 | 10" | 10% | 10° E | NV | N/A | Mars, Spica |
| 5th | Vir | 13h 14m | -11° 6' | 3.1 | 10" | 5% | 7° E | NV | N/A | Mars, Spica |
| 7th | Vir | 13h 6m | -9° 51' | 4.2 | 10" | 2% | 3° E | NV | N/A | Moon, Mars, Spica |
| 9th | Vir | 12h 58m | -8° 24' | 5.2 | 10" | 0% | 0° W | NV | N/A | Mars, Spica |

## Venus

| Date | Con. | R.A. | Dec. | Mag. | Diam. | Ill. | Elon. | Vis. | Rat. | Close To |
|------|------|------|------|------|-------|------|-------|------|------|----------|
| 1st | Lib | 15h 21m | -20° 47' | -4.2 | 19" | 62% | 43° E | PM | *** | |
| 3rd | Lib | 15h 30m | -21° 27' | -4.2 | 19" | 61% | 43° E | PM | *** | |
| 5th | Lib | 15h 39m | -22° 6' | -4.2 | 20" | 60% | 43° E | PM | *** | |
| 7th | Sco | 15h 49m | -22° 42' | -4.2 | 20" | 60% | 44° E | PM | *** | Antares |
| 9th | Sco | 15h 58m | -23° 17' | -4.2 | 20" | 59% | 44° E | PM | *** | Moon, Antares |

## Mars and the Outer Planets

Mars
5th

Jupiter
5th

Saturn
5th

### Mars

| Date | Con. | R.A. | Dec. | Mag. | Diam. | Ill. | Elon. | Vis. | Rat. | Close To |
|------|------|------|------|------|-------|------|-------|------|------|----------|
| 1st | Vir | 12h 40m | -3° 36' | 1.7 | 4" | 100% | 2° E | NV | N/A | |
| 5th | Vir | 12h 50m | -4° 39' | 1.7 | 4" | 100% | 1° E | NV | N/A | Moon, Mercury, Spica |
| 10th | Vir | 13h 2m | -5° 57' | 1.6 | 4" | 100% | 0° W | NV | N/A | Mercury, Spica |

### The Outer Planets

| Planet | Date | Con. | R.A. | Dec. | Mag. | Diam. | Elon. | Vis. | Rat. | Close To |
|--------|------|------|------|------|------|-------|-------|------|------|----------|
| Jupiter | 5th | Cap | 21h 41m | -15° 5' | -2.7 | 46" | 134° E | PM | **** | |
| Saturn | 5th | Cap | 20h 38m | -19° 20' | 0.5 | 18" | 118° E | PM | *** | |
| Uranus | 5th | Ari | 2h 46m | 15° 37' | 5.7 | 4" | 150° W | AM | *** | |
| Neptune | 5th | Aqr | 23h 29m | -4° 33' | 7.8 | 2" | 161° E | PM | ***** | |

## Highlights

| Date | Time (UT) | Event |
|------|-----------|-------|
| 1st | 14:22 | The waning crescent Moon is north of the Praesepe star cluster. (Cancer, morning sky.) |
| 3rd | 03:56 | The waning crescent Moon is north of the bright star Regulus. (Leo, morning sky.) |
| | N/A | Good opportunity to see Earthshine on the waning crescent Moon. (Morning sky.) |
| 6th | 00:42 | Dwarf planet Pluto is stationary prior to resuming prograde motion. (Evening sky.) |
| | 11:06 | New Moon. (Not visible.) |
| 8th | 22:56 | Dwarf planet Ceres is stationary prior to beginning retrograde motion. (Morning sky.) |
| | 23:57 | Mars is in conjunction with the Sun. (Not visible.) |
| | N/A | The Draconid meteor shower is at its maximum. (ZHR: Variable.) |
| 9th | 16:12 | Mercury is at inferior conjunction with the Sun. (Not visible.) |
| | 19:52 | The waxing crescent Moon is north of Venus. (Evening sky.) |
| | N/A | Good opportunity to see Earthshine on the waxing crescent Moon. (Evening sky.) |
| 10th | 05:42 | The waxing crescent Moon is north of the bright star Antares. (Scorpius, evening sky.) |
| | N/A | The Southern Taurid meteor shower is at its maximum. (ZHR: 5) |

# October 11th to 20th, 2021

## The Moon

| 11th | 13th | 15th | 17th | 19th |

| Date | Con | R.A. | Dec | Mag | Diam | Ill. | Elon. | Phase | Close To |
|------|-----|------|-----|-----|------|------|-------|-------|----------|
| 11th | Sgr | 17h 46m | -25° 32' | -9.2 | 32' | 33% | 70° E | +Cr | |
| 12th | Sgr | 18h 49m | -26° 11' | -9.8 | 32' | 43% | 84° E | FQ | |
| 13th | Sgr | 19h 50m | -25° 10' | -10.3 | 32' | 54% | 99° E | FQ | Saturn |
| 14th | Cap | 20h 49m | -22° 40' | -10.7 | 32' | 65% | 113° E | FQ | Saturn |
| 15th | Cap | 21h 44m | -18° 57' | -11.1 | 31' | 75% | 126° E | +G | Jupiter |
| 16th | Aqr | 22h 36m | -14° 23' | -11.5 | 31' | 83% | 137° E | +G | |
| 17th | Aqr | 23h 24m | -9° 13' | -11.8 | 31' | 90% | 149° E | +G | Neptune |
| 18th | Psc | 0h 10m | -3° 46' | -12.1 | 30' | 95% | 159° E | FM | Neptune |
| 19th | Cet | 0h 56m | 1° 46' | -12.4 | 30' | 99% | 170° E | FM | |
| 20th | Psc | 1h 40m | 7° 10' | -12.6 | 30' | 100% | 180° E | FM | |

## Mercury and Venus

Mercury
15th

Venus
15th

### Mercury

| Date | Con. | R.A. | Dec. | Mag. | Diam. | Ill. | Elon. | Vis. | Rat. | Close To |
|------|------|------|------|------|-------|------|-------|------|------|----------|
| 11th | Vir | 12h 51m | -6° 55' | 4.4 | 10" | 1% | 4° W | NV | N/A | Mars, Spica |
| 13th | Vir | 12h 45m | -5° 30' | 3.1 | 10" | 5% | 8° W | NV | N/A | Mars, Spica |
| 15th | Vir | 12h 40m | -4° 19' | 2.0 | 9" | 11% | 11° W | NV | N/A | Mars |
| 17th | Vir | 12h 38m | -3° 28' | 1.1 | 9" | 19% | 13° W | NV | N/A | |
| 19th | Vir | 12h 39m | -2° 59' | 0.4 | 8" | 28% | 15° W | NV | N/A | |

## Venus

| Date | Con. | R.A. | Dec. | Mag. | Diam. | Ill. | Elon. | Vis. | Rat. | Close To |
|------|------|------|------|------|-------|------|-------|------|------|----------|
| 11th | Sco | 16h 7m | -23° 49' | -4.3 | 21" | 58% | 45° E | PM | *** | Antares |
| 13th | Sco | 16h 16m | -24° 19' | -4.3 | 21" | 57% | 45° E | PM | *** | Antares |
| 15th | Oph | 16h 25m | -24° 47' | -4.3 | 21" | 56% | 46° E | PM | *** | Antares |
| 17th | Sco | 16h 35m | -25° 13' | -4.3 | 22" | 55% | 46° E | PM | *** | Antares |
| 19th | Sco | 16h 44m | -25° 36' | -4.3 | 22" | 54% | 47° E | PM | *** | Antares |

## Mars and the Outer Planets

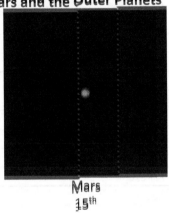

Mars
15th

Jupiter
15th

Saturn
15th

## Mars

| Date | Con. | R.A. | Dec. | Mag. | Diam. | Ill. | Elon. | Vis. | Rat. | Close To |
|------|------|------|------|------|-------|------|-------|------|------|----------|
| 11th | Vir | 13h 4m | -6° 13' | 1.6 | 4" | 100% | 1° W | NV | N/A | Mercury, Spica |
| 15th | Vir | 13h 14m | -7° 15' | 1.6 | 4" | 100% | 2° W | NV | N/A | Mercury, Spica |
| 20th | Vir | 13h 26m | -8° 31' | 1.6 | 4" | 100% | 4° W | NV | N/A | Spica |

## The Outer Planets

| Planet | Date | Con. | R.A. | Dec. | Mag. | Diam. | Elon. | Vis. | Rat. | Close To |
|--------|------|------|------|------|------|-------|-------|------|------|----------|
| Jupiter | 15th | Cap | 21h 40m | -15° 9' | -2.6 | 44" | 124° E | PM | **** | Moon |
| Saturn | 15th | Cap | 20h 38m | -19° 20' | 0.5 | 17" | 109° E | PM | *** | |
| Uranus | 15th | Ari | 2h 45m | 15° 31' | 5.7 | 4" | 159° W | AM | **** | |
| Neptune | 15th | Aqr | 23h 28m | -4° 39' | 7.8 | 2" | 152° E | PM | **** | |

## Highlights

| Date | Time (UT) | Event |
|------|-----------|-------|
| 11th | 01:34 | Saturn is stationary prior to resuming prograde motion. (Evening sky.) |
| 12th | N/A | The Delta Aurigid meteor shower is at its maximum. (ZHR: 2) |
| 13th | 03:26 | First Quarter Moon. (Evening sky.) |
| | 08:48 | The first quarter Moon is south of the dwarf planet Pluto. (Evening sky.) |
| 14th | 07:06 | The just-past first quarter Moon is south of Saturn. (Evening sky.) |
| 15th | 09:32 | The waxing gibbous Moon is south of Jupiter. (Evening sky.) |
| 16th | 06:35 | Venus is 1.5° north of the bright star Antares. (Scorpius, evening sky.) |
| 17th | 12:56 | The waxing gibbous Moon is south of Neptune. (Evening sky.) |
| 18th | 00:53 | Mercury is stationary prior to resuming prograde motion. (Not visible.) |
| | 09:54 | Jupiter is stationary prior to resuming prograde motion. (Evening sky.) |
| | N/A | The Epsilon Geminid meteor shower is at its maximum. (ZHR: 3) |
| 20th | 14:57 | Full Moon. (Visible all night.) |

# October 21st to 31st, 2021

## The Moon

| 21st | 23rd | 25th | 27th | 29th | 31st |

| Date | Con | R.A. | Dec | Mag | Diam | Ill. | Elon. | Phase | Close To |
|------|-----|------|-----|-----|------|------|-------|-------|----------|
| 21st | Ari | 2h 25m | 12° 13' | -12.5 | 30' | 99% | 170° W | FM | Uranus |
| 22nd | Ari | 3h 11m | 16° 44' | -12.2 | 30' | 97% | 159° W | FM | Uranus, Pleiades |
| 23rd | Tau | 3h 59m | 20° 34' | -11.9 | 30' | 93% | 148° W | -G | Pleiades, Hyades, Aldebaran |
| 24th | Tau | 4h 49m | 23° 32' | -11.6 | 29' | 87% | 137° W | -G | Hyades, Aldebaran |
| 25th | Tau | 5h 40m | 25° 28' | -11.3 | 29' | 80% | 125° W | -G | |
| 26th | Gem | 6h 32m | 26° 16' | -11.0 | 30' | 72% | 113° W | -G | |
| 27th | Gem | 7h 26m | 25° 52' | -10.7 | 30' | 63% | 100° W | LQ | |
| 28th | Cnc | 8h 19m | 24° 15' | -10.3 | 30' | 53% | 88° W | LQ | Praesepe |
| 29th | Cnc | 9h 11m | 21° 26' | -9.8 | 30' | 43% | 76° W | LQ | Praesepe |
| 30th | Leo | 10h 2m | 17° 32' | -9.3 | 31' | 33% | 64° W | -Cr | Regulus |
| 31st | Leo | 10h 53m | 12° 41' | -8.7 | 31' | 24% | 53° W | -Cr | Regulus |

## Mercury and Venus

Mercury
25th

Venus
25th

**Mercury**

| Date | Con. | R.A. | Dec. | Mag. | Diam. | Ill. | Elon. | Vis. | Rat. | Close To |
|------|------|------|------|------|-------|------|-------|------|------|----------|
| 21st | Vir | 12h 42m | -2° 53' | 0.0 | 8" | 38% | 16° W | AM | ** | |
| 23rd | Vir | 12h 47m | -3° 8' | -0.3 | 7" | 48% | 16° W | AM | ** | Spica |
| 25th | Vir | 12h 54m | -3° 40' | -0.6 | 7" | 57% | 16° W | AM | ** | Spica |
| 27th | Vir | 13h 3m | -4° 27' | -0.7 | 6" | 65% | 16° W | AM | ** | Spica |
| 29th | Vir | 13h 12m | -5° 25' | -0.8 | 6" | 72% | 16° W | AM | ** | Mars, Spica |
| 31st | Vir | 13h 23m | -6° 30' | -0.8 | 6" | 78% | 15° W | AM | ** | Mars, Spica |

## Venus

| Date | Con. | R.A. | Dec. | Mag. | Diam. | Ill. | Elon. | Vis. | Rat. | Close To |
|------|------|------|------|------|-------|------|-------|------|------|----------|
| 21st | Sco | 16h 53m | -25° 56' | -4.3 | 23" | 53% | 47° E | PM | *** | Antares |
| 23rd | Oph | 17h 2m | -26° 15' | -4.4 | 23" | 53% | 47° E | PM | *** | Antares |
| 25th | Oph | 17h 11m | -26° 30' | -4.4 | 24" | 52% | 48° E | PM | *** | |
| 27th | Oph | 17h 20m | -26° 44' | -4.4 | 24" | 51% | 48° E | PM | **** | |
| 29th | Oph | 17h 29m | -26° 55' | -4.4 | 25" | 50% | 48° E | PM | **** | |
| 31st | Oph | 17h 38m | -27° 3' | -4.4 | 25" | 49% | 49° E | PM | **** | |

## Mars and the Outer Planets

Mars
25th

Jupiter
25th

Saturn
25th

### Mars

| Date | Con. | R.A. | Dec. | Mag. | Diam. | Ill. | Elon. | Vis. | Rat. | Close To |
|------|------|------|------|------|-------|------|-------|------|------|----------|
| 21st | Vir | 13h 29m | -8° 46' | 1.6 | 4" | 100% | 4° W | NV | N/A | Spica |
| 25th | Vir | 13h 39m | -9° 46' | 1.7 | 4" | 100% | 5° W | NV | N/A | Spica |
| 31st | Vir | 13h 54m | -11° 15' | 1.7 | 4" | 100% | 7° W | NV | N/A | Mercury, Spica |

### The Outer Planets

| Planet | Date | Con. | R.A. | Dec. | Mag. | Diam. | Elon. | Vis. | Rat. | Close To |
|--------|------|------|------|------|------|-------|-------|------|------|----------|
| Jupiter | 25th | Cap | 21h 40m | -15° 6' | -2.5 | 43" | 115° E | PM | **** | |
| Saturn | 25th | Cap | 20h 38m | -19° 18' | 0.6 | 17" | 100° E | PM | *** | |
| Uranus | 25th | Ari | 2h 43m | 15° 24' | 5.7 | 4" | 169° W | AM | **** | |
| Neptune | 25th | Aqr | 23h 27m | -4° 44' | 7.8 | 2" | 142° E | PM | **** | |

## Highlights

| Date | Time (UT) | Event |
|------|-----------|-------|
| 21st | 20:27 | The just-past full Moon is south of Uranus. (Morning sky.) |
| | N/A | The Orionid meteor shower is at its maximum. (ZHR: 25) |
| 23rd | 07:07 | The waning gibbous Moon is south of the Pleiades star cluster. (Taurus, morning sky.) |
| 24th | 06:56 | The waning gibbous Moon is north of the bright star Aldebaran. (Taurus, morning sky.) |
| | N/A | The Leo Minorid meteor shower is at its maximum. (ZHR: 2) |
| 25th | 05:18 | Mercury is at greatest western elongation from the Sun. (Morning sky.) |
| 28th | 20:06 | Last Quarter Moon. (Morning sky.) |
| | 20:42 | The last quarter Moon is north of the Praesepe star cluster. (Cancer, morning sky.) |
| 29th | 20:38 | Venus is at greatest eastern elongation from the Sun. (Evening sky.) |
| 30th | 15:41 | The waning crescent Moon is north of the bright star Regulus. (Leo, morning sky.) |
| 31st | 20:32 | Mercury is 4.5° north of the bright star Spica. (Virgo, morning sky.) |

# November 1st to 10th, 2021

## The Moon

| | 1st | | 3rd | | 5th | | 7th | | 9th |

| Date | Con | R.A. | Dec | Mag | Diam | Ill. | Elon. | Phase | Close To |
|------|-----|------|-----|-----|------|------|-------|-------|----------|
| 1st | Vir | 11h 43m | -7° 4' | -8.0 | 32' | 15% | 41° W | -Cr | |
| 2nd | Vir | 12h 34m | 0° 55' | -7.0 | 32' | 8% | 29° W | NM | |
| 3rd | Vir | 13h 26m | -5° 29' | -5.9 | 33' | 3% | 17° W | NM | Mercury, Mars, Spica |
| 4th | Vir | 14h 21m | -11° 43' | -4.5 | 33' | 0% | 5° W | NM | Mercury, Mars |
| 5th | Lib | 15h 18m | -17° 22' | -4.8 | 33' | 1% | 9° E | NM | |
| 6th | Sco | 16h 20m | -21° 57' | -6.2 | 33' | 4% | 23° E | NM | Antares |
| 7th | Oph | 17h 24m | -25° 1' | -7.3 | 33' | 10% | 38° E | NM | Venus |
| 8th | Sgr | 18h 29m | -26° 18' | -8.2 | 33' | 18% | 54° E | +Cr | Venus |
| 9th | Sgr | 19h 34m | -25° 45' | -9.0 | 32' | 28% | 69° E | +Cr | |
| 10th | Cap | 20h 35m | -23° 33' | -9.6 | 32' | 39% | 83° E | FQ | Saturn |

## Mercury and Venus

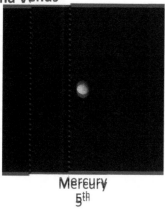

Mercury
5th

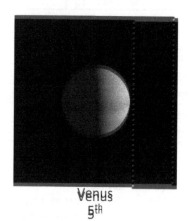

Venus
5th

### Mercury

| Date | Con. | R.A. | Dec. | Mag. | Diam. | Ill. | Elon. | Vis. | Rat. | Close To |
|------|------|------|------|------|-------|------|-------|------|------|----------|
| 1st | Vir | 13h 28m | -7° 5' | -0.8 | 6" | 80% | 15° W | NV | N/A | Mars, Spica |
| 3rd | Vir | 13h 39m | -8° 17' | -0.9 | 6" | 85% | 14° W | NV | N/A | Moon, Mars, Spica |
| 5th | Vir | 13h 51m | -9° 32' | -0.9 | 5" | 88% | 13° W | NV | N/A | Mars, Spica |
| 7th | Vir | 14h 3m | -10° 48' | -0.9 | 5" | 91% | 12° W | NV | N/A | Mars, Spica |
| 9th | Vir | 14h 15m | -12° 3' | -0.9 | 5" | 93% | 11° W | NV | N/A | Mars |

## Venus

| Date | Con. | R.A. | Dec. | Mag. | Diam. | Ill. | Elon. | Vis. | Rat. | Close To |
|------|------|------|------|------|-------|------|-------|------|------|----------|
| 1st | Oph | 17h 42m | -27° 7' | -4.4 | 26" | 48% | 49° E | PM | **** | |
| 3rd | Sgr | 17h 51m | -27° 12' | -4.4 | 26" | 47% | 49° E | PM | **** | |
| 5th | Sgr | 18h 0m | -27° 14' | -4.5 | 27" | 46% | 49° E | PM | **** | |
| 7th | Sgr | 18h 8m | -27° 14' | -4.5 | 28" | 45% | 49° E | PM | **** | Moon |
| 9th | Sgr | 18h 16m | -27° 12' | -4.5 | 28" | 44% | 49° E | PM | **** | |

## Mars and the Outer Planets

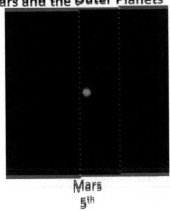

Mars
5th

Jupiter
5th

Saturn
5th

## Mars

| Date | Con. | R.A. | Dec. | Mag. | Diam. | Ill. | Elon. | Vis. | Rat. | Close To |
|------|------|------|------|------|-------|------|-------|------|------|----------|
| 1st | Vir | 13h 57m | -11° 29' | 1.7 | 4" | 100% | 8° W | NV | N/A | Mercury, Spica |
| 5th | Vir | 14h 7m | -12° 27' | 1.6 | 4" | 100% | 9° W | NV | N/A | Mercury |
| 10th | Vir | 14h 20m | -13° 36' | 1.6 | 4" | 100% | 11° W | NV | N/A | Mercury |

## The Outer Planets

| Planet | Date | Con. | R.A. | Dec. | Mag. | Diam. | Elon. | Vis. | Rat. | Close To |
|--------|------|------|------|------|------|-------|-------|------|------|----------|
| Jupiter | 5th | Cap | 21h 42m | -14° 56' | -2.5 | 42" | 105° E | PM | *** | |
| Saturn | 5th | Cap | 20h 40m | -19° 12' | 0.6 | 17" | 89° E | PM | *** | |
| Uranus | 5th | Ari | 2h 41m | 15° 16' | 5.6 | 4" | 180° E | AN | **** | |
| Neptune | 5th | Aqr | 23h 27m | -4° 48' | 7.8 | 2" | 131° E | PM | **** | |

## Highlights

| Date | Time (UT) | Event |
|------|-----------|-------|
| 1st | N/A | Good opportunity to see Earthshine on the waning crescent Moon. (Morning sky.) |
| 4th | 21:15 | New Moon. (Not visible.) |
| 5th | 04:31 | Uranus is at opposition. (Visible all night.) |
| 8th | 04:49 | The waxing crescent Moon is north of Venus. (Evening sky.) |
| | N/A | Good opportunity to see Earthshine on the waxing crescent Moon. (Evening sky.) |
| 9th | 16:35 | The waxing crescent Moon is south of the dwarf planet Pluto. (Evening sky.) |
| 10th | 13:16 | The nearly first quarter Moon is south of Saturn. (Evening sky.) |

# November 11<sup>th</sup> to 20<sup>th</sup>, 2021

## The Moon

11<sup>th</sup>

13<sup>th</sup>

15<sup>th</sup>

17<sup>th</sup>

19<sup>th</sup>

| Date | Con | R.A. | Dec | Mag | Diam | Ill. | Elon. | Phase | Close To |
|------|-----|------|-----|-----|------|------|-------|-------|----------|
| 11th | Cap | 21h 32m | -20° 3' | -10.1 | 32' | 50% | 96° E | FQ | Jupiter |
| 12th | Aqr | 22h 24m | -15° 36' | -10.6 | 31' | 60% | 108° E | FQ | Jupiter |
| 13th | Aqr | 23h 13m | -10° 32' | -10.9 | 31' | 70% | 120° E | +G | Neptune |
| 14th | Psc | 0h 0m | -5° 9' | -11.3 | 30' | 79% | 130° E | +G | Neptune |
| 15th | Cet | 0h 45m | 0° 21' | -11.6 | 30' | 87% | 140° E | +G | |
| 16th | Psc | 1h 29m | 5° 46' | -11.9 | 30' | 93% | 150° E | +G | |
| 17th | Ari | 2h 13m | 10° 53' | -12.2 | 30' | 97% | 160° E | FM | Uranus |
| 18th | Ari | 2h 59m | 15° 34' | -12.5 | 30' | 99% | 171° E | FM | Uranus, Pleiades |
| 19th | Tau | 3h 46m | 19° 37' | -12.7 | 29' | 100% | 179° W | FM | Pleiades, Hyades |
| 20th | Tau | 4h 35m | 22° 50' | -12.4 | 29' | 99% | 167° W | FM | Hyades, Aldebaran |

## Mercury and Venus

Mercury
15<sup>th</sup>

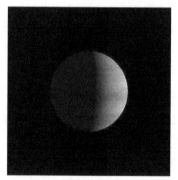
Venus
15<sup>th</sup>

### Mercury

| Date | Con. | R.A. | Dec. | Mag. | Diam. | Ill. | Elon. | Vis. | Rat. | Close To |
|------|------|------|------|------|-------|------|-------|------|------|----------|
| 11th | Lib | 14h 27m | -13° 18' | -0.9 | 5" | 95% | 10° W | NV | N/A | Mars |
| 13th | Lib | 14h 40m | -14° 30' | -0.9 | 5" | 96% | 9° W | NV | N/A | Mars |
| 15th | Lib | 14h 52m | -15° 40' | -0.9 | 5" | 97% | 8° W | NV | N/A | Mars |
| 17th | Lib | 15h 5m | -16° 48' | -1.0 | 5" | 98% | 7° W | NV | N/A | Mars |
| 19th | Lib | 15h 17m | -17° 52' | -1.0 | 5" | 99% | 6° W | NV | N/A | Mars |

## Venus

| Date | Con. | R.A. | Dec. | Mag. | Diam. | Ill. | Elon. | Vis. | Rat. | Close To |
|------|------|------|------|------|-------|------|-------|------|------|----------|
| 11th | Sgr | 18h 25m | -27° 8' | -4.5 | 29" | 42% | 49° E | PM | **** | |
| 13th | Sgr | 18h 33m | -27° 1' | -4.5 | 30" | 41% | 49° E | PM | **** | |
| 15th | Sgr | 18h 40m | -26° 53' | -4.5 | 31" | 40% | 49° E | PM | **** | |
| 17th | Sgr | 18h 48m | -26° 42' | -4.6 | 32" | 39% | 49° E | PM | **** | |
| 19th | Sgr | 18h 55m | -26° 30' | -4.6 | 33" | 37% | 49° E | PM | **** | |

## Mars and the Outer Planets

Mars
15<sup>th</sup>

Jupiter
15<sup>th</sup>

Saturn
15<sup>th</sup>

### Mars

| Date | Con. | R.A. | Dec. | Mag. | Diam. | Ill. | Elon. | Vis. | Rat. | Close To |
|------|------|------|------|------|-------|------|-------|------|------|----------|
| 11th | Lib | 14h 23m | -13° 50' | 1.6 | 4" | 100% | 11° W | NV | N/A | Mercury |
| 15th | Lib | 14h 33m | -14° 44' | 1.6 | 4" | 100% | 13° W | NV | N/A | Mercury |
| 20th | Lib | 14h 47m | -15° 49' | 1.6 | 4" | 99% | 14° W | NV | N/A | Mercury |

### The Outer Planets

| Planet | Date | Con. | R.A. | Dec. | Mag. | Diam. | Elon. | Vis. | Rat. | Close To |
|--------|------|------|------|------|------|-------|-------|------|------|----------|
| Jupiter | 15th | Cap | 21h 45m | -14° 40' | -2.4 | 40" | 95° E | PM | *** | |
| Saturn | 15th | Cap | 20h 42m | -19° 5' | 0.7 | 16" | 80° E | PM | ** | |
| Uranus | 15th | Ari | 2h 40m | 15° 9' | 5.7 | 4" | 169° E | PM | **** | |
| Neptune | 15th | Aqr | 23h 26m | -4° 50' | 7.9 | 2" | 121° E | PM | **** | |

## Highlights

| Date | Time (UT) | Event |
|------|-----------|-------|
| 11<sup>th</sup> | 12:47 | First Quarter Moon. (Evening sky.) |
| | 16:39 | The first quarter Moon is south of Jupiter. (Evening sky.) |
| 12<sup>th</sup> | N/A | The Northern Taurid meteor shower is at its maximum. (ZHR: 5) |
| 13<sup>th</sup> | 17:52 | The waxing gibbous Moon is south of Neptune. (Evening sky.) |
| 18<sup>th</sup> | 03:08 | The nearly full Moon is south of Uranus. (Evening sky.) |
| | N/A | The Leonid meteor shower is at its maximum. (ZHR: 20) |
| 19<sup>th</sup> | 08:58 | Full Moon. (Visible all night.) |
| | 09:03 | Partial lunar eclipse. Visible from the Arctic, Asia, the northern Atlantic, Australia, far western Europe, North America, the Pacific, and South America. |
| | 12:14 | The full Moon is south of the Pleiades star cluster. (Taurus, evening sky.) |
| 20<sup>th</sup> | 12:23 | The just-past full Moon is north of the bright star Aldebaran. (Taurus, evening sky.) |

# November 21st to 30th, 2021

## The Moon

| 21st | 23rd | 25th | 27th | 29th |

| Date | Con | R.A. | Dec | Mag | Diam | Ill. | Elon. | Phase | Close To |
|------|-----|------|-----|-----|------|------|-------|-------|----------|
| 21st | Tau | 5h 26m | 25° 5' | -12.1 | 29' | 96% | 156° W | FM | |
| 22nd | Gem | 6h 18m | 26° 13' | -11.8 | 29' | 92% | 144° W | -G | |
| 23rd | Gem | 7h 11m | 26° 9' | -11.6 | 30' | 86% | 131° W | -G | |
| 24th | Cnc | 8h 4m | 24° 53' | -11.2 | 30' | 78% | 119° W | -G | Praesepe |
| 25th | Cnc | 8h 56m | 22° 27' | -10.9 | 30' | 70% | 107° W | -G | Praesepe |
| 26th | Leo | 9h 46m | 18° 56' | -10.5 | 30' | 60% | 96° W | LQ | Regulus |
| 27th | Leo | 10h 36m | 14° 31' | -10.1 | 31' | 50% | 84° W | LQ | Regulus |
| 28th | Leo | 11h 24m | 9° 18' | -9.7 | 31' | 40% | 73° W | LQ | |
| 29th | Vir | 12h 13m | 3° 31' | -9.1 | 32' | 30% | 62° W | -Cr | |
| 30th | Vir | 13h 3m | -2° 39' | -8.4 | 32' | 20% | 51° W | -Cr | Spica |

## Mercury and Venus

 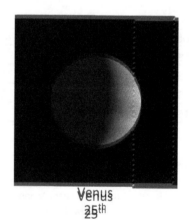

Mercury
25th

Venus
25th

### Mercury

| Date | Con. | R.A. | Dec. | Mag. | Diam. | Ill. | Elon. | Vis. | Rat. | Close To |
|------|------|------|------|------|-------|------|-------|------|------|----------|
| 21st | Lib | 15h 30m | -18° 54' | -1.0 | 5" | 99% | 5° W | NV | N/A | |
| 23rd | Lib | 15h 43m | -19° 51' | -1.1 | 5" | 100% | 3° W | NV | N/A | |
| 25th | Sco | 15h 56m | -20° 45' | -1.1 | 5" | 100% | 2° W | NV | N/A | Antares |
| 27th | Sco | 16h 9m | -21° 34' | -1.2 | 5" | 100% | 1° W | NV | N/A | Antares |
| 29th | Sco | 16h 22m | -22° 19' | -1.2 | 5" | 100% | 0° E | NV | N/A | Antares |

## Venus

| Date | Con. | R.A. | Dec. | Mag. | Diam. | Ill. | Elon. | Vis. | Rat. | Close To |
|------|------|------|------|------|-------|------|-------|------|------|----------|
| 21st | Sgr | 19h 2m | -26° 16' | -4.6 | 34" | 36% | 48° E | PM | **** | |
| 23rd | Sgr | 19h 8m | -25° 60' | -4.6 | 35" | 34% | 48° E | PM | **** | |
| 25th | Sgr | 19h 15m | -25° 42' | -4.6 | 36" | 33% | 47° E | PM | **** | |
| 27th | Sgr | 19h 21m | -25° 23' | -4.6 | 37" | 31% | 47° E | PM | **** | |
| 29th | Sgr | 19h 26m | -25° 3' | -4.6 | 38" | 30% | 46° E | PM | **** | |

## Mars and the Outer Planets

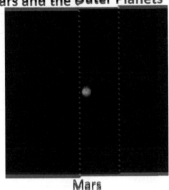

Mars
25th

Jupiter
25th

Saturn
25th

## Mars

| Date | Con. | R.A. | Dec. | Mag. | Diam. | Ill. | Elon. | Vis. | Rat. | Close To |
|------|------|------|------|------|-------|------|-------|------|------|----------|
| 21st | Lib | 14h 49m | -16° 1' | 1.6 | 4" | 99% | 15° W | NV | N/A | |
| 25th | Lib | 15h 0m | -16° 51' | 1.6 | 4" | 99% | 16° W | AM | * | |
| 30th | Lib | 15h 14m | -17° 50' | 1.6 | 4" | 99% | 18° W | AM | * | |

## The Outer Planets

| Planet | Date | Con. | R.A. | Dec. | Mag. | Diam. | Elon. | Vis. | Rat. | Close To |
|--------|------|------|------|------|------|-------|-------|------|------|----------|
| Jupiter | 25th | Cap | 21h 49m | -14° 17' | -2.3 | 39" | 86° E | PM | *** | |
| Saturn | 25th | Cap | 20h 44m | -18° 55' | 0.7 | 16" | 70° E | PM | ** | |
| Uranus | 25th | Ari | 2h 38m | 15° 2' | 5.7 | 4" | 158° E | PM | **** | |
| Neptune | 25th | Aqr | 23h 26m | -4° 52' | 7.9 | 2" | 110° E | PM | *** | |

## Highlights

| Date | Time (UT) | Event |
|------|-----------|-------|
| 22nd | N/A | The Alpha Monocerotid meteor shower is at its maximum. (ZHR: Variable.) |
| 25th | 04:35 | The waning gibbous Moon is north of the Praesepe star cluster. (Cancer, morning sky.) |
| 26th | 21:24 | The nearly last quarter Moon is north of the bright star Regulus. (Leo, morning sky.) |
| 27th | 12:28 | Last Quarter Moon. (Morning sky.) |
| | 22:38 | Dwarf planet Ceres is at opposition. (Visible all night.) |
| 29th | 02:47 | Asteroid Vesta is in conjunction with the Sun. (Not visible.) |
| | 04:25 | Mercury is at superior conjunction with the Sun. (Not visible.) |
| 30th | 21:44 | The waning crescent Moon is north of the bright star Spica. (Virgo, morning sky.) |

# December 1st to 10th, 2021

## The Moon

 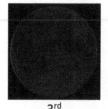   

| 1st | 3rd | 5th | 7th | 9th |

| Date | Con | R.A. | Dec | Mag | Diam | Ill. | Elon. | Phase | Close To |
|------|-----|------|-----|-----|------|------|-------|-------|----------|
| 1st | Vir | 13h 55m | -8° 53' | -7.5 | 33' | 11% | 39° W | NM | Spica |
| 2nd | Lib | 14h 51m | -14° 48' | -6.5 | 33' | 5% | 26° W | NM | Mars |
| 3rd | Lib | 15h 50m | -19° 57' | -5.1 | 33' | 1% | 12° W | NM | Mars, Antares |
| 4th | Oph | 16h 54m | -23° 49' | -4.2 | 33' | 0% | 3° E | NM | Mercury, Antares |
| 5th | Sgr | 18h 1m | -25° 58' | -5.7 | 33' | 2% | 18° E | NM | |
| 6th | Sgr | 19h 8m | -26° 11' | -6.9 | 33' | 7% | 34° E | NM | Venus |
| 7th | Cap | 20h 13m | -24° 31' | -7.9 | 33' | 14% | 49° E | +Cr | Venus, Saturn |
| 8th | Cap | 21h 14m | -21° 19' | -8.7 | 32' | 23% | 63° E | +Cr | Jupiter, Saturn |
| 9th | Aqr | 22h 9m | -16° 59' | -9.3 | 32' | 34% | 76° E | +Cr | Jupiter |
| 10th | Aqr | 23h 0m | -11° 56' | -9.9 | 31' | 44% | 88° E | FQ | Neptune |

## Mercury and Venus

Mercury
5th

Venus
5th

### Mercury

| Date | Con. | R.A. | Dec. | Mag. | Diam. | Ill. | Elon. | Vis. | Rat. | Close To |
|------|------|------|------|------|-------|------|-------|------|------|----------|
| 1st | Oph | 16h 36m | -23° 0' | -1.1 | 5" | 100% | 1° E | NV | N/A | Antares |
| 3rd | Oph | 16h 49m | -23° 36' | -1.1 | 5" | 100% | 2° E | NV | N/A | Antares |
| 5th | Oph | 17h 3m | -24° 8' | -1.0 | 5" | 100% | 4° E | NV | N/A | Antares |
| 7th | Oph | 17h 16m | -24° 34' | -0.9 | 5" | 99% | 5° E | NV | N/A | |
| 9th | Oph | 17h 30m | -24° 55' | -0.9 | 5" | 99% | 6° E | NV | N/A | |

## Venus

| Date | Con. | R.A. | Dec. | Mag. | Diam. | Ill. | Elon. | Vis. | Rat. | Close To |
|------|------|------|------|------|-------|------|-------|------|------|----------|
| 1st | Sgr | 19h 31m | -24° 42' | -4.7 | 39" | 28% | 45° E | PM | **** | |
| 3rd | Sgr | 19h 36m | -24° 20' | -4.7 | 40" | 27% | 44° E | PM | **** | |
| 5th | Sgr | 19h 40m | -23° 56' | -4.7 | 42" | 25% | 43° E | PM | **** | |
| 7th | Sgr | 19h 44m | -23° 32' | -4.7 | 43" | 23% | 42° E | PM | **** | Moon |
| 9th | Sgr | 19h 47m | -23° 8' | -4.7 | 45" | 21% | 40° E | PM | **** | |

## Mars and the Outer Planets

Mars
5th

Jupiter
5th

Saturn
5th

### Mars

| Date | Con. | R.A. | Dec. | Mag. | Diam. | Ill. | Elon. | Vis. | Rat. | Close To |
|------|------|------|------|------|-------|------|-------|------|------|----------|
| 1st | Lib | 15h 17m | -18° 1' | 1.6 | 4" | 99% | 18° W | AM | * | |
| 5th | Lib | 15h 28m | -18° 45' | 1.6 | 4" | 99% | 20° W | AM | * | |
| 10th | Lib | 15h 42m | -19° 37' | 1.6 | 4" | 99% | 22° W | AM | * | |

### The Outer Planets

| Planet | Date | Con. | R.A. | Dec. | Mag. | Diam. | Elon. | Vis. | Rat. | Close To |
|--------|------|------|------|------|------|-------|-------|------|------|----------|
| Jupiter | 5th | Cap | 21h 54m | -13° 50' | -2.3 | 38" | 76° E | PM | *** | |
| Saturn | 5th | Cap | 20h 47m | -18° 42' | 0.7 | 16" | 60° E | PM | ** | |
| Uranus | 5th | Ari | 2h 37m | 14° 55' | 5.7 | 4" | 147° E | PM | *** | |
| Neptune | 5th | Aqr | 23h 26m | -4° 51' | 7.9 | 2" | 99° E | PM | *** | |

## Highlights

| Date | Time (UT) | Event |
|------|-----------|-------|
| 1st | 18:14 | Neptune is stationary prior to resuming prograde motion. (Evening sky.) |
| | N/A | Good opportunity to see Earthshine on the waning crescent Moon. (Morning sky.) |
| 3rd | 00:00 | The nearly new Moon is north of Mars. (Morning sky.) |
| 4th | 07:35 | Total solar eclipse. Visible from Antarctica and the southern Atlantic. |
| | 07:44 | New Moon. (Not visible.) |
| 7th | 01:17 | The waxing crescent Moon is south of Venus. (Evening sky.) |
| | N/A | Good opportunity to see Earthshine on the waxing crescent Moon. (Evening sky.) |
| 8th | 02:17 | The waxing crescent Moon is south of Saturn. (Evening sky.) |
| 9th | 05:47 | The waxing crescent Moon is south of Jupiter. (Evening sky.) |
| | N/A | The Monocerotid meteor shower is at its maximum. (ZHR: 2) |

# December 11th to 20th, 2021

## The Moon

   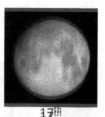 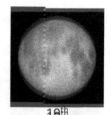

| 11th | 13th | 15th | 17th | 19th |

| Date | Con | R.A. | Dec | Mag | Diam | Ill. | Elon. | Phase | Close To |
|------|-----|------|-----|-----|------|------|-------|-------|----------|
| 11th | Aqr | 23h 48m | -6° 31' | -10.3 | 31' | 54% | 98° E | FQ | Neptune |
| 12th | Cet | 0h 34m | 0° 58' | -10.7 | 30' | 64% | 109° E | FQ | |
| 13th | Psc | 1h 18m | 4° 29' | -11.1 | 30' | 74% | 119° E | +G | |
| 14th | Psc | 2h 2m | 9° 41' | -11.4 | 30' | 82% | 129° E | +G | Uranus |
| 15th | Ari | 2h 47m | 14° 27' | -11.7 | 30' | 88% | 139° E | +G | Uranus |
| 16th | Tau | 3h 34m | 18° 39' | -12.0 | 29' | 94% | 149° E | +G | Pleiades |
| 17th | Tau | 4h 22m | 22° 5' | -12.3 | 29' | 97% | 160° E | FM | Pleiades, Hyades, Aldebaran |
| 18th | Tau | 5h 12m | 24° 36' | -12.5 | 29' | 100% | 172° E | FM | Hyades, Aldebaran |
| 19th | Gem | 6h 5m | 26° 1' | -12.6 | 29' | 100% | 176° W | FM | |
| 20th | Gem | 6h 58m | 26° 15' | -12.3 | 30' | 98% | 164° W | FM | |

## Mercury and Venus

 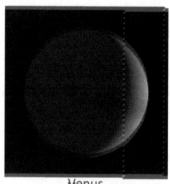

Mercury
15th

Venus
15th

### Mercury

| Date | Con. | R.A. | Dec. | Mag. | Diam. | Ill. | Elon. | Vis. | Rat. | Close To |
|------|------|------|------|------|-------|------|-------|------|------|----------|
| 11th | Oph | 17h 44m | -25° 11' | -0.8 | 5" | 98% | 7° E | NV | N/A | |
| 13th | Sgr | 17h 58m | -25° 22' | -0.8 | 5" | 98% | 9° E | NV | N/A | |
| 15th | Sgr | 18h 12m | -25° 26' | -0.8 | 5" | 97% | 10° E | NV | N/A | |
| 17th | Sgr | 18h 26m | -25° 25' | -0.8 | 5" | 96% | 11° E | NV | N/A | |
| 19th | Sgr | 18h 40m | -25° 18' | -0.7 | 5" | 95% | 12° E | NV | N/A | |

## Venus

| Date | Con. | R.A. | Dec. | Mag. | Diam. | Ill. | Elon. | Vis. | Rat. | Close To |
|------|------|------|------|------|-------|------|-------|------|------|----------|
| 11th | Sgr | 19h 50m | -22° 43' | -4.7 | 46" | 20% | 39° E | PM | **** | |
| 13th | Sgr | 19h 52m | -22° 18' | -4.7 | 48" | 18% | 37° E | PM | **** | |
| 15th | Sgr | 19h 53m | -21° 52' | -4.6 | 49" | 16% | 35° E | PM | **** | |
| 17th | Sgr | 19h 54m | -21° 27' | -4.6 | 51" | 14% | 33° E | PM | **** | |
| 19th | Sgr | 19h 54m | -21° 2' | -4.6 | 52" | 12% | 31° E | PM | **** | |

## Mars and the Outer Planets

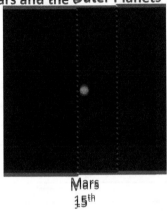

Mars
15th

Jupiter
15th

Saturn
15th

### Mars

| Date | Con. | R.A. | Dec. | Mag. | Diam. | Ill. | Elon. | Vis. | Rat. | Close To |
|------|------|------|------|------|-------|------|-------|------|------|----------|
| 11th | Lib | 15h 45m | -19° 47' | 1.6 | 4" | 99% | 22° W | AM | * | |
| 15th | Lib | 15h 57m | -20° 25' | 1.6 | 4" | 99% | 24° W | AM | * | Antares |
| 20th | Sco | 16h 12m | -21° 8' | 1.6 | 4" | 98% | 26° W | AM | * | Antares |

### The Outer Planets

| Planet | Date | Con. | R.A. | Dec. | Mag. | Diam. | Elon. | Vis. | Rat. | Close To |
|--------|------|------|------|------|------|-------|-------|------|------|----------|
| Jupiter | 15th | Aqr | 22h 0m | -13° 17' | -2.3 | 37" | 67° E | PM | ** | |
| Saturn | 15th | Cap | 20h 51m | -18° 28' | 0.7 | 16" | 50° E | PM | ** | |
| Uranus | 15th | Ari | 2h 36m | 14° 50' | 5.7 | 4" | 136° E | PM | *** | Moon |
| Neptune | 15th | Aqr | 23h 26m | -4° 50' | 7.9 | 2" | 89° E | PM | *** | |

## Highlights

| Date | Time (UT) | Event |
|------|-----------|-------|
| 11th | 01:36 | First Quarter Moon. (Evening sky.) |
| | 01:53 | The first quarter Moon is south of Neptune. (Evening sky.) |
| 12th | N/A | The Sigma Hydrid meteor shower is at its maximum. (ZHR: 3) |
| 14th | N/A | The Geminid meteor shower is at its maximum. (ZHR: 120) |
| 15th | 06:42 | The waxing gibbous Moon is south of Uranus. (Evening sky.) |
| 16th | 17:14 | The waxing gibbous Moon is south of the Pleiades star cluster. (Taurus, evening sky.) |
| | N/A | The Comae Berenicid meteor shower is at its maximum. (ZHR: 3) |
| 17th | 17:11 | The waxing gibbous Moon is north of the bright star Aldebaran. (Taurus, evening sky.) |
| 18th | 08:33 | Venus is stationary prior to beginning retrograde motion. (Evening sky.) |
| 19th | 04:36 | Full Moon. (Visible all night.) |
| 20th | N/A | The Leo Minorid meteor shower is at its maximum. (ZHR: 5) |

# December 21st to 31st, 2021

## The Moon

| 21st | 23rd | 25th | 27th | 29th | 31st |

| Date | Con | R.A. | Dec | Mag | Diam | Ill. | Elon. | Phase | Close To |
|------|-----|------|-----|-----|------|------|-------|-------|----------|
| 21st | Gem | 7h 51m | 25° 15' | -12.1 | 30' | 95% | 152° W | FM | |
| 22nd | Cnc | 8h 43m | 23° 5' | -11.8 | 30' | 90% | 140° W | -G | Praesepe |
| 23rd | Leo | 9h 34m | 19° 50' | -11.5 | 30' | 84% | 129° W | -G | Regulus |
| 24th | Leo | 10h 23m | 15° 40' | -11.1 | 30' | 76% | 117° W | -G | Regulus |
| 25th | Leo | 11h 11m | 10° 44' | -10.8 | 31' | 66% | 106° W | -G | |
| 26th | Vir | 11h 58m | 5° 15' | -10.4 | 31' | 56% | 96° W | LQ | |
| 27th | Vir | 12h 46m | 0° 38' | -9.9 | 32' | 46% | 85° W | LQ | Spica |
| 28th | Vir | 13h 35m | -6° 38' | -9.4 | 32' | 35% | 74° W | -Cr | Spica |
| 29th | Lib | 14h 27m | -12° 30' | -8.7 | 32' | 24% | 62° W | -Cr | |
| 30th | Lib | 15h 23m | -17° 51' | -7.9 | 33' | 15% | 49° W | -Cr | |
| 31st | Sco | 16h 24m | -22° 14' | -7.0 | 33' | 8% | 35° W | NM | Mars, Antares |

## Mercury and Venus

Mercury
25th

Venus
25th

### Mercury

| Date | Con. | R.A. | Dec. | Mag. | Diam. | Ill. | Elon. | Vis. | Rat. | Close To |
|------|------|------|------|------|-------|------|-------|------|------|----------|
| 21st | Sgr | 18h 53m | -25° 5' | -0.7 | 5" | 93% | 14° E | NV | N/A | |
| 23rd | Sgr | 19h 7m | -24° 46' | -0.7 | 5" | 91% | 15° E | NV | N/A | |
| 25th | Sgr | 19h 21m | -24° 20' | -0.7 | 5" | 89% | 16° E | PM | ** | Venus |
| 27th | Sgr | 19h 34m | -23° 49' | -0.7 | 5" | 86% | 17° E | PM | ** | Venus |
| 29th | Sgr | 19h 47m | -23° 12' | -0.7 | 6" | 83% | 18° E | PM | ** | Venus |
| 31st | Sgr | 19h 59m | -22° 30' | -0.7 | 6" | 79% | 19° E | PM | *** | Venus |

### Venus

| Date | Con. | R.A. | Dec. | Mag. | Diam. | Ill. | Elon. | Vis. | Rat. | Close To |
|------|------|------|------|------|-------|------|-------|------|------|----------|
| 21st | Sgr | 19h 53m | -20° 37' | -4.6 | 54" | 10% | 29° E | PM | **** | |
| 23rd | Sgr | 19h 52m | -20° 12' | -4.5 | 55" | 9% | 26° E | PM | **** | |
| 25th | Sgr | 19h 49m | -19° 48' | -4.5 | 57" | 7% | 23° E | PM | *** | Mercury |
| 27th | Sgr | 19h 47m | -19° 25' | -4.4 | 58" | 5% | 20° E | PM | *** | Mercury |
| 29th | Sgr | 19h 43m | -19° 2' | -4.4 | 60" | 4% | 17° E | PM | *** | Mercury |
| 31st | Sgr | 19h 39m | -18° 40' | -4.3 | 61" | 3% | 14° E | NV | N/A | Mercury |

## Mars and the Outer Planets

Mars
25th

Jupiter
25th

Saturn
25th

### Mars

| Date | Con. | R.A. | Dec. | Mag. | Diam. | Ill. | Elon. | Vis. | Rat. | Close To |
|------|------|------|------|------|-------|------|-------|------|------|----------|
| 21st | Sco | 16h 15m | -21° 16' | 1.6 | 4" | 98% | 26° W | AM | * | Antares |
| 25th | Oph | 16h 26m | -21° 47' | 1.6 | 4" | 98% | 28° W | AM | * | Antares |
| 31st | Oph | 16h 45m | -22° 27' | 1.5 | 4" | 98% | 30° W | AM | * | Moon, Antares |

### The Outer Planets

| Planet | Date | Con. | R.A. | Dec. | Mag. | Diam. | Elon. | Vis. | Rat. | Close To |
|--------|------|------|------|------|------|-------|-------|------|------|----------|
| Jupiter | 25th | Aqr | 22h 7m | -12° 39' | -2.2 | 36" | 58° E | PM | ** | |
| Saturn | 25th | Cap | 20h 55m | -18° 12' | 0.7 | 16" | 40° E | PM | ** | |
| Uranus | 25th | Ari | 2h 35m | 14° 46' | 5.7 | 4" | 125° E | PM | *** | |
| Neptune | 25th | Aqr | 23h 27m | -4° 47' | 7.9 | 2" | 78° E | PM | ** | |

## Highlights

| Date | Time (UT) | Event |
|------|-----------|-------|
| 21st | 16:00 | Winter solstice. |
| 22nd | 11:24 | The waning gibbous Moon is north of the Praesepe star cluster. (Cancer, morning sky.) |
| 23rd | N/A | The Ursid meteor shower is at its maximum. (ZHR: 10) |
| 24th | 05:06 | The waning gibbous Moon is north of the bright star Regulus. (Leo, morning sky.) |
| 26th | 06:55 | Mars is 4.6° north of the bright star Antares. (Scorpius, morning sky.) |
| 27th | 02:24 | Last Quarter Moon. (Morning sky.) |
| 28th | 07:02 | The just-past last quarter Moon is north of the bright star Spica. (Virgo, morning sky.) |
| 29th | 00:56 | Mercury is 4.2° south of Venus. (Evening sky.) |
| 30th | N/A | Good opportunity to see Earthshine on the waning crescent Moon. (Morning sky.) |
| 31st | 15:02 | The waning crescent Moon is north of the bright star Antares. (Scorpius, morning sky.) |
| | 20:38 | The waning crescent Moon is south of Mars. (Morning sky.) |

# Planet Visibility Ratings

| | | Morning Sky | | | | | | | Evening Sky | | | | | | |
|---|---|---|---|---|---|---|---|---|---|---|---|---|---|---|---|
| | | Me | Ve | Ma | Ju | Sa | Ur | Ne | Me | Ve | Ma | Ju | Sa | Ur | Ne |
| Jan | 5th | | | * | | | | | *** | | | ** | ** | *** | ** |
| | 15th | | | * | | | | | ** | | | ** | ** | *** | ** |
| | 25th | | *** | * | | | | | | | | ** | | ** | ** |
| Feb | 5th | *** | **** | * | | | | | | | | * | | ** | * |
| | 15th | **** | **** | * | | | | | | | | | | ** | * |
| | 25th | *** | **** | * | | * | | | | | | | | ** | * |
| Mar | 5th | *** | **** | * | | ** | | | | | | | | ** | |
| | 15th | | **** | * | | ** | | | | | | | | * | |
| | 25th | | *** | * | | ** | | | | | | | | * | |
| Apr | 5th | | *** | * | * | ** | | * | | | | | | * | |
| | 15th | | *** | * | ** | ** | | * | | | | | | * | |
| | 25th | | *** | * | ** | ** | | * | *** | | | | | * | |
| May | 5th | | ** | * | ** | ** | | ** | *** | | | | | | |
| | 15th | | ** | * | ** | *** | | ** | | | | | | | |
| | 25th | | ** | ** | ** | *** | * | ** | | | | | | | |
| Jun | 5th | *** | ** | ** | *** | *** | * | *** | | | | | | | |
| | 15th | *** | ** | ** | *** | *** | * | *** | | | | | | | |
| | 25th | *** | ** | ** | *** | **** | * | *** | | | | | | | |
| Jul | 5th | | ** | ** | *** | **** | ** | *** | | | | | | | |
| | 15th | | * | ** | **** | **** | ** | **** | | | | | | | |
| | 25th | | * | ** | **** | **** | ** | **** | | | | | | | |
| Aug | 5th | | * | ** | **** | **** | ** | **** | ** | | | | | | |
| | 15th | | * | ** | ***** | | ** | **** | *** | | | | ***** | | |
| | 25th | | | ** | ***** | | *** | ***** | *** | | | | **** | | |
| Sep | 5th | | | ** | ***** | | *** | ***** | *** | | | | **** | | |
| | 15th | | | *** | ***** | | *** | ***** | | | | | **** | | |
| | 25th | | | *** | ***** | | *** | | | | | | **** | | ***** |
| Oct | 5th | ** | | *** | | | *** | | | | | ***** | **** | | ***** |
| | 15th | | | *** | | | **** | | | | | ***** | *** | | ***** |
| | 25th | | | *** | | | **** | | | | | ***** | *** | | **** |
| Nov | 5th | | | **** | | | **** | | | | | ***** | *** | | **** |
| | 15th | | | **** | | | | | | | | **** | *** | **** | **** |
| | 25th | | | **** | | | | | | | | **** | ** | **** | *** |
| Dec | 5th | | | **** | | | | | ** | | | **** | ** | **** | *** |
| | 15th | | | | | | | | *** | | **** | *** | ** | *** | *** |
| | 25th | | | | | | | | *** | * | **** | *** | ** | *** | *** |

# Solar and Lunar Eclipses

| Date | Time (UT) | Type | Visible From |
|---|---|---|---|
| Apr 30th | 20:42 | Partial Solar | Antarctica, the Pacific and south-western South America. |
| May 16th | 04:10 | Total Lunar | Africa, western Europe, eastern and central North America and South America. |
| Oct 25th | 11:02 | Partial Solar | North-eastern Africa, western and central Asia, the North Atlantic and Europe. |
| Nov 8th | 10:59 | Total Lunar | The Arctic, eastern Asia, Australia, North America and the Pacific. |

# Planetary Highlights

| Date | Time (UT) | Elon. | Vis. | Description |
|---|---|---|---|---|
| Feb 16th | 20:50 | 26° W | AM | Mercury is at greatest western elongation from the Sun. (Capricornus) |
| Mar 2nd | 12:31 | 22° W | AM | Mercury appears 0.7° south of Saturn. (Capricornus) |
| Mar 12th | 15:09 | 45° W | AM | Venus appears 4.0° north of Mars. (Capricornus) |
| Mar 20th | 09:20 | 45° W | AM | Venus is at greatest eastern elongation from the Sun. (Capricornus) |
| Mar 29th | 13:16 | 44° W | AM | Venus appears 2.2° north of Saturn. (Capricornus) |
| Apr 4th | 21:57 | 49° W | AM | Mars is 0.3° south of Saturn. (Capricornus) |
| Apr 12th | 20:08 | 28° W | AM | Jupiter appears 0.1° north of Neptune. (Aquarius) |
| Apr 30th | 18:42 | 40° W | AM | Venus appears 0.2° south of Jupiter. (Pisces) |
| May 17th | 23:00 | 58° W | AM | Mars appears 0.6° south of Neptune. (Aquarius) |
| May 28th | 23:55 | 62° W | AM | Mars appears 0.6° south of Jupiter. (Pisces) |
| Jun 11th | 13:15 | 35° W | AM | Venus is 1.6° south of Uranus. (Aries) |
| Jun 21st | 20:05 | 34° W | AM | Venus is 5.9° south of the Pleiades star cluster. (Taurus) |
| Aug 1st | 09:24 | 85° W | AM | Mars is 1.4° south of Uranus. (Aries) |
| Aug 15th | 03:05 | 180° | AN | Saturn is at opposition. (Capricornus) |
| Aug 18th | 00:11 | 91° W | AM | Mars is 5.7° south of the Pleiades star cluster. (Taurus) |
| Aug 27th | 15:59 | 24° E | PM | Mercury is at greatest eastern elongation. (Virgo) |
| Sep 8th | 10:00 | 98° W | AM | Mars is 4.3° north of the bright star Aldebaran. (Taurus) |
| Sep 17th | 11:20 | 180° | AN | Neptune is at opposition. (Aquarius) |
| Sep 27th | 10:13 | 180° | AN | Jupiter is at opposition. (Pisces) |
| Nov 9th | 11:17 | 180° | AN | Uranus is at opposition. (Aries) |
| Dec 8th | 01:32 | 180° | AN | Mars is at opposition. (Taurus) |

# Major Meteor Showers

| Shower Name | Start Date | End Date | Peak | ZHR | Speed | Brightness | Moon |
|---|---|---|---|---|---|---|---|
| Quadrantids | Dec 28th | Jan 12th | Jan 3rd | 120 | *** | ***** | ● |
| Lyrids | Apr 18th | Apr 25th | Apr 22nd | 18 | *** | ***** | ◑ |
| Eta Aquariids | Apr 24th | May 19th | May 7th | 40 | * | **** | ◐ |
| June Bootids | Jun 23rd | Jun 25th | Jun 24th | Var | ***** | ***** | ◕ |
| Alpha Capricornids | Jul 8th | Aug 10th | Jul 27th | 5 | ***** | **** | ◑ |
| Southern Delta Aquariids | Jul 21st | Aug 23rd | Jul 30th | 16 | *** | * | ● |
| Perseids | Jul 13th | Aug 26th | Aug 12th | 100 | * | ***** | ○ |
| Kappa Cygnids | Aug 6th | Aug 31st | Aug 17th | 3 | ***** | ** | ◑ |
| Aurigids | Aug 29th | Sep 4th | Sep 1st | 6 | * | **** | ◐ |
| September Epsilon Perseids | Sep 5th | Sep 28th | Sep 9th | 5 | * | ** | ○ |
| Draconids | Oct 6th | Oct 10th | Oct 8th | Var | ***** | *** | ○ |
| Southern Taurids | Sep 7th | Nov 19th | Oct 10th | 5 | **** | **** | ○ |
| Orionids | Aug 25th | Nov 19th | Oct 22nd | 15 | * | **** | ◕ |
| Andromedids | Oct 26th | Nov 20th | Nov 8th | Var | ***** | **** | ○ |
| Northern Taurids | Oct 25th | Dec 4th | Nov 11th | 5 | **** | **** | ◑ |
| Leonids | Nov 5th | Dec 3rd | Nov 18th | 15 | * | **** | ◕ |
| Alpha Monocerotids | Nov 21st | Nov 23rd | Nov 21st | Var | * | **** | ● |
| Geminids | Nov 30th | Dec 17th | Dec 13th | 120 | **** | *** | ◑ |
| December Leonis Minorids | Dec 6th | Jan 18th | Dec 20th | 5 | * | ** | ◕ |
| Ursids | Dec 17th | Dec 24th | Dec 22nd | 10 | **** | ** | ● |
| Coma Berenicids | Dec 24th | Jan 3rd | Dec 31st | 5 | * | ** | ◐ |

# January 1st to 10th, 2022

## The Moon

| 1st | 3rd | 5th | 7th | 9th |

| Date | Con | R.A. | Dec | Mag | Diam | Ill. | Elon. | Phase | Close To |
|------|-----|------|-----|-----|------|------|-------|-------|----------|
| 1st | Oph | 17h 29m | -25° 11' | -5.8 | 33' | 2% | 20° W | NM | Mars |
| 2nd | Sgr | 18h 36m | -26° 18' | -4.4 | 33' | 0% | 4° W | NM | |
| 3rd | Sgr | 19h 43m | -25° 27' | -5.1 | 33' | 1% | 12° E | NM | Mercury, Venus |
| 4th | Cap | 20h 47m | -22° 48' | -6.4 | 33' | 5% | 27° E | NM | Mercury, Saturn |
| 5th | Cap | 21h 47m | -18° 45' | -7.4 | 32' | 11% | 40° E | NM | Jupiter, Saturn |
| 6th | Aqr | 22h 42m | -13° 45' | -8.3 | 32' | 18% | 53° E | +Cr | Jupiter, Neptune |
| 7th | Aqr | 23h 32m | -8° 15' | -9.0 | 31' | 28% | 65° E | +Cr | Neptune |
| 8th | Psc | 0h 19m | -2° 34' | -9.5 | 31' | 37% | 75° E | +Cr | |
| 9th | Psc | 1h 5m | 3° 4' | -10.0 | 30' | 47% | 86° E | FQ | |
| 10th | Psc | 1h 50m | 8° 25' | -10.4 | 30' | 57% | 96° E | FQ | Uranus |

## Mercury and Venus

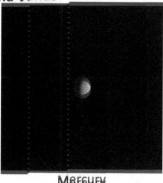

Mercury
5th

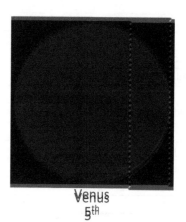

Venus
5th

### Mercury

| Date | Con. | R.A. | Dec. | Mag. | Diam. | Ill. | Elon. | Vis. | Rat. | Close To |
|------|------|------|------|------|-------|------|-------|------|------|----------|
| 1st | Sgr | 20h 5m | -22° 7' | -0.7 | 6" | 77% | 19° E | PM | *** | Venus |
| 3rd | Cap | 20h 16m | -21° 18' | -0.7 | 6" | 72% | 20° E | PM | *** | Moon |
| 5th | Cap | 20h 27m | -20° 26' | -0.6 | 6" | 66% | 20° E | PM | *** | Saturn |
| 7th | Cap | 20h 35m | -19° 32' | -0.5 | 7" | 59% | 20° E | PM | *** | Saturn |
| 9th | Cap | 20h 42m | -18° 39' | -0.4 | 7" | 50% | 20° E | PM | *** | Saturn |

## Venus

| Date | Con. | R.A. | Dec. | Mag. | Diam. | Ill. | Elon. | Vis. | Rat. | Close To |
|------|------|------|------|------|-------|------|-------|------|------|----------|
| 1st | Sgr | 19h 37m | -18° 30' | -4.3 | 61" | 2% | 12° E | NV | N/A | Mercury |
| 3rd | Sgr | 19h 32m | -18° 9' | -4.2 | 62" | 1% | 9° E | NV | N/A | Moon |
| 5th | Sgr | 19h 27m | -17° 50' | -4.1 | 62" | 1% | 5° E | NV | N/A | |
| 7th | Sgr | 19h 22m | -17° 32' | -4.1 | 63" | 0% | 2° E | NV | N/A | |
| 9th | Sgr | 19h 17m | -17° 16' | -4.0 | 63" | 0% | 2° W | NV | N/A | |

## Mars and the Outer Planets

Mars
5th

Jupiter
5th

Saturn
5th

### Mars

| Date | Con. | R.A. | Dec. | Mag. | Diam. | Ill. | Elon. | Vis. | Rat. | Close To |
|------|------|------|------|------|-------|------|-------|------|------|----------|
| 1st | Oph | 16h 48m | -22° 33' | 1.5 | 4" | 98% | 30° W | AM | * | Moon, Antares |
| 5th | Oph | 17h 0m | -22° 55' | 1.5 | 4" | 98% | 31° W | AM | * | Antares |
| 10th | Oph | 17h 15m | -23° 17' | 1.5 | 4" | 97% | 33° W | AM | * | |

### The Outer Planets

| Planet | Date | Con. | R.A. | Dec. | Mag. | Diam. | Elon. | Vis. | Rat. | Close To |
|--------|------|------|------|------|------|-------|-------|------|------|----------|
| Jupiter | 5th | Aqr | 22h 15m | -11° 53' | -2.1 | 35" | 47° E | PM | ** | Moon |
| Saturn | 5th | Cap | 21h 0m | -17° 52' | 0.7 | 15" | 29° E | PM | ** | Moon, Mercury |
| Uranus | 5th | Ari | 2h 34m | 14° 43' | 5.7 | 4" | 112° E | PM | *** | |
| Neptune | 5th | Aqr | 23h 27m | -4° 43' | 7.9 | 2" | 66° E | PM | ** | |

## Highlights

| Date | Time (UT) | Event |
|------|-----------|-------|
| 2nd | 18:34 | New Moon. (Not visible.) |
| 3rd | 07:04 | The just-past new Moon is south of Venus. (Not visible.) |
| | 15:56 | The just-past new Moon is south of dwarf planet Pluto. (Not visible.) |
| 4th | 01:24 | The waxing crescent Moon is south of Mercury. (Evening sky.) |
| | 17:50 | The waxing crescent Moon is south of Saturn (Evening sky.) |
| | N/A | The Quadrantid meteor shower is at its maximum. (ZHR: 120.) |
| 6th | 00:56 | The waxing crescent Moon is south of Jupiter. (Evening sky.) |
| | N/A | Good opportunity to see Earthshine on the waxing crescent Moon. (Evening sky.) |
| 7th | 08:33 | The waxing crescent Moon appears south of Neptune. (Evening sky.) |
| | 10:50 | Mercury is at greatest elongation from the Sun. (Evening sky.) |
| 9th | 00:41 | Venus is at inferior conjunction with the Sun. (Not visible.) |
| | 18:12 | First Quarter Moon. (Evening sky.) |

# January 11<sup>th</sup> to 20<sup>th</sup>, 2022

## The Moon

11<sup>th</sup>

13<sup>th</sup>

15<sup>th</sup>

17<sup>th</sup>

19<sup>th</sup>

| Date | Con | R.A. | Dec | Mag | Diam | Ill. | Elon. | Phase | Close To |
|------|-----|------|-----|-----|------|------|-------|-------|----------|
| 11th | Ari | 2h 35m | 13° 21' | -10.8 | 30' | 67% | 106° E | +G | Uranus |
| 12th | Ari | 3h 21m | 17° 43' | -11.1 | 30' | 75% | 116° E | +G | Uranus, Pleiades |
| 13th | Tau | 4h 9m | 21° 21' | -11.4 | 29' | 83% | 127° E | +G | Pleiades, Hyades, Aldebaran |
| 14th | Tau | 4h 59m | 24° 5' | -11.7 | 29' | 89% | 139° E | +G | Hyades, Aldebaran |
| 15th | Tau | 5h 50m | 25° 47' | -12.0 | 29' | 94% | 150° E | +G | |
| 16th | Gem | 6h 43m | 26° 18' | -12.3 | 30' | 98% | 163° E | FM | |
| 17th | Gem | 7h 37m | 25° 36' | -12.6 | 30' | 100% | 175° E | FM | |
| 18th | Cnc | 8h 30m | 23° 40' | -12.5 | 30' | 100% | 173° W | FM | Praesepe |
| 19th | Cnc | 9h 21m | 20° 37' | -12.3 | 30' | 98% | 161° W | FM | Praesepe, Regulus |
| 20th | Leo | 10h 11m | 16° 35' | -12.0 | 30' | 94% | 150° W | -G | Regulus |

## Mercury and Venus

Mercury
15<sup>th</sup>

Venus
15<sup>th</sup>

**Mercury**

| Date | Con. | R.A. | Dec. | Mag. | Diam. | Ill. | Elon. | Vis. | Rat. | Close To |
|------|------|------|------|------|-------|------|-------|------|------|----------|
| 11th | Cap | 20h 47m | -17° 49' | -0.1 | 8" | 41% | 19° E | PM | *** | Saturn |
| 13th | Cap | 20h 50m | -17° 5' | 0.2 | 8" | 31% | 17° E | PM | *** | Saturn |
| 15th | Cap | 20h 49m | -16° 30' | 0.8 | 9" | 22% | 15° E | PM | ** | Saturn |
| 17th | Cap | 20h 45m | -16° 7' | 1.6 | 9" | 13% | 12° E | NV | N/A | Saturn |
| 19th | Cap | 20h 38m | -15° 56' | 2.7 | 10" | 6% | 8° E | NV | N/A | Saturn |

## Venus

| Date | Con. | R.A. | Dec. | Mag. | Diam. | Ill. | Elon. | Vis. | Rat. | Close To |
|------|------|------|------|------|-------|------|-------|------|------|----------|
| 11th | Sgr | 19h 11m | -17° 1' | -4.1 | 62" | 1% | 5° W | NV | N/A | |
| 13th | Sgr | 19h 6m | -16° 48' | -4.2 | 62" | 1% | 8° W | NV | N/A | |
| 15th | Sgr | 19h 2m | -16° 37' | -4.2 | 61" | 2% | 12° W | NV | N/A | |
| 17th | Sgr | 18h 57m | -16° 27' | -4.3 | 60" | 3% | 15° W | NV | N/A | |
| 19th | Sgr | 18h 54m | -16° 20' | -4.4 | 59" | 4% | 18° W | AM | *** | |

## Mars and the Outer Planets

Mars
15th

Jupiter
15th

Saturn
15th

### Mars

| Date | Con. | R.A. | Dec. | Mag. | Diam. | Ill. | Elon. | Vis. | Rat. | Close To |
|------|------|------|------|------|-------|------|-------|------|------|----------|
| 11th | Oph | 17h 18m | -23° 21' | 1.5 | 4" | 97% | 33° W | AM | * | |
| 15th | Oph | 17h 31m | -23° 34' | 1.5 | 4" | 97% | 34° W | AM | * | |
| 20th | Sgr | 17h 47m | -23° 46' | 1.5 | 4" | 97% | 36° W | AM | * | |

### The Outer Planets

| Planet | Date | Con. | R.A. | Dec. | Mag. | Diam. | Elon. | Vis. | Rat. | Close To |
|--------|------|------|------|------|------|-------|-------|------|------|----------|
| Jupiter | 15th | Aqr | 22h 23m | -11° 8' | -2.1 | 34" | 39° E | PM | ** | |
| Saturn | 15th | Cap | 21h 4m | -17° 33' | 0.7 | 15" | 19° E | PM | ** | Mercury |
| Uranus | 15th | Ari | 2h 34m | 14° 42' | 5.7 | 4" | 101° E | PM | *** | |
| Neptune | 15th | Aqr | 23h 28m | -4° 37' | 7.9 | 2" | 55° E | PM | ** | |

## Highlights

| Date | Time (UT) | Event |
|------|-----------|-------|
| 11th | 10:25 | The waxing gibbous Moon is south of Uranus. (Evening sky.) |
| 13th | 02:22 | The waxing gibbous Moon is south of the Pleiades. (Taurus, evening sky.) |
| 14th | 01:01 | Mercury is stationary prior to beginning retrograde motion. (Evening sky.) |
| | 02:18 | The waxing gibbous Moon is north of the star Aldebaran. (Taurus, evening sky.) |
| 15th | 15:36 | Dwarf planet Pluto is in conjunction with the Sun. (Not visible.) |
| 17th | 23:49 | Full Moon. (Visible all night.) |
| 18th | 15:38 | The just-past full Moon is north of the Praesepe. (Cancer, visible all night.) |
| | 17:34 | Uranus is stationary prior to resuming prograde motion. (Evening sky.) |
| 20th | 11:19 | The waning gibbous Moon is north of the star Regulus. (Leo, morning sky.) |

# January 21st to 31st, 2022

## The Moon

| 21st | 23rd | 25th | 27th | 29th | 31st |

| Date | Con | R.A. | Dec | Mag | Diam | Ill. | Elon. | Phase | Close To |
|------|-----|------|-----|-----|------|------|-------|-------|----------|
| 21st | Leo | 10h 59m | 11° 46' | -11.7 | 31' | 88% | 139° W | -G | |
| 22nd | Vir | 11h 47m | 6° 23' | -11.3 | 31' | 81% | 128° W | -G | |
| 23rd | Vir | 12h 34m | 0° 37' | -11.0 | 31' | 72% | 117° W | -G | |
| 24th | Vir | 13h 22m | -5° 17' | -10.6 | 32' | 62% | 106° W | LQ | Spica |
| 25th | Vir | 14h 12m | -11° 3' | -10.2 | 32' | 51% | 95° W | LQ | Spica |
| 26th | Lib | 15h 5m | -16° 25' | -9.6 | 32' | 40% | 83° W | LQ | |
| 27th | Sco | 16h 2m | -20° 59' | -9.0 | 33' | 29% | 69° W | -Cr | Antares |
| 28th | Oph | 17h 3m | -24° 22' | -8.3 | 33' | 19% | 55° W | -Cr | Antares |
| 29th | Sgr | 18h 7m | -26° 9' | -7.4 | 33' | 11% | 40° W | NM | Venus, Mars |
| 30th | Sgr | 19h 13m | -26° 6' | -6.3 | 33' | 4% | 25° W | NM | Mercury, Venus |
| 31st | Cap | 20h 18m | -24° 12' | -5.1 | 33' | 1% | 9° W | NM | Mercury |

## Mercury and Venus

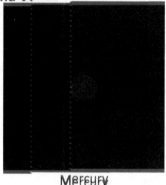 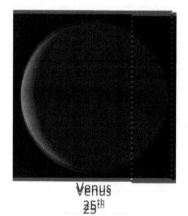

Mercury
25th

Venus
25th

### Mercury

| Date | Con. | R.A. | Dec. | Mag. | Diam. | Ill. | Elon. | Vis. | Rat. | Close To |
|------|------|------|------|------|-------|------|-------|------|------|----------|
| 21st | Cap | 20h 29m | -15° 59' | 3.9 | 10" | 2% | 4° E | NV | N/A | Saturn |
| 23rd | Cap | 20h 19m | -16° 12' | 4.6 | 10" | 1% | 1° W | NV | N/A | |
| 25th | Cap | 20h 8m | -16° 33' | 3.8 | 10" | 2% | 6° W | NV | N/A | |
| 27th | Sgr | 19h 59m | -16° 59' | 2.8 | 10" | 6% | 10° W | NV | N/A | |
| 29th | Sgr | 19h 51m | -17° 26' | 2.0 | 10" | 12% | 14° W | NV | N/A | |
| 31st | Sgr | 19h 46m | -17° 53' | 1.4 | 9" | 18% | 17° W | AM | *** | Moon |

## Venus

| Date | Con. | R.A. | Dec. | Mag. | Diam. | Ill. | Elon. | Vis. | Rat. | Close To |
|------|------|------|------|------|-------|------|-------|------|------|----------|
| 21st | Sgr | 18h 51m | -16° 15' | -4.4 | 57" | 6% | 21° W | AM | *** | |
| 23rd | Sgr | 18h 48m | -16° 12' | -4.5 | 56" | 8% | 24° W | AM | *** | |
| 25th | Sgr | 18h 47m | -16° 10' | -4.5 | 54" | 9% | 26° W | AM | *** | |
| 27th | Sgr | 18h 46m | -16° 10' | -4.6 | 53" | 11% | 28° W | AM | **** | Mars |
| 29th | Sgr | 18h 45m | -16° 12' | -4.6 | 51" | 13% | 31° W | AM | **** | Moon, Mars |
| 31st | Sgr | 18h 46m | -16° 14' | -4.6 | 50" | 15% | 32° W | AM | **** | Mars |

## Mars and the Outer Planets

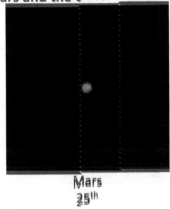

Mars
25th

Jupiter
25th

Saturn
25th

### Mars

| Date | Con. | R.A. | Dec. | Mag. | Diam. | Ill. | Elon. | Vis. | Rat. | Close To |
|------|------|------|------|------|-------|------|-------|------|------|----------|
| 21st | Sgr | 17h 50m | -23° 47' | 1.5 | 4" | 97% | 36° W | AM | * | |
| 25th | Sgr | 18h 3m | -23° 51' | 1.4 | 4" | 96% | 37° W | AM | * | |
| 31st | Sgr | 18h 22m | -23° 49' | 1.4 | 4" | 96% | 38° W | AM | * | Venus |

### The Outer Planets

| Planet | Date | Con. | R.A. | Dec. | Mag. | Diam. | Elon. | Vis. | Rat. | Close To |
|--------|------|------|------|------|------|-------|-------|------|------|----------|
| Jupiter | 25th | Aqr | 22h 31m | -10° 19' | -2.1 | 34" | 30° E | PM | ** | |
| Saturn | 25th | Cap | 21h 9m | -17° 13' | 0.7 | 15" | 10° E | NV | N/A | |
| Uranus | 25th | Ari | 2h 34m | 14° 42' | 5.7 | 4" | 91° E | PM | ** | |
| Neptune | 25th | Aqr | 23h 29m | -4° 30' | 7.9 | 2" | 45° E | PM | ** | |

## Highlights

| Date | Time (UT) | Event |
|------|-----------|-------|
| 23rd | 10:21 | Mercury is at inferior conjunction with the Sun. (Not visible.) |
| 24th | 14:21 | The nearly last quarter Moon appears north of the star Spica. (Virgo, morning sky.) |
| 25th | 13:42 | Last Quarter Moon. (Morning sky.) |
| 27th | 22:17 | The waning crescent Moon is north of the star Antares. (Scorpius, morning sky.) |
| 28th | 18:59 | Venus is stationary prior to resuming prograde motion. (Morning sky.) |
| 29th | N/A | Good opportunity to see Earthshine on the waning crescent Moon. (Morning sky.) |
| | 16:15 | The waning crescent Moon is south of Mars. (Morning sky.) |
| 31st | 00:01 | The nearly new Moon is south of Mercury. (Morning sky.) |

# February 1st to 10th, 2022

## The Moon

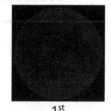
1st

3rd

5th

7th

9th

| Date | Con | R.A. | Dec | Mag | Diam | Ill. | Elon. | Phase | Close To |
|------|-----|------|-----|-----|------|------|-------|-------|----------|
| 1st | Cap | 21h 20m | -20° 40' | -4.5 | 33' | 0% | 5° E | NM | Saturn |
| 2nd | Aqr | 22h 18m | -15° 57' | -5.7 | 32' | 2% | 18° E | NM | Jupiter |
| 3rd | Aqr | 23h 11m | -10° 28' | -6.8 | 32' | 7% | 31° E | NM | Jupiter, Neptune |
| 4th | Psc | 0h 0m | -4° 39' | -7.8 | 31' | 13% | 42° E | +Cr | Neptune |
| 5th | Cet | 0h 48m | 1° 12' | -8.5 | 31' | 21% | 53° E | +Cr | |
| 6th | Psc | 1h 34m | 6° 50' | -9.1 | 30' | 30% | 63° E | +Cr | |
| 7th | Ari | 2h 20m | 12° 2' | -9.6 | 30' | 40% | 74° E | FQ | Uranus |
| 8th | Ari | 3h 6m | 16° 40' | -10.1 | 30' | 49% | 85° E | FQ | Uranus, Pleiades |
| 9th | Tau | 3h 54m | 20° 33' | -10.5 | 30' | 59% | 95° E | FQ | Pleiades, Hyades, Aldebaran |
| 10th | Tau | 4h 43m | 23° 34' | -10.8 | 30' | 68% | 107° E | +G | Hyades, Aldebaran |

## Mercury and Venus

Mercury
5th

Venus
5th

**Mercury**

| Date | Con. | R.A. | Dec. | Mag. | Diam. | Ill. | Elon. | Vis. | Rat. | Close To |
|------|------|------|------|------|-------|------|-------|------|------|----------|
| 1st | Sgr | 19h 44m | -18° 6' | 1.1 | 9" | 21% | 19° W | AM | *** | |
| 3rd | Sgr | 19h 43m | -18° 30' | 0.8 | 9" | 28% | 21° W | AM | *** | |
| 5th | Sgr | 19h 43m | -18° 51' | 0.5 | 9" | 34% | 23° W | AM | *** | |
| 7th | Sgr | 19h 46m | -19° 8' | 0.3 | 8" | 40% | 25° W | AM | **** | |
| 9th | Sgr | 19h 50m | -19° 21' | 0.2 | 8" | 45% | 25° W | AM | **** | |

**Venus**

| Date | Con. | R.A. | Dec. | Mag. | Diam. | Ill. | Elon. | Vis. | Rat. | Close To |
|------|------|------|------|------|-------|------|-------|------|------|----------|
| 1st | Sgr | 18h 46m | -16° 16' | -4.6 | 49" | 16% | 33° W | AM | **** | Mars |
| 3rd | Sgr | 18h 48m | -16° 20' | -4.6 | 47" | 17% | 35° W | AM | **** | Mars |
| 5th | Sgr | 18h 49m | -16° 24' | -4.6 | 46" | 19% | 37° W | AM | **** | Mars |
| 7th | Sgr | 18h 52m | -16° 29' | -4.6 | 44" | 21% | 38° W | AM | **** | Mars |
| 9th | Sgr | 18h 55m | -16° 35' | -4.6 | 43" | 23% | 39° W | AM | **** | Mars |

## Mars and the Outer Planets

Mars
5th

Jupiter
5th

Saturn
5th

**Mars**

| Date | Con. | R.A. | Dec. | Mag. | Diam. | Ill. | Elon. | Vis. | Rat. | Close To |
|------|------|------|------|------|-------|------|-------|------|------|----------|
| 1st | Sgr | 18h 25m | -23° 48' | 1.4 | 4" | 96% | 39° W | AM | * | Venus |
| 5th | Sgr | 18h 38m | -23° 42' | 1.4 | 4" | 96% | 40° W | AM | * | Venus |
| 10th | Sgr | 18h 54m | -23° 28' | 1.4 | 4" | 95% | 41° W | AM | * | Venus |

**The Outer Planets**

| Planet | Date | Con. | R.A. | Dec. | Mag. | Diam. | Elon. | Vis. | Rat. | Close To |
|--------|------|------|------|------|------|-------|-------|------|------|----------|
| Jupiter | 5th | Aqr | 22h 41m | -9° 22' | -2.0 | 33" | 21° E | PM | * | |
| Saturn | 5th | Cap | 21h 14m | -16° 51' | 0.7 | 15" | 0° W | NV | N/A | |
| Uranus | 5th | Ari | 2h 34m | 14° 45' | 5.8 | 4" | 80° E | PM | ** | |
| Neptune | 5th | Aqr | 23h 30m | -4° 22' | 7.9 | 2" | 34° E | PM | * | |

## Highlights

| Date | Time (UT) | Event |
|------|-----------|-------|
| 1st | 05:47 | New Moon. (Not visible.) |
| 2nd | 22:12 | The just-past new Moon is south of Jupiter. (Evening sky.) |
| 3rd | 22:09 | Mercury is stationary prior to resuming prograde motion. (Morning sky.) |
| | 22:24 | The just-past new Moon is south of Neptune. (Evening sky.) |
| 4th | N/A | Good opportunity to see Earthshine on the waxing crescent Moon. (Evening sky.) |
| 5th | 02:13 | Saturn is in conjunction with the Sun. (Not visible.) |
| 7th | 20:46 | The nearly first quarter Moon is south of Uranus. (Evening sky.) |
| 8th | 13:51 | First quarter Moon. (Evening sky.) |
| 9th | 07:57 | The just-past first quarter Moon is south of the Pleiades. (Taurus, evening sky.) |
| 10th | 08:02 | The waxing gibbous Moon is north of the star Aldebaran. (Taurus, evening sky.) |

# February 11th to 20th, 2022

## The Moon

| 11th | 13th | 15th | 17th | 19th |

| Date | Con | R.A. | Dec | Mag | Diam | Ill. | Elon. | Phase | Close To |
|------|-----|------|-----|-----|------|------|-------|-------|----------|
| 11th | Tau | 5h 35m | 25° 34' | -11.2 | 30' | 76% | 119° E | +G | |
| 12th | Gem | 6h 27m | 26° 25' | -11.5 | 30' | 84% | 131° E | +G | |
| 13th | Gem | 7h 21m | 26° 3' | -11.8 | 30' | 90% | 143° E | +G | |
| 14th | Cnc | 8h 14m | 24° 25' | -12.1 | 30' | 95% | 156° E | FM | Praesepe |
| 15th | Cnc | 9h 6m | 21° 37' | -12.3 | 30' | 98% | 168° E | FM | Praesepe |
| 16th | Leo | 9h 57m | 17° 46' | -12.6 | 30' | 100% | 179° E | FM | Regulus |
| 17th | Leo | 10h 46m | 13° 2' | -12.5 | 31' | 99% | 169° W | FM | Regulus |
| 18th | Leo | 11h 35m | 7° 38' | -12.2 | 31' | 96% | 158° W | FM | |
| 19th | Vir | 12h 22m | 1° 50' | -11.8 | 31' | 91% | 147° W | -G | |
| 20th | Vir | 13h 10m | -4° 9' | -11.5 | 31' | 85% | 136° W | -G | Spica |

## Mercury and Venus

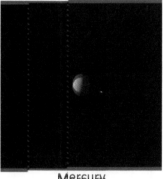 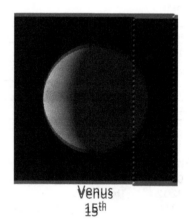

Mercury
15th

Venus
15th

### Mercury

| Date | Con. | R.A. | Dec. | Mag. | Diam. | Ill. | Elon. | Vis. | Rat. | Close To |
|------|------|------|------|------|-------|------|-------|------|------|----------|
| 11th | Sgr | 19h 56m | -19° 29' | 0.1 | 8" | 50% | 26° W | AM | **** | |
| 13th | Sgr | 20h 2m | -19° 33' | 0.1 | 7" | 54% | 26° W | AM | **** | |
| 15th | Cap | 20h 10m | -19° 31' | 0.1 | 7" | 58% | 26° W | AM | **** | |
| 17th | Cap | 20h 19m | -19° 25' | 0.0 | 7" | 61% | 26° W | AM | **** | |
| 19th | Cap | 20h 28m | -19° 14' | 0.0 | 7" | 64% | 26° W | AM | *** | |

## Venus

| Date | Con. | R.A. | Dec. | Mag. | Diam. | Ill. | Elon. | Vis. | Rat. | Close To |
|------|------|------|------|------|-------|------|-------|------|------|----------|
| 11th | Sgr | 18h 59m | -16° 40' | -4.6 | 41" | 25% | 40° W | AM | **** | Mars |
| 13th | Sgr | 19h 3m | -16° 45' | -4.6 | 40" | 26% | 41° W | AM | **** | Mars |
| 15th | Sgr | 19h 7m | -16° 49' | -4.6 | 39" | 28% | 42° W | AM | **** | Mars |
| 17th | Sgr | 19h 12m | -16° 53' | -4.6 | 38" | 30% | 43° W | AM | **** | Mars |
| 19th | Sgr | 19h 17m | -16° 55' | -4.6 | 36" | 31% | 43° W | AM | **** | Mars |

## Mars and the Outer Planets

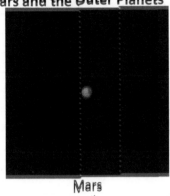

Mars
15th

Jupiter
15th

Saturn
15th

## Mars

| Date | Con. | R.A. | Dec. | Mag. | Diam. | Ill. | Elon. | Vis. | Rat. | Close To |
|------|------|------|------|------|-------|------|-------|------|------|----------|
| 11th | Sgr | 18h 57m | -23° 24' | 1.4 | 4" | 95% | 41° W | AM | * | Venus |
| 15th | Sgr | 19h 10m | -23° 8' | 1.3 | 5" | 95% | 41° W | AM | * | Venus |
| 20th | Sgr | 19h 26m | -22° 42' | 1.3 | 5" | 95% | 42° W | AM | * | Venus |

## The Outer Planets

| Planet | Date | Con. | R.A. | Dec. | Mag. | Diam. | Elon. | Vis. | Rat. | Close To |
|--------|------|------|------|------|------|-------|-------|------|------|----------|
| Jupiter | 15th | Aqr | 22h 49m | -8° 29' | -2.0 | 33" | 13° E | NV | N/A | |
| Saturn | 15th | Cap | 21h 19m | -16° 30' | 0.7 | 15" | 9° W | NV | N/A | |
| Uranus | 15th | Ari | 2h 35m | 14° 49' | 5.8 | 4" | 70° E | PM | ** | |
| Neptune | 15th | Aqr | 23h 32m | -4° 14' | 8.0 | 2" | 24° E | PM | * | |

## Highlights

| Date | Time (UT) | Event |
|------|-----------|-------|
| 13th | 00:59 | Venus appears 6.6° north of Mars. (Morning sky.) |
| 15th | 00:02 | The nearly full Moon is north of the Praesepe. (Cancer, all night.) |
| 16th | 16:06 | The almost full Moon is north of the star Regulus. (Leo, all night.) |
| | 16:57 | Full Moon. (Visible all night.) |
| | 20:50 | Mercury is at greatest western elongation from the Sun. (Morning sky.) |
| 20th | 18:05 | The waning gibbous Moon is north of the star Spica. (Virgo, morning sky.) |

# February 21st to 28th, 2022

## The Moon

21st

23rd

25th

27th

| Date | Con | R.A. | Dec | Mag | Diam | Ill. | Elon. | Phase | Close To |
|------|-----|------|-----|-----|------|------|-------|-------|----------|
| 21st | Vir | 14h 0m | -10° 2' | -11.2 | 32' | 76% | 125° W | -G | Spica |
| 22nd | Lib | 14h 52m | -15° 30' | -10.8 | 32' | 66% | 113° W | LQ | |
| 23rd | Lib | 15h 47m | -20° 13' | -10.3 | 32' | 55% | 100° W | LQ | Antares |
| 24th | Oph | 16h 46m | -23° 51' | -9.8 | 32' | 44% | 86° W | LQ | Antares |
| 25th | Sgr | 17h 48m | -26° 2' | -9.3 | 32' | 33% | 72° W | -Cr | |
| 26th | Sgr | 18h 52m | -26° 31' | -8.6 | 32' | 22% | 57° W | -Cr | Venus |
| 27th | Sgr | 19h 55m | -25° 13' | -7.8 | 32' | 13% | 42° W | -Cr | Venus, Mars |
| 28th | Cap | 20h 57m | -22° 17' | -6.8 | 32' | 7% | 27° W | NM | Mercury, Saturn |

## Mercury and Venus

Mercury
24th

Venus
24th

**Mercury**

| Date | Con. | R.A. | Dec. | Mag. | Diam. | Ill. | Elon. | Vis. | Rat. | Close To |
|------|------|------|------|------|-------|------|-------|------|------|----------|
| 21st | Cap | 20h 37m | -18° 58' | 0.0 | 6" | 67% | 25° W | AM | *** | |
| 23rd | Cap | 20h 48m | -18° 36' | 0.0 | 6" | 70% | 25° W | AM | *** | Saturn |
| 25th | Cap | 20h 58m | -18° 10' | 0.0 | 6" | 72% | 24° W | AM | *** | Saturn |
| 27th | Cap | 21h 9m | -17° 38' | -0.1 | 6" | 74% | 23° W | AM | *** | Saturn |

## Venus

| Date | Con. | R.A. | Dec. | Mag. | Diam. | Ill. | Elon. | Vis. | Rat. | Close To |
|------|------|------|------|------|-------|------|-------|------|------|----------|
| 21st | Sgr | 19h 23m | -16° 57' | -4.6 | 35" | 33% | 44° W | AM | **** | Mars |
| 23rd | Sgr | 19h 29m | -16° 58' | -4.6 | 34" | 34% | 44° W | AM | **** | Mars |
| 25th | Sgr | 19h 35m | -16° 57' | -4.6 | 33" | 36% | 45° W | AM | **** | Mars |
| 27th | Sgr | 19h 41m | -16° 55' | -4.6 | 32" | 37% | 45° W | AM | **** | Moon, Mars |

## Mars and the Outer Planets

Mars
24th

Jupiter
24th

Saturn
24th

### Mars

| Date | Con. | R.A. | Dec. | Mag. | Diam. | Ill. | Elon. | Vis. | Rat. | Close To |
|------|------|------|------|------|-------|------|-------|------|------|----------|
| 21st | Sgr | 19h 29m | -22° 36' | 1.3 | 5" | 95% | 42° W | AM | * | Venus |
| 24th | Sgr | 19h 38m | -22° 17' | 1.3 | 5" | 94% | 43° W | AM | * | Venus |
| 28th | Sgr | 19h 51m | -21° 49' | 1.3 | 5" | 94% | 44° W | AM | * | Venus |

### The Outer Planets

| Planet | Date | Con. | R.A. | Dec. | Mag. | Diam. | Elon. | Vis. | Rat. | Close To |
|--------|------|------|------|------|------|-------|-------|------|------|----------|
| Jupiter | 24th | Aqr | 22h 57m | -7° 40' | -2.0 | 33" | 7° E | NV | N/A | |
| Saturn | 24th | Cap | 21h 23m | -16° 11' | 0.8 | 15" | 17° W | AM | * | Mercury |
| Uranus | 24th | Ari | 2h 36m | 14° 53' | 5.8 | 4" | 61° E | PM | ** | |
| Neptune | 24th | Aqr | 23h 33m | -4° 7' | 8.0 | 2" | 16° E | PM | * | Jupiter |

## Highlights

| Date | Time (UT) | Event |
|------|-----------|-------|
| 23rd | 22:33 | Last Quarter Moon. (Morning sky.) |
| 24th | 04:56 | The just-past last quarter Moon is north of the star Antares. (Scorpius, morning sky.) |
| 27th | 05:30 | The waning crescent Moon is south of Venus. (Morning sky.) |
| | 08:55 | The waning crescent Moon is south of Mars. (Morning sky.) |
| | 14:39 | The waning crescent Moon is south of dwarf planet Pluto. (Morning sky.) |
| | N/A | Good opportunity to see Earthshine on the waning crescent Moon. (Morning sky.) |
| 28th | 20:47 | The waning crescent Moon is south of Mercury (Morning sky.) |
| | 23:33 | The waning crescent Moon is south of Saturn. (Morning sky.) |

# March 1st to 10th, 2022

## The Moon

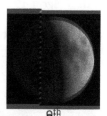

| | 1st | 3rd | 5th | 7th | 9th |

| Date | Con. | R.A. | Dec. | Mag. | Diam | Ill. | Elon. | Phase | Close To |
|------|------|------|------|------|------|------|-------|-------|----------|
| 1st | Cap | 21h 55m | -17° 60' | -5.7 | 32' | 3% | 14° W | NM | Mercury, Saturn |
| 2nd | Aqr | 22h 49m | -13° 46' | -4.5 | 32' | 0% | 1° W | NM | Jupiter, Neptune |
| 3rd | Aqr | 23h 40m | -6° 59' | -5.1 | 32' | 1% | 11° E | NM | Jupiter, Neptune |
| 4th | Cet | 0h 28m | -1° 1' | -6.2 | 31' | 4% | 22° E | NM | |
| 5th | Psc | 1h 15m | 4° 51' | -7.2 | 31' | 9% | 33° E | NM | |
| 6th | Psc | 2h 2m | 10° 22' | -8.0 | 30' | 15% | 44° E | ±Cr | Uranus |
| 7th | Ari | 2h 49m | 15° 19' | -8.7 | 30' | 23% | 54° E | ±Cr | Uranus |
| 8th | Tau | 3h 37m | 19° 34' | -9.2 | 30' | 32% | 65° E | ±Cr | Pleiades |
| 9th | Tau | 4h 26m | 23° 56' | -9.7 | 30' | 41% | 77° E | FQ | Pleiades, Hyades, Aldebaran |
| 10th | Tau | 5h 17m | 25° 18' | -10.1 | 30' | 50% | 89° E | FQ | Aldebaran |

## Mercury and Venus

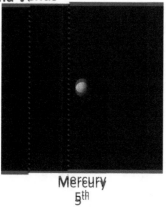

Mercury
5th

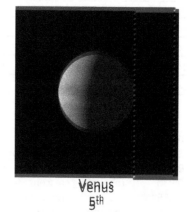

Venus
5th

### Mercury

| Date | Con. | R.A. | Dec. | Mag. | Diam. | Ill. | Elon. | Vis. | Rat. | Close To |
|------|------|------|------|------|-------|------|-------|------|------|----------|
| 1st | Cap | 21h 20m | -17° 1' | -0.1 | 6" | 76% | 22° W | AM | *** | Moon, Saturn |
| 3rd | Cap | 21h 32m | -16° 19' | -0.1 | 6" | 78% | 21° W | AM | *** | Saturn |
| 5th | Cap | 21h 44m | -15° 32' | -0.2 | 6" | 80% | 20° W | AM | *** | Saturn |
| 7th | Cap | 21h 55m | -14° 40' | -0.2 | 6" | 82% | 19° W | AM | ** | Saturn |
| 9th | Aqr | 22h 7m | -13° 42' | -0.2 | 5" | 84% | 18° W | AM | ** | Saturn |

## Venus

| Date | Con. | R.A. | Dec. | Mag. | Diam. | Ill. | Elon. | Vis. | Rat. | Close To |
|------|------|------|------|------|-------|------|-------|------|------|----------|
| 1st | Sgr | 19h 48m | -16° 52' | -4.5 | 31" | 38% | 45° W | AM | **** | Mars |
| 3rd | Sgr | 19h 55m | -16° 47' | -4.5 | 30" | 40% | 45° W | AM | **** | Mars |
| 5th | Sgr | 20h 2m | -16° 40' | -4.5 | 30" | 41% | 45° W | AM | **** | Mars |
| 7th | Cap | 20h 10m | -16° 32' | -4.5 | 29" | 42% | 45° W | AM | **** | Mars |
| 9th | Cap | 20h 17m | -16° 21' | -4.5 | 28" | 43% | 45° W | AM | **** | Mars |

## Mars and the Outer Planets

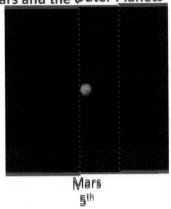

Mars
5th

Jupiter
5th

Saturn
5th

## Mars

| Date | Con. | R.A. | Dec. | Mag. | Diam. | Ill. | Elon. | Vis. | Rat. | Close To |
|------|------|------|------|------|-------|------|-------|------|------|----------|
| 1st | Sgr | 19h 54m | -21° 41' | 1.3 | 5" | 94% | 44° W | AM | * | Venus |
| 5th | Sgr | 20h 7m | -21° 8' | 1.2 | 5" | 94% | 44° W | AM | * | Venus |
| 10th | Cap | 20h 22m | -20° 33' | 1.2 | 5" | 93% | 45° W | AM | * | Venus |

## The Outer Planets

| Planet | Date | Con. | R.A. | Dec. | Mag. | Diam. | Elon. | Vis. | Rat. | Close To |
|--------|------|------|------|------|------|-------|-------|------|------|----------|
| Jupiter | 5th | Aqr | 23h 6m | -6° 50' | -2.0 | 33" | 0° E | NV | N/A | |
| Saturn | 5th | Cap | 21h 28m | -15° 53' | 0.8 | 15" | 24° W | AM | ** | Mercury |
| Uranus | 5th | Ari | 2h 37m | 14° 59' | 5.8 | 3" | 53° E | PM | ** | |
| Neptune | 5th | Aqr | 23h 34m | -3° 59' | 8.0 | 2" | 8° E | NV | N/A | Jupiter |

## Highlights

| Date | Time (UT) | Event |
|------|-----------|-------|
| 2nd | 12:31 | Mercury appears 0.7° south of Saturn. (Morning sky.) |
| | 17:35 | New Moon. (Not visible.) |
| 6th | 03:03 | Jupiter is in conjunction with the Sun. (Not visible.) |
| | N/A | Good opportunity to see Earthshine on the waxing crescent Moon. (Evening sky.) |
| 7th | 05:32 | The waxing crescent Moon is south of Uranus. (Evening sky.) |
| 8th | 16:54 | The waxing crescent Moon is south of the Pleiades star cluster. (Taurus, evening sky.) |
| 9th | 15:55 | The nearly first quarter Moon is north of the star Aldebaran. (Taurus, evening sky.) |
| 10th | 10:46 | First Quarter Moon. (Evening sky.) |

# March 11<sup>th</sup> to 20<sup>th</sup>, 2022

## The Moon

| 11<sup>th</sup> | 13<sup>th</sup> | 15<sup>th</sup> | 17<sup>th</sup> | 19<sup>th</sup> |

| Date | Con | R.A. | Dec | Mag | Diam | Ill. | Elon. | Phase | Close To |
|------|-----|------|-----|-----|------|------|-------|-------|----------|
| 11th | Gem | 6h 10m | 26° 31' | -10.5 | 30' | 60% | 101° E | FQ | |
| 12th | Gem | 7h 3m | 26° 32' | -10.9 | 30' | 69% | 113° E | +G | |
| 13th | Gem | 7h 56m | 25° 18' | -11.2 | 30' | 77% | 126° E | +G | Praesepe |
| 14th | Cnc | 8h 48m | 22° 51' | -11.5 | 30' | 85% | 138° E | +G | Praesepe |
| 15th | Leo | 9h 40m | 19° 18' | -11.8 | 30' | 91% | 150° E | +G | Regulus |
| 16th | Leo | 10h 30m | 14° 47' | -12.1 | 31' | 96% | 161° E | FM | Regulus |
| 17th | Leo | 11h 19m | 9° 29' | -12.4 | 31' | 99% | 173° E | FM | |
| 18th | Vir | 12h 7m | 3° 38' | -12.6 | 31' | 100% | 176° W | FM | |
| 19th | Vir | 12h 56m | -2° 30' | -12.3 | 32' | 98% | 165° W | FM | Spica |
| 20th | Vir | 13h 46m | -8° 37' | -12.0 | 32' | 94% | 153° W | -G | Spica |

## Mercury and Venus

Mercury
15<sup>th</sup>

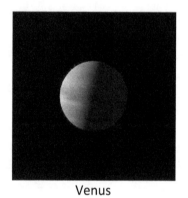

Venus
15<sup>th</sup>

**Mercury**

| Date | Con. | R.A. | Dec. | Mag. | Diam. | Ill. | Elon. | Vis. | Rat. | Close To |
|------|------|------|------|------|-------|------|-------|------|------|----------|
| 11th | Aqr | 22h 20m | -12° 40' | -0.3 | 5" | 85% | 17° W | AM | ** | |
| 13th | Aqr | 22h 32m | -11° 33' | -0.4 | 5" | 87% | 15° W | AM | ** | |
| 15th | Aqr | 22h 45m | -10° 21' | -0.5 | 5" | 89% | 14° W | NV | N/A | Jupiter |
| 17th | Aqr | 22h 57m | -9° 4' | -0.6 | 5" | 90% | 13° W | NV | N/A | Jupiter |
| 19th | Aqr | 23h 10m | -7° 42' | -0.7 | 5" | 92% | 11° W | NV | N/A | Jupiter |

**Venus**

| Date | Con. | R.A. | Dec. | Mag. | Diam. | Ill. | Elon. | Vis. | Rat. | Close To |
|------|------|------|------|------|-------|------|-------|------|------|----------|
| 11th | Cap | 20h 25m | -16° 10' | -4.5 | 27" | 45% | 45° W | AM | **** | Mars |
| 13th | Cap | 20h 32m | -15° 56' | -4.4 | 27" | 46% | 45° W | AM | **** | Mars |
| 15th | Cap | 20h 40m | -15° 40' | -4.4 | 26" | 47% | 45° W | AM | **** | Mars |
| 17th | Cap | 20h 48m | -15° 23' | -4.4 | 25" | 48% | 45° W | AM | *** | Mars |
| 19th | Cap | 20h 56m | -15° 3' | -4.4 | 25" | 49% | 45° W | AM | *** | Mars, Saturn |

## Mars and the Outer Planets

Mars
15th

Jupiter
15th

Saturn
15th

**Mars**

| Date | Con. | R.A. | Dec. | Mag. | Diam. | Ill. | Elon. | Vis. | Rat. | Close To |
|------|------|------|------|------|-------|------|-------|------|------|----------|
| 11th | Cap | 20h 25m | -20° 13' | 1.2 | 5" | 93% | 45° W | AM | * | Venus |
| 15th | Cap | 20h 38m | -19° 32' | 1.2 | 5" | 93% | 46° W | AM | * | Venus |
| 20th | Cap | 20h 53m | -18° 36' | 1.1 | 5" | 93% | 46° W | AM | * | Venus |

**The Outer Planets**

| Planet | Date | Con. | R.A. | Dec. | Mag. | Diam. | Elon. | Vis. | Rat. | Close To |
|--------|------|------|------|------|------|-------|-------|------|------|----------|
| Jupiter | 15th | Aqr | 23h 15m | -5° 54' | -2.0 | 33" | 7° W | NV | N/A | Mercury |
| Saturn | 15th | Cap | 21h 32m | -15° 33' | 0.8 | 16" | 32° W | AM | ** | |
| Uranus | 15th | Ari | 2h 39m | 15° 7' | 5.8 | 3" | 44° E | PM | * | |
| Neptune | 15th | Aqr | 23h 35m | -3° 50' | 8.0 | 2" | 1° W | NV | N/A | Jupiter |

## Highlights

| Date | Time (UT) | Event |
|------|-----------|-------|
| 12th | 15:09 | Venus appears 4.0° north of Mars. (Morning sky.) |
| 13th | 23:49 | Neptune is in conjunction with the Sun. (Not visible.) |
| 14th | 08:13 | The waxing gibbous Moon is north of the Praesepe. (Cancer, evening sky.) |
| 16th | 02:39 | The waxing gibbous Moon is north of the star Regulus. (Leo, evening sky.) |
| 18th | 07:18 | Full Moon. (Visible all night.) |
| 20th | 01:57 | The waning gibbous Moon is north of the star Spica. (Virgo, morning sky.) |
| | 09:20 | Venus is at greatest eastern elongation from the Sun. (Morning sky.) |
| | 15:34 | Spring equinox |

# March 21st to 31st, 2022

## The Moon

| 21st | 23rd | 25th | 27th | 29th | 31st |

| Date | Con | R.A. | Dec | Mag | Diam | Ill. | Elon. | Phase | Close To |
|------|-----|------|-----|-----|------|------|-------|-------|----------|
| 21st | Lib | 14h 38m | -14° 23' | -11.7 | 32' | 88% | 141° W | -G | |
| 22nd | Lib | 15h 34m | -19° 25' | -11.3 | 32' | 79% | 128° W | -G | |
| 23rd | Oph | 16h 32m | -23° 23' | -10.9 | 32' | 69% | 114° W | -G | Antares |
| 24th | Oph | 17h 34m | -25° 55' | -10.5 | 32' | 58% | 100° W | LQ | |
| 25th | Sgr | 18h 37m | -26° 47' | -10.0 | 32' | 47% | 85° W | LQ | |
| 26th | Sgr | 19h 40m | -25° 54' | -9.4 | 32' | 36% | 70° W | -Cr | |
| 27th | Cap | 20h 40m | -23° 24' | -8.8 | 32' | 25% | 56° W | -Cr | Mars |
| 28th | Cap | 21h 38m | -19° 32' | -8.1 | 32' | 16% | 43° W | -Cr | Venus, Mars, Saturn |
| 29th | Aqr | 22h 32m | -14° 38' | -7.2 | 32' | 9% | 30° W | NM | |
| 30th | Aqr | 23h 23m | -9° 4' | -6.2 | 31' | 4% | 18° W | NM | Jupiter, Neptune |
| 31st | Psc | 0h 11m | -3° 10' | -5.0 | 31' | 1% | 7° W | NM | Mercury, Jupiter, Neptune |

## Mercury and Venus

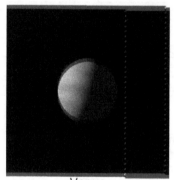

Mercury
25th

Venus
25th

### Mercury

| Date | Con. | R.A. | Dec. | Mag. | Diam. | Ill. | Elon. | Vis. | Rat. | Close To |
|------|------|------|------|------|-------|------|-------|------|------|----------|
| 21st | Aqr | 23h 23m | -6° 16' | -0.8 | 5" | 93% | 10° W | NV | N/A | Jupiter, Neptune |
| 23rd | Aqr | 23h 36m | -4° 45' | -0.9 | 5" | 95% | 8° W | NV | N/A | Jupiter, Neptune |
| 25th | Psc | 23h 50m | -3° 9' | -1.1 | 5" | 96% | 7° W | NV | N/A | Jupiter, Neptune |
| 27th | Psc | 0h 4m | -1° 30' | -1.3 | 5" | 98% | 5° W | NV | N/A | Jupiter |
| 29th | Psc | 0h 18m | 0° 13' | -1.5 | 5" | 99% | 4° W | NV | N/A | |
| 31st | Cet | 0h 32m | 1° 60' | -1.7 | 5" | 99% | 3° W | NV | N/A | Moon |

## Venus

| Date | Con. | R.A. | Dec. | Mag. | Diam. | Ill. | Elon. | Vis. | Rat. | Close To |
|------|------|------|------|------|-------|------|-------|------|------|----------|
| 21st | Cap | 21h 4m | -14° 42' | -4.4 | 24" | 50% | 45° W | AM | *** | Mars, Saturn |
| 23rd | Aqr | 21h 12m | -14° 19' | -4.4 | 24" | 51% | 44° W | AM | *** | Mars, Saturn |
| 25th | Aqr | 21h 21m | -13° 54' | -4.3 | 23" | 52% | 44° W | AM | *** | Mars, Saturn |
| 27th | Cap | 21h 29m | -13° 27' | -4.3 | 23" | 53% | 44° W | AM | *** | Mars, Saturn |
| 29th | Cap | 21h 37m | -12° 59' | -4.3 | 22" | 54% | 44° W | AM | *** | Mars, Saturn |
| 31st | Cap | 21h 46m | -12° 28' | -4.3 | 22" | 55% | 43° W | AM | *** | Mars, Saturn |

## Mars and the Outer Planets

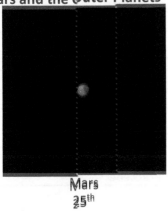
Mars
25th

Jupiter
25th

Saturn
25th

## Mars

| Date | Con. | R.A. | Dec. | Mag. | Diam. | Ill. | Elon. | Vis. | Rat. | Close To |
|------|------|------|------|------|-------|------|-------|------|------|----------|
| 21st | Cap | 20h 56m | -18° 24' | 1.1 | 5" | 93% | 47° W | AM | * | Venus, Saturn |
| 25th | Cap | 21h 8m | -17° 35' | 1.1 | 5" | 93% | 47° W | AM | * | Venus, Saturn |
| 31st | Cap | 21h 27m | -16° 17' | 1.1 | 5" | 93% | 48° W | AM | * | Venus, Saturn |

## The Outer Planets

| Planet | Date | Con. | R.A. | Dec. | Mag. | Diam. | Elon. | Vis. | Rat. | Close To |
|--------|------|------|------|------|------|-------|-------|------|------|----------|
| Jupiter | 25th | Aqr | 23h 23m | -4° 58' | -2.0 | 33" | 13° W | NV | N/A | Mercury, Neptune |
| Saturn | 25th | Cap | 21h 36m | -15° 15' | 0.9 | 16" | 40° W | AM | ** | Venus, Mars |
| Uranus | 25th | Ari | 2h 40m | 15° 15' | 5.8 | 3" | 36° E | PM | * | |
| Neptune | 25th | Aqr | 23h 37m | -3° 41' | 8.0 | 2" | 10° W | NV | N/A | Mercury, Jupiter |

## Highlights

| Date | Time (UT) | Event |
|------|-----------|-------|
| 23rd | 11:50 | The waning gibbous Moon is north of the star Antares. (Scorpius, morning sky.) |
| 25th | 05:38 | Last Quarter Moon. (Morning sky.) |
| 26th | 20:25 | The just-past last quarter Moon is south of dwarf planet Pluto. (Morning sky.) |
| 28th | 01:43 | The waning crescent Moon is south of Mars. (Morning sky.) |
| | 10:18 | The waning crescent Moon is south of Venus. (Morning sky.) |
| | 12:48 | The waning crescent Moon is south of Saturn. (Morning sky.) |
| 29th | N/A | Good opportunity to see Earthshine on the waning crescent Moon. (Morning sky.) |
| | 13:16 | Venus appears 2.2° north of Saturn. (Morning sky.) |

# April 1<sup>st</sup> to 10<sup>th</sup>, 2022

## The Moon

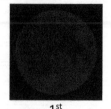
1<sup>st</sup>

3<sup>rd</sup>

5<sup>th</sup>

7<sup>th</sup>

9<sup>th</sup>

| Date | Con | R.A. | Dec | Mag | Diam | Ill. | Elon. | Phase | Close To |
|------|-----|------|-----|-----|------|------|-------|-------|----------|
| 1st | Cet | 0h 58m | 2° 46' | -4.3 | 31' | 0% | 4° E | NM | Mercury |
| 2nd | Psc | 1h 45m | 8° 29' | -5.5 | 31' | 2% | 15° E | NM | |
| 3rd | Ari | 2h 32m | 13° 44' | -6.5 | 30' | 5% | 25° E | NM | Uranus |
| 4th | Ari | 3h 19m | 18° 19' | -7.4 | 30' | 10% | 36° E | NM | Uranus, Pleiades |
| 5th | Tau | 4h 9m | 22° 4' | -8.1 | 30' | 17% | 48° E | +Cr | Pleiades, Hyades, Aldebaran |
| 6th | Tau | 5h 0m | 24° 49' | -8.8 | 30' | 25% | 60° E | +Cr | Hyades, Aldebaran |
| 7th | Tau | 5h 52m | 26° 27' | -9.3 | 30' | 33% | 72° E | +Cr | |
| 8th | Gem | 6h 45m | 26° 52' | -9.8 | 30' | 43% | 84° E | FQ | |
| 9th | Gem | 7h 38m | 26° 3' | -10.2 | 30' | 52% | 96° E | FQ | |
| 10th | Cnc | 8h 30m | 24° 2' | -10.6 | 30' | 62% | 109° E | FQ | Praesepe |

## Mercury and Venus

Mercury
5<sup>th</sup>

Venus
5<sup>th</sup>

**Mercury**

| Date | Con. | R.A. | Dec. | Mag. | Diam. | Ill. | Elon. | Vis. | Rat. | Close To |
|------|------|------|------|------|-------|------|-------|------|------|----------|
| 1st | Psc | 0h 39m | 2° 54' | -1.9 | 5" | 100% | 1° W | NV | N/A | Moon |
| 3rd | Psc | 0h 54m | 4° 45' | -2.0 | 5" | 100% | 1° E | NV | N/A | |
| 5th | Psc | 1h 8m | 6° 37' | -1.9 | 5" | 99% | 3° E | NV | N/A | |
| 7th | Psc | 1h 23m | 8° 29' | -1.8 | 5" | 98% | 5° E | NV | N/A | |
| 9th | Psc | 1h 39m | 10° 20' | -1.6 | 5" | 96% | 7° E | NV | N/A | |

**Venus**

| Date | Con. | R.A. | Dec. | Mag. | Diam. | Ill. | Elon. | Vis. | Rat. | Close To |
|------|------|------|------|------|-------|------|-------|------|------|----------|
| **1st** | Cap | 21h 50m | -12° 13' | -4.3 | 22" | 56% | 43° W | AM | *** | Mars, Saturn |
| **3rd** | Cap | 21h 58m | -11° 40' | -4.3 | 21" | 56% | 43° W | AM | *** | Mars, Saturn |
| **5th** | Aqr | 22h 7m | -11° 6' | -4.3 | 21" | 57% | 43° W | AM | *** | Mars, Saturn |
| **7th** | Aqr | 22h 15m | -10° 30' | -4.2 | 20" | 58% | 42° W | AM | *** | Mars, Saturn |
| **9th** | Aqr | 22h 24m | -9° 53' | -4.2 | 20" | 59% | 42° W | AM | *** | Mars |

## Mars and the Outer Planets

Mars
5th

Jupiter
5th

Saturn
5th

**Mars**

| Date | Con. | R.A. | Dec. | Mag. | Diam. | Ill. | Elon. | Vis. | Rat. | Close To |
|------|------|------|------|------|-------|------|-------|------|------|----------|
| **1st** | Cap | 21h 30m | -16° 4' | 1.1 | 5" | 92% | 48° W | AM | * | Venus, Saturn |
| **5th** | Cap | 21h 41m | -15° 8' | 1.0 | 5" | 91% | 49° W | AM | * | Venus, Saturn |
| **10th** | Cap | 21h 56m | -13° 55' | 1.0 | 5" | 91% | 50° W | AM | * | Venus, Saturn |

**The Outer Planets**

| Planet | Date | Con. | R.A. | Dec. | Mag. | Diam. | Elon. | Vis. | Rat. | Close To |
|--------|------|------|------|------|------|-------|-------|------|------|----------|
| Jupiter | 5th | Aqr | 23h 33m | -3° 58' | -2.0 | 34" | 21° W | AM | * | Neptune |
| Saturn | 5th | Cap | 21h 40m | -14° 56' | 0.9 | 16" | 49° W | AM | ** | Venus, Mars |
| Uranus | 5th | Ari | 2h 43m | 15° 26' | 5.9 | 3" | 26° E | PM | * | |
| Neptune | 5th | Aqr | 23h 38m | -3° 31' | 8.0 | 2" | 20° W | AM | * | Jupiter |

## Highlights

| Date | Time (UT) | Event |
|------|-----------|-------|
| 1st | 06:25 | New Moon. (Not visible.) |
| 2nd | 22:58 | Mercury is at superior conjunction with the Sun. (Not visible.) |
| 3rd | 18:49 | The waxing crescent Moon is south of Uranus. (Evening sky.) |
| 4th | N/A | Good opportunity to see Earthshine on the waxing crescent Moon. (Evening sky.) |
| | 21:57 | Mars is 0.3° south of Saturn. (Morning sky.) |
| 5th | 01:43 | The waxing crescent Moon is south of the Pleiades. (Taurus, evening sky.) |
| 6th | 01:27 | The waxing crescent Moon is north of the star Aldebaran. (Taurus, evening sky.) |
| 9th | 06:48 | First Quarter Moon. (Evening sky.) |
| 10th | 15:13 | The just-past first quarter Moon is north of the Praesepe. (Cancer, evening sky.) |

# April 11th to 20th, 2022

## The Moon

   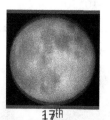 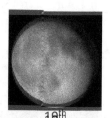

| 11th | 13th | 15th | 17th | 19th |

| Date | Con | R.A. | Dec | Mag | Diam | Ill. | Elon. | Phase | | Close To |
|------|-----|------|-----|-----|------|------|-------|-------|--|----------|
| 11th | Cnc | 9h 21m | 20° 53' | -11.0 | 30' | 71% | 120° E | +G | | Praesepe, Regulus |
| 12th | Leo | 10h 11m | 16° 43' | -11.3 | 31' | 79% | 132° E | +G | | Regulus |
| 13th | Leo | 11h 0m | 11° 42' | -11.6 | 31' | 87% | 143° E | +G | | |
| 14th | Vir | 11h 48m | 6° 1' | -12.0 | 31' | 93% | 155° E | +G | | |
| 15th | Vir | 12h 37m | 0° 7' | -12.3 | 32' | 98% | 166° E | FM | | Spica |
| 16th | Vir | 13h 27m | -6° 24' | -12.6 | 32' | 100% | 177° E | FM | | Spica |
| 17th | Vir | 14h 20m | -12° 29' | -12.5 | 32' | 99% | 170° W | FM | | |
| 18th | Lib | 15h 15m | -17° 60' | -12.1 | 33' | 96% | 157° W | FM | | |
| 19th | Sco | 16h 15m | -22° 29' | -11.8 | 33' | 90% | 144° W | -G | | Antares |
| 20th | Oph | 17h 17m | -25° 33' | -11.4 | 33' | 82% | 129° W | -G | | |

## Mercury and Venus

Mercury
15th

Venus
15th

### Mercury

| Date | Con. | R.A. | Dec. | Mag. | Diam. | Ill. | Elon. | Vis. | Rat. | Close To |
|------|------|------|------|------|-------|------|-------|------|------|----------|
| 11th | Ari | 1h 54m | 12° 9' | -1.5 | 5" | 93% | 9° E | NV | N/A | |
| 13th | Ari | 2h 9m | 13° 53' | -1.3 | 6" | 88% | 11° E | NV | N/A | |
| 15th | Ari | 2h 24m | 15° 31' | -1.2 | 6" | 83% | 12° E | NV | N/A | |
| 17th | Ari | 2h 38m | 17° 2' | -1.0 | 6" | 77% | 14° E | NV | N/A | Uranus |
| 19th | Ari | 2h 52m | 18° 24' | -0.8 | 6" | 70% | 16° E | PM | ** | Uranus |

## Venus

| Date | Con. | R.A. | Dec. | Mag. | Diam. | Ill. | Elon. | Vis. | Rat. | Close To |
|------|------|------|------|------|-------|------|-------|------|------|----------|
| 11th | Aqr | 22h 32m | -9° 14' | -4.2 | 20" | 60% | 43° W | AM | *** | Mars |
| 13th | Aqr | 22h 40m | -8° 34' | -4.2 | 19" | 61% | 42° W | AM | *** | Mars |
| 15th | Aqr | 22h 49m | -7° 53' | -4.2 | 19" | 62% | 41° W | AM | *** | Mars |
| 17th | Aqr | 22h 57m | -7° 10' | -4.2 | 19" | 62% | 41° W | AM | *** | |
| 19th | Aqr | 23h 6m | -6° 26' | -4.2 | 18" | 63% | 41° W | AM | *** | Jupiter |

## Mars and the Outer Planets

Mars
15th

Jupiter
15th

Saturn
15th

## Mars

| Date | Con. | R.A. | Dec. | Mag. | Diam. | Ill. | Elon. | Vis. | Rat. | Close To |
|------|------|------|------|------|-------|------|-------|------|------|----------|
| 11th | Cap | 21h 59m | -13° 40' | 1.0 | 5" | 91% | 50° W | AM | * | Venus, Saturn |
| 15th | Aqr | 22h 11m | -12° 40' | 1.0 | 5" | 91% | 51° W | AM | * | Venus, Saturn |
| 20th | Aqr | 22h 25m | -11° 21' | 0.9 | 6" | 90% | 52° W | AM | * | |

## The Outer Planets

| Planet | Date | Con. | R.A. | Dec. | Mag. | Diam. | Elon. | Vis. | Rat. | Close To |
|--------|------|------|------|------|------|-------|-------|------|------|----------|
| Jupiter | 15th | Psc | 23h 41m | -3° 5' | -2.1 | 34" | 28° W | AM | ** | Neptune |
| Saturn | 15th | Cap | 21h 43m | -14° 41' | 0.9 | 16" | 58° W | AM | ** | Mars |
| Uranus | 15th | Ari | 2h 45m | 15° 36' | 5.9 | 3" | 18° E | PM | * | Mercury |
| Neptune | 15th | Aqr | 23h 40m | -3° 23' | 8.0 | 2" | 29° W | AM | * | Jupiter |

## Highlights

| Date | Time (UT) | Event |
|------|-----------|-------|
| 12th | 10:02 | The waxing gibbous Moon is north of the star Regulus. (Leo, evening sky.) |
| | 20:08 | Jupiter appears 0.1° north of Neptune. (Morning sky.) |
| 16th | 11:02 | The almost full Moon is north of the star Spica. (Virgo, visible all night.) |
| | 18:56 | Full Moon. (Visible all night.) |
| 18th | 13:41 | Mercury appears 2.1° north of Uranus. (Evening sky.) |
| 19th | 16:54 | The waning gibbous Moon is north of the star Antares. (Scorpius, morning sky.) |

# April 21st to 30th, 2022

## The Moon

| 21st | 23rd | 25th | 27th | 29th |
| --- | --- | --- | --- | --- |

| Date | Con | R.A. | Dec | Mag | Diam | Ill. | Elon. | Phase | Close To |
| --- | --- | --- | --- | --- | --- | --- | --- | --- | --- |
| 21st | Sgr | 18h 21m | -26° 53' | -11.0 | 33' | 72% | 114° W | -G | |
| 22nd | Sgr | 19h 25m | -26° 24' | -10.6 | 32' | 61% | 99° W | LQ | |
| 23rd | Cap | 20h 27m | -24° 14' | -10.1 | 32' | 50% | 84° W | LQ | |
| 24th | Cap | 21h 25m | -20° 38' | -9.6 | 32' | 39% | 71° W | LQ | Saturn |
| 25th | Aqr | 22h 19m | -15° 59' | -9.0 | 32' | 28% | 58° W | -Cr | Mars, Saturn |
| 26th | Aqr | 23h 10m | -10° 37' | -8.3 | 31' | 19% | 46° W | -Cr | Venus, Mars, Jupiter, Neptune |
| 27th | Psc | 23h 58m | -4° 51' | -7.5 | 31' | 12% | 35° W | NM | Venus, Jupiter, Neptune |
| 28th | Cet | 0h 44m | 1° 1' | -6.6 | 31' | 6% | 25° W | NM | |
| 29th | Psc | 1h 30m | 6° 46' | -5.6 | 30' | 2% | 14° W | NM | |
| 30th | Ari | 2h 17m | 12° 9' | -4.4 | 30' | 0% | 3° W | NM | Uranus |

## Mercury and Venus

Mercury
25th

Venus
25th

**Mercury**

| Date | Con. | R.A. | Dec. | Mag. | Diam. | Ill. | Elon. | Vis. | Rat. | Close To |
| --- | --- | --- | --- | --- | --- | --- | --- | --- | --- | --- |
| 21st | Ari | 3h 5m | 19° 36' | -0.6 | 6" | 63% | 17° E | PM | ** | Uranus |
| 23rd | Ari | 3h 17m | 20° 38' | -0.4 | 7" | 57% | 18° E | PM | *** | Pleiades |
| 25th | Ari | 3h 28m | 21° 30' | -0.2 | 7" | 50% | 19° E | PM | *** | Pleiades |
| 27th | Tau | 3h 38m | 22° 11' | 0.1 | 8" | 43% | 20° E | PM | *** | Pleiades |
| 29th | Tau | 3h 47m | 22° 43' | 0.3 | 8" | 37% | 20° E | PM | *** | Pleiades, Hyades |

## Venus

| Date | Con. | R.A. | Dec. | Mag. | Diam. | Ill. | Elon. | Vis. | Rat. | Close To |
|------|------|------|------|------|-------|------|-------|------|------|----------|
| 21st | Aqr | 23h 14m | -5° 42' | -4.2 | 18" | 64% | 41° W | AM | *** | Jupiter |
| 23rd | Aqr | 23h 23m | -4° 56' | -4.1 | 18" | 65% | 40° W | AM | *** | Jupiter, Neptune |
| 25th | Aqr | 23h 31m | -4° 9' | -4.1 | 17" | 66% | 40° W | AM | *** | Jupiter, Neptune |
| 27th | Aqr | 23h 40m | -3° 22' | -4.1 | 17" | 66% | 40° W | AM | ** | Moon, Jupiter, Neptune |
| 29th | Psc | 23h 48m | -2° 34' | -4.1 | 17" | 67% | 40° W | AM | ** | Jupiter, Neptune |

## Mars and the Outer Planets

Mars
25th

Jupiter
25th

Saturn
25th

### Mars

| Date | Con. | R.A. | Dec. | Mag. | Diam. | Ill. | Elon. | Vis. | Rat. | Close To |
|------|------|------|------|------|-------|------|-------|------|------|----------|
| 21st | Aqr | 22h 28m | -11° 5' | 0.9 | 6" | 90% | 52° W | AM | * | |
| 25th | Aqr | 22h 40m | -9° 60' | 0.9 | 6" | 90% | 53° W | AM | * | Moon |
| 30th | Aqr | 22h 54m | -8° 37' | 0.9 | 6" | 89% | 54° W | AM | * | |

### The Outer Planets

| Planet | Date | Con. | R.A. | Dec. | Mag. | Diam. | Elon. | Vis. | Rat. | Close To |
|--------|------|------|------|------|------|-------|-------|------|------|----------|
| Jupiter | 25th | Psc | 23h 50m | -2° 14' | -2.1 | 34" | 35° W | AM | ** | Venus, Neptune |
| Saturn | 25th | Cap | 21h 46m | -14° 29' | 0.9 | 16" | 66° W | AM | ** | Moon |
| Uranus | 25th | Ari | 2h 47m | 15° 46' | 5.9 | 3" | 9° E | NV | N/A | |
| Neptune | 25th | Aqr | 23h 41m | -3° 16' | 7.9 | 2" | 38° W | AM | * | Venus, Jupiter |

## Highlights

| Date | Time (UT) | Event |
|------|-----------|-------|
| 22nd | N/A | The Lyrid meteor shower is at its maximum. (ZHR: 18.) |
| 23rd | 11:57 | Last Quarter Moon. (Morning sky.) |
| 24th | 20:37 | The just-past last quarter Moon is south of Saturn. (Morning sky.) |
| 25th | 21:41 | The waning crescent Moon is south of Mars. (Morning sky.) |
| 27th | 00:43 | The waning crescent Moon is south of Venus. (Morning sky.) |
| | 02:05 | The waning crescent Moon is south of Neptune. (Morning sky.) |
| | 07:54 | The waning crescent Moon is south of Jupiter. (Morning sky.) |
| | N/A | Good opportunity to see Earthshine on the waning crescent Moon. (Morning sky.) |
| 28th | 07:58 | Mercury is at greatest eastern elongation. (Evening sky.) |
| | 11:22 | Mercury appears 1.4° south of the Pleiades. (Taurus, evening sky.) |
| 30th | 18:42 | Venus appears 0.2° south of Jupiter. (Morning sky.) |
| | 20:29 | New Moon (Not visible.) |
| | 20:42 | Partial Solar Eclipse. Visible from Antarctica, the Pacific and south-western South America. |

# May 1st to 10th, 2022

## The Moon

| 1st | 3rd | 5th | 7th | 9th |

| Date | Con | R.A. | Dec | Mag | Diam | Ill. | Elon. | Phase | | Close To |
|------|-----|------|-----|-----|------|------|-------|-------|--|----------|
| 1st | Ari | 3h 4m | 16° 58' | -4.7 | 30' | 0% | 7° E | NM | | Uranus, Pleiades |
| 2nd | Tau | 3h 53m | 21° 1' | -5.8 | 30' | 3% | 19° E | NM | | Mercury, Pleiades, Hyades, Aldeb |
| 3rd | Tau | 4h 43m | 24° 7' | -6.8 | 30' | 7% | 30° E | NM | | Mercury, Hyades, Aldebaran |
| 4th | Tau | 5h 35m | 26° 8' | -7.6 | 30' | 12% | 42° E | NM | | |
| 5th | Gem | 6h 28m | 26° 57' | -8.3 | 29' | 19% | 55° E | +Cr | | |
| 6th | Gem | 7h 21m | 26° 32' | -8.9 | 30' | 27% | 67° E | +Cr | | |
| 7th | Cnc | 8h 13m | 24° 54' | -9.4 | 30' | 36% | 79° E | +Cr | | Praesepe |
| 8th | Cnc | 9h 4m | 22° 9' | -9.9 | 30' | 45% | 91° E | FQ | | Praesepe |
| 9th | Leo | 9h 54m | 18° 23' | -10.3 | 30' | 55% | 102° E | FQ | | Regulus |
| 10th | Leo | 10h 42m | 13° 44' | -10.7 | 31' | 65% | 113° E | FQ | | Regulus |

## Mercury and Venus

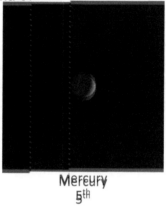

Mercury
5th

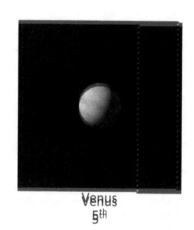

Venus
5th

### Mercury

| Date | Con. | R.A. | Dec. | Mag. | Diam. | Ill. | Elon. | Vis. | Rat. | | Close To |
|------|------|------|------|------|-------|------|-------|------|------|--|----------|
| 1st | Tau | 3h 54m | 23° 4' | 0.6 | 8" | 31% | 20° E | PM | *** | | Pleiades, Hyades |
| 3rd | Tau | 4h 0m | 23° 16' | 0.9 | 9" | 26% | 20° E | PM | *** | | Moon, Pleiades, Hyades, Al |
| 5th | Tau | 4h 5m | 23° 19' | 1.3 | 9" | 21% | 19° E | PM | *** | | Pleiades, Hyades, Aldebara |
| 7th | Tau | 4h 8m | 23° 13' | 1.7 | 10" | 17% | 18° E | PM | *** | | Pleiades, Hyades, Aldebara |
| 9th | Tau | 4h 9m | 22° 59' | 2.1 | 10" | 13% | 16° E | PM | ** | | Pleiades, Hyades, Aldebara |

## Venus

| Date | Con. | R.A. | Dec. | Mag. | Diam. | Ill. | Elon. | Vis. | Rat. | Close To |
|------|------|------|------|------|-------|------|-------|------|------|----------|
| 1st | Psc | 23h 57m | -1° 45' | -4.1 | 17" | 68% | 39° W | AM | ** | Jupiter, Neptune |
| 3rd | Psc | 0h 5m | 0° 56' | -4.1 | 16" | 69% | 39° W | AM | ** | Jupiter |
| 5th | Psc | 0h 14m | 0° 6' | -4.1 | 16" | 69% | 39° W | AM | ** | Jupiter |
| 7th | Psc | 0h 22m | 0° 44' | -4.1 | 16" | 70% | 39° W | AM | ** | Jupiter |
| 9th | Cet | 0h 31m | 1° 35' | -4.1 | 16" | 71% | 39° W | AM | ** | Jupiter |

## Mars and the Outer Planets

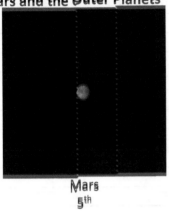

Mars
5th

Jupiter
5th

Saturn
5th

## Mars

| Date | Con. | R.A. | Dec. | Mag. | Diam. | Ill. | Elon. | Vis. | Rat. | Close To |
|------|------|------|------|------|-------|------|-------|------|------|----------|
| 1st | Aqr | 22h 57m | -8° 20' | 0.9 | 6" | 89% | 54° W | AM | * | |
| 5th | Aqr | 23h 8m | -7° 12' | 0.8 | 6" | 89% | 55° W | AM | * | |
| 10th | Aqr | 23h 22m | -5° 46' | 0.8 | 6" | 89% | 57° W | AM | * | Jupiter |

## The Outer Planets

| Planet | Date | Con. | R.A. | Dec. | Mag. | Diam. | Elon. | Vis. | Rat. | Close To |
|--------|------|------|------|------|------|-------|-------|------|------|----------|
| Jupiter | 5th | Psc | 23h 57m | -1° 26' | -2.1 | 35" | 43° W | AM | ** | Venus, Neptune |
| Saturn | 5th | Cap | 21h 48m | -14° 20' | 0.8 | 17" | 75° W | AM | ** | |
| Uranus | 5th | Ari | 2h 49m | 15° 56' | 5.9 | 3" | 0° W | NV | N/A | |
| Neptune | 5th | Psc | 23h 42m | -3° 9' | 7.9 | 2" | 47° W | AM | ** | Venus, Mars, Jupiter |

## Highlights

| Date | Time (UT) | Event |
|------|-----------|-------|
| 2nd | 07:43 | The waxing crescent Moon is south of the Pleiades. (Taurus, evening sky.) |
| | 14:58 | The waxing crescent Moon is south of Mercury. (Taurus, evening sky.) |
| 3rd | 07:07 | The waxing crescent Moon is north of the star Aldebaran. (Taurus, evening sky.) |
| 4th | N/A | Good opportunity to see Earthshine on the waxing crescent Moon. (Evening sky.) |
| 5th | 10:16 | Uranus is in conjunction with the Sun. (Not visible.) |
| 6th | N/A | The Eta Aquariid meteor shower is at its maximum. (ZHR: 70) |
| 8th | 01:29 | The nearly first quarter Moon is north of the Praesepe. (Cancer, evening sky.) |
| 9th | 00:22 | First Quarter Moon. (Evening sky.) |
| | 19:09 | The just-past first quarter Moon is north of the star Regulus. (Leo, evening sky.) |
| | N/A | The Eta Lyrid meteor shower is at its maximum. (ZHR: 3) |
| 10th | 22:39 | Mercury is stationary prior to beginning retrograde motion. (Evening sky.) |

# May 11<sup>th</sup> to 20<sup>th</sup>, 2022

## The Moon

11<sup>th</sup>

13<sup>th</sup>

15<sup>th</sup>

17<sup>th</sup>

19<sup>th</sup>

| Date | Con | R.A. | Dec | Mag | Diam | Ill. | Elon. | Phase | Close To |
|------|-----|------|-----|-----|------|------|-------|-------|----------|
| 11th | Leo | 11h 29m | 8° 23' | -11.1 | 31' | 74% | 124° E | +G | |
| 12th | Vir | 12h 17m | 2° 29' | -11.4 | 31' | 83% | 135° E | +G | |
| 13th | Vir | 13h 6m | -3° 43' | -11.8 | 32' | 90% | 146° E | +G | Spica |
| 14th | Vir | 13h 57m | -9° 56' | -12.1 | 32' | 96% | 158° E | FM | Spica |
| 15th | Lib | 14h 52m | -15° 48' | -12.5 | 33' | 99% | 171° E | FM | |
| 16th | Sco | 15h 51m | -20° 51' | -12.6 | 33' | 100% | 176° W | FM | Antares |
| 17th | Oph | 16h 53m | -24° 37' | -12.2 | 33' | 97% | 161° W | FM | Antares |
| 18th | Sgr | 17h 59m | -26° 40' | -11.9 | 33' | 92% | 145° W | -G | |
| 19th | Sgr | 19h 6m | -26° 46' | -11.5 | 33' | 84% | 130° W | -G | |
| 20th | Cap | 20h 10m | -25° 1' | -11.1 | 33' | 75% | 115° W | -G | |

## Mercury and Venus

Mercury
15<sup>th</sup>

Venus
15<sup>th</sup>

**Mercury**

| Date | Con. | R.A. | Dec. | Mag. | Diam. | Ill. | Elon. | Vis. | Rat. | Close To |
|------|------|------|------|------|-------|------|-------|------|------|----------|
| 11th | Tau | 4h 10m | 22° 36' | 2.6 | 11" | 9% | 14° E | NV | N/A | Pleiades, Hyades, Aldebara |
| 13th | Tau | 4h 9m | 22° 6' | 3.2 | 11" | 6% | 12° E | NV | N/A | Pleiades, Hyades, Aldebara |
| 15th | Tau | 4h 7m | 21° 29' | 3.8 | 12" | 3% | 9° E | NV | N/A | Pleiades, Hyades, Aldebara |
| 17th | Tau | 4h 3m | 20° 47' | 4.5 | 12" | 2% | 7° E | NV | N/A | Pleiades, Hyades, Aldebara |
| 19th | Tau | 4h 0m | 20° 1' | 5.3 | 12" | 0% | 4° E | NV | N/A | Pleiades, Hyades, Aldebara |

**Venus**

| Date | Con. | R.A. | Dec. | Mag. | Diam. | Ill. | Elon. | Vis. | Rat. | Close To |
|------|------|------|------|------|-------|------|-------|------|------|----------|
| 11th | Cet | 0h 39m | 2° 25' | -4.0 | 16" | 71% | 38° W | AM | ** | Jupiter |
| 13th | Psc | 0h 48m | 3° 16' | -4.0 | 15" | 72% | 38° W | AM | ** | |
| 15th | Psc | 0h 57m | 4° 7' | -4.0 | 15" | 73% | 38° W | AM | ** | |
| 17th | Psc | 1h 5m | 4° 58' | -4.0 | 15" | 73% | 38° W | AM | ** | |
| 19th | Psc | 1h 14m | 5° 49' | -4.0 | 15" | 74% | 38° W | AM | ** | |

## Mars and the Outer Planets

Mars
15th

Jupiter
15th

Saturn
15th

**Mars**

| Date | Con. | R.A. | Dec. | Mag. | Diam. | Ill. | Elon. | Vis. | Rat. | Close To |
|------|------|------|------|------|-------|------|-------|------|------|----------|
| 11th | Aqr | 23h 25m | -5° 29' | 0.8 | 6" | 89% | 57° W | AM | * | Jupiter, Neptune |
| 15th | Aqr | 23h 36m | -4° 20' | 0.8 | 6" | 88% | 58° W | AM | * | Jupiter, Neptune |
| 20th | Psc | 23h 50m | -2° 53' | 0.7 | 6" | 88% | 60° W | AM | * | Jupiter, Neptune |

**The Outer Planets**

| Planet | Date | Con. | R.A. | Dec. | Mag. | Diam. | Elon. | Vis. | Rat. | Close To |
|--------|------|------|------|------|------|-------|-------|------|------|----------|
| Jupiter | 15th | Psc | 0h 4m | 0° 40' | -2.2 | 36" | 51° W | AM | ** | Mars |
| Saturn | 15th | Cap | 21h 50m | -14° 13' | 0.8 | 17" | 85° W | AM | *** | |
| Uranus | 15th | Ari | 2h 52m | 16° 6' | 5.9 | 3" | 9° W | NV | N/A | |
| Neptune | 15th | Psc | 23h 43m | -3° 4' | 7.9 | 2" | 56° W | AM | ** | Mars, Jupiter |

## Highlights

| Date | Time (UT) | Event |
|------|-----------|-------|
| 13th | 20:24 | The waxing gibbous Moon is north of the star Spica. (Virgo, evening sky.) |
| 16th | 04:10 | Total lunar eclipse. Visible from western Africa, the Atlantic, Central America, western Europe, eastern and central North America, the Pacific and South America. |
| | 04:15 | Full Moon. (Visible all night.) |
| 17th | 03:35 | The just-past full Moon is north of the star Antares. (Scorpius, visible all night.) |
| | 23:00 | Mars appears 0.6° south of Neptune. (Morning sky.) |
| 20th | 10:07 | The waning gibbous Moon is south of dwarf planet Pluto. (Morning sky.) |

# May 21st to 31st, 2022

## The Moon

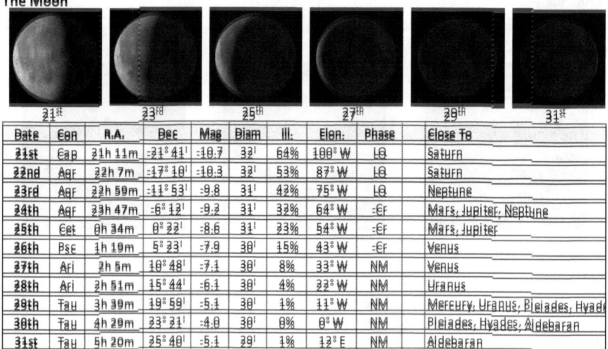

21st     23rd     25th     27th     29th     31st

| Date | Con | R.A. | Dec | Mag | Diam | Ill. | Elon. | Phase | Close To |
|------|-----|------|-----|-----|------|------|-------|-------|----------|
| 21st | Cap | 21h 11m | -21° 41' | -10.7 | 32' | 64% | 100° W | LQ | Saturn |
| 22nd | Aqr | 22h 7m | -17° 10' | -10.3 | 32' | 53% | 87° W | LQ | Saturn |
| 23rd | Aqr | 22h 59m | -11° 53' | -9.8 | 31' | 42% | 75° W | LQ | Neptune |
| 24th | Aqr | 23h 47m | -6° 12' | -9.2 | 31' | 32% | 64° W | -Cr | Mars, Jupiter, Neptune |
| 25th | Cet | 0h 34m | 0° 22' | -8.6 | 31' | 23% | 54° W | -Cr | Mars, Jupiter |
| 26th | Psc | 1h 19m | 5° 23' | -7.9 | 30' | 15% | 43° W | -Cr | Venus |
| 27th | Ari | 2h 5m | 10° 48' | -7.1 | 30' | 8% | 33° W | NM | Venus |
| 28th | Ari | 2h 51m | 15° 44' | -6.1 | 30' | 4% | 22° W | NM | Uranus |
| 29th | Tau | 3h 39m | 19° 59' | -5.1 | 30' | 1% | 11° W | NM | Mercury, Uranus, Pleiades, Hyades |
| 30th | Tau | 4h 29m | 23° 21' | -4.0 | 30' | 0% | 0° W | NM | Pleiades, Hyades, Aldebaran |
| 31st | Tau | 5h 20m | 25° 40' | -5.1 | 29' | 1% | 12° E | NM | Aldebaran |

## Mercury and Venus

Mercury
25th

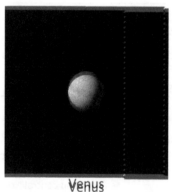

Venus
25th

### Mercury

| Date | Con. | R.A. | Dec. | Mag. | Diam. | Ill. | Elon. | Vis. | Rat. | Close To |
|------|------|------|------|------|-------|------|-------|------|------|----------|
| 21st | Tau | 3h 56m | 19° 13' | 5.9 | 12" | 0% | 1° E | NV | N/A | Pleiades, Hyades, Aldebara |
| 23rd | Tau | 3h 51m | 18° 25' | 5.4 | 12" | 0% | 2° W | NV | N/A | Pleiades, Hyades |
| 25th | Tau | 3h 47m | 17° 40' | 4.7 | 12" | 1% | 5° W | NV | N/A | Pleiades, Hyades |
| 27th | Tau | 3h 44m | 16° 59' | 4.0 | 12" | 3% | 8° W | NV | N/A | Pleiades |
| 29th | Tau | 3h 41m | 16° 24' | 3.4 | 12" | 5% | 11° W | NV | N/A | Moon, Pleiades |
| 31st | Tau | 3h 39m | 15° 58' | 2.9 | 11" | 8% | 13° W | NV | N/A | Pleiades |

## Venus

| Date | Con. | R.A. | Dec. | Mag. | Diam. | Ill. | Elon. | Vis. | Rat. | Close To |
|------|------|------|------|------|-------|------|-------|------|------|----------|
| 21st | Psc | 1h 23m | 6° 39' | -4.0 | 15" | 75% | 37° W | AM | ** | |
| 23rd | Psc | 1h 31m | 7° 29' | -4.0 | 14" | 75% | 37° W | AM | ** | |
| 25th | Psc | 1h 40m | 8° 19' | -4.0 | 14" | 76% | 37° W | AM | ** | |
| 27th | Psc | 1h 49m | 9° 9' | -4.0 | 14" | 76% | 37° W | AM | ** | Moon |
| 29th | Psc | 1h 58m | 9° 57' | -4.0 | 14" | 77% | 37° W | AM | ** | |
| 31st | Ari | 2h 7m | 10° 46' | -4.0 | 14" | 78% | 37° W | AM | ** | |

## Mars and the Outer Planets

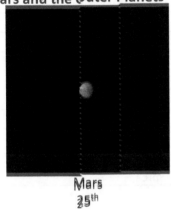

Mars
25th

Jupiter
25th

Saturn
25th

## Mars

| Date | Con. | R.A. | Dec. | Mag. | Diam. | Ill. | Elon. | Vis. | Rat. | Close To |
|------|------|------|------|------|-------|------|-------|------|------|----------|
| 21st | Psc | 23h 53m | -2° 35' | 0.7 | 6" | 88% | 60° W | AM | * | Jupiter, Neptune |
| 25th | Psc | 0h 4m | -1° 25' | 0.7 | 6" | 88% | 61° W | AM | ** | Moon, Jupiter |
| 31st | Psc | 0h 20m | 0° 19' | 0.7 | 6" | 87% | 63° W | AM | ** | Jupiter |

## The Outer Planets

| Planet | Date | Con. | R.A. | Dec. | Mag. | Diam. | Elon. | Vis. | Rat. | Close To |
|--------|------|------|------|------|------|-------|-------|------|------|----------|
| Jupiter | 25th | Psc | 0h 11m | 0° 1' | -2.2 | 37" | 59° W | AM | ** | Moon, Mars |
| Saturn | 25th | Cap | 21h 51m | -14° 10' | 0.8 | 17" | 94° W | AM | *** | |
| Uranus | 25th | Ari | 2h 54m | 16° 16' | 5.9 | 3" | 19° W | AM | * | |
| Neptune | 25th | Psc | 23h 43m | -2° 60' | 7.9 | 2" | 66° W | AM | ** | Mars, Jupiter |

## Highlights

| Date | Time (UT) | Event |
|------|-----------|-------|
| 21st | 19:13 | Mercury is at inferior conjunction with the Sun. (Not visible.) |
| 22nd | 04:12 | The almost last quarter Moon is south of Saturn. (Morning sky.) |
| | 18:44 | Last Quarter Moon. (Morning sky.) |
| 24th | 11:07 | The waning crescent Moon is south of Neptune. (Morning sky.) |
| | 19:32 | The waning crescent Moon is south of Mars. (Morning sky.) |
| | 23:02 | The waning crescent Moon is south of Jupiter. (Morning sky.) |
| | N/A | The May Camelopardalid meteor shower is at its maximum. (ZHR: Variable.) |
| 27th | 01:31 | The waning crescent Moon appears close to Venus. (Morning sky.) |
| | N/A | Good opportunity to see Earthshine on the waning crescent Moon. (Morning sky.) |
| 28th | 15:00 | The waning crescent Moon is south of Uranus. (Morning sky.) |
| | 23:55 | Mars appears 0.6° south of Jupiter. (Morning sky.) |
| 30th | 11:31 | New Moon. (Not visible.) |

# June 1ˢᵗ to 10ᵗʰ, 2022

## The Moon

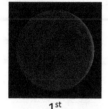    

| | 1ˢᵗ | | 3ʳᵈ | | 5ᵗʰ | | 7ᵗʰ | | 9ᵗʰ | |

| Date | Con | R.A. | Dec | Mag | Diam | Ill. | Elon. | Phase | Close To |
|------|-----|------|-----|-----|------|------|-------|-------|----------|
| 1st | Gem | 6h 13m | 26° 49' | -6.2 | 29' | 4% | 24° E | NM | |
| 2nd | Gem | 7h 6m | 26° 45' | -7.1 | 29' | 8% | 36° E | NM | |
| 3rd | Gem | 7h 58m | 25° 27' | -7.8 | 29' | 14% | 48° E | +Cr | Praesepe |
| 4th | Cnc | 8h 49m | 23° 1' | -8.5 | 30' | 21% | 60° E | +Cr | Praesepe |
| 5th | Leo | 9h 39m | 19° 34' | -9.1 | 30' | 30% | 71° E | +Cr | Regulus |
| 6th | Leo | 10h 27m | 15° 15' | -9.6 | 30' | 39% | 82° E | FQ | Regulus |
| 7th | Leo | 11h 13m | 10° 13' | -10.1 | 30' | 49% | 93° E | FQ | |
| 8th | Vir | 12h 0m | 4° 38' | -10.5 | 31' | 59% | 103° E | FQ | |
| 9th | Vir | 12h 47m | -1° 20' | -10.9 | 31' | 69% | 114° E | +G | Spica |
| 10th | Vir | 13h 36m | -7° 25' | -11.3 | 32' | 79% | 125° E | +G | Spica |

## Mercury and Venus

Mercury
5ᵗʰ

Venus
5ᵗʰ

**Mercury**

| Date | Con. | R.A. | Dec. | Mag. | Diam. | Ill. | Elon. | Vis. | Rat. | Close To |
|------|------|------|------|------|-------|------|-------|------|------|----------|
| 1st | Tau | 3h 39m | 15° 47' | 2.7 | 11" | 9% | 15° W | NV | N/A | Pleiades |
| 3rd | Tau | 3h 38m | 15° 33' | 2.3 | 11" | 12% | 17° W | AM | *** | Pleiades |
| 5th | Tau | 3h 39m | 15° 27' | 1.9 | 10" | 15% | 19° W | AM | *** | Pleiades |
| 7th | Tau | 3h 41m | 15° 30' | 1.6 | 10" | 19% | 20° W | AM | *** | Pleiades |
| 9th | Tau | 3h 44m | 15° 41' | 1.3 | 10" | 23% | 21° W | AM | *** | Pleiades |

## Venus

| Date | Con. | R.A. | Dec. | Mag. | Diam. | Ill. | Elon. | Vis. | Rat. | Close To |
|------|------|------|------|------|-------|------|-------|------|------|----------|
| 1st | Ari | 2h 11m | 11° 9' | -4.0 | 14" | 78% | 36° W | AM | ** | |
| 3rd | Ari | 2h 20m | 11° 56' | -4.0 | 14" | 79% | 36° W | AM | ** | |
| 5th | Ari | 2h 30m | 12° 42' | -4.0 | 13" | 79% | 36° W | AM | ** | |
| 7th | Ari | 2h 39m | 13° 27' | -4.0 | 13" | 80% | 36° W | AM | ** | Uranus |
| 9th | Ari | 2h 48m | 14° 11' | -3.9 | 13" | 80% | 36° W | AM | ** | Uranus |

## Mars and the Outer Planets

Mars
5th

Jupiter
5th

Saturn
5th

### Mars

| Date | Con. | R.A. | Dec. | Mag. | Diam. | Ill. | Elon. | Vis. | Rat. | Close To |
|------|------|------|------|------|-------|------|-------|------|------|----------|
| 1st | Psc | 0h 23m | 0° 37' | 0.7 | 6" | 87% | 64° W | AM | ** | Jupiter |
| 5th | Cet | 0h 34m | 1° 46' | 0.6 | 7" | 87% | 65° W | AM | ** | Jupiter |
| 10th | Psc | 0h 47m | 3° 11' | 0.6 | 7" | 87% | 67° W | AM | ** | Jupiter |

### The Outer Planets

| Planet | Date | Con. | R.A. | Dec. | Mag. | Diam. | Elon. | Vis. | Rat. | Close To |
|--------|------|------|------|------|------|-------|-------|------|------|----------|
| Jupiter | 5th | Psc | 0h 18m | 0° 40' | -2.3 | 38" | 69° W | AM | *** | Mars |
| Saturn | 5th | Cap | 21h 51m | -14° 10' | 0.7 | 17" | 106° W | AM | *** | |
| Uranus | 5th | Ari | 2h 56m | 16° 26' | 5.9 | 3" | 29° W | AM | * | Venus |
| Neptune | 5th | Psc | 23h 44m | -2° 56' | 7.9 | 2" | 77° W | AM | *** | Jupiter |

## Highlights

| Date | Time (UT) | Event |
|------|-----------|-------|
| 2nd | N/A | Good opportunity to see Earthshine on the waxing crescent Moon. (Evening sky.) |
| 3rd | 00:30 | Mercury is stationary prior to resuming prograde motion. (Morning sky.) |
| 4th | 06:24 | The waxing crescent Moon is north of the Praesepe star cluster. (Cancer, evening sky.) |
| 5th | 12:29 | Saturn is stationary prior to beginning retrograde motion. (Morning sky.) |
| 6th | 03:12 | The nearly first quarter Moon is north of the star Regulus. (Leo, evening sky.) |
| 7th | 14:49 | First Quarter Moon. (Evening sky.) |
| 10th | 07:07 | The waxing gibbous Moon is north of the star Spica. (Virgo, evening sky.) |
| | 15:29 | Mercury is 8.3° south of the Pleiades. (Taurus, morning sky.) |

# June 11th to 20th, 2022

## The Moon

  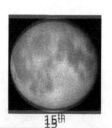  

11th      13th      15th      17th      19th

| Date | Con | R.A. | Dec | Mag | Diam | Ill. | Elon. | Phase | Close To |
|------|-----|------|-----|-----|------|------|-------|-------|----------|
| 11th | Lib | 14h 28m | -13° 22' | -11.6 | 32' | 87% | 137° E | +G | |
| 12th | Lib | 15h 24m | -18° 46' | -12.0 | 33' | 94% | 150° E | +G | |
| 13th | Oph | 16h 25m | -23° 8' | -12.4 | 33' | 98% | 165° E | FM | Antares |
| 14th | Oph | 17h 31m | -25° 59' | -12.7 | 33' | 100% | 180° E | FM | |
| 15th | Sgr | 18h 39m | -26° 55' | -12.3 | 33' | 98% | 164° W | FM | |
| 16th | Sgr | 19h 46m | -25° 50' | -12.0 | 33' | 94% | 148° W | -G | |
| 17th | Cap | 20h 50m | -22° 55' | -11.6 | 33' | 87% | 133° W | -G | |
| 18th | Cap | 21h 50m | -18° 36' | -11.2 | 32' | 78% | 119° W | -G | Saturn |
| 19th | Aqr | 22h 44m | -13° 21' | -10.8 | 32' | 68% | 107° W | -G | |
| 20th | Aqr | 23h 35m | -7° 36' | -10.4 | 32' | 57% | 95° W | LQ | Neptune |

## Mercury and Venus

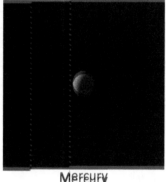 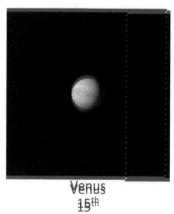

Mercury                     Venus
15th                          15th

### Mercury

| Date | Con. | R.A. | Dec. | Mag. | Diam. | Ill. | Elon. | Vis. | Rat. | Close To |
|------|------|------|------|------|-------|------|-------|------|------|----------|
| 11th | Tau | 3h 48m | 15° 59' | 1.1 | 9" | 26% | 23° W | AM | *** | Pleiades, Hyades |
| 13th | Tau | 3h 54m | 16° 23' | 0.9 | 9" | 30% | 23° W | AM | *** | Pleiades, Hyades |
| 15th | Tau | 4h 0m | 16° 54' | 0.7 | 8" | 34% | 24° W | AM | *** | Pleiades, Hyades, Aldebara |
| 17th | Tau | 4h 8m | 17° 29' | 0.5 | 8" | 39% | 24° W | AM | *** | Pleiades, Hyades, Aldebara |
| 19th | Tau | 4h 16m | 18° 7' | 0.3 | 8" | 43% | 24° W | AM | *** | Pleiades, Hyades, Aldebara |

## Venus

| Date | Con. | R.A. | Dec. | Mag. | Diam. | Ill. | Elon. | Vis. | Rat. | Close To |
|------|------|------|------|------|-------|------|-------|------|------|----------|
| 11th | Ari | 2h 57m | 14° 54' | -3.9 | 13" | 81% | 35° W | AM | ** | Uranus |
| 13th | Ari | 3h 7m | 15° 36' | -3.9 | 13" | 81% | 35° W | AM | ** | Uranus, Pleiades |
| 15th | Ari | 3h 16m | 16° 16' | -3.9 | 13" | 82% | 35° W | AM | ** | Uranus, Pleiades |
| 17th | Tau | 3h 26m | 16° 55' | -3.9 | 13" | 83% | 34° W | AM | ** | Pleiades |
| 19th | Tau | 3h 35m | 17° 33' | -3.9 | 12" | 83% | 34° W | AM | ** | Pleiades |

## Mars and the Outer Planets

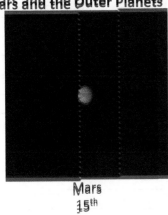

Mars
15th

Jupiter
15th

Saturn
15th

## Mars

| Date | Con. | R.A. | Dec. | Mag. | Diam. | Ill. | Elon. | Vis. | Rat. | Close To |
|------|------|------|------|------|-------|------|-------|------|------|----------|
| 11th | Psc | 0h 50m | 3° 28' | 0.6 | 7" | 87% | 67° W | AM | ** | Jupiter |
| 15th | Psc | 1h 1m | 4° 35' | 0.6 | 7" | 86% | 69° W | AM | ** | Jupiter |
| 20th | Psc | 1h 14m | 5° 57' | 0.5 | 7" | 86% | 70° W | AM | ** | |

## The Outer Planets

| Planet | Date | Con. | R.A. | Dec. | Mag. | Diam. | Elon. | Vis. | Rat. | Close To |
|--------|------|------|------|------|------|-------|-------|------|------|----------|
| Jupiter | 15th | Psc | 0h 23m | 1° 11' | -2.3 | 39" | 78° W | AM | *** | Mars |
| Saturn | 15th | Cap | 21h 51m | -14° 14' | 0.7 | 18" | 116° W | AM | *** | |
| Uranus | 15th | Ari | 2h 58m | 16° 35' | 5.8 | 3" | 39° W | AM | * | Venus |
| Neptune | 15th | Psc | 23h 44m | -2° 54' | 7.9 | 2" | 88° W | AM | *** | Jupiter |

## Highlights

| Date | Time (UT) | Event |
|------|-----------|-------|
| 11th | 13:15 | Venus is 1.6° south of Uranus. (Morning sky.) |
| 13th | 12:51 | The nearly full Moon is north of the star Antares. (Scorpius, evening sky.) |
| 14th | 11:52 | Full Moon. (Visible all night.) |
| 16th | 14:41 | Mercury is at greatest western elongation. (Morning sky.) |
| | 16:41 | The waning gibbous Moon is south of dwarf planet Pluto. (Morning sky.) |
| 18th | 13:16 | The waning gibbous Moon is south of Saturn. (Morning sky.) |
| 20th | 17:06 | The nearly last quarter Moon is south of Neptune. (Morning sky.) |

# June 21st to 30th, 2022

## The Moon

21st

23rd

25th

27th

29th

| Date | Con | R.A. | Dec | Mag | Diam | Ill. | Elon. | Phase | Close To |
|------|-----|------|-----|-----|------|------|-------|-------|----------|
| 21st | Psc | 0h 22m | -1° 41' | -10.0 | 31' | 46% | 84° W | LQ | Jupiter, Neptune |
| 22nd | Psc | 1h 8m | 4° 8' | -9.5 | 31' | 36% | 74° W | -Cr | Mars, Jupiter |
| 23rd | Psc | 1h 54m | 9° 40' | -8.9 | 30' | 27% | 64° W | -Cr | Mars |
| 24th | Ari | 2h 40m | 14° 42' | -8.2 | 30' | 18% | 53° W | -Cr | Uranus |
| 25th | Tau | 3h 27m | 19° 5' | -7.5 | 30' | 11% | 42° W | NM | Venus, Uranus, Pleiades |
| 26th | Tau | 4h 16m | 22° 38' | -6.7 | 30' | 6% | 31° W | NM | Mercury, Venus, Pleiades, Hyades |
| 27th | Tau | 5h 7m | 25° 12' | -5.7 | 29' | 2% | 19° W | NM | Mercury, Hyades, Aldebaran |
| 28th | Tau | 6h 0m | 26° 37' | -4.7 | 29' | 0% | 7° W | NM | |
| 29th | Gem | 6h 53m | 26° 50' | -4.5 | 29' | 0% | 5° E | NM | |
| 30th | Gem | 7h 45m | 25° 49' | -5.5 | 29' | 2% | 17° E | NM | |

## Mercury and Venus

Mercury
25th

Venus
25th

**Mercury**

| Date | Con. | R.A. | Dec. | Mag. | Diam. | Ill. | Elon. | Vis. | Rat. | Close To |
|------|------|------|------|------|-------|------|-------|------|------|----------|
| 21st | Tau | 4h 26m | 18° 49' | 0.1 | 7" | 48% | 24° W | AM | *** | Pleiades, Hyades, Aldebara |
| 23rd | Tau | 4h 36m | 19° 32' | -0.1 | 7" | 53% | 23° W | AM | *** | Hyades, Aldebaran |
| 25th | Tau | 4h 48m | 20° 16' | -0.2 | 7" | 58% | 22° W | AM | *** | Hyades, Aldebaran |
| 27th | Tau | 5h 0m | 20° 59' | -0.4 | 6" | 63% | 21° W | AM | *** | Moon, Hyades, Aldebaran |
| 29th | Tau | 5h 14m | 21° 41' | -0.6 | 6" | 68% | 20° W | AM | *** | Aldebaran |

## Venus

| Date | Con. | R.A. | Dec. | Mag. | Diam. | Ill. | Elon. | Vis. | Rat. | Close To |
|------|------|------|------|------|-------|------|-------|------|------|----------|
| 21st | Tau | 3h 45m | 18° 7' | -3.9 | 12" | 84% | 34° W | AM | ** | Pleiades |
| 23rd | Tau | 3h 55m | 18° 41' | -3.9 | 12" | 84% | 33° W | AM | ** | Pleiades, Hyades, Aldebaran |
| 25th | Tau | 4h 4m | 19° 13' | -3.9 | 12" | 85% | 33° W | AM | ** | Moon, Pleiades, Hyades, Aldebar: |
| 27th | Tau | 4h 14m | 19° 43' | -3.9 | 12" | 85% | 33° W | AM | ** | Pleiades, Hyades, Aldebaran |
| 29th | Tau | 4h 24m | 20° 12' | -3.9 | 12" | 86% | 32° W | AM | ** | Pleiades, Hyades, Aldebaran |

## Mars and the Outer Planets

Mars
25th

Jupiter
25th

Saturn
25th

### Mars

| Date | Con. | R.A. | Dec. | Mag. | Diam. | Ill. | Elon. | Vis. | Rat. | Close To |
|------|------|------|------|------|-------|------|-------|------|------|----------|
| 21st | Psc | 1h 17m | 6° 13' | 0.5 | 7" | 86% | 71° W | AM | ** | |
| 25th | Psc | 1h 28m | 7° 17' | 0.5 | 7" | 86% | 72° W | AM | ** | |
| 30th | Psc | 1h 41m | 8° 35' | 0.5 | 7" | 86% | 74° W | AM | ** | |

### The Outer Planets

| Planet | Date | Con. | R.A. | Dec. | Mag. | Diam. | Elon. | Vis. | Rat. | Close To |
|--------|------|------|------|------|------|-------|-------|------|------|----------|
| Jupiter | 25th | Cet | 0h 27m | 1° 36' | -2.4 | 40" | 87° W | AM | *** | |
| Saturn | 25th | Cap | 21h 50m | -14° 20' | 0.6 | 18" | 127° W | AM | **** | |
| Uranus | 25th | Ari | 3h 0m | 16° 43' | 5.8 | 3" | 49° W | AM | * | Moon |
| Neptune | 25th | Psc | 23h 45m | -2° 54' | 7.9 | 2" | 98° W | AM | *** | |

## Highlights

| Date | Time (UT) | Event |
|------|-----------|-------|
| 21st | 03:11 | Last Quarter Moon. (Morning sky.) |
| | 09:14 | Summer Solstice. |
| | 20:05 | Venus is 5.9° south of the Pleiades star cluster. (Taurus, morning sky.) |
| 22nd | 18:33 | The just-past last quarter Moon is south of Mars. (Morning sky.) |
| 23rd | 08:39 | Mercury is 3.0° north of the star Aldebaran. (Taurus, morning sky.) |
| 24th | 21:52 | The waning crescent Moon is south of Uranus. (Morning sky.) |
| 25th | 21:28 | The waning crescent Moon is south of the Pleiades star cluster. (Taurus, morning sky.) |
| | N/A | Good opportunity to see Earthshine on the waning crescent Moon. |
| 26th | 07:18 | The waning crescent Moon is north of Venus. (Morning sky.) |
| | 21:28 | The waning crescent Moon is north of the star Aldebaran. (Taurus, morning sky.) |
| 27th | 07:11 | The waning crescent Moon is north of Mercury. (Morning sky.) |
| | N/A | The Boötid meteor shower is at its maximum. (ZHR: Variable.) |
| 29th | 02:53 | New Moon. (Not visible.) |

# July 1st to 10th, 2022

## The Moon

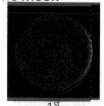
1st

3rd

5th

7th

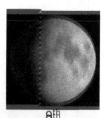
9th

| Date | Con | R.A. | Dec | Mag | Diam | Ill. | Elon. | Phase | Close To |
|------|-----|------|-----|-----|------|------|-------|-------|----------|
| 1st | Cnc | 8h 37m | 23° 38' | -6.5 | 30' | 5% | 29° E | NM | Praesepe |
| 2nd | Leo | 9h 26m | 20° 24' | -7.4 | 30' | 10% | 40° E | NM | Praesepe, Regulus |
| 3rd | Leo | 10h 14m | 16° 17' | -8.1 | 30' | 17% | 51° E | +Cr | Regulus |
| 4th | Leo | 11h 1m | 11° 28' | -8.8 | 30' | 25% | 62° E | +Cr | |
| 5th | Vir | 11h 46m | 6° 6' | -9.3 | 30' | 34% | 72° E | +Cr | |
| 6th | Vir | 12h 32m | 0° 22' | -9.8 | 31' | 44% | 82° E | FQ | |
| 7th | Vir | 13h 19m | -5° 32' | -10.3 | 31' | 54% | 93° E | FQ | Spica |
| 8th | Vir | 14h 8m | -11° 22' | -10.7 | 32' | 65% | 104° E | FQ | Spica |
| 9th | Lib | 15h 1m | -16° 51' | -11.1 | 32' | 75% | 117° E | +G | |
| 10th | Sco | 15h 59m | -21° 33' | -11.5 | 33' | 85% | 130° E | +G | Antares |

## Mercury and Venus

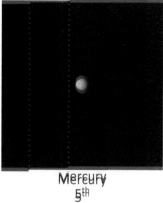
Mercury
5th

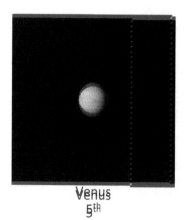
Venus
5th

### Mercury

| Date | Con. | R.A. | Dec. | Mag. | Diam. | Ill. | Elon. | Vis. | Rat. | Close To |
|------|------|------|------|------|-------|------|-------|------|------|----------|
| 1st | Tau | 5h 29m | 22° 19' | -0.8 | 6" | 74% | 18° W | AM | ** | |
| 3rd | Tau | 5h 45m | 22° 51' | -0.9 | 6" | 79% | 16° W | AM | ** | |
| 5th | Gem | 6h 1m | 23° 18' | -1.1 | 6" | 85% | 14° W | NV | N/A | |
| 7th | Gem | 6h 19m | 23° 36' | -1.3 | 5" | 89% | 13° W | NV | N/A | |

## Venus

| Date | Con. | R.A. | Dec. | Mag. | Diam. | Ill. | Elon. | Vis. | Rat. | Close To |
|------|------|------|------|------|-------|------|-------|------|------|----------|
| 1st | Tau | 4h 34m | 20° 38' | -3.9 | 12" | 86% | 32° W | AM | ** | Hyades, Aldebaran |
| 3rd | Tau | 4h 44m | 21° 3' | -3.9 | 12" | 87% | 31° W | AM | ** | Hyades, Aldebaran |
| 5th | Tau | 4h 55m | 21° 23' | -3.9 | 12" | 87% | 31° W | AM | ** | Hyades, Aldebaran |
| 7th | Tau | 5h 5m | 21° 43' | -3.9 | 12" | 87% | 30° W | AM | ** | Hyades, Aldebaran |
| 9th | Tau | 5h 15m | 22° 0' | -3.9 | 12" | 88% | 30° W | AM | * | |

## Mars and the Outer Planets

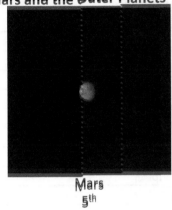

Mars
5th

Jupiter
5th

Saturn
5th

### Mars

| Date | Con. | R.A. | Dec. | Mag. | Diam. | Ill. | Elon. | Vis. | Rat. | Close To |
|------|------|------|------|------|-------|------|-------|------|------|----------|
| 1st | Psc | 1h 44m | 8° 50' | 0.5 | 7" | 86% | 74° W | AM | ** | |
| 5th | Psc | 1h 54m | 9° 50' | 0.4 | 7" | 85% | 76° W | AM | ** | |
| 10th | Ari | 2h 8m | 11° 3' | 0.4 | 8" | 85% | 78° W | AM | ** | |

### The Outer Planets

| Planet | Date | Con. | R.A. | Dec. | Mag. | Diam. | Elon. | Vis. | Rat. | Close To |
|--------|------|------|------|------|------|-------|-------|------|------|----------|
| Jupiter | 5th | Cet | 0h 30m | 1° 54' | -2.5 | 41" | 97° W | AM | *** | |
| Saturn | 5th | Cap | 21h 48m | -14° 30' | 0.5 | 18" | 137° W | AM | **** | |
| Uranus | 5th | Ari | 3h 2m | 16° 50' | 5.8 | 3" | 59° W | AM | ** | |
| Neptune | 5th | Psc | 23h 45m | -2° 54' | 7.9 | 2" | 108° W | AM | *** | |

## Highlights

| Date | Time (UT) | Event |
|------|-----------|-------|
| 1st | 17:17 | Venus is 4.2° north of the star Aldebaran. (Taurus, morning sky.) |
| 2nd | N/A | Good opportunity to see Earthshine on the waxing crescent Moon. (Evening sky.) |
| 3rd | 07:34 | The waxing crescent Moon is north of the star Regulus. (Leo, evening sky.) |
| 7th | 02:15 | First Quarter Moon. (Evening sky.) |
| | 13:44 | The first quarter Moon is north of the star Spica. (Virgo, evening sky.) |

# July 11<sup>th</sup> to 20<sup>th</sup>, 2022

## The Moon

| 11<sup>th</sup> | 13<sup>th</sup> | 15<sup>th</sup> | 17<sup>th</sup> | 19<sup>th</sup> |

| Date | Con | R.A. | Dec | Mag | Diam | Ill. | Elon. | Phase | Close To |
|------|-----|------|-----|-----|------|------|-------|-------|----------|
| 11th | Oph | 17h 1m | -25° 0' | -11.9 | 33' | 92% | 145° E | +G | Antares |
| 12th | Sgr | 18h 8m | -26° 46' | -12.2 | 33' | 97% | 160° E | FM | |
| 13th | Sgr | 19h 16m | -26° 32' | -12.6 | 33' | 100% | 176° E | FM | |
| 14th | Cap | 20h 23m | -24° 19' | -12.4 | 33' | 99% | 168° W | FM | |
| 15th | Cap | 21h 26m | -20° 24' | -12.1 | 33' | 95% | 153° W | FM | Saturn |
| 16th | Aqr | 22h 24m | -15° 18' | -11.7 | 33' | 89% | 140° W | -G | Saturn |
| 17th | Aqr | 23h 17m | -9° 29' | -11.4 | 32' | 81% | 127° W | -G | Neptune |
| 18th | Psc | 0h 7m | -3° 24' | -11.0 | 32' | 71% | 116° W | -G | Jupiter, Neptune |
| 19th | Cet | 0h 55m | 2° 39' | -10.6 | 31' | 61% | 105° W | LQ | Jupiter |
| 20th | Psc | 1h 41m | 8° 23' | -10.2 | 31' | 51% | 94° W | LQ | |

## Mercury and Venus

Mercury
15<sup>th</sup>

Venus
15<sup>th</sup>

**Mercury**

| Date | Con. | R.A. | Dec. | Mag. | Diam. | Ill. | Elon. | Vis. | Rat. | Close To |
|------|------|------|------|------|-------|------|-------|------|------|----------|
| 11th | Gem | 6h 56m | 23° 43' | -1.7 | 5" | 96% | 7° W | NV | N/A | |
| 13th | Gem | 7h 14m | 23° 30' | -1.9 | 5" | 99% | 4° W | NV | N/A | |
| 15th | Gem | 7h 33m | 23° 7' | -2.1 | 5" | 100% | 1° W | NV | N/A | |
| 17th | Gem | 7h 52m | 22° 33' | -2.1 | 5" | 100% | 1° E | NV | N/A | |
| 19th | Cnc | 8h 10m | 21° 49' | -1.8 | 5" | 99% | 4° E | NV | N/A | Praesepe |

**Venus**

| Date | Con. | R.A. | Dec. | Mag. | Diam. | Ill. | Elon. | Vis. | Rat. | Close To |
|------|------|------|------|------|-------|------|-------|------|------|----------|
| 11th | Tau | 5h 25m | 22° 15' | -3.9 | 11" | 88% | 29° W | AM | * | |
| 13th | Tau | 5h 36m | 22° 27' | -3.9 | 11" | 89% | 29° W | AM | * | |
| 15th | Tau | 5h 46m | 22° 37' | -3.9 | 11" | 89% | 28° W | AM | * | |
| 17th | Ori | 5h 57m | 22° 44' | -3.9 | 11" | 90% | 28° W | AM | * | |
| 19th | Gem | 6h 7m | 22° 49' | -3.9 | 11" | 90% | 27° W | AM | * | |

## Mars and the Outer Planets

Mars
15th

Jupiter
15th

Saturn
15th

**Mars**

| Date | Con. | R.A. | Dec. | Mag. | Diam. | Ill. | Elon. | Vis. | Rat. | Close To |
|------|------|------|------|------|-------|------|-------|------|------|----------|
| 11th | Ari | 2h 10m | 11° 17' | 0.4 | 8" | 85% | 78° W | AM | ** | |
| 15th | Ari | 2h 21m | 12° 12' | 0.3 | 8" | 85% | 79° W | AM | ** | |
| 20th | Ari | 2h 34m | 13° 18' | 0.3 | 8" | 85% | 81° W | AM | ** | |

**The Outer Planets**

| Planet | Date | Con. | R.A. | Dec. | Mag. | Diam. | Elon. | Vis. | Rat. | Close To |
|--------|------|------|------|------|------|-------|-------|------|------|----------|
| Jupiter | 15th | Cet | 0h 33m | 2° 5' | -2.5 | 43" | 107° W | AM | **** | |
| Saturn | 15th | Cap | 21h 46m | -14° 42' | 0.5 | 18" | 148° W | AM | **** | Moon |
| Uranus | 15th | Ari | 3h 3m | 16° 55' | 5.8 | 4" | 69° W | AM | ** | |
| Neptune | 15th | Psc | 23h 44m | -2° 56' | 7.9 | 2" | 119° W | AM | **** | |

## Highlights

| Date | Time (UT) | Event |
|------|-----------|-------|
| 11th | 00:41 | The waxing gibbous Moon is north of the star Antares. (Scorpius, evening sky.) |
| 13th | 18:38 | Full Moon. (Visible all night.) |
| 14th | 03:51 | The just-past full Moon is south of dwarf planet Pluto. (Morning sky.) |
| 15th | 19:09 | The waning gibbous Moon is south of Saturn. (Morning sky.) |
| 16th | 19:24 | Mercury is at superior conjunction with the Sun. (Not visible.) |
| 17th | 23:40 | The waning gibbous Moon is south of Neptune. (Morning sky.) |
| 18th | 23:41 | The waning gibbous Moon is south of Jupiter. (Morning sky.) |
| 20th | 05:34 | Dwarf planet Pluto is at opposition. (Visible all night.) |
| | 14:19 | Last Quarter Moon. (Morning sky.) |

# July 21ˢᵗ to 31ˢᵗ, 2022

## The Moon

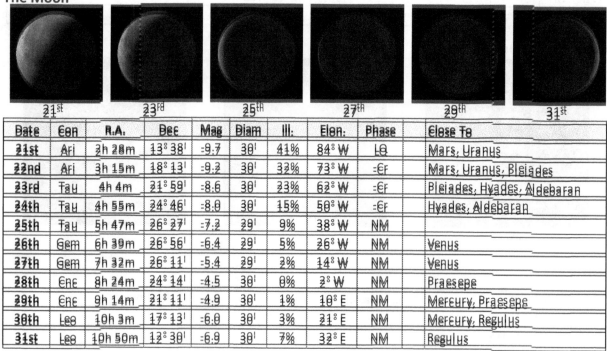

| | 21st | | 23rd | | 25th | | 27th | | 29th | | 31st |

| Date | Con | R.A. | Dec | Mag | Diam | Ill. | Elon. | Phase | Close To |
|------|-----|------|-----|-----|------|------|-------|-------|----------|
| 21st | Ari | 2h 28m | 13° 38' | -9.7 | 30' | 41% | 84° W | LQ | Mars, Uranus |
| 22nd | Ari | 3h 15m | 18° 13' | -9.2 | 30' | 32% | 73° W | Cr | Mars, Uranus, Pleiades |
| 23rd | Tau | 4h 4m | 21° 59' | -8.6 | 30' | 23% | 62° W | Cr | Pleiades, Hyades, Aldebaran |
| 24th | Tau | 4h 55m | 24° 46' | -8.0 | 30' | 15% | 50° W | Cr | Hyades, Aldebaran |
| 25th | Tau | 5h 47m | 26° 27' | -7.2 | 29' | 9% | 38° W | NM | |
| 26th | Gem | 6h 39m | 26° 56' | -6.4 | 29' | 5% | 26° W | NM | Venus |
| 27th | Gem | 7h 32m | 26° 11' | -5.4 | 29' | 2% | 14° W | NM | Venus |
| 28th | Cnc | 8h 24m | 24° 14' | -4.5 | 30' | 0% | 2° W | NM | Praesepe |
| 29th | Cnc | 9h 14m | 21° 11' | -4.9 | 30' | 1% | 10° E | NM | Mercury, Praesepe |
| 30th | Leo | 10h 3m | 17° 13' | -6.0 | 30' | 3% | 21° E | NM | Mercury, Regulus |
| 31st | Leo | 10h 50m | 12° 30' | -6.9 | 30' | 7% | 32° E | NM | Regulus |

## Mercury and Venus

**Mercury**
**25th**

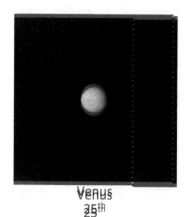

**Venus**
**25th**

### Mercury

| Date | Con. | R.A. | Dec. | Mag. | Diam. | Ill. | Elon. | Vis. | Rat. | Close To |
|------|------|------|------|------|-------|------|-------|------|------|----------|
| 21st | Cnc | 8h 27m | 20° 57' | -1.6 | 5" | 98% | 6° E | NV | N/A | Praesepe |
| 23rd | Cnc | 8h 44m | 19° 57' | -1.3 | 5" | 96% | 8° E | NV | N/A | Praesepe |
| 25th | Cnc | 9h 0m | 18° 51' | -1.1 | 5" | 94% | 10° E | NV | N/A | Praesepe |
| 27th | Cnc | 9h 16m | 17° 40' | -0.9 | 5" | 93% | 12° E | NV | N/A | Praesepe |
| 29th | Leo | 9h 30m | 16° 25' | -0.8 | 5" | 89% | 14° E | NV | N/A | Moon, Regulus |
| 31st | Leo | 9h 44m | 15° 7' | -0.6 | 5" | 87% | 16° E | PM | ** | Regulus |

## Venus

| Date | Con. | R.A. | Dec. | Mag. | Diam. | Ill. | Elon. | Vis. | Rat. | Close To |
|------|------|------|------|------|-------|------|-------|------|------|----------|
| 21st | Gem | 6h 18m | 22°51' | -3.9 | 11" | 91% | 26° W | AM | * | |
| 23rd | Gem | 6h 28m | 22°51' | -3.9 | 11" | 91% | 26° W | AM | * | |
| 25th | Gem | 6h 39m | 22°48' | -3.9 | 11" | 91% | 25° W | AM | * | |
| 27th | Gem | 6h 49m | 22°42' | -3.9 | 11" | 92% | 24° W | AM | * | Moon |
| 29th | Gem | 7h 0m | 22°34' | -3.9 | 11" | 92% | 24° W | AM | * | |
| 31st | Gem | 7h 10m | 22°23' | -3.9 | 11" | 92% | 23° W | AM | * | |

## Mars and the Outer Planets

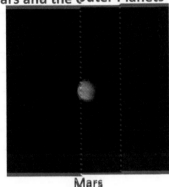

Mars
25th

Jupiter
25th

Saturn
25th

### Mars

| Date | Con. | R.A. | Dec. | Mag. | Diam. | Ill. | Elon. | Vis. | Rat. | Close To |
|------|------|------|------|------|-------|------|-------|------|------|----------|
| 21st | Ari | 2h 37m | 13°30' | 0.3 | 8" | 85% | 82° W | AM | ** | Moon |
| 25th | Ari | 2h 47m | 14°20' | 0.3 | 8" | 85% | 83° W | AM | ** | Uranus |
| 31st | Ari | 3h 3m | 15°30' | 0.2 | 8" | 85% | 85° W | AM | ** | Uranus |

### The Outer Planets

| Planet | Date | Con. | R.A. | Dec. | Mag. | Diam. | Ill. | Elon. | Vis. | Rat. | Close To |
|--------|------|------|------|------|------|-------|------|-------|------|------|----------|
| Jupiter | 25th | Cet | 0h 34m | 2°9' | -2.6 | 44" | | 116° W | AM | **** | |
| Saturn | 25th | Cap | 21h 44m | -14°56' | 0.4 | 19" | | 159° W | AM | **** | |
| Uranus | 25th | Ari | 3h 4m | 16°60' | 5.8 | 4" | | 79° W | AM | ** | Mars |
| Neptune | 25th | Psc | 23h 44m | -2°59' | 7.8 | 2" | | 139° W | AM | **** | |

## Highlights

| Date | Time (UT) | Event |
|------|-----------|-------|
| 21st | 17:18 | The just-past last quarter Moon is north of Mars. (Morning sky.) |
| 22nd | 06:02 | The waning crescent Moon is south of Uranus. (Morning sky.) |
| 23rd | 02:11 | The waning crescent Moon is south of the Pleiades star cluster. (Taurus, morning sky.) |
| | 21:50 | Dwarf planet Ceres is in conjunction with the Sun. (Not visible.) |
| 24th | 01:45 | The waning crescent Moon is north of the bright star Aldebaran. (Taurus, morning sky.) |
| 25th | N/A | Good opportunity to see Earthshine on the waning crescent Moon. (Morning sky.) |
| 26th | 15:38 | The waning crescent Moon is north of Venus. (Morning sky.) |
| 28th | 17:56 | New Moon (Not visible.) |
| 29th | 11:11 | Jupiter is stationary prior to beginning retrograde motion. (Morning sky.) |
| 30th | 15:04 | The waxing crescent Moon is north of the bright star Regulus. (Evening sky.) |
| | N/A | The Delta Aquariid meteor shower is at its maximum. (ZHR: 16) |
| 31st | N/A | Good opportunity to see Earthshine on the waxing crescent Moon. (Evening sky.) |

# August 1st to 10th, 2022

## The Moon

 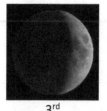   

| 1st | 3rd | 5th | 7th | 9th |

| Date | Con | R.A. | Dec | Mag | Diam | Ill. | Elon. | Phase | Close To |
|------|-----|------|-----|-----|------|------|-------|-------|----------|
| 1st | Leo | 11h 35m | 7° 13' | -7.7 | 30' | 13% | 42° E | +Cr | |
| 2nd | Vir | 12h 20m | 1° 34' | -8.4 | 31' | 21% | 53° E | +Cr | |
| 3rd | Vir | 13h 6m | -4° 16' | -9.1 | 31' | 30% | 63° E | +Cr | Spica |
| 4th | Vir | 13h 54m | -10° 2' | -9.6 | 31' | 40% | 74° E | FQ | Spica |
| 5th | Lib | 14h 44m | -15° 30' | -10.1 | 32' | 50% | 86° E | FQ | |
| 6th | Lib | 15h 39m | -20° 19' | -10.6 | 32' | 61% | 98° E | FQ | |
| 7th | Oph | 16h 38m | -24° 6' | -11.0 | 33' | 72% | 112° E | +G | Antares |
| 8th | Oph | 17h 41m | -26° 27' | -11.4 | 33' | 82% | 127° E | +G | |
| 9th | Sgr | 18h 48m | -26° 58' | -11.8 | 33' | 90% | 143° E | +G | |

## Mercury and Venus

Mercury
5th

Venus
5th

**Mercury**

| Date | Con. | R.A. | Dec. | Mag. | Diam. | Ill. | Elon. | Vis. | Rat. | Close To |
|------|------|------|------|------|-------|------|-------|------|------|----------|
| 1st | Leo | 9h 51m | 14° 27' | -0.6 | 5" | 86% | 16° E | PM | ** | Regulus |
| 3rd | Leo | 10h 4m | 13° 6' | -0.5 | 5" | 83% | 18° E | PM | ** | Regulus |
| 5th | Leo | 10h 17m | 11° 43' | -0.4 | 6" | 81% | 19° E | PM | ** | Regulus |
| 7th | Leo | 10h 29m | 10° 19' | -0.3 | 6" | 79% | 20° E | PM | *** | Regulus |

### Venus

| Date | Con. | R.A. | Dec. | Mag. | Diam. | Ill. | Elon. | Vis. | Rat. | Close To |
|------|------|------|------|------|-------|------|-------|------|------|----------|
| 1st | Gem | 7h 15m | 22° 16' | -3.9 | 11" | 93% | 23° W | AM | * | |
| 3rd | Gem | 7h 26m | 22° 1' | -3.9 | 11" | 93% | 22° W | AM | * | |
| 5th | Gem | 7h 36m | 21° 44' | -3.9 | 11" | 93% | 21° W | AM | * | |
| 7th | Gem | 7h 47m | 21° 24' | -3.9 | 11" | 94% | 21° W | AM | * | |
| 9th | Gem | 7h 57m | 21° 2' | -3.9 | 11" | 94% | 20° W | AM | * | |

## Mars and the Outer Planets

Mars
5th

Jupiter
5th

Saturn
5th

### Mars

| Date | Con. | R.A. | Dec. | Mag. | Diam. | Ill. | Elon. | Vis. | Rat. | Close To |
|------|------|------|------|------|-------|------|-------|------|------|----------|
| 1st | Ari | 3h 5m | 15° 41' | 0.2 | 8" | 85% | 85° W | AM | ** | Uranus |
| 5th | Ari | 3h 15m | 16° 24' | 0.2 | 8" | 85% | 87° W | AM | ** | Uranus, Pleiades |
| 10th | Tau | 3h 28m | 17° 15' | 0.1 | 9" | 85% | 88° W | AM | ** | Pleiades |

### The Outer Planets

| Planet | Date | Con. | R.A. | Dec. | Mag. | Diam. | Elon. | Vis. | Rat. | Close To |
|--------|------|------|------|------|------|-------|-------|------|------|----------|
| Jupiter | 5th | Cet | 0h 33m | 2° 4' | -2.7 | 46" | 127° W | AM | **** | |
| Saturn | 5th | Cap | 21h 41m | -15° 12' | 0.3 | 19" | 170° W | AN | **** | |
| Uranus | 5th | Ari | 3h 5m | 17° 3' | 5.8 | 4" | 89° W | AM | ** | Mars |
| Neptune | 5th | Psc | 23h 43m | -3° 4' | 7.8 | 2" | 140° W | AM | **** | |

## Highlights

| Date | Time (UT) | Event |
|------|-----------|-------|
| 1st | 09:24 | Mars is 1.4° south of Uranus. (Morning sky.) |
| 3rd | 22:47 | The waxing crescent Moon is to the north of the bright star Spica. (Virgo, evening sky.) |
| 4th | 00:33 | Mercury is 0.8° north of the bright star Regulus. (Leo, evening sky.) |
| 5th | 11:07 | First Quarter Moon. (Evening sky.) |
| 7th | 08:11 | The waxing gibbous Moon is north of the bright star Antares. (Scorpius, evening sky.) |
| 10th | 11:59 | The waxing gibbous Moon is south of dwarf planet Pluto. (Evening sky.) |

# August 11th to 20th, 2022

## The Moon

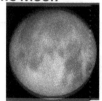    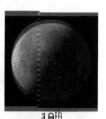

| 11th | 13th | 15th | 17th | 19th |

| Date | Con | R.A. | Dec | Mag | Diam | Ill. | Elon. | Phase | Close To |
|------|-----|------|-----|-----|------|------|-------|-------|----------|
| 11th | Cap | 20h 59m | -22° 17' | -12.5 | 33' | 99% | 174° E | FM | Saturn |
| 12th | Cap | 21h 59m | -17° 34' | -12.5 | 33' | 100% | 172° W | FM | Saturn |
| 13th | Aqr | 22h 55m | -11° 52' | -12.2 | 33' | 97% | 159° W | FM | Neptune |
| 14th | Aqr | 23h 47m | -5° 41' | -11.9 | 32' | 93% | 147° W | =G | Jupiter, Neptune |
| 15th | Cet | 0h 37m | 0° 36' | -11.5 | 32' | 85% | 136° W | =G | Jupiter |
| 16th | Psc | 1h 25m | 6° 40' | -11.2 | 31' | 76% | 125° W | =G | |
| 17th | Ari | 2h 12m | 12° 15' | -10.8 | 31' | 67% | 114° W | =G | |
| 18th | Ari | 3h 0m | 17° 9' | -10.4 | 30' | 57% | 103° W | LQ | Mars, Uranus, Pleiades |
| 19th | Tau | 3h 50m | 21° 14' | -10.0 | 30' | 47% | 91° W | LQ | Mars, Uranus, Pleiades, Hyades, |
| 20th | Tau | 4h 40m | 24° 20' | -9.5 | 30' | 37% | 80° W | =Cr | Mars, Hyades, Aldebaran |

## Mercury and Venus

Mercury
15th

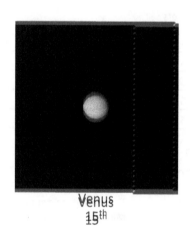

Venus
15th

### Mercury

| Date | Con. | R.A. | Dec. | Mag. | Diam. | Ill. | Elon. | Vis. | Rat. | Close To |
|------|------|------|------|------|-------|------|-------|------|------|----------|
| 11th | Leo | 10h 52m | 7° 32' | -0.1 | 6" | 74% | 22° E | PM | *** | |
| 13th | Leo | 11h 2m | 6° 9' | 0.0 | 6" | 72% | 23° E | PM | *** | |
| 15th | Leo | 11h 12m | 4° 47' | 0.0 | 6" | 69% | 23° E | PM | *** | |
| 17th | Leo | 11h 21m | 3° 27' | 0.1 | 6" | 67% | 24° E | PM | *** | |
| 19th | Leo | 11h 30m | 2° 9' | 0.1 | 6" | 65% | 24° E | PM | *** | |

## Venus

| Date | Con. | R.A. | Dec. | Mag. | Diam. | Ill. | Elon. | Vis. | Rat. | Close To |
|------|------|------|------|------|-------|------|-------|------|------|----------|
| 11th | Cnc | 8h 7m | 20° 37' | -3.9 | 10" | 94% | 19° W | AM | * | Praesepe |
| 13th | Cnc | 8h 17m | 20° 10' | -3.9 | 10" | 95% | 19° W | AM | * | Praesepe |
| 15th | Cnc | 8h 28m | 19° 40' | -3.9 | 10" | 95% | 18° W | AM | * | Praesepe |
| 17th | Cnc | 8h 38m | 19° 8' | -3.9 | 10" | 95% | 17° W | AM | * | Praesepe |
| 19th | Cnc | 8h 48m | 18° 34' | -3.9 | 10" | 96% | 17° W | AM | * | Praesepe |

## Mars and the Outer Planets

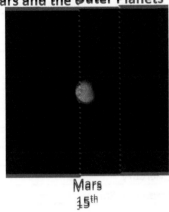

Mars
15th

Jupiter
15th

Saturn
15th

## Mars

| Date | Con. | R.A. | Dec. | Mag. | Diam. | Ill. | Elon. | Vis. | Rat. | Close To |
|------|------|------|------|------|-------|------|-------|------|------|----------|
| 11th | Tau | 3h 30m | 17° 24' | 0.1 | 9" | 85% | 89° W | AM | ** | Pleiades |
| 15th | Tau | 3h 40m | 18° 1' | 0.1 | 9" | 85% | 90° W | AM | ** | Pleiades |
| 20th | Tau | 3h 53m | 18° 44' | 0.0 | 9" | 85% | 91° W | AM | ** | Moon, Pleiades, Hyades |

## The Outer Planets

| Planet | Date | Con. | R.A. | Dec. | Mag. | Diam. | Elon. | Vis. | Rat. | Close To |
|--------|------|------|------|------|------|-------|-------|------|------|----------|
| Jupiter | 15th | Cet | 0h 32m | 1° 52' | -2.8 | 47" | 137° W | AM | ***** | Moon |
| Saturn | 15th | Cap | 21h 38m | -15° 28' | 0.3 | 19" | 180° E | AN | ***** | |
| Uranus | 15th | Ari | 3h 6m | 17° 5' | 5.7 | 4" | 98° W | AM | ** | Mars |
| Neptune | 15th | Psc | 23h 43m | -3° 9' | 7.8 | 2" | 149° W | AM | **** | |

## Highlights

| Date | Time (UT) | Event |
|------|-----------|-------|
| 12th | 01:36 | Full Moon. (Visible all night.) |
| | 05:04 | The full Moon appears to the south of Saturn. (Visible all night.) |
| 13th | N/A | The Perseid meteor shower is at its maximum. (ZHR: 100) |
| 14th | 10:54 | The just-past full Moon is south of Neptune. (Morning sky.) |
| 15th | 03:05 | Saturn is at opposition. (Visible all night.) |
| | 10:54 | The waning gibbous Moon is south of Jupiter. (Morning sky.) |
| 17th | 19:44 | Venus is 0.6° south of the Praesepe star cluster. (Cancer, morning sky.) |
| 18th | 00:11 | Mars is 5.7° south of the Pleiades star cluster. (Taurus, morning sky.) |
| | 15:21 | The nearly last quarter Moon is south of Uranus. (Morning sky.) |
| 19th | 04:37 | Last Quarter Moon. (Morning sky.) |
| | 1156 | The last quarter Moon is south of the Pleiades star cluster. (Taurus, morning sky.) |
| | 13:38 | The last quarter Moon is north of Mars. (Morning sky.) |
| 20th | 11:05 | The just-past last quarter Moon is north of the bright star Aldebaran. (Taurus, morning sky.) |

# August 21st to 31st, 2022

## The Moon

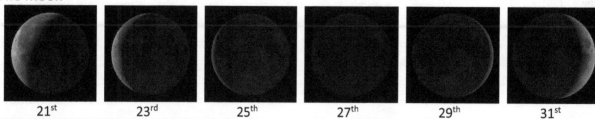

| 21st | 23rd | 25th | 27th | 29th | 31st |

| Date | Con | R.A. | Dec | Mag | Diam | Ill. | Elon. | Phase | Close To |
|------|-----|------|-----|-----|------|------|-------|-------|----------|
| 21st | Tau | 5h 32m | 26° 19' | -9.0 | 30' | 29% | 67° W | -Cr | |
| 22nd | Gem | 6h 25m | 27° 6' | -8.4 | 29' | 20% | 55° W | -Cr | |
| 23rd | Gem | 7h 18m | 26° 38' | -7.8 | 29' | 13% | 43° W | -Cr | |
| 24th | Cnc | 8h 10m | 24° 57' | -7.0 | 30' | 8% | 31° W | NM | Praesepe |
| 25th | Cnc | 9h 1m | 22° 8' | -6.1 | 30' | 3% | 19° W | NM | Venus, Praesepe |
| 26th | Leo | 9h 50m | 18° 20' | -5.1 | 30' | 1% | 7° W | NM | Venus, Regulus |
| 27th | Leo | 10h 38m | 13° 42' | -4.4 | 30' | 0% | 3° E | NM | Regulus |
| 28th | Leo | 11h 24m | 8° 27' | -5.4 | 30' | 1% | 14° E | NM | Mercury |
| 29th | Vir | 12h 10m | 2° 46' | -6.4 | 31' | 5% | 25° E | NM | Mercury |
| 30th | Vir | 12h 55m | -3° 7' | -7.3 | 31' | 10% | 35° E | NM | Mercury, Spica |
| 31st | Vir | 13h 43m | -8° 58' | -8.2 | 31' | 17% | 46° E | +Cr | Spica |

## Mercury and Venus

Mercury
25th

Venus
25th

**Mercury**

| Date | Con. | R.A. | Dec. | Mag. | Diam. | Ill. | Elon. | Vis. | Rat. | Close To |
|------|------|------|------|------|-------|------|-------|------|------|----------|
| 21st | Vir | 11h 39m | 0° 53' | 0.1 | 7" | 62% | 24° E | PM | *** | |
| 23rd | Vir | 11h 47m | 0° 20' | 0.2 | 7" | 59% | 24° E | PM | *** | |
| 25th | Vir | 11h 54m | -1° 30' | 0.2 | 7" | 57% | 24° E | PM | *** | |
| 27th | Vir | 12h 1m | -2° 36' | 0.3 | 7" | 54% | 24° E | PM | *** | |
| 29th | Vir | 12h 8m | -3° 37' | 0.3 | 7" | 50% | 24° E | PM | *** | Moon |
| 31st | Vir | 12h 13m | -4° 33' | 0.4 | 8" | 47% | 24° E | PM | *** | |

**Venus**

| Date | Con. | R.A. | Dec. | Mag. | Diam. | Ill. | Elon. | Vis. | Rat. | Close To |
|------|------|------|------|------|-------|------|-------|------|------|----------|
| 21st | Cnc | 8h 58m | 17° 58' | -3.9 | 10" | 96% | 16° W | AM | * | Praesepe |
| 23rd | Cnc | 9h 8m | 17° 20' | -3.9 | 10" | 96% | 15° W | AM | * | Praesepe |
| 25th | Cnc | 9h 18m | 16° 40' | -3.9 | 10" | 96% | 15° W | NV | N/A | Moon, Praesepe |
| 27th | Leo | 9h 28m | 15° 58' | -3.9 | 10" | 97% | 14° W | NV | N/A | Regulus |
| 29th | Leo | 9h 37m | 15° 14' | -3.9 | 10" | 97% | 13° W | NV | N/A | Regulus |
| 31st | Leo | 9h 47m | 14° 28' | -3.9 | 10" | 97% | 13° W | NV | N/A | Regulus |

## Mars and the Outer Planets

Mars
25th

Jupiter
25th

Saturn
25th

**Mars**

| Date | Con. | R.A. | Dec. | Mag. | Diam. | Ill. | Elon. | Vis. | Rat. | Close To |
|------|------|------|------|------|-------|------|-------|------|------|----------|
| 21st | Tau | 3h 55m | 18° 52' | 0.0 | 9" | 85% | 92° W | AM | ** | Pleiades, Hyades, Aldebaran |
| 25th | Tau | 4h 4m | 19° 22' | 0.0 | 9" | 85% | 93° W | AM | ** | Pleiades, Hyades, Aldebaran |
| 31st | Tau | 4h 18m | 20° 4' | -0.1 | 10" | 85% | 95° W | AM | ** | Pleiades, Hyades, Aldebaran |

**The Outer Planets**

| Planet | Date | Con. | R.A. | Dec. | Mag. | Diam. | Elon. | Vis. | Rat. | Close To |
|--------|------|------|------|------|------|-------|-------|------|------|----------|
| Jupiter | 25th | Cet | 0h 29m | 1° 33' | -2.8 | 48" | 147° W | AM | ***** | |
| Saturn | 25th | Cap | 21h 35m | -15° 43' | 0.3 | 19" | 170° E | PM | **** | |
| Uranus | 25th | Ari | 3h 6m | 17° 6' | 5.7 | 4" | 108° W | AM | *** | |
| Neptune | 25th | Aqr | 23h 42m | -3° 15' | 7.8 | 2" | 159° W | AM | ***** | |

## Highlights

| Date | Time (UT) | Event |
|------|-----------|-------|
| 23rd | N/A | Good opportunity to see Earthshine on the waning crescent Moon. (Morning sky.) |
| 24th | 12:20 | Uranus is stationary prior to beginning retrograde motion. (Morning sky.) |
| 25th | 00:55 | The waning crescent Moon is north of the Praesepe star cluster. (Cancer, morning sky.) |
| | 21:34 | The waning crescent Moon is north of Venus. (Morning sky.) |
| | 21:37 | Asteroid Vesta is at opposition. (Visible all night.) |
| 27th | 08:18 | New Moon. (Not visible.) |
| | 15:59 | Mercury is at greatest eastern elongation. (Evening sky.) |
| 29th | 09:33 | The waxing crescent Moon is north of Mercury. (Evening sky.) |
| 30th | N/A | Good opportunity to see Earthshine on the waxing crescent Moon. (Evening sky.) |
| 31st | 02:54 | The waxing crescent Moon is north of the bright star Spica. (Virgo, evening sky.) |

# September 1st to 10th, 2022

## The Moon

  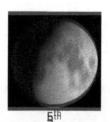  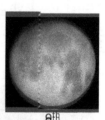

1st     3rd     5th     7th     9th

| Date | Con | R.A. | Dec | Mag | Diam | Ill. | Elon. | Phase | Close To |
|------|-----|------|-----|-----|------|------|-------|-------|----------|
| 1st | Lib | 14h 32m | -14° 31' | -8.9 | 31' | 26% | 58° E | +Cr | |
| 2nd | Lib | 15h 25m | -19° 28' | -9.5 | 32' | 36% | 70° E | +Cr | |
| 3rd | Sco | 16h 22m | -23° 29' | -10.0 | 32' | 47% | 83° E | FQ | Antares |
| 4th | Oph | 17h 22m | -26° 10' | -10.5 | 32' | 58% | 97° E | FQ | |
| 5th | Sgr | 18h 26m | -27° 13' | -10.9 | 33' | 70% | 112° E | +G | |
| 6th | Sgr | 19h 31m | -26° 26' | -11.3 | 33' | 80% | 128° E | +G | |
| 7th | Cap | 20h 35m | -23° 50' | -11.7 | 33' | 88% | 143° E | +G | |
| 8th | Cap | 21h 35m | -19° 40' | -12.0 | 33' | 95% | 157° E | +G | Saturn |
| 9th | Aqr | 22h 32m | -14° 20' | -12.4 | 33' | 99% | 170° E | FM | |
| 10th | Aqr | 23h 25m | -8° 16' | -12.6 | 32' | 100% | 177° W | FM | Neptune |

## Mercury and Venus

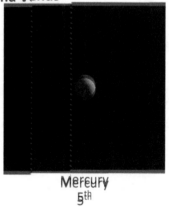 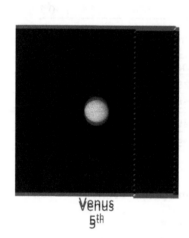

Mercury
5th

Venus
5th

### Mercury

| Date | Con. | R.A. | Dec. | Mag. | Diam. | Ill. | Elon. | Vis. | Rat. | Close To |
|------|------|------|------|------|-------|------|-------|------|------|----------|
| 1st | Vir | 12h 15m | -4° 59' | 0.4 | 8" | 45% | 23° E | PM | *** | |
| 3rd | Vir | 12h 20m | -5° 45' | 0.5 | 8" | 41% | 23° E | PM | *** | |
| 5th | Vir | 12h 23m | -6° 23' | 0.6 | 8" | 37% | 22° E | PM | *** | |
| 7th | Vir | 12h 25m | -6° 52' | 0.8 | 9" | 33% | 20° E | PM | *** | |
| 9th | Vir | 12h 26m | -7° 10' | 1.0 | 9" | 28% | 19° E | PM | *** | |

## Venus

| Date | Con. | R.A. | Dec. | Mag. | Diam. | Ill. | Elon. | Vis. | Rat. | Close To |
|------|------|------|------|------|-------|------|-------|------|------|----------|
| 1st | Leo | 9h 52m | 14° 5' | -3.9 | 10" | 97% | 13° W | NV | N/A | Regulus |
| 3rd | Leo | 10h 2m | 13° 17' | -3.9 | 10" | 97% | 12° W | NV | N/A | Regulus |
| 5th | Leo | 10h 11m | 12° 28' | -3.9 | 10" | 98% | 11° W | NV | N/A | Regulus |
| 7th | Leo | 10h 21m | 11° 37' | -3.9 | 10" | 98% | 11° W | NV | N/A | Regulus |
| 9th | Leo | 10h 30m | 10° 45' | -3.9 | 10" | 98% | 10° W | NV | N/A | Regulus |

## Mars and the Outer Planets

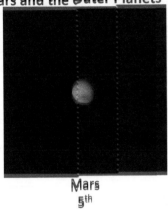

Mars
5th

Jupiter
5th

Saturn
5th

## Mars

| Date | Con. | R.A. | Dec. | Mag. | Diam. | Ill. | Elon. | Vis. | Rat. | Close To |
|------|------|------|------|------|-------|------|-------|------|------|----------|
| 1st | Tau | 4h 20m | 20° 10' | -0.1 | 10" | 85% | 95° W | AM | ** | Pleiades, Hyades, Aldebaran |
| 5th | Tau | 4h 29m | 20° 34' | -0.2 | 10" | 85% | 97° W | AM | ** | Hyades, Aldebaran |
| 10th | Tau | 4h 40m | 21° 2' | -0.3 | 10" | 86% | 99° W | AM | *** | Hyades, Aldebaran |

## The Outer Planets

| Planet | Date | Con. | R.A. | Dec. | Mag. | Diam. | Elon. | Vis. | Rat. | Close To |
|--------|------|------|------|------|------|-------|-------|------|------|----------|
| Jupiter | 5th | Psc | 0h 25m | 1° 5' | -2.9 | 49" | 158° W | AM | ***** | |
| Saturn | 5th | Cap | 21h 32m | -15° 58' | 0.4 | 19" | 159° E | PM | **** | |
| Uranus | 5th | Ari | 3h 6m | 17° 5' | 5.7 | 4" | 118° W | AM | *** | |
| Neptune | 5th | Aqr | 23h 41m | -3° 22' | 7.8 | 2" | 169° W | AM | ***** | |

## Highlights

| Date | Time (UT) | Event |
|------|-----------|-------|
| 1st | N/A | The Alpha Aurigid meteor shower is at its maximum. (ZHR: 6) |
| 3rd | 13:52 | The almost first quarter Moon is north of the bright star Antares. (Scorpius, evening sky.) |
| | 18:08 | First quarter Moon. (Evening sky.) |
| 6th | 20:07 | The waxing gibbous Moon is south of dwarf planet Pluto. (Evening sky.) |
| 8th | 10:00 | Mars is 4.3° north of the bright star Aldebaran. (Taurus, morning sky.) |
| | 16:28 | The waxing gibbous Moon is south of Saturn. (Evening sky.) |
| 9th | 19:36 | Mercury is stationary prior to beginning retrograde motion. (Evening sky.) |
| 10th | 16:00 | Full Moon. (Visible all night.) |
| | 17:35 | The full Moon is south of Neptune. (Visible all night.) |
| | N/A | The Epsilon Perseid meteor shower is at its maximum. (ZHR: 5) |

# September 11ᵗʰ to 20ᵗʰ, 2022

## The Moon

| | 11th | 13th | 15th | 17th | 19th |

| Date | Con | R.A. | Dec | Mag | Diam | Ill. | Elon. | Phase | Close To |
|------|-----|------|-----|-----|------|------|-------|-------|----------|
| 11th | Psc | 0h 16m | -1° 54' | -12.3 | 32' | 98% | 166° W | FM | Jupiter, Neptune |
| 12th | Psc | 1h 5m | 4° 24' | -12.0 | 32' | 94% | 154° W | -G | Jupiter |
| 13th | Psc | 1h 54m | 10° 20' | -11.7 | 31' | 89% | 143° W | -G | |
| 14th | Ari | 2h 42m | 15° 39' | -11.4 | 31' | 81% | 132° W | -G | Uranus |
| 15th | Tau | 3h 32m | 20° 9' | -11.0 | 30' | 73% | 120° W | -G | Uranus, Pleiades |
| 16th | Tau | 4h 23m | 23° 40' | -10.7 | 30' | 63% | 108° W | LQ | Mars, Pleiades, Hyades, Aldebara |
| 17th | Tau | 5h 15m | 26° 2' | -10.3 | 30' | 54% | 96° W | LQ | Mars, Aldebaran |
| 18th | Gem | 6h 8m | 27° 12' | -9.9 | 30' | 44% | 84° W | LQ | |
| 19th | Gem | 7h 2m | 27° 7' | -9.4 | 30' | 35% | 71° W | -Cr | |
| 20th | Gem | 7h 54m | 25° 46' | -8.9 | 30' | 26% | 59° W | -Cr | Praesepe |

## Mercury and Venus

Mercury
15th

Venus
15th

**Mercury**

| Date | Con. | R.A. | Dec. | Mag. | Diam. | Ill. | Elon. | Vis. | Rat. | Close To |
|------|------|------|------|------|-------|------|-------|------|------|----------|
| 11th | Vir | 12h 25m | -7° 16' | 1.2 | 9" | 23% | 17° E | PM | *** | |
| 13th | Vir | 12h 23m | -7° 7' | 1.6 | 10" | 18% | 15° E | NV | N/A | |
| 15th | Vir | 12h 20m | -6° 43' | 2.1 | 10" | 13% | 12° E | NV | N/A | |
| 17th | Vir | 12h 15m | -6° 2' | 2.7 | 10" | 8% | 9° E | NV | N/A | |
| 19th | Vir | 12h 9m | -5° 5' | 3.5 | 10" | 4% | 6° E | NV | N/A | |

**Venus**

| Date | Con. | R.A. | Dec. | Mag. | Diam. | Ill. | Elon. | Vis. | Rat. | Close To |
|------|------|------|------|------|-------|------|-------|------|------|----------|
| 11th | Leo | 10h 40m | 9° 51' | -3.9 | 10" | 98% | 10° W | NV | N/A | Regulus |
| 13th | Leo | 10h 49m | 8° 57' | -3.9 | 10" | 98% | 9° W | NV | N/A | |
| 15th | Leo | 10h 58m | 8° 2' | -3.9 | 10" | 99% | 9° W | NV | N/A | |
| 17th | Leo | 11h 7m | 7° 5' | -3.9 | 10" | 99% | 8° W | NV | N/A | |
| 19th | Leo | 11h 17m | 6° 8' | -3.9 | 10" | 99% | 8° W | NV | N/A | |

## Mars and the Outer Planets

Mars
15th

Jupiter
15th

Saturn
15th

**Mars**

| Date | Con. | R.A. | Dec. | Mag. | Diam. | Ill. | Elon. | Vis. | Rat. | Close To |
|------|------|------|------|------|-------|------|-------|------|------|----------|
| 11th | Tau | 4h 42m | 21° 7' | -0.3 | 10" | 86% | 99° W | AM | *** | Hyades, Aldebaran |
| 15th | Tau | 4h 50m | 21° 26' | -0.3 | 11" | 86% | 101° W | AM | *** | Hyades, Aldebaran |
| 20th | Tau | 4h 59m | 21° 48' | -0.4 | 11" | 86% | 103° W | AM | *** | Hyades, Aldebaran |

**The Outer Planets**

| Planet | Date | Con. | R.A. | Dec. | Mag. | Diam. | Elon. | Vis. | Rat. | Close To |
|--------|------|------|------|------|------|-------|-------|------|------|----------|
| Jupiter | 15th | Psc | 0h 21m | 0° 35' | -2.9 | 50" | 168° W | AM | ***** | |
| Saturn | 15th | Cap | 21h 30m | -16° 9' | 0.4 | 18" | 149° E | PM | **** | |
| Uranus | 15th | Ari | 3h 5m | 17° 2' | 5.7 | 4" | 127° W | AM | *** | Moon |
| Neptune | 15th | Aqr | 23h 40m | -3° 29' | 7.8 | 2" | 178° W | AN | ***** | |

## Highlights

| Date | Time (UT) | Event |
|------|-----------|-------|
| 11th | 14:42 | The just-past full Moon is south of Jupiter. (Morning sky.) |
| 14th | 21:40 | The waning gibbous Moon is south of Uranus. (Morning sky.) |
| 15th | 18:12 | The waning gibbous Moon is south of the Pleiades star cluster. (Taurus, morning sky.) |
| 16th | 17:33 | The nearly last quarter Moon is north of the bright star Aldebaran. (Taurus, morning sky.) |
| 17th | 00:31 | The almost last quarter Moon is north of Mars. (Morning sky.) |
| | 11:20 | Neptune is at opposition. (Visible all night.) |
| | 21:53 | Last Quarter Moon. (Visible all night.) |

# September 21st to 30th, 2022

## The Moon

| 21st | 23rd | 25th | 27th | 29th |

| Date | Con | R.A. | Dec | Mag | Diam | Ill. | Elon. | Phase | Close To |
|------|-----|------|-----|-----|------|------|-------|-------|----------|
| 21st | Cnc | 8h 46m | 23° 16' | -8.3 | 30' | 19% | 47° W | =Cr | Praesepe |
| 22nd | Leo | 9h 35m | 19° 43' | -7.5 | 30' | 12% | 36° W | NM | Regulus |
| 23rd | Leo | 10h 23m | 15° 17' | -6.7 | 30' | 6% | 24° W | NM | Regulus |
| 24th | Leo | 11h 10m | 10° 7' | -5.7 | 30' | 2% | 14° W | NM | Mercury, Venus |
| 25th | Vir | 11h 56m | 4° 26' | -4.6 | 31' | 0% | 3° W | NM | Mercury, Venus |
| 26th | Vir | 12h 43m | -1° 32' | -4.7 | 31' | 0% | 8° E | NM | Spica |
| 27th | Vir | 13h 30m | -7° 33' | -5.9 | 31' | 3% | 19° E | NM | Spica |
| 28th | Vir | 14h 20m | -13° 20' | -6.9 | 32' | 8% | 30° E | NM | |
| 29th | Lib | 15h 12m | -18° 34' | -7.9 | 32' | 14% | 42° E | ±Cr | |
| 30th | Sco | 16h 8m | -22° 52' | -8.6 | 33' | 23% | 56° E | ±Cr | Antares |

## Mercury and Venus

 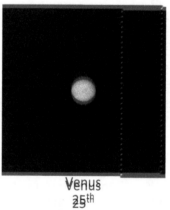

Mercury
25th

Venus
25th

## Mercury

| Date | Con. | R.A. | Dec. | Mag. | Diam. | Ill. | Elon. | Vis. | Rat. | Close To |
|------|------|------|------|------|-------|------|-------|------|------|----------|
| 21st | Vir | 12h 2m | -3° 54' | 4.4 | 10'' | 1% | 2° E | NV | N/A | Venus |
| 23rd | Vir | 11h 55m | -2° 34' | 5.0 | 10'' | 0% | 2° W | NV | N/A | Venus |
| 25th | Vir | 11h 48m | -1° 12' | 4.2 | 10'' | 2% | 5° W | NV | N/A | Moon, Venus |
| 27th | Vir | 11h 43m | 0° 7' | 3.1 | 10'' | 5% | 8° W | NV | N/A | Venus |
| 29th | Vir | 11h 39m | 1° 14' | 2.1 | 9'' | 11% | 11° W | NV | N/A | Venus |

## Venus

| Date | Con. | R.A. | Dec. | Mag. | Diam. | Ill. | Elon. | Vis. | Rat. | Close To |
|------|------|------|------|------|-------|------|-------|------|------|----------|
| 21st | Leo | 11h 26m | 5° 10' | -3.9 | 10" | 99% | 7° W | NV | N/A | Mercury |
| 23rd | Leo | 11h 35m | 4° 13' | -3.9 | 10" | 99% | 7° W | NV | N/A | Mercury |
| 25th | Vir | 11h 44m | 3° 13' | -3.9 | 10" | 99% | 6° W | NV | N/A | Moon, Mercury |
| 27th | Vir | 11h 53m | 2° 13' | -3.9 | 10" | 99% | 6° W | NV | N/A | Mercury |
| 29th | Vir | 12h 3m | 1° 13' | -3.9 | 10" | 99% | 5° W | NV | N/A | Mercury |

## Mars and the Outer Planets

Mars
25th

Jupiter
25th

Saturn
25th

## Mars

| Date | Con. | R.A. | Dec. | Mag. | Diam. | Ill. | Elon. | Vis. | Rat. | Close To |
|------|------|------|------|------|-------|------|-------|------|------|----------|
| 21st | Tau | 5h 1m | 21° 52' | -0.4 | 11" | 87% | 103° W | AM | *** | Hyades, Aldebaran |
| 25th | Tau | 5h 8m | 22° 7' | -0.5 | 11" | 87% | 105° W | AM | *** | Aldebaran |
| 30th | Tau | 5h 16m | 22° 24' | -0.6 | 12" | 88% | 108° W | AM | *** | |

## The Outer Planets

| Planet | Date | Con. | R.A. | Dec. | Mag. | Diam. | Elon. | Vis. | Rat. | Close To |
|--------|------|------|------|------|------|-------|-------|------|------|----------|
| Jupiter | 25th | Psc | 0h 16m | 0° 3' | -2.9 | 50" | 178° W | AN | ***** | |
| Saturn | 25th | Cap | 21h 28m | -16° 19' | 0.5 | 18" | 140° E | PM | **** | |
| Uranus | 25th | Ari | 3h 4m | 16° 59' | 5.7 | 4" | 136° W | AM | *** | |
| Neptune | 25th | Aqr | 23h 39m | -3° 35' | 7.8 | 2" | 173° E | AN | ***** | Jupiter |

## Highlights

| Date | Time (UT) | Event |
|------|-----------|-------|
| 21st | 09:24 | The waning crescent Moon is north of the Praesepe star cluster. (Cancer, morning sky.) |
| 22nd | N/A | Good opportunity to see Earthshine on the waning crescent Moon. (Morning sky.) |
| 23rd | 01:05 | Autumn Equinox. |
| | 02:57 | The waning crescent Moon is north of the bright star Regulus. (Leo, morning sky) |
| | 06:44 | Mercury is at inferior conjunction with the Sun. (Not visible.) |
| 25th | 21:55 | New Moon. (Not visible.) |
| 27th | 08:01 | The waxing crescent Moon is north of the bright star Spica. (Virgo, evening sky.) |
| | 10:13 | Jupiter is at opposition. (Visible all night.) |
| 28th | N/A | Good opportunity to see Earthshine on the waxing crescent Moon. (Evening sky.) |
| 30th | 21:28 | The waxing crescent Moon is north of the bright star Antares. (Scorpius, evening sky.) |

# October 1st to 10th, 2022

## The Moon

| 1st | 3rd | 5th | 7th | 9th |

| Date | Con | R.A. | Dec | Mag | Diam | Ill. | Elon. | Phase | Close To |
|------|-----|------|-----|-----|------|------|-------|-------|----------|
| 1st | Oph | 17h 8m | -25° 54' | -9.3 | 32' | 33% | 70° E | +Cr | Antares |
| 2nd | Sgr | 18h 11m | -27° 20' | -9.9 | 32' | 44% | 84° E | FQ | |
| 3rd | Sgr | 19h 15m | -26° 59' | -10.4 | 32' | 56% | 99° E | FQ | |
| 4th | Cap | 20h 17m | -24° 53' | -10.8 | 32' | 67% | 114° E | +G | |
| 5th | Cap | 21h 17m | -21° 13' | -11.2 | 32' | 77% | 128° E | +G | Saturn |
| 6th | Aqr | 22h 13m | -16° 19' | -11.6 | 32' | 86% | 141° E | +G | Saturn |
| 7th | Aqr | 23h 7m | -10° 34' | -11.9 | 32' | 93% | 154° E | +G | Neptune |
| 8th | Psc | 23h 57m | -4° 20' | -12.3 | 32' | 98% | 165° E | FM | Jupiter, Neptune |
| 9th | Cet | 0h 46m | 1° 60' | -12.6 | 32' | 100% | 177° E | FM | Jupiter |
| 10th | Psc | 1h 35m | 8° 8' | -12.5 | 31' | 100% | 172° W | FM | |

## Mercury and Venus

Mercury
5th

Venus
5th

### Mercury

| Date | Con. | R.A. | Dec. | Mag. | Diam. | Ill. | Elon. | Vis. | Rat. | Close To |
|------|------|------|------|------|-------|------|-------|------|------|----------|
| 1st | Vir | 11h 38m | 2° 5' | 1.2 | 9" | 18% | 13° W | NV | N/A | Venus |
| 3rd | Vir | 11h 39m | 2° 36' | 0.6 | 8" | 27% | 15° W | NV | N/A | |
| 5th | Vir | 11h 43m | 2° 46' | 0.1 | 8" | 36% | 15° W | AM | ** | |
| 7th | Vir | 11h 48m | 2° 36' | -0.3 | 7" | 46% | 16° W | AM | ** | |
| 9th | Vir | 11h 56m | 2° 8' | -0.6 | 7" | 55% | 16° W | AM | ** | |

**Venus**

| Date | Con. | R.A. | Dec. | Mag. | Diam. | Ill. | Elon. | Vis. | Rat. | Close To |
|------|------|------|------|------|-------|------|-------|------|------|----------|
| 1st | Vir | 12h 12m | 0° 13' | -3.9 | 10" | 100% | 5° W | NV | N/A | Mercury |
| 3rd | Vir | 12h 21m | 0° 47' | -3.9 | 10" | 100% | 4° W | NV | N/A | |
| 5th | Vir | 12h 30m | -1° 47' | -3.9 | 10" | 100% | 4° W | NV | N/A | |
| 7th | Vir | 12h 39m | -2° 47' | -3.9 | 10" | 100% | 3° W | NV | N/A | |
| 9th | Vir | 12h 48m | -3° 47' | -3.9 | 10" | 100% | 3° W | NV | N/A | Spica |

## Mars and the Outer Planets

Mars
5th

Jupiter
5th

Saturn
5th

**Mars**

| Date | Con. | R.A. | Dec. | Mag. | Diam. | Ill. | Elon. | Vis. | Rat. | Close To |
|------|------|------|------|------|-------|------|-------|------|------|----------|
| 1st | Tau | 5h 17m | 22° 27' | -0.6 | 12" | 88% | 108° W | AM | *** | |
| 5th | Tau | 5h 23m | 22° 40' | -0.7 | 12" | 88% | 110° W | AM | *** | |
| 10th | Tau | 5h 29m | 22° 54' | -0.8 | 13" | 89% | 113° W | AM | *** | |

**The Outer Planets**

| Planet | Date | Con. | R.A. | Dec. | Mag. | Diam. | Elon. | Vis. | Rat. | Close To |
|--------|------|------|------|------|------|-------|-------|------|------|----------|
| Jupiter | 5th | Psc | 0h 11m | 0° 28' | -2.9 | 50" | 172° E | AN | ***** | |
| Saturn | 5th | Cap | 21h 26m | -16° 25' | 0.5 | 18" | 130° E | PM | **** | Moon |
| Uranus | 5th | Ari | 3h 3m | 16° 54' | 5.7 | 4" | 145° W | AM | *** | |
| Neptune | 5th | Aqr | 23h 38m | -3° 42' | 7.8 | 2" | 163° E | PM | ***** | Jupiter |

## Highlights

| Date | Time (UT) | Event |
|------|-----------|-------|
| 1st | 14:46 | Mercury is stationary prior to resuming prograde motion. (Not visible.) |
| 3rd | 00:15 | First Quarter Moon. (Evening sky.) |
| 4th | 03:21 | The just-past first quarter Moon is south of dwarf planet Pluto. (Evening sky.) |
| 5th | 14:39 | The waxing gibbous Moon is south of Saturn. (Evening sky.) |
| 7th | 04:51 | Asteroid Vesta is stationary prior to resuming prograde motion. (Evening sky.) |
| 8th | 03:53 | The nearly full Moon is south of Neptune. (Evening sky.) |
| | 05:24 | Dwarf planet Pluto is stationary prior to resuming prograde motion. (Evening sky.) |
| | 16:53 | The nearly full Moon is south of Jupiter. (Evening sky.) |
| | 21:01 | Mercury is at greatest western elongation from the Sun. (Morning sky.) |
| | N/A | The Draconid meteor shower is at its maximum. (ZHR: Variable.) |
| 9th | 20:56 | Full Moon. (Visible all night.) |
| 10th | N/A | The Southern Taurid meteor shower is at its maximum. (ZHR: 5) |

# October 11th to 20th, 2022

## The Moon

11th

13th

15th

17th

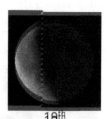
19th

| Date | Con | R.A. | Dec | Mag | Diam | Ill. | Elon. | Phase | Close To |
|------|-----|------|-----|-----|------|------|-------|-------|----------|
| 11th | Ari | 2h 23m | 13° 46' | -12.2 | 31' | 97% | 161° W | FM | Uranus |
| 12th | Ari | 3h 13m | 18° 41' | -11.9 | 31' | 93% | 149° W | =G | Uranus, Pleiades |
| 13th | Tau | 4h 4m | 22° 38' | -11.6 | 30' | 86% | 137° W | =G | Pleiades, Hyades, Aldebaran |
| 14th | Tau | 4h 57m | 25° 29' | -11.3 | 30' | 79% | 125° W | =G | Mars, Hyades, Aldebaran |
| 15th | Tau | 5h 50m | 27° 6' | -10.9 | 30' | 70% | 113° W | =G | Mars |
| 16th | Gem | 6h 44m | 27° 26' | -10.6 | 30' | 61% | 100° W | LQ | |
| 17th | Gem | 7h 37m | 26° 30' | -10.2 | 30' | 52% | 88° W | LQ | |
| 18th | Cnc | 8h 28m | 24° 23' | -9.8 | 30' | 43% | 76° W | LQ | Praesepe |
| 19th | Cnc | 9h 18m | 21° 11' | -9.3 | 30' | 33% | 64° W | =Cr | Praesepe |
| 20th | Leo | 10h 7m | 17° 2' | -8.8 | 30' | 25% | 53° W | =Cr | Regulus |

## Mercury and Venus

Mercury
15th

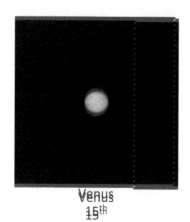
Venus
15th

### Mercury

| Date | Con. | R.A. | Dec. | Mag. | Diam. | Ill. | Elon. | Vis. | Rat. | Close To |
|------|------|------|------|------|-------|------|-------|------|------|----------|
| 11th | Vir | 12h 5m | 1° 34' | -0.7 | 6" | 64% | 15° W | AM | ** | |
| 13th | Vir | 12h 15m | 0° 27' | -0.8 | 6" | 71% | 15° W | NV | N/A | |
| 15th | Vir | 12h 26m | 0° 40' | -0.9 | 6" | 78% | 14° W | NV | N/A | |
| 17th | Vir | 12h 37m | -1° 54' | -1.0 | 6" | 83% | 13° W | NV | N/A | |
| 19th | Vir | 12h 49m | -3° 13' | -1.0 | 5" | 87% | 13° W | NV | N/A | Spica |

## Venus

| Date | Con. | R.A. | Dec. | Mag. | Diam. | Ill. | Elon. | Vis. | Rat. | Close To |
|------|------|------|------|------|-------|------|-------|------|------|----------|
| 11th | Vir | 12h 58m | -4° 47' | -3.9 | 10" | 100% | 2° W | NV | N/A | Spica |
| 13th | Vir | 13h 7m | -5° 47' | -3.9 | 10" | 100% | 2° W | NV | N/A | Spica |
| 15th | Vir | 13h 16m | -6° 45' | -3.9 | 10" | 100% | 1° W | NV | N/A | Spica |
| 17th | Vir | 13h 25m | -7° 44' | -3.9 | 10" | 100% | 1° W | NV | N/A | Spica |
| 19th | Vir | 13h 35m | -8° 42' | -3.9 | 10" | 100% | 0° W | NV | N/A | Spica |

## Mars and the Outer Planets

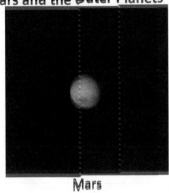

Mars
15th

Jupiter
15th

Saturn
15th

### Mars

| Date | Con. | R.A. | Dec. | Mag. | Diam. | Ill. | Elon. | Vis. | Rat. | Close To |
|------|------|------|------|------|-------|------|-------|------|------|----------|
| 11th | Tau | 5h 30m | 22° 57' | -0.8 | 13" | 89% | 114° W | AM | *** | |
| 15th | Tau | 5h 34m | 23° 8' | -0.9 | 13" | 90% | 117° W | AM | *** | Moon |
| 20th | Tau | 5h 37m | 23° 21' | -1.0 | 14" | 91% | 121° W | AM | *** | |

### The Outer Planets

| Planet | Date | Con. | R.A. | Dec. | Mag. | Diam. | Elon. | Vis. | Rat. | Close To |
|--------|------|------|------|------|------|-------|-------|------|------|----------|
| Jupiter | 15th | Psc | 0h 7m | 0° 57' | -2.9 | 49" | 161° E | PM | ***** | |
| Saturn | 15th | Cap | 21h 25m | -16° 29' | 0.6 | 18" | 121° E | PM | *** | |
| Uranus | 15th | Ari | 3h 2m | 16° 48' | 5.6 | 4" | 155° W | AM | **** | |
| Neptune | 15th | Aqr | 23h 37m | -3° 47' | 7.8 | 2" | 154° E | PM | ***** | Jupiter |

## Highlights

| Date | Time (UT) | Event |
|------|-----------|-------|
| 12th | 08:06 | The waning gibbous Moon is north of Uranus. (Morning sky.) |
| | N/A | The Delta Aurigid meteor shower is at its maximum. (ZHR: 2) |
| 13th | 04:30 | The waning gibbous Moon is south of the Pleiades star cluster. (Taurus, morning sky.) |
| 14th | 02:01 | The waning gibbous Moon is north of the bright star Aldebaran. (Taurus, morning sky.) |
| 15th | 04:53 | The waning gibbous Moon is north of Mars. (Morning sky.) |
| 17th | 17:16 | Last Quarter Moon. (Morning sky.) |
| 18th | 17:39 | The just-past last quarter Moon is north of the Praesepe star cluster. (Cancer, morning sky.) |
| | N/A | The Epsilon Geminid meteor shower is at its maximum. (ZHR: 3) |
| 20th | 14:03 | The waning crescent Moon is north of the bright star Regulus. (Leo, morning sky.) |

# October 21st to 31st, 2022

## The Moon

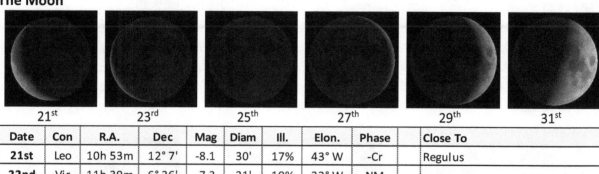

| 21st | 23rd | 25th | 27th | 29th | 31st |

| Date | Con | R.A. | Dec | Mag | Diam | Ill. | Elon. | Phase | Close To |
|------|-----|------|-----|-----|------|------|-------|-------|----------|
| 21st | Leo | 10h 53m | 12° 7' | -8.1 | 30' | 17% | 43° W | -Cr | Regulus |
| 22nd | Vir | 11h 39m | 6° 36' | -7.3 | 31' | 10% | 32° W | NM | |
| 23rd | Vir | 12h 26m | 0° 40' | -6.3 | 31' | 4% | 21° W | NM | Mercury |
| 24th | Vir | 13h 13m | -5° 27' | -5.2 | 31' | 1% | 10° W | NM | Mercury, Venus, Spica |
| 25th | Vir | 14h 3m | -11° 29' | -4.0 | 32' | 0% | 1° E | NM | Mercury, Venus, Spica |
| 26th | Lib | 14h 56m | -17° 4' | -5.3 | 32' | 1% | 13° E | NM | Venus |
| 27th | Sco | 15h 52m | -21° 49' | -6.5 | 32' | 5% | 26° E | NM | Antares |
| 28th | Sco | 16h 52m | -25° 20' | -7.5 | 32' | 12% | 40° E | NM | Antares |
| 29th | Sgr | 17h 56m | -27° 14' | -8.4 | 32' | 20% | 55° E | +Cr | |
| 30th | Sgr | 19h 0m | -27° 19' | -9.1 | 32' | 30% | 71° E | +Cr | |
| 31st | Sgr | 20h 3m | -25° 35' | -9.7 | 32' | 41% | 85° E | FQ | |

## Mercury and Venus

Mercury
25th

Venus
25th

**Mercury**

| Date | Con. | R.A. | Dec. | Mag. | Diam. | Ill. | Elon. | Vis. | Rat. | Close To |
|------|------|------|------|------|-------|------|-------|------|------|----------|
| 21st | Vir | 13h 1m | -4° 35' | -1.0 | 5" | 91% | 11° W | NV | N/A | Spica |
| 23rd | Vir | 13h 14m | -5° 59' | -1.0 | 5" | 93% | 9° W | NV | N/A | Moon, Venus, Spica |
| 25th | Vir | 13h 26m | -7° 23' | -1.1 | 5" | 95% | 8° W | NV | N/A | Moon, Venus, Spica |
| 27th | Vir | 13h 38m | -8° 47' | -1.1 | 5" | 97% | 7° W | NV | N/A | Venus, Spica |
| 29th | Vir | 13h 51m | -10° 10' | -1.1 | 5" | 98% | 6° W | NV | N/A | Venus, Spica |
| 31st | Vir | 14h 3m | -11° 31' | -1.2 | 5" | 99% | 5° W | NV | N/A | Venus, Spica |

## Venus

| Date | Con. | R.A. | Dec. | Mag. | Diam. | Ill. | Elon. | Vis. | Rat. | Close To |
|------|------|------|------|------|-------|------|-------|------|------|----------|
| 21st | Vir | 13h 44m | -9° 39' | -3.9 | 10" | 100% | 0° E | NV | N/A | Spica |
| 23rd | Vir | 13h 53m | -10° 35' | -3.9 | 10" | 100% | 1° E | NV | N/A | Mercury, Spica |
| 25th | Vir | 14h 3m | -11° 30' | -3.9 | 10" | 100% | 1° E | NV | N/A | Moon, Mercury, Spica |
| 27th | Vir | 14h 13m | -12° 24' | -3.9 | 10" | 100% | 1° E | NV | N/A | Mercury |
| 29th | Vir | 14h 22m | -13° 17' | -3.9 | 10" | 100% | 2° E | NV | N/A | Mercury |
| 31st | Lib | 14h 32m | -14° 8' | -3.9 | 10" | 100% | 2° E | NV | N/A | Mercury |

## Mars and the Outer Planets

Mars
25th

Jupiter
25th

Saturn
25th

### Mars

| Date | Con. | R.A. | Dec. | Mag. | Diam. | Ill. | Elon. | Vis. | Rat. | Close To |
|------|------|------|------|------|-------|------|-------|------|------|----------|
| 21st | Tau | 5h 38m | 23° 24' | -1.0 | 14" | 91% | 121° W | AM | *** | |
| 25th | Tau | 5h 40m | 23° 35' | -1.1 | 14" | 92% | 125° W | AM | *** | |
| 31st | Tau | 5h 40m | 23° 51' | -1.2 | 15" | 93% | 130° W | AM | **** | |

### The Outer Planets

| Planet | Date | Con. | R.A. | Dec. | Mag. | Diam. | Elon. | Vis. | Rat. | Close To |
|--------|------|------|------|------|------|-------|-------|------|------|----------|
| Jupiter | 25th | Psc | 0h 3m | -1° 21' | -2.8 | 48" | 151° E | PM | ***** | |
| Saturn | 25th | Cap | 21h 25m | -16° 29' | 0.6 | 17" | 112° E | PM | *** | |
| Uranus | 25th | Ari | 3h 0m | 16° 42' | 5.6 | 4" | 165° W | AM | **** | |
| Neptune | 25th | Aqr | 23h 36m | -3° 52' | 7.8 | 2" | 144° E | PM | **** | Jupiter |

## Highlights

| Date | Time (UT) | Event |
|------|-----------|-------|
| 21st | N/A | The Orionid meteor shower is at its maximum. (ZHR: 25) |
| | N/A | Good opportunity to see Earthshine on the waning crescent Moon. (Morning sky.) |
| 22nd | 20:40 | Venus is at superior conjunction with the Sun. (Not visible.) |
| 23rd | 06:43 | Saturn is stationary prior to resuming prograde motion. (Evening sky.) |
| 24th | N/A | The Leo Minorid meteor shower is at its maximum. (ZHR: 2) |
| 25th | 10:49 | New Moon. (Not visible) |
| | 11:02 | Partial solar eclipse. Visible from north-eastern Africa, western and central Asia, the North Atlantic and Europe. |
| 28th | 02:35 | The waxing crescent Moon is north of the bright star Antares. (Evening sky.) |
| | N/A | Good opportunity to see Earthshine on the waxing crescent Moon. (Evening sky.) |
| 30th | 10:47 | Mars is stationary prior to beginning retrograde motion. (Morning sky.) |
| 31st | 07:15 | The nearly first quarter Moon is south of Pluto. (Evening sky.) |

# November 1st to 10th, 2022

## The Moon

  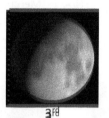  

| 1st | 3rd | 5th | 7th | 9th |

| Date | Con | R.A. | Dec | Mag | Diam | Ill. | Elon. | Phase | Close To |
|------|-----|------|-----|-----|------|------|-------|-------|----------|
| 1st | Cap | 21h 4m | -22° 16' | -10.2 | 32' | 53% | 99° E | FQ | Saturn |
| 2nd | Aqr | 22h 0m | -17° 41' | -10.7 | 32' | 64% | 112° E | FQ | Saturn |
| 3rd | Aqr | 22h 53m | -12° 13' | -11.1 | 32' | 74% | 125° E | +G | Neptune |
| 4th | Aqr | 23h 43m | -6° 12' | -11.4 | 32' | 83% | 136° E | +G | Jupiter, Neptune |
| 5th | Cet | 0h 31m | 0° 0' | -11.8 | 31' | 90% | 147° E | +G | Jupiter |
| 6th | Psc | 1h 19m | 6° 8' | -12.1 | 31' | 96% | 158° E | FM | |
| 7th | Ari | 2h 6m | 11° 54' | -12.4 | 31' | 99% | 169° E | FM | |
| 8th | Ari | 2h 55m | 17° 3' | -12.7 | 31' | 100% | 180° W | FM | Uranus |
| 9th | Tau | 3h 46m | 21° 22' | -12.4 | 30' | 99% | 168° W | FM | Pleiades, Hyades |
| 10th | Tau | 4h 38m | 24° 39' | -12.1 | 30' | 96% | 156° W | FM | Hyades, Aldebaran |

## Mercury and Venus

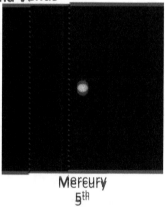 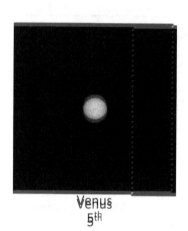

Mercury
5th

Venus
5th

### Mercury

| Date | Con. | R.A. | Dec. | Mag. | Diam. | Ill. | Elon. | Vis. | Rat. | Close To |
|------|------|------|------|------|-------|------|-------|------|------|----------|
| 1st | Vir | 14h 10m | -12° 11' | -1.2 | 5" | 99% | 4° W | NV | N/A | Venus |
| 3rd | Vir | 14h 22m | -13° 28' | -1.2 | 5" | 100% | 3° W | NV | N/A | Venus |
| 5th | Lib | 14h 35m | -14° 43' | -1.3 | 5" | 100% | 2° W | NV | N/A | Venus |
| 7th | Lib | 14h 47m | -15° 55' | -1.3 | 5" | 100% | 1° W | NV | N/A | Venus |
| 9th | Lib | 15h 0m | -17° 4' | -1.3 | 5" | 100% | 0° E | NV | N/A | Venus |

## Venus

| Date | Con. | R.A. | Dec. | Mag. | Diam. | Ill. | Elon. | Vis. | Rat. | Close To |
|------|------|------|------|------|-------|------|-------|------|------|----------|
| 1st | Lib | 14h 37m | -14° 33' | -3.9 | 10" | 100% | 3° E | NV | N/A | Mercury |
| 3rd | Lib | 14h 47m | -15° 23' | -3.9 | 10" | 100% | 3° E | NV | N/A | Mercury |
| 5th | Lib | 14h 56m | -16° 11' | -3.9 | 10" | 100% | 4° E | NV | N/A | Mercury |
| 7th | Lib | 15h 6m | -16° 57' | -3.9 | 10" | 100% | 4° E | NV | N/A | Mercury |
| 9th | Lib | 15h 16m | -17° 41' | -3.9 | 10" | 100% | 5° E | NV | N/A | Mercury |

## Mars and the Outer Planets

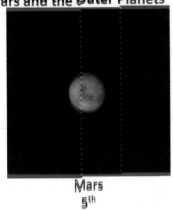

Mars
5th

Jupiter
5th

Saturn
5th

### Mars

| Date | Con. | R.A. | Dec. | Mag. | Diam. | Ill. | Elon. | Vis. | Rat. | Close To |
|------|------|------|------|------|-------|------|-------|------|------|----------|
| 1st | Tau | 5h 40m | 23° 54' | -1.2 | 15" | 94% | 131° W | AM | **** | |
| 5th | Tau | 5h 39m | 24° 5' | -1.3 | 16" | 95% | 136° W | AM | **** | |
| 10th | Tau | 5h 36m | 24° 18' | -1.4 | 16" | 96% | 141° W | AM | **** | |

### The Outer Planets

| Planet | Date | Con. | R.A. | Dec. | Mag. | Diam. | Elon. | Vis. | Rat. | Close To |
|--------|------|------|------|------|------|-------|-------|------|------|----------|
| Jupiter | 5th | Psc | 0h 0m | -1° 39' | -2.8 | 47" | 139° E | PM | ***** | Moon |
| Saturn | 5th | Cap | 21h 26m | -16° 26' | 0.7 | 17" | 101° E | PM | *** | |
| Uranus | 5th | Ari | 2h 58m | 16° 34' | 5.6 | 4" | 176° W | AN | **** | |
| Neptune | 5th | Aqr | 23h 35m | -3° 57' | 7.8 | 2" | 133° E | PM | **** | Jupiter |

## Highlights

| Date | Time (UT) | Event |
|------|-----------|-------|
| 1st | 06:38 | First Quarter Moon. (Evening sky.) |
| | 22:06 | The just-past first quarter Moon is south of Saturn. (Evening sky.) |
| 4th | 08:27 | The waxing gibbous Moon is south of Neptune. (Evening sky.) |
| | 20:07 | The waxing gibbous Moon is south of Jupiter. (Evening sky.) |
| 8th | 10:59 | Total lunar eclipse. Visible from the Arctic, eastern Asia, Australia, North America and the Pacific. |
| | 11:03 | Full Moon. (Visible all night.) |
| | 12:51 | The full Moon is south of Uranus. (Visible all night.) |
| | 16:26 | Mercury is at superior conjunction with the Sun. (Not visible.) |
| 9th | 11:17 | Uranus is at opposition. (Visible all night.) |
| | 12:31 | The just-past full Moon is south of the Pleiades star cluster. (Taurus, morning sky.) |
| 10th | 11:24 | The waning gibbous Moon is north of the bright star Aldebaran. (Taurus, morning sky.) |

# November 11th to 20th, 2022

## The Moon

| 11th | 13th | 15th | 17th | 19th |

| Date | Con | R.A. | Dec | Mag | Diam | Ill. | Elon. | Phase | Close To |
|------|-----|------|-----|-----|------|------|-------|-------|----------|
| 11th | Tau | 5h 32m | 26° 43' | -11.8 | 30' | 91% | 144° W | -G | Mars |
| 12th | Gem | 6h 26m | 27° 30' | -11.5 | 30' | 85% | 131° W | -G | |
| 13th | Gem | 7h 19m | 26° 59' | -11.2 | 30' | 77% | 119° W | -G | |
| 14th | Cnc | 8h 11m | 25° 14' | -10.9 | 30' | 69% | 107° W | -G | Praesepe |
| 15th | Cnc | 9h 2m | 22° 24' | -10.5 | 30' | 60% | 95° W | LQ | Praesepe |
| 16th | Leo | 9h 50m | 18° 36' | -10.2 | 30' | 51% | 84° W | LQ | Regulus |
| 17th | Leo | 10h 36m | 14° 1' | -9.7 | 30' | 41% | 74° W | LQ | Regulus |
| 18th | Leo | 11h 22m | 8° 47' | -9.2 | 30' | 32% | 63° W | -Cr | |
| 19th | Vir | 12h 7m | 3° 4' | -8.6 | 31' | 23% | 53° W | -Cr | |
| 20th | Vir | 12h 54m | -2° 56' | -7.9 | 31' | 14% | 42° W | -Cr | Spica |

## Mercury and Venus

Mercury
15th

Venus
15th

**Mercury**

| Date | Con. | R.A. | Dec. | Mag. | Diam. | Ill. | Elon. | Vis. | Rat. | Close To |
|------|------|------|------|------|-------|------|-------|------|------|----------|
| 11th | Lib | 15h 12m | -18° 9' | -1.2 | 5" | 100% | 2° E | NV | N/A | Venus |
| 13th | Lib | 15h 25m | -19° 11' | -1.1 | 5" | 100% | 3° E | NV | N/A | Venus |
| 15th | Lib | 15h 38m | -20° 9' | -1.0 | 5" | 99% | 4° E | NV | N/A | Venus |
| 17th | Sco | 15h 51m | -21° 3' | -0.9 | 5" | 99% | 5° E | NV | N/A | Venus, Antares |
| 19th | Sco | 16h 4m | -21° 52' | -0.8 | 5" | 99% | 6° E | NV | N/A | Venus, Antares |

## Venus

| Date | Con. | R.A. | Dec. | Mag. | Diam. | Ill. | Elon. | Vis. | Rat. | Close To |
|------|------|------|------|------|-------|------|-------|------|------|----------|
| 11th | Lib | 15h 27m | -18° 24' | -3.9 | 10" | 100% | 5° E | NV | N/A | Mercury |
| 13th | Lib | 15h 37m | -19° 4' | -3.9 | 10" | 100% | 6° E | NV | N/A | Mercury |
| 15th | Lib | 15h 47m | -19° 42' | -3.9 | 10" | 99% | 6° E | NV | N/A | Mercury |
| 17th | Lib | 15h 57m | -20° 19' | -3.9 | 10" | 99% | 7° E | NV | N/A | Mercury, Antares |
| 19th | Sco | 16h 8m | -20° 53' | -3.9 | 10" | 99% | 7° E | NV | N/A | Mercury, Antares |

## Mars and the Outer Planets

Mars
15th

Jupiter
15th

Saturn
15th

### Mars

| Date | Con. | R.A. | Dec. | Mag. | Diam. | Ill. | Elon. | Vis. | Rat. | Close To |
|------|------|------|------|------|-------|------|-------|------|------|----------|
| 11th | Tau | 5h 36m | 24° 21' | -1.5 | 16" | 96% | 143° W | AM | **** | Moon |
| 15th | Tau | 5h 32m | 24° 31' | -1.6 | 17" | 97% | 148° W | AM | **** | |
| 20th | Tau | 5h 26m | 24° 42' | -1.7 | 17" | 98% | 154° W | AM | **** | |

### The Outer Planets

| Planet | Date | Con. | R.A. | Dec. | Mag. | Diam. | Elon. | Vis. | Rat. | Close To |
|--------|------|------|------|------|------|-------|-------|------|------|----------|
| Jupiter | 15th | Psc | 23h 58m | -1° 48' | -2.7 | 46" | 129° E | PM | **** | |
| Saturn | 15th | Cap | 21h 27m | -16° 19' | 0.7 | 17" | 91° E | PM | *** | |
| Uranus | 15th | Ari | 2h 57m | 16° 28' | 5.6 | 4" | 174° E | AN | **** | |
| Neptune | 15th | Aqr | 23h 35m | -3° 60' | 7.9 | 2" | 123° E | PM | **** | Jupiter |

## Highlights

| Date | Time (UT) | Event |
|------|-----------|-------|
| 11th | 13:54 | The waning gibbous Moon is north of Mars. (Morning sky.) |
| 12th | N/A | The Northern Taurid meteor shower is at its maximum. (ZHR: 5) |
| 15th | 00:03 | The nearly last quarter Moon is north of the Praesepe star cluster. (Cancer, morning sky.) |
| 16th | 13:28 | Last Quarter Moon. (Morning sky.) |
| | 20:30 | The last quarter Moon is north of the bright star Regulus. (Leo, morning sky.) |
| 18th | N/A | The Leonid meteor shower is at its maximum. (ZHR: 20) |
| 20th | N/A | Good opportunity to see Earthshine on the waning crescent Moon. (Morning sky.) |

# November 21st to 30th, 2022

## The Moon

21st

23rd

25th

27th
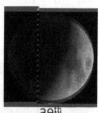
29th

| Date | Con | R.A. | Dec | Mag | Diam | Ill. | Elon. | Phase | Close To |
|------|-----|------|-----|-----|------|------|-------|-------|----------|
| 21st | Vir | 13h 42m | -9° 1' | -7.0 | 32' | 8% | 31° W | NM | Spica |
| 22nd | Lib | 14h 33m | -14° 51' | -5.9 | 32' | 3% | 19° W | NM | |
| 23rd | Lib | 15h 29m | -20° 3' | -4.6 | 32' | 0% | 7° W | NM | |
| 24th | Oph | 16h 29m | -24° 11' | -4.7 | 33' | 0% | 7° E | NM | Mercury, Venus, Antares |
| 25th | Oph | 17h 34m | -26° 45' | -6.1 | 33' | 3% | 22° E | NM | |
| 26th | Sgr | 18h 40m | -27° 27' | -7.2 | 33' | 9% | 38° E | NM | |
| 27th | Sgr | 19h 46m | -26° 12' | -8.1 | 33' | 17% | 53° E | +Cr | |
| 28th | Cap | 20h 49m | -23° 11' | -8.9 | 33' | 27% | 68° E | +Cr | Saturn |
| 29th | Cap | 21h 47m | -18° 48' | -9.5 | 32' | 38% | 81° E | FQ | Saturn |
| 30th | Aqr | 22h 41m | -13° 27' | -10.1 | 32' | 49% | 94° E | FQ | |

## Mercury and Venus

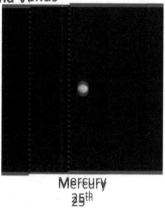
Mercury
25th

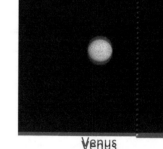
Venus
25th

### Mercury

| Date | Con. | R.A. | Dec. | Mag. | Diam. | Ill. | Elon. | Vis. | Rat. | Close To |
|------|------|------|------|------|-------|------|-------|------|------|----------|
| 21st | Sco | 16h 17m | -22° 38' | -0.8 | 5" | 98% | 7° E | NV | N/A | Venus, Antares |
| 23rd | Oph | 16h 30m | -23° 19' | -0.7 | 5" | 97% | 9° E | NV | N/A | Venus, Antares |
| 25th | Oph | 16h 43m | -23° 55' | -0.7 | 5" | 97% | 10° E | NV | N/A | Venus, Antares |
| 27th | Oph | 16h 57m | -24° 26' | -0.6 | 5" | 96% | 11° E | NV | N/A | Venus, Antares |
| 29th | Oph | 17h 10m | -24° 53' | -0.6 | 5" | 95% | 12° E | NV | N/A | Venus |

## Venus

| Date | Con. | R.A. | Dec. | Mag. | Diam. | Ill. | Elon. | Vis. | Rat. | Close To |
|------|------|------|------|------|-------|------|-------|------|------|----------|
| 21st | Sco | 16h 18m | -21° 24' | -3.9 | 10" | 99% | 8° E | NV | N/A | Mercury, Antares |
| 23rd | Oph | 16h 29m | -21° 53' | -3.9 | 10" | 99% | 8° E | NV | N/A | Mercury, Antares |
| 25th | Oph | 16h 40m | -22° 20' | -3.9 | 10" | 99% | 9° E | NV | N/A | Mercury, Antares |
| 27th | Oph | 16h 50m | -22° 44' | -3.9 | 10" | 99% | 9° E | NV | N/A | Mercury, Antares |
| 29th | Oph | 17h 1m | -23° 5' | -3.9 | 10" | 99% | 10° E | NV | N/A | Mercury, Antares |

## Mars and the Outer Planets

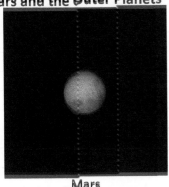

Mars
25th

Jupiter
25th

Saturn
25th

## Mars

| Date | Con. | R.A. | Dec. | Mag. | Diam. | Ill. | Elon. | Vis. | Rat. | Close To |
|------|------|------|------|------|-------|------|-------|------|------|----------|
| 21st | Tau | 5h 25m | 24° 44' | -1.7 | 17" | 98% | 156° W | AM | **** | |
| 25th | Tau | 5h 19m | 24° 51' | -1.7 | 17" | 99% | 161° W | AM | **** | |
| 30th | Tau | 5h 11m | 24° 57' | -1.8 | 17" | 100% | 169° W | AM | **** | Aldebaran |

## The Outer Planets

| Planet | Date | Con. | R.A. | Dec. | Mag. | Diam. | Elon. | Vis. | Rat. | Close To |
|--------|------|------|------|------|------|-------|-------|------|------|----------|
| Jupiter | 25th | Psc | 23h 57m | -1° 48' | -2.6 | 44" | 118° E | PM | **** | |
| Saturn | 25th | Cap | 21h 39m | -16° 10' | 0.8 | 17" | 81° E | PM | ** | |
| Uranus | 25th | Ari | 2h 55m | 16° 21' | 5.6 | 4" | 163° E | PM | **** | |
| Neptune | 25th | Aqr | 23h 34m | -4° 1' | 7.9 | 2" | 113° E | PM | *** | Jupiter |

## Highlights

| Date | Time (UT) | Event |
|------|-----------|-------|
| 21st | 02:19 | The waning crescent Moon is north of the bright star Spica. (Virgo, morning sky.) |
| 22nd | N/A | The Alpha Monocerotid meteor shower is at its maximum. (ZHR: Variable.) |
| 23rd | 22:58 | New Moon. (Not visible.) |
| 24th | 12:24 | Jupiter is stationary prior to resuming prograde motion. |
| 27th | 15:13 | The waxing crescent Moon appears south of dwarf planet Pluto. (Evening sky.) |
| | N/A | Good opportunity to see Earthshine on the waxing crescent Moon. (Evening sky.) |
| 29th | 04:43 | The nearly first quarter Moon is south of Saturn. (Evening sky.) |
| 30th | 14:37 | First Quarter Moon. (Evening sky.) |

# December 1st to 10th, 2022

## The Moon

1st

3rd

5th

7th

9th

| Date | Con | R.A. | Dec | Mag | Diam | Ill. | Elon. | Phase | Close To |
|------|-----|------|-----|-----|------|------|-------|-------|----------|
| 1st | Aqr | 23h 32m | -7° 33' | -10.5 | 32' | 60% | 105° E | FQ | Jupiter, Neptune |
| 2nd | Psc | 0h 20m | -1° 25' | -10.9 | 31' | 70% | 116° E | +G | Jupiter, Neptune |
| 3rd | Psc | 1h 7m | 4° 39' | -11.3 | 31' | 79% | 127° E | +G | |
| 4th | Psc | 1h 54m | 10° 26' | -11.6 | 31' | 87% | 138° E | +G | |
| 5th | Ari | 2h 41m | 15° 41' | -11.9 | 30' | 93% | 149° E | +G | Uranus |
| 6th | Tau | 3h 31m | 20° 12' | -12.2 | 30' | 97% | 160° E | FM | Uranus, Pleiades |
| 7th | Tau | 4h 22m | 23° 45' | -12.5 | 30' | 100% | 172° E | FM | Mars, Pleiades, Hyades, Aldebara |
| 8th | Tau | 5h 15m | 26° 10' | -12.6 | 30' | 100% | 176° W | FM | Mars, Aldebaran |
| 9th | Gem | 6h 9m | 27° 20' | -12.3 | 30' | 98% | 164° W | FM | |
| 10th | Gem | 7h 3m | 27° 12' | -12.1 | 30' | 95% | 152° W | -G | |

## Mercury and Venus

Mercury
5th

Venus
5th

**Mercury**

| Date | Con. | R.A. | Dec. | Mag. | Diam. | Ill. | Elon. | Vis. | Rat. | Close To |
|------|------|------|------|------|-------|------|-------|------|------|----------|
| 1st | Oph | 17h 23m | -25° 14' | -0.6 | 5" | 94% | 13° E | NV | N/A | Venus |
| 3rd | Oph | 17h 37m | -25° 29' | -0.6 | 5" | 92% | 15° E | NV | N/A | Venus |
| 5th | Sgr | 17h 50m | -25° 39' | -0.6 | 5" | 91% | 16° E | PM | ** | Venus |
| 7th | Sgr | 18h 3m | -25° 44' | -0.6 | 5" | 89% | 17° E | PM | ** | Venus |
| 9th | Sgr | 18h 16m | -25° 42' | -0.6 | 5" | 87% | 18° E | PM | ** | Venus |

**Venus**

| Date | Con. | R.A. | Dec. | Mag. | Diam. | Ill. | Elon. | Vis. | Rat. | Close To |
|------|------|------|------|------|-------|------|-------|------|------|----------|
| 1st | Oph | 17h 12m | -23° 23' | -3.9 | 10" | 99% | 11° E | NV | N/A | Mercury |
| 3rd | Oph | 17h 23m | -23° 39' | -3.9 | 10" | 98% | 11° E | NV | N/A | Mercury |
| 5th | Oph | 17h 34m | -23° 52' | -3.9 | 10" | 98% | 12° E | NV | N/A | Mercury |
| 7th | Sgr | 17h 45m | -24° 2' | -3.9 | 10" | 98% | 12° E | NV | N/A | Mercury |
| 9th | Sgr | 17h 56m | -24° 8' | -3.9 | 10" | 98% | 13° E | NV | N/A | Mercury |

## Mars and the Outer Planets

Mars
5th

Jupiter
5th

Saturn
5th

**Mars**

| Date | Con. | R.A. | Dec. | Mag. | Diam. | Ill. | Elon. | Vis. | Rat. | Close To |
|------|------|------|------|------|-------|------|-------|------|------|----------|
| 1st | Tau | 5h 10m | 24° 58' | -1.8 | 17" | 100% | 170° W | AN | **** | Aldebaran |
| 5th | Tau | 5h 3m | 24° 60' | -1.9 | 17" | 100% | 176° W | AN | **** | Hyades, Aldebaran |
| 10th | Tau | 4h 54m | 24° 59' | -1.8 | 17" | 100% | 176° E | AN | **** | Hyades, Aldebaran |

**The Outer Planets**

| Planet | Date | Con. | R.A. | Dec. | Mag. | Diam. | Elon. | Vis. | Rat. | Close To |
|--------|------|------|------|------|------|-------|-------|------|------|----------|
| Jupiter | 5th | Psc | 23h 58m | -1° 40' | -2.6 | 43" | 108° E | PM | **** | |
| Saturn | 5th | Cap | 21h 31m | -15° 58' | 0.8 | 16" | 71° E | PM | ** | |
| Uranus | 5th | Ari | 2h 54m | 16° 14' | 5.6 | 4" | 152° E | PM | **** | Moon |
| Neptune | 5th | Aqr | 23h 34m | -4° 1' | 7.9 | 2" | 102° E | PM | *** | Jupiter |

## Highlights

| Date | Time (UT) | Event |
|------|-----------|-------|
| 1st | 12:02 | The just-past first quarter Moon is south of Neptune. (Evening sky.) |
| 2nd | 02:15 | The waxing gibbous Moon is south of Jupiter. (Evening sky.) |
| 4th | 05:40 | Neptune is stationary prior to resuming prograde motion. (Evening sky.) |
| 5th | 16:43 | The waxing gibbous Moon is south of Uranus. (Evening sky.) |
| 6th | 18:17 | The waxing gibbous Moon is south of the Pleiades star cluster. (Taurus, evening sky.) |
| 7th | 16:53 | The nearly full Moon is north of the bright star Aldebaran. (Taurus, evening sky.) |
| 8th | 01:32 | Mars is at opposition. (Visible all night.) |
| | 04:09 | Full Moon. (Visible all night.) |
| | 05:43 | The full Moon is north of Mars. (Visible all night.) |
| 9th | N/A | The Monocerotid meteor shower is at its maximum. (ZHR: 2) |

# December 11th to 20th, 2022

## The Moon

  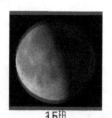  

| 11th | 13th | 15th | 17th | 19th |

| Date | Con | R.A. | Dec | Mag | Diam | Ill. | Elon. | Phase | Close To |
|------|-----|------|-----|-----|------|------|-------|-------|----------|
| 11th | Gem | 7h 55m | 25° 48' | =11.8 | 29' | 90% | 139° W | =G | Praesepe |
| 12th | Cnc | 8h 46m | 23° 17' | =11.5 | 29' | 84% | 128° W | =G | Praesepe |
| 13th | Leo | 9h 35m | 19° 46' | =11.3 | 30' | 76% | 117° W | =G | Regulus |
| 14th | Leo | 10h 21m | 15° 27' | =10.8 | 30' | 68% | 106° W | =G | Regulus |
| 15th | Leo | 11h 6m | 10° 30' | =10.5 | 30' | 58% | 96° W | LQ | |
| 16th | Vir | 11h 51m | 5° 3' | =10.1 | 30' | 49% | 86° W | LQ | |
| 17th | Vir | 12h 35m | 0° 43' | =9.6 | 31' | 39% | 76° W | LQ | |
| 18th | Vir | 13h 22m | -6° 38' | =9.0 | 31' | 29% | 66° W | =Cr | Spica |
| 19th | Vir | 14h 10m | =12° 27' | =8.4 | 32' | 20% | 55° W | =Cr | Spica |
| 20th | Lib | 15h 3m | =17° 53' | =7.6 | 32' | 13% | 42° W | NM | |

## Mercury and Venus

 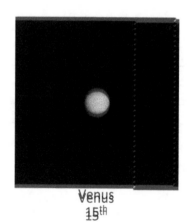

Mercury
15th

Venus
15th

### Mercury

| Date | Con. | R.A. | Dec. | Mag. | Diam. | Ill. | Elon. | Vis. | Rat. | Close To |
|------|------|------|------|------|-------|------|-------|------|------|----------|
| 11th | Sgr | 18h 29m | =25° 35' | =0.6 | 6" | 84% | 19° E | PM | ** | Venus |
| 13th | Sgr | 18h 42m | =25° 22' | =0.6 | 6" | 81% | 20° E | PM | *** | Venus |
| 15th | Sgr | 18h 54m | =25° 3' | =0.6 | 6" | 77% | 21° E | PM | *** | Venus |
| 17th | Sgr | 19h 6m | =24° 39' | =0.6 | 6" | 73% | 21° E | PM | *** | Venus |
| 19th | Sgr | 19h 16m | =24° 10' | =0.5 | 6" | 68% | 22° E | PM | *** | Venus |

## Venus

| Date | Con. | R.A. | Dec. | Mag. | Diam. | Ill. | Elon. | Vis. | Rat. | Close To |
|------|------|------|------|------|-------|------|-------|------|------|----------|
| 11th | Sgr | 18h 7m | -24° 12' | -3.9 | 10" | 98% | 13° E | NV | N/A | Mercury |
| 13th | Sgr | 18h 18m | -24° 13' | -3.9 | 10" | 98% | 14° E | NV | N/A | Mercury |
| 15th | Sgr | 18h 29m | -24° 11' | -3.9 | 10" | 98% | 14° E | NV | N/A | Mercury |
| 17th | Sgr | 18h 40m | -24° 6' | -3.9 | 10" | 97% | 15° E | NV | N/A | Mercury |
| 19th | Sgr | 18h 51m | -23° 58' | -3.9 | 10" | 97% | 16° E | PM | * | Mercury |

## Mars and the Outer Planets

Mars
15th

Jupiter
15th

Saturn
15th

## Mars

| Date | Con. | R.A. | Dec. | Mag. | Diam. | Ill. | Elon. | Vis. | Rat. | Close To |
|------|------|------|------|------|-------|------|-------|------|------|----------|
| 11th | Tau | 4h 53m | 24° 58' | -1.8 | 17" | 100% | 175° E | AN | **** | Hyades, Aldebaran |
| 15th | Tau | 4h 46m | 24° 55' | -1.7 | 17" | 100% | 169° E | PM | **** | Hyades, Aldebaran |
| 20th | Tau | 4h 39m | 24° 50' | -1.6 | 16" | 99% | 161° E | PM | **** | Hyades, Aldebaran |

## The Outer Planets

| Planet | Date | Con. | R.A. | Dec. | Mag. | Diam. | Elon. | Vis. | Rat. | Close To |
|--------|------|------|------|------|------|-------|-------|------|------|----------|
| Jupiter | 15th | Psc | 0h 0m | -1° 25' | -2.5 | 43" | 97° E | PM | *** | |
| Saturn | 15th | Cap | 21h 34m | -15° 43' | 0.8 | 16" | 61° E | PM | ** | |
| Uranus | 15th | Ari | 2h 52m | 16° 9' | 5.7 | 4" | 140° E | PM | *** | |
| Neptune | 15th | Aqr | 23h 35m | -4° 0' | 7.9 | 2" | 91° E | PM | *** | Jupiter |

## Highlights

| Date | Time (UT) | Event |
|------|-----------|-------|
| 12th | 10:04 | The waning gibbous Moon is north of the Praesepe star cluster. (Cancer, morning sky.) |
| | N/A | The Sigma Hydrid meteor shower is at its maximum. (ZHR: 3) |
| 14th | 05:11 | The waning gibbous Moon is north of the bright star Regulus. (Leo, morning sky.) |
| | N/A | The Geminid meteor shower is at its maximum. (ZHR: 120) |
| 16th | 08:57 | Last Quarter Moon. (Morning sky.) |
| | N/A | The Comae Berenicid meteor shower is at its maximum. (ZHR: 3) |
| 18th | 14:55 | The waning crescent Moon is north of the bright star Spica. (Virgo, morning sky.) |
| 19th | N/A | Good opportunity to see Earthshine on the waning crescent Moon. (Morning sky.) |
| 20th | N/A | The Leonis Minorid meteor shower is at its maximum. (ZHR: 5) |

# December 21<sup>st</sup> to 31<sup>st</sup>, 2022

## The Moon

| 21<sup>st</sup> | 23<sup>rd</sup> | 25<sup>th</sup> | 27<sup>th</sup> | 29<sup>th</sup> | 31<sup>st</sup> |

| Date | Con | R.A. | Dec | Mag | Diam | Ill. | Elon. | Phase | Close To |
|------|-----|------|-----|-----|------|------|-------|-------|----------|
| 21st | Sco | 16h 1m | -22° 30' | -6.5 | 33' | 5% | 29° W | NM | Antares |
| 22nd | Oph | 17h 4m | -25° 48' | -5.3 | 33' | 1% | 14° W | NM | Antares |
| 23rd | Sgr | 18h 11m | -27° 21' | -4.3 | 33' | 0% | 1° E | NM | |
| 24th | Sgr | 19h 20m | -26° 52' | -5.6 | 33' | 2% | 17° E | NM | Mercury, Venus |
| 25th | Cap | 20h 26m | -24° 23' | -6.8 | 33' | 7% | 33° E | NM | Mercury |
| 26th | Cap | 21h 28m | -20° 15' | -7.8 | 33' | 14% | 47° E | +Cr | Saturn |
| 27th | Aqr | 22h 25m | -14° 58' | -8.6 | 33' | 23% | 60° E | +Cr | Saturn |
| 28th | Aqr | 23h 18m | -8° 60' | -9.3 | 32' | 33% | 72° E | +Cr | Jupiter, Neptune |
| 29th | Psc | 0h 8m | -2° 46' | -9.9 | 32' | 44% | 84° E | FQ | Jupiter, Neptune |
| 30th | Psc | 0h 55m | 3° 25' | -10.3 | 31' | 55% | 94° E | FQ | |
| 31st | Psc | 1h 43m | 9° 18' | -10.7 | 31' | 65% | 105° E | FQ | |

## Mercury and Venus

Mercury
25<sup>th</sup>

Venus
25<sup>th</sup>

### Mercury

| Date | Con. | R.A. | Dec. | Mag. | Diam. | Ill. | Elon. | Vis. | Rat. | Close To |
|------|------|------|------|------|-------|------|-------|------|------|----------|
| 21st | Sgr | 19h 26m | -23° 37' | -0.5 | 7" | 62% | 22° E | PM | *** | Venus |
| 23rd | Sgr | 19h 34m | -23° 1' | -0.4 | 7" | 54% | 22° E | PM | *** | Venus |
| 25th | Sgr | 19h 40m | -22° 23' | -0.2 | 7" | 46% | 21° E | PM | *** | Moon, Venus |
| 27th | Sgr | 19h 44m | -21° 45' | 0.0 | 8" | 37% | 20° E | PM | *** | Venus |
| 29th | Sgr | 19h 44m | -21° 9' | 0.5 | 8" | 28% | 18° E | PM | *** | Venus |
| 31st | Sgr | 19h 42m | -20° 37' | 1.1 | 9" | 19% | 15° E | PM | ** | Venus |

## Venus

| Date | Con. | R.A. | Dec. | Mag. | Diam. | Ill. | Elon. | Vis. | Rat. | Close To |
|------|------|------|------|------|-------|------|-------|------|------|----------|
| 21st | Sgr | 19h 2m | -23° 47' | -3.9 | 10" | 97% | 16° E | PM | * | Mercury |
| 23rd | Sgr | 19h 13m | -23° 33' | -3.9 | 10" | 97% | 17° E | PM | * | Mercury |
| 25th | Sgr | 19h 24m | -23° 16' | -3.9 | 10" | 97% | 17° E | PM | * | Mercury |
| 27th | Sgr | 19h 34m | -22° 57' | -3.9 | 10" | 96% | 18° E | PM | * | Mercury |
| 29th | Sgr | 19h 45m | -22° 34' | -3.9 | 10" | 96% | 18° E | PM | * | Mercury |
| 31st | Sgr | 19h 56m | -22° 9' | -3.9 | 10" | 96% | 18° E | PM | * | Mercury |

## Mars and the Outer Planets

Mars
25th

Jupiter
25th

Saturn
25th

**Mars**

| Date | Con. | R.A. | Dec. | Mag. | Diam. | Ill. | Elon. | Vis. | Rat. | Close To |
|------|------|------|------|------|-------|------|-------|------|------|----------|
| 21st | Tau | 4h 38m | 24° 48' | -1.5 | 16" | 99% | 160° E | PM | **** | Hyades, Aldebaran |
| 25th | Tau | 4h 33m | 24° 43' | -1.4 | 16" | 98% | 154° E | PM | **** | Hyades, Aldebaran |
| 31st | Tau | 4h 27m | 24° 36' | -1.2 | 15" | 97% | 146° E | PM | **** | Hyades, Aldebaran |

**The Outer Planets**

| Planet | Date | Con. | R.A. | Dec. | Mag. | Diam. | Elon. | Vis. | Rat. | Close To |
|--------|------|------|------|------|------|-------|-------|------|------|----------|
| Jupiter | 25th | Psc | 0h 3m | -1° 1' | -2.4 | 40" | 87° E | PM | *** | |
| Saturn | 25th | Cap | 21h 38m | -15° 25' | 0.8 | 16" | 51° E | PM | ** | |
| Uranus | 25th | Ari | 2h 51m | 16° 4' | 5.7 | 4" | 129° E | PM | *** | |
| Neptune | 25th | Aqr | 23h 35m | -3° 58' | 7.9 | 2" | 80° E | PM | *** | Jupiter |

## Highlights

| Date | Time (UT) | Event |
|------|-----------|-------|
| 21st | 15:17 | Mercury is at greatest eastern elongation from the Sun. (Evening sky.) |
| | 21:49 | Winter Solstice. |
| 23rd | 10:18 | New Moon. (Not visible.) |
| | N/A | The Ursid meteor shower is at its maximum. (ZHR: 10) |
| 24th | 10:50 | The waxing crescent Moon is south of Venus. (Evening sky.) |
| | 19:46 | The waxing crescent Moon is south of Mercury. (Evening sky.) |
| 26th | 16:42 | The waxing crescent Moon is south of Saturn. (Evening sky.) |
| | N/A | Good opportunity to see Earthshine on the waxing crescent Moon. (Evening sky.) |
| 29th | 02:49 | Mercury is stationary prior to beginning retrograde motion. (Evening sky.) |
| | 09:03 | Mercury is 1.4° north of Venus. (Evening sky.) |
| | 09:23 | The nearly first quarter Moon is south of Jupiter. (Evening sky.) |
| 30th | 01:21 | First quarter Moon. (Evening sky.) |

# Planet Visibility Ratings

| | | Morning Sky | | | | | | | Evening Sky | | | | | | |
|---|---|---|---|---|---|---|---|---|---|---|---|---|---|---|---|
| | | Me | Ve | Ma | Ju | Sa | Ur | Ne | Me | Ve | Ma | Ju | Sa | Ur | Ne |
| Jan | 5th | | | | | | | | | * | *** | ** | ** | ** | *** |
| | 15th | *** | | | | | | | | * | *** | ** | * | ** | *** |
| | 25th | **** | | | | | | | | * | *** | ** | * | * | *** |
| Feb | 5th | *** | | | | | | | | * | *** | ** | | * | *** |
| | 15th | *** | | | | | | | | * | ** | * | | * | *** |
| | 25th | | | | | | | | | * | ** | * | | * | ** |
| Mar | 5th | | | | | | | | | * | ** | * | | * | |
| | 15th | | | | | * | | | | * | ** | * | | * | |
| | 25th | | | | | * | | | | * | ** | | | * | |
| Apr | 5th | | | | | * | | ** | ** | * | ** | | | * | |
| | 15th | | | | | ** | | ** | *** | ** | * | | | * | |
| | 25th | | | | | ** | | *** | | ** | * | | | | |
| May | 5th | | | | * | ** | | *** | | ** | * | | | | |
| | 15th | *** | | | * | ** | | *** | | ** | * | | | | |
| | 25th | *** | | | * | ** | | *** | | *** | * | | | | |
| Jun | 5th | *** | | | * | ** | * | **** | | *** | * | | | | |
| | 15th | ** | | | * | *** | * | **** | | *** | * | | | | |
| | 25th | | | | ** | *** | * | **** | | *** | * | | | | |
| Jul | 5th | | | | ** | *** | * | ***** | | *** | * | | | | |
| | 15th | | | | ** | *** | * | ***** | ** | *** | * | | | | |
| | 25th | | | | ** | **** | * | ***** | *** | *** | * | | | | |
| Aug | 5th | | | | *** | **** | * | ***** | *** | | * | | | | |
| | 15th | | | | *** | **** | * | ***** | *** | | * | | | | |
| | 25th | | ** | | *** | **** | ** | ***** | *** | | * | | | | |
| Sep | 5th | | *** | | **** | **** | ** | ***** | | | * | | | | |
| | 15th | | *** | | **** | **** | ** | ***** | | | * | | | | |
| | 25th | ** | *** | | **** | **** | ** | ***** | | | * | | | | |
| Oct | 5th | | *** | | **** | **** | ** | ***** | | | | | | | |
| | 15th | | *** | | ***** | *** | *** | ***** | | | | | | | |
| | 25th | | *** | | ***** | *** | *** | ***** | | | | | | | |
| Nov | 5th | | ** | | ***** | *** | *** | ***** | | | | | | | |
| | 15th | ** | ** | | ***** | *** | *** | ***** | | | | | | | |
| | 25th | *** | ** | | **** | ** | *** | ***** | | | | | | | |
| Dec | 5th | *** | ** | | **** | ** | *** | ***** | | | | | | | |
| | 15th | ** | ** | | **** | ** | *** | **** | | | | | | | |
| | 25th | | ** | | **** | ** | ** | **** | | | | | | | |

# Solar and Lunar Eclipses

| Date | Time (UT) | Type | Visible From |
|---|---|---|---|
| Apr 20th | 04:18 | Total Solar | South-east Asia, Australia and the Indian and Pacific oceans. |
| May 5th | 17:22 | Penumbral Lunar | Africa, Asia, the Atlantic, Australia, Europe and the Indian and Pacific oceans. |
| Oct 14th | 18:01 | Annular Solar | Western Africa, the Atlantic, North America, the Pacific and South America. |
| Oct 28th | 20:14 | Partial Lunar | Africa, Asia, the Atlantic, Australia, Europe, north-eastern North America, the Pacific and north-eastern South America |

# Planetary Highlights

| Date | Time (UT) | Elon. | Vis. | Description |
|------|-----------|-------|------|-------------|
| Jan 22nd | 19:32 | 23° E | PM | Venus is 0.4° south of Saturn. (Capricornus) |
| Feb 4th | 00:36 | 111° E | PM | Mars is 8.2° north of the bright star Aldebaran. (Taurus) |
| Feb 15th | 12:14 | 26° E | PM | Venus is 0.0° south of Neptune. (Aquarius) |
| Mar 2nd | 10:37 | 29° E | PM | Venus is 0.5° north of Jupiter. (Pisces) |
| Mar 31st | 06:04 | 34° E | PM | Venus is 1.3° north of Uranus. (Aries) |
| Apr 10th | 14:16 | 38° E | PM | Venus is 2.7° south of the Pleaides star cluster. (Taurus) |
| May 29th | 05:27 | 24° W | AM | Mercury is at greatest western elongation from the Sun. (Aries) |
| Jun 2nd | 03:36 | 60° E | PM | Mars is 0.2° north of the Praesepe star cluster. (Cancer) |
| Jun 4th | 10:45 | 49° E | PM | Venus is at greatest eastern elongation from the Sun. (Cancer) |
| Jun 13th | 05:16 | 49° E | PM | Venus is 0.9° north of the Praesepe star cluster. (Cancer) |
| Jul 9th | 18:54 | 39° E | PM | Mars is 0.7° north of the bright star Regulus. (Leo) |
| Jul 26th | 12:33 | 24° E | PM | Mercury is 5.3° north of Venus. (Leo) |
| Jul 28th | 18:56 | 24° E | PM | Mercury is 0.1° south of the bright star Regulus. (Leo) |
| Aug 10th | 01:39 | 26° E | PM | Mercury is at greatest eastern elongation from the Sun. (Leo) |
| Aug 28th | 00:09 | 180° E | AN | Saturn is at opposition. (Aquarius) |
| Sep 20th | 00:30 | 180° E | AN | Neptune is at opposition. (Pisces) |
| Oct 23rd | 22:59 | 43° W | AM | Venus is at greatest western elongation from the Sun. (Leo) |
| Nov 3rd | 14:34 | 180° E | AN | Jupiter is at opposition. (Aries) |
| Nov 13th | 16:58 | 180° E | AN | Uranus is at opposition. (Aries) |
| Dec 4th | 14:22 | 23° E | PM | Mercury is at greatest eastern elongation from the Sun. (Sagittarius) |

# Major Meteor Showers

| Shower Name | Start Date | End Date | Peak | ZHR | Speed | Brightness | Moon |
|-------------|-----------|----------|------|-----|-------|------------|------|
| Quadrantids | Dec 28th | Jan 12th | Jan 3rd | 120 | *** | ***** | ○ |
| Lyrids | Apr 18th | Apr 25th | Apr 22nd | 18 | *** | ***** | ● |
| Eta Aquariids | Apr 24th | May 19th | May 7th | 40 | * | **** | ○ |
| June Bootids | Jun 23rd | Jun 25th | Jun 24th | Var | ***** | ***** | ◗ |
| Alpha Capricornids | Jul 8th | Aug 10th | Jul 27th | 5 | ***** | **** | ○ |
| Southern Delta Aquariids | Jul 21st | Aug 23rd | Jul 30th | 16 | *** | * | ○ |
| Perseids | Jul 13th | Aug 26th | Aug 12th | 100 | * | ***** | ◑ |
| Kappa Cygnids | Aug 6th | Aug 31st | Aug 17th | 3 | ***** | ** | ● |
| Aurigids | Aug 29th | Sep 4th | Sep 1st | 6 | * | **** | ○ |
| September Epsilon Perseids | Sep 5th | Sep 28th | Sep 9th | 5 | * | ** | ◗ |
| Draconids | Oct 6th | Oct 10th | Oct 8th | Var | ***** | *** | ◗ |
| Southern Taurids | Sep 7th | Nov 19th | Oct 10th | 5 | **** | **** | ◗ |
| Orionids | Aug 25th | Nov 19th | Oct 22nd | 15 | * | **** | ◐ |
| Andromedids | Oct 26th | Nov 20th | Nov 8th | Var | ***** | **** | ◗ |
| Northern Taurids | Oct 25th | Dec 4th | Nov 11th | 5 | **** | **** | ◑ |
| Leonids | Nov 5th | Dec 3rd | Nov 18th | 15 | * | **** | ◑ |
| Alpha Monocerotids | Nov 21st | Nov 23rd | Nov 21st | Var | * | **** | ◐ |
| Geminids | Nov 30th | Dec 17th | Dec 13th | 120 | **** | *** | ◑ |
| December Leonis Minorids | Dec 6th | Jan 18th | Dec 20th | 5 | * | ** | ◐ |
| Ursids | Dec 17th | Dec 24th | Dec 22nd | 10 | **** | ** | ○ |
| Coma Berenicids | Dec 24th | Jan 3rd | Dec 31st | 5 | * | ** | ◖ |

# January 1st to 10th, 2023

## The Moon

    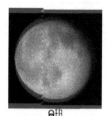

1ˢᵗ     3ʳᵈ     5ᵗʰ     7ᵗʰ     9ᵗʰ

| Date | Con | R.A. | Dec | Mag | Diam | Ill. | Elon. | Phase | Close To |
|------|-----|------|-----|-----|------|------|-------|-------|----------|
| 1st | Ari | 2h 30m | 14° 39' | -11.1 | 30' | 74% | 116° E | +G | Uranus |
| 2nd | Ari | 3h 19m | 19° 18' | -11.4 | 30' | 83% | 127° E | +G | Uranus, Pleiades |
| 3rd | Tau | 4h 9m | 23° 2' | -11.7 | 30' | 89% | 138° E | +G | Mars, Pleiades, Hyades, Aldebaran |
| 4th | Tau | 5h 1m | 25° 42' | -12.0 | 30' | 94% | 150° E | +G | Mars, Hyades, Aldebaran |
| 5th | Tau | 5h 54m | 27° 9' | -12.3 | 30' | 98% | 163° E | FM | |
| 6th | Gem | 6h 48m | 27° 19' | -12.6 | 29' | 100% | 175° E | FM | |
| 7th | Gem | 7h 41m | 26° 13' | -12.5 | 29' | 100% | 173° W | FM | |
| 8th | Cnc | 8h 33m | 23° 57' | -12.3 | 29' | 98% | 161° W | FM | Praesepe |
| 9th | Cnc | 9h 22m | 20° 39' | -12.0 | 29' | 94% | 150° W | -G | Praesepe, Regulus |
| 10th | Leo | 10h 9m | 16° 31' | -11.7 | 30' | 89% | 139° W | -G | Regulus |

## Mercury and Venus

Mercury
5ᵗʰ

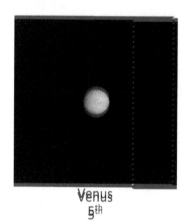

Venus
5ᵗʰ

### Mercury

| Date | Con. | R.A. | Dec. | Mag. | Diam. | Ill. | Elon. | Vis. | Rat. | Close To |
|------|------|------|------|------|-------|------|-------|------|------|----------|
| 1st | Sgr | 19h 40m | -20° 24' | 1.1 | 9" | 14% | 13° E | NV | N/A | Venus |
| 3rd | Sgr | 19h 33m | -20° 1' | 1.7 | 10" | 7% | 9° E | NV | N/A | Venus |
| 5th | Sgr | 19h 23m | -19° 45' | 2.3 | 10" | 3% | 5° E | NV | N/A | |
| 7th | Sgr | 19h 12m | -19° 36' | 2.7 | 10" | 1% | 0° W | NV | N/A | |
| 9th | Sgr | 19h 0m | -19° 33' | 2.3 | 10" | 3% | 5° W | NV | N/A | |

## Venus

| Date | Con. | R.A. | Dec. | Mag. | Diam. | Ill. | Elon. | Vis. | Rat. | Close To |
|------|------|------|------|------|-------|------|-------|------|------|----------|
| 1st | Sgr | 20h 1m | -21° 55' | -3.3 | 10" | 96% | 19° E | PM | * | Mercury |
| 3rd | Cap | 20h 12m | -21° 26' | -3.3 | 10" | 96% | 19° E | PM | * | Mercury |
| 5th | Cap | 20h 22m | -20° 55' | -3.3 | 10" | 95% | 20° E | PM | * | |
| 7th | Cap | 20h 33m | -20° 20' | -3.3 | 11" | 95% | 20° E | PM | * | |
| 9th | Cap | 20h 43m | -19° 44' | -3.3 | 11" | 95% | 20° E | PM | * | |

## Mars and the Outer Planets

Mars
5th

Jupiter
5th

Saturn
5th

## Mars

| Date | Con. | R.A. | Dec. | Mag. | Diam. | Ill. | Elon. | Vis. | Rat. | Close To |
|------|------|------|------|------|-------|------|-------|------|------|----------|
| 1st | Tau | 4h 27m | 24° 35' | -1.0 | 15" | 97% | 145° E | PM | **** | Hyades, Aldebaran |
| 5th | Tau | 4h 25m | 24° 31' | -0.9 | 14" | 96% | 140° E | PM | *** | Pleiades, Hyades, Aldebaran |
| 10th | Tau | 4h 23m | 24° 28' | -0.7 | 13" | 95% | 134° E | PM | *** | Pleiades, Hyades, Aldebaran |

## The Outer Planets

| Planet | Date | Con. | R.A. | Dec. | Mag. | Diam. | Elon. | Vis. | Rat. | Close To |
|--------|------|------|------|------|------|-------|-------|------|------|----------|
| Jupiter | 5th | Psc | 0h 8m | 0° 28' | -1.9 | 39" | 76° E | PM | ** | |
| Saturn | 5th | Cap | 21h 42m | -15° 4' | 1.0 | 16" | 40° E | PM | ** | |
| Uranus | 5th | Ari | 2h 50m | 16° 1' | 6.0 | 4" | 117° E | PM | ** | |
| Neptune | 5th | Aqr | 23h 35m | -3° 53' | 7.7 | 2" | 68° E | PM | *** | Jupiter |

## Highlights

| Date | Time (UT) | Event |
|------|-----------|-------|
| 1st | 23:52 | The waxing gibbous Moon is north of Uranus. (Evening sky.) |
| 3rd | 03:19 | The waxing gibbous Moon is south of the Pleiades star cluster. (Taurus, evening sky.) |
| | 18:14 | The waxing gibbous Moon is south of Mars. (Evening sky.) |
| | N/A | The Quadrantid meteor shower is at its maximum. (ZHR: 120) |
| 4th | 02:00 | The waxing gibbous Moon is north of the bright star Aldebaran. (Taurus, evening sky.) |
| 6th | 23:09 | Full Moon. (Visible all night.) |
| 7th | 12:50 | Mercury is at inferior conjunction with the Sun. (Not visible.) |
| 10th | 12:39 | The waning gibbous Moon is north of the bright star Regulus. (Leo, morning sky.) |

# January 11<sup>th</sup> to 20<sup>th</sup>, 2023

## The Moon

| 11<sup>th</sup> | 13<sup>th</sup> | 15<sup>th</sup> | 17<sup>th</sup> | 19<sup>th</sup> |

| Date | Con | R.A. | Dec | Mag | Diam | Ill. | Elon. | Phase | Close To |
|------|-----|------|-----|-----|------|------|-------|-------|----------|
| 11th | Leo | 10h 54m | 11° 43' | -11.4 | 30' | 82% | 129° W | -G | Regulus |
| 12th | Leo | 11h 38m | 6° 27' | -11.1 | 30' | 75% | 119° W | -G | |
| 13th | Vir | 12h 21m | 0° 51' | -10.8 | 30' | 66% | 109° W | LQ | |
| 14th | Vir | 13h 6m | -4° 54' | -10.4 | 31' | 56% | 99° W | LQ | Spica |
| 15th | Vir | 13h 52m | -10° 37' | -9.9 | 31' | 46% | 89° W | LQ | Spica |
| 16th | Lib | 14h 42m | -16° 2' | -9.4 | 32' | 35% | 77° W | -Cr | |
| 17th | Lib | 15h 36m | -20° 51' | -8.8 | 32' | 25% | 65° W | -Cr | |
| 18th | Oph | 16h 35m | -24° 38' | -8.0 | 33' | 16% | 51° W | -Cr | Antares |
| 19th | Oph | 17h 40m | -26° 57' | -7.1 | 33' | 8% | 36° W | NM | |
| 20th | Sgr | 18h 47m | -27° 23' | -6.0 | 33' | 3% | 20° W | NM | Mercury |

## Mercury and Venus

Mercury
15<sup>th</sup>

Venus
15<sup>th</sup>

### Mercury

| Date | Con. | R.A. | Dec. | Mag. | Diam. | Ill. | Elon. | Vis. | Rat. | Close To |
|------|------|------|------|------|-------|------|-------|------|------|----------|
| 11th | Sgr | 18h 50m | -19° 35' | 1.8 | 10" | 7% | 10° W | NV | N/A | |
| 13th | Sgr | 18h 42m | -19° 42' | 1.3 | 10" | 13% | 14° W | NV | N/A | |
| 15th | Sgr | 18h 37m | -19° 54' | 1.0 | 9" | 20% | 18° W | AM | *** | |
| 17th | Sgr | 18h 34m | -20° 8' | 0.7 | 9" | 27% | 20° W | AM | *** | |
| 19th | Sgr | 18h 34m | -20° 24' | 0.5 | 8" | 34% | 23° W | AM | *** | |

**Venus**

| Date | Con. | R.A. | Dec. | Mag. | Diam. | Ill. | Elon. | Vis. | Rat. | Close To |
|------|------|------|------|------|-------|------|-------|------|------|----------|
| 11th | Cap | 20h 53m | -19° 5' | -3.3 | 11" | 95% | 21° E | PM | * | |
| 13th | Cap | 21h 3m | -18° 24' | -3.3 | 11" | 94% | 21° E | PM | * | |
| 15th | Cap | 21h 13m | -17° 41' | -3.3 | 11" | 94% | 21° E | PM | * | Saturn |
| 17th | Cap | 21h 23m | -16° 56' | -3.3 | 11" | 94% | 22° E | PM | * | Saturn |
| 19th | Cap | 21h 33m | -16° 10' | -3.3 | 11" | 93% | 22° E | PM | * | Saturn |

## Mars and the Outer Planets

Mars
15th

Jupiter
15th

Saturn
15th

**Mars**

| Date | Con. | R.A. | Dec. | Mag. | Diam. | Ill. | Elon. | Vis. | Rat. | Close To |
|------|------|------|------|------|-------|------|-------|------|------|----------|
| 11th | Tau | 4h 23m | 24° 28' | -0.7 | 13" | 95% | 133° E | PM | *** | Pleiades, Hyades, Aldebaran |
| 15th | Tau | 4h 23m | 24° 28' | -0.6 | 13" | 95% | 129° E | PM | *** | Pleiades, Hyades, Aldebaran |
| 20th | Tau | 4h 25m | 24° 29' | -0.4 | 12" | 94% | 124° E | PM | *** | Pleiades, Hyades, Aldebaran |

**The Outer Planets**

| Planet | Date | Con. | R.A. | Dec. | Mag. | Diam. | Elon. | Vis. | Rat. | Close To |
|--------|------|------|------|------|------|-------|-------|------|------|----------|
| Jupiter | 15th | Psc | 0h 13m | 0° 9' | -1.9 | 38" | 66° E | PM | ** | |
| Saturn | 15th | Cap | 21h 46m | -14° 43' | 1.0 | 16" | 30° E | PM | * | Venus |
| Uranus | 15th | Ari | 2h 50m | 15° 59' | 6.0 | 4" | 106° E | PM | ** | |
| Neptune | 15th | Aqr | 23h 36m | -3° 48' | 7.7 | 2" | 57° E | PM | *** | Jupiter |

## Highlights

| Date | Time (UT) | Event |
|------|-----------|-------|
| 12th | 20:09 | Mars is stationary prior to resuming prograde motion. (Evening sky.) |
| 14th | 20:41 | The nearly last quarter Moon is north of the bright star Spica. (Virgo, morning sky.) |
| 15th | 02:11 | Last Quarter Moon. (Morning sky.) |
| 18th | 10:08 | The waning crescent Moon is north of the bright star Antares. (Scorpius, morning sky.) |
| | 11:42 | Mercury is stationary prior to resuming prograde motion. (Morning sky.) |
| | N/A | Earthshine on the waning crescent Moon. (Morning sky.) |
| 20th | 06:15 | The waning crescent Moon is south of Mercury. (Morning sky.) |

# January 21st to 31st, 2023

## The Moon

| 21st | 23rd | 25th | 27th | 29th | 31st |

| Date | Con | R.A. | Dec | Mag | Diam | Ill. | Elon. | Phase | Close To |
|------|-----|------|-----|-----|------|------|-------|-------|----------|
| 21st | Sgr | 19h 55m | -25° 45' | -4.7 | 33' | 0% | 5° W | NM | |
| 22nd | Cap | 21h 0m | -22° 12' | -5.0 | 33' | 1% | 11° E | NM | Venus |
| 23rd | Aqr | 22h 1m | -17° 11' | -6.3 | 33' | 4% | 25° E | NM | Venus, Saturn |
| 24th | Aqr | 22h 58m | -11° 11' | -7.4 | 33' | 10% | 38° E | NM | Neptune |
| 25th | Aqr | 23h 50m | -4° 45' | -8.3 | 32' | 18% | 50° E | +Cr | Jupiter, Neptune |
| 26th | Cet | 0h 40m | 1° 44' | -9.0 | 32' | 28% | 62° E | +Cr | Jupiter |
| 27th | Psc | 1h 29m | 7° 55' | -9.6 | 31' | 38% | 73° E | FQ | |
| 28th | Ari | 2h 17m | 13° 33' | -10.1 | 31' | 49% | 84° E | FQ | Uranus |
| 29th | Ari | 3h 6m | 18° 27' | -10.5 | 30' | 59% | 95° E | FQ | Uranus, Pleiades |
| 30th | Tau | 3h 56m | 22° 26' | -10.9 | 30' | 68% | 106° E | +G | Mars, Pleiades, Hyades, Aldebaran |
| 31st | Tau | 4h 48m | 25° 21' | -11.2 | 30' | 77% | 118° E | +G | Mars, Hyades, Aldebaran |

## Mercury and Venus

Mercury
25th

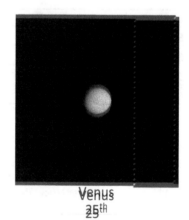

Venus
25th

### Mercury

| Date | Con. | R.A. | Dec. | Mag. | Diam. | Ill. | Elon. | Vis. | Rat. | Close To |
|------|------|------|------|------|-------|------|-------|------|------|----------|
| 21st | Sgr | 18h 37m | -20° 41' | 0.4 | 8" | 41% | 24° W | AM | *** | |
| 23rd | Sgr | 18h 41m | -20° 57' | 0.3 | 8" | 47% | 25° W | AM | **** | |
| 25th | Sgr | 18h 46m | -21° 12' | 0.2 | 7" | 52% | 26° W | AM | **** | |
| 27th | Sgr | 18h 53m | -21° 24' | 0.2 | 7" | 57% | 26° W | AM | **** | |
| 29th | Sgr | 19h 1m | -21° 34' | 0.1 | 7" | 61% | 26° W | AM | **** | |
| 31st | Sgr | 19h 10m | -21° 40' | 0.1 | 7" | 65% | 26° W | AM | *** | |

## Venus

| Date | Con. | R.A. | Dec. | Mag. | Diam. | Ill. | Elon. | Vis. | Rat. | Close To |
|------|------|------|------|------|-------|------|-------|------|------|----------|
| 21st | Cap | 21h 43m | -15° 21' | -3.3 | 11" | 93% | 22° E | PM | * | Saturn |
| 23rd | Cap | 21h 53m | -14° 31' | -3.3 | 11" | 93% | 23° E | PM | * | Moon, Saturn |
| 25th | Aqr | 22h 2m | -13° 39' | -3.3 | 11" | 93% | 23° E | PM | * | Saturn |
| 27th | Aqr | 22h 12m | -12° 46' | -3.3 | 11" | 92% | 23° E | PM | * | Saturn |
| 29th | Aqr | 22h 21m | -11° 51' | -3.3 | 11" | 92% | 24° E | PM | * | Saturn |
| 31st | Aqr | 22h 31m | -10° 55' | -3.3 | 11" | 92% | 24° E | PM | * | Saturn |

## Mars and the Outer Planets

Mars
25th

Jupiter
25th

Saturn
25th

## Mars

| Date | Con. | R.A. | Dec. | Mag. | Diam. | Ill. | Elon. | Vis. | Rat. | Close To |
|------|------|------|------|------|-------|------|-------|------|------|----------|
| 21st | Tau | 4h 25m | 24° 30' | -0.4 | 12" | 94% | 123° E | PM | *** | Pleiades, Hyades, Aldebaran |
| 25th | Tau | 4h 27m | 24° 33' | -0.3 | 11" | 93% | 119° E | PM | *** | Hyades, Aldebaran |
| 31st | Tau | 4h 32m | 24° 40' | -0.1 | 11" | 92% | 114° E | PM | *** | Moon, Hyades, Aldebaran |

## The Outer Planets

| Planet | Date | Con. | R.A. | Dec. | Mag. | Diam. | Elon. | Vis. | Rat. | Close To |
|--------|------|------|------|------|------|-------|-------|------|------|----------|
| Jupiter | 25th | Psc | 0h 19m | 0° 50' | -1.8 | 37" | 57° E | PM | ** | Moon |
| Saturn | 25th | Cap | 21h 51m | -14° 20' | 1.0 | 15" | 20° E | PM | * | Venus |
| Uranus | 25th | Ari | 2h 50m | 15° 59' | 6.1 | 4" | 95° E | PM | * | |
| Neptune | 25th | Aqr | 23h 37m | -3° 42' | 7.8 | 2" | 47° E | PM | *** | Moon |

## Highlights

| Date | Time (UT) | Event |
|------|-----------|-------|
| 21st | 20:54 | New Moon. (Not visible.) |
| 22nd | 19:32 | Venus is 0.4° south of Saturn. (Evening sky.) |
| 23rd | 05:52 | The waxing crescent Moon is south of Saturn. (Evening sky.) |
| | 06:35 | The waxing crescent Moon is south of Venus. (Evening sky.) |
| 25th | 05:08 | The waxing crescent Moon is south of Neptune. (Evening sky.) |
| | N/A | Good opportunity to see Earthshine on the waxing crescent Moon. (Evening sky.) |
| 26th | 02:43 | The waxing crescent Moon is south of Jupiter. (Evening sky.) |
| 28th | 15:20 | First Quarter Moon. (Evening sky.) |
| 29th | 04:50 | The just-past first quarter Moon is north of Uranus. (Evening sky.) |
| 30th | 05:46 | Mercury is at greatest western elongation from the Sun. (Morning sky.) |
| 31st | 05:32 | The waxing gibbous Moon is south of Mars. (Evening sky.) |
| | 06:44 | The waxing gibbous Moon is north of the bright star Aldebaran. (Taurus, evening sky.) |

# February 1st to 10th, 2023

## The Moon

| | 1st | | 3rd | | 5th | | 7th | | 9th |

| Date | Con | R.A. | Dec | Mag | Diam | Ill. | Elon. | Phase | Close To |
|------|-----|------|-----|-----|------|------|-------|-------|----------|
| 1st | Tau | 5h 41m | 27° 3' | -11.5 | 30' | 84% | 131° E | +G | |
| 2nd | Gem | 6h 35m | 27° 30' | -11.8 | 29' | 91% | 143° E | +G | |
| 3rd | Gem | 7h 28m | 26° 40' | -12.1 | 29' | 95% | 155° E | FM | |
| 4th | Cnc | 8h 20m | 24° 38' | -12.3 | 29' | 98% | 167° E | FM | Praesepe |
| 5th | Cnc | 9h 9m | 21° 32' | -12.6 | 29' | 100% | 179° E | FM | Praesepe |
| 6th | Leo | 9h 57m | 17° 32' | -12.5 | 30' | 99% | 171° W | FM | Regulus |
| 7th | Leo | 10h 42m | 12° 49' | -12.2 | 30' | 97% | 160° W | FM | Regulus |
| 8th | Leo | 11h 27m | 7° 35' | -11.9 | 30' | 93% | 150° W | -G | |
| 9th | Vir | 12h 10m | 2° 1' | -11.6 | 30' | 88% | 140° W | -G | |
| 10th | Vir | 12h 54m | -3° 42' | -11.3 | 30' | 80% | 130° W | -G | Spica |

## Mercury and Venus

Mercury
5th

Venus
5th

**Mercury**

| Date | Con. | R.A. | Dec. | Mag. | Diam. | Ill. | Elon. | Vis. | Rat. | Close To |
|------|------|------|------|------|-------|------|-------|------|------|----------|
| 1st | Sgr | 19h 15m | -21° 41' | 0.1 | 6" | 66% | 26° W | AM | *** | |
| 3rd | Sgr | 19h 25m | -21° 41' | 0.1 | 6" | 70% | 25° W | AM | *** | |
| 5th | Sgr | 19h 35m | -21° 37' | 0.0 | 6" | 72% | 25° W | AM | *** | |
| 7th | Sgr | 19h 46m | -21° 28' | 0.0 | 6" | 75% | 24° W | AM | *** | |
| 9th | Sgr | 19h 58m | -21° 15' | 0.0 | 6" | 77% | 23° W | AM | *** | |

**Venus**

| Date | Con. | R.A. | Dec. | Mag. | Diam. | Ill. | Elon. | Vis. | Rat. | Close To |
|------|------|------|------|------|-------|------|-------|------|------|----------|
| **1st** | Aqr | 22h 36m | -10° 27' | -3.4 | 11" | 91% | 24° E | PM | * | |
| **3rd** | Aqr | 22h 45m | -9° 29' | -3.4 | 11" | 91% | 24° E | PM | * | |
| **5th** | Aqr | 22h 54m | -8° 31' | -3.4 | 11" | 91% | 25° E | PM | * | |
| **7th** | Aqr | 23h 3m | -7° 31' | -3.4 | 11" | 90% | 25° E | PM | * | |
| **9th** | Aqr | 23h 12m | -6° 31' | -3.4 | 11" | 90% | 25° E | PM | * | |

## Mars and the Outer Planets

Mars
5th

Jupiter
5th

Saturn
5th

**Mars**

| Date | Con. | R.A. | Dec. | Mag. | Diam. | Ill. | Elon. | Vis. | Rat. | Close To |
|------|------|------|------|------|-------|------|-------|------|------|----------|
| **1st** | Tau | 4h 33m | 24° 41' | -0.1 | 11" | 92% | 114° E | PM | *** | Hyades, Aldebaran |
| **5th** | Tau | 4h 37m | 24° 47' | 0.0 | 10" | 92% | 111° E | PM | *** | Hyades, Aldebaran |
| **10th** | Tau | 4h 43m | 24° 55' | 0.2 | 10" | 91% | 107° E | PM | ** | Hyades, Aldebaran |

**The Outer Planets**

| Planet | Date | Con. | R.A. | Dec. | Mag. | Diam. | Elon. | Vis. | Rat. | Close To |
|--------|------|------|------|------|------|-------|-------|------|------|----------|
| Jupiter | 5th | Psc | 0h 27m | 1° 41' | -1.7 | 36" | 48° E | PM | ** | |
| Saturn | 5th | Cap | 21h 56m | -13° 54' | 1.0 | 15" | 10° E | NV | N/A | |
| Uranus | 5th | Ari | 2h 50m | 16° 1' | 6.1 | 4" | 84° E | PM | * | |
| Neptune | 5th | Aqr | 23h 38m | -3° 34' | 7.8 | 2" | 36° E | PM | *** | |

## Highlights

| Date | Time (UT) | Event |
|------|-----------|-------|
| 4th | 00:36 | Mars is 8.2° north of the bright star Aldebaran. (Taurus, evening sky.) |
| 5th | 18:29 | Full Moon. (Visible all night.) |
| 6th | 16:14 | The just-past full Moon is north of the bright star Regulus. (Leo, visible all night.) |

# February 11th to 20th, 2023

## The Moon

11th

13th

15th

17th

19th

| Date | Con | R.A. | Dec | Mag | Diam | Ill. | Elon. | Phase | Close To |
|------|-----|------|-----|-----|------|------|-------|-------|----------|
| 11th | Vir | 13h 39m | -9° 23' | -11.0 | 31' | 72% | 120° W | -G | Spica |
| 12th | Lib | 14h 27m | -14° 48' | -10.6 | 31' | 62% | 109° W | LQ | |
| 13th | Lib | 15h 18m | -19° 41' | -10.2 | 31' | 52% | 97° W | LQ | |
| 14th | Sco | 16h 14m | -23° 43' | -9.7 | 32' | 41% | 84° W | LQ | Antares |
| 15th | Oph | 17h 14m | -26° 29' | -9.1 | 32' | 30% | 70° W | -Cr | Antares |
| 16th | Sgr | 18h 19m | -27° 37' | -8.4 | 33' | 20% | 55° W | -Cr | |
| 17th | Sgr | 19h 25m | -26° 51' | -7.5 | 33' | 12% | 39° W | NM | |
| 18th | Cap | 20h 30m | -24° 8' | -6.5 | 33' | 5% | 24° W | NM | Mercury |
| 19th | Cap | 21h 33m | -19° 42' | -5.2 | 33' | 1% | 9° W | NM | Mercury, Saturn |
| 20th | Aqr | 22h 31m | -14° 1' | -4.5 | 33' | 0% | 4° E | NM | Saturn |

## Mercury and Venus

Mercury
15th

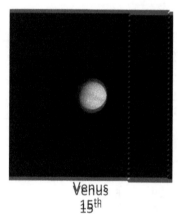
Venus
15th

### Mercury

| Date | Con. | R.A. | Dec. | Mag. | Diam. | Ill. | Elon. | Vis. | Rat. | Close To |
|------|------|------|------|------|-------|------|-------|------|------|----------|
| 11th | Cap | 20h 9m | -20° 56' | 0.0 | 6" | 79% | 22° W | AM | *** | |
| 13th | Cap | 20h 21m | -20° 33' | 0.0 | 6" | 81% | 21° W | AM | *** | |
| 15th | Cap | 20h 33m | -20° 4' | -0.1 | 5" | 83% | 20° W | AM | *** | |
| 17th | Cap | 20h 45m | -19° 31' | -0.1 | 5" | 84% | 19° W | AM | ** | |
| 19th | Cap | 20h 58m | -18° 52' | -0.2 | 5" | 86% | 18° W | AM | ** | Moon |

## Venus

| Date | Con. | R.A. | Dec. | Mag. | Diam. | Ill. | Elon. | Vis. | Rat. | Close To |
|------|------|------|------|------|-------|------|-------|------|------|----------|
| 11th | Aqr | 23h 21m | -5° 30' | -3.4 | 11" | 90% | 26° E | PM | * | Neptune |
| 13th | Aqr | 23h 31m | -4° 29' | -3.4 | 12" | 89% | 26° E | PM | * | Neptune |
| 15th | Aqr | 23h 40m | -3° 27' | -3.4 | 12" | 89% | 26° E | PM | * | Neptune |
| 17th | Psc | 23h 48m | -2° 25' | -3.4 | 12" | 88% | 27° E | PM | * | Neptune |
| 19th | Psc | 23h 57m | -1° 22' | -3.4 | 12" | 88% | 27° E | PM | * | Jupiter, Neptune |

## Mars and the Outer Planets

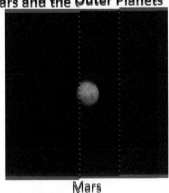

Mars
15th

Jupiter
15th

Saturn
15th

### Mars

| Date | Con. | R.A. | Dec. | Mag. | Diam. | Ill. | Elon. | Vis. | Rat. | Close To |
|------|------|------|------|------|-------|------|-------|------|------|----------|
| 11th | Tau | 4h 44m | 24° 56' | 0.2 | 10" | 91% | 106° E | PM | ** | Hyades, Aldebaran |
| 15th | Tau | 4h 50m | 25° 3' | 0.3 | 9" | 91% | 104° E | PM | ** | Hyades, Aldebaran |
| 20th | Tau | 4h 57m | 25° 11' | 0.4 | 9" | 90% | 101° E | PM | ** | Hyades, Aldebaran |

### The Outer Planets

| Planet | Date | Con. | R.A. | Dec. | Mag. | Diam. | Elon. | Vis. | Rat. | Close To |
|--------|------|------|------|------|------|-------|-------|------|------|----------|
| Jupiter | 15th | Cet | 0h 34m | 2° 30' | -1.7 | 35" | 40° E | PM | * | |
| Saturn | 15th | Aqr | 22h 0m | -13° 29' | 1.0 | 15" | 1° E | NV | N/A | |
| Uranus | 15th | Ari | 2h 51m | 16° 4' | 6.1 | 4" | 74° E | PM | * | |
| Neptune | 15th | Aqr | 23h 40m | -3° 26' | 7.8 | 2" | 26° E | PM | ** | Venus |

## Highlights

| Date | Time (UT) | Event |
|------|-----------|-------|
| 11th | 04:49 | The waning gibbous Moon is north of the bright star Spica. (Virgo, morning sky.) |
| 13th | 16:01 | Last Quarter Moon. (Morning sky.) |
| 14th | 18:24 | The just-past last quarter Moon is north of the bright star Antares. (Scorpius, morning sky.) |
| 15th | 12:14 | Venus is less than 6 arc seconds south of Neptune. (Evening sky.) |
| 17th | 12:07 | Saturn is in conjunction with the Sun. (Not visible.) |
| | N/A | Good opportunity to see Earthshine on the waning crescent Moon. (Morning sky.) |
| 18th | 21:31 | The waning crescent Moon is south of Mercury. (Morning sky.) |
| 20th | 07:06 | New Moon. (Not visible.) |

# February 21st to 28th, 2023

## The Moon

21st

23rd

25th

27th

| Date | Con | R.A. | Dec | Mag | Diam | Ill. | Elon. | Phase | Close To |
|------|-----|------|-----|-----|------|------|-------|-------|----------|
| 21st | Aqr | 23h 27m | -7° 33' | -5.7 | 33' | 2% | 17° E | NM | Venus, Neptune |
| 22nd | Psc | 0h 19m | 0° 50' | -6.9 | 33' | 7% | 29° E | NM | Venus, Jupiter, Neptune |
| 23rd | Psc | 1h 9m | 5° 44' | -7.8 | 32' | 14% | 41° E | +Cr | Jupiter |
| 24th | Ari | 1h 59m | 11° 50' | -8.6 | 32' | 22% | 53° E | +Cr | |
| 25th | Ari | 2h 49m | 17° 11' | -9.2 | 31' | 32% | 64° E | +Cr | Uranus |
| 26th | Tau | 3h 41m | 21° 36' | -9.7 | 31' | 42% | 76° E | FQ | Pleiades, Hyades |
| 27th | Tau | 4h 33m | 24° 53' | -10.2 | 30' | 52% | 88° E | FQ | Mars, Pleiades, Hyades, Aldebara |
| 28th | Tau | 5h 27m | 26° 57' | -10.6 | 30' | 61% | 101° E | FQ | Mars |

## Mercury and Venus

Mercury
24th

Venus
24th

### Mercury

| Date | Con. | R.A. | Dec. | Mag. | Diam. | Ill. | Elon. | Vis. | Rat. | Close To |
|------|------|------|------|------|-------|------|-------|------|------|----------|
| 21st | Cap | 21h 10m | -18° 8' | -0.2 | 5" | 88% | 17° W | AM | ** | |
| 23rd | Cap | 21h 23m | -17° 19' | -0.3 | 5" | 89% | 16° W | AM | ** | |
| 25th | Cap | 21h 36m | -16° 24' | -0.3 | 5" | 90% | 14° W | NV | N/A | Saturn |
| 27th | Cap | 21h 48m | -15° 25' | -0.4 | 5" | 92% | 13° W | NV | N/A | Saturn |

**Venus**

| Date | Con. | R.A. | Dec. | Mag. | Diam. | Ill. | Elon. | Vis. | Rat. | Close To |
|------|------|------|------|------|-------|------|-------|------|------|----------|
| 21st | Psc | 0h 6m | 0° 20' | -3.4 | 12" | 88% | 27° E | PM | * | Moon, Jupiter |
| 23rd | Psc | 0h 15m | 0° 43' | -3.4 | 12" | 87% | 27° E | PM | * | Jupiter |
| 25th | Psc | 0h 24m | 1° 46' | -3.4 | 12" | 87% | 28° E | PM | * | Jupiter |
| 27th | Cet | 0h 33m | 2° 49' | -3.4 | 12" | 86% | 28° E | PM | * | Jupiter |

## Mars and the Outer Planets

Mars
24th

Jupiter
24th

Saturn
24th

**Mars**

| Date | Con. | R.A. | Dec. | Mag. | Diam. | Ill. | Elon. | Vis. | Rat. | Close To |
|------|------|------|------|------|-------|------|-------|------|------|----------|
| 21st | Tau | 4h 59m | 25° 13' | 0.4 | 9" | 90% | 100° E | PM | ** | Hyades, Aldebaran |
| 24th | Tau | 5h 4m | 25° 17' | 0.5 | 9" | 90% | 99° E | PM | ** | Hyades, Aldebaran |
| 28th | Tau | 5h 10m | 25° 23' | 0.6 | 8" | 90% | 97° E | PM | ** | Moon, Aldebaran |

**The Outer Planets**

| Planet | Date | Con. | R.A. | Dec. | Mag. | Diam. | Elon. | Vis. | Rat. | Close To |
|--------|------|------|------|------|------|-------|-------|------|------|----------|
| Jupiter | 24th | Psc | 0h 41m | 3° 16' | -1.7 | 34" | 33° E | PM | * | Venus |
| Saturn | 24th | Aqr | 22h 5m | -13° 7' | 1.0 | 15" | 6° W | NV | N/A | Mercury |
| Uranus | 24th | Ari | 2h 52m | 16° 8' | 6.1 | 4" | 66° E | PM | * | |
| Neptune | 24th | Aqr | 23h 41m | -3° 18' | 7.8 | 2" | 18° E | PM | ** | Venus |

## Highlights

| Date | Time (UT) | Event |
|------|-----------|-------|
| 22nd | 06:11 | The waxing crescent Moon is south of Venus. (Evening sky.) |
| | 23:26 | The waxing crescent Moon is south of Jupiter. (Evening sky.) |
| 23rd | N/A | Good opportunity to see Earthshine on the waxing crescent Moon. (Evening sky.) |
| 25th | 11:16 | The waxing crescent Moon is north of Uranus. (Evening sky.) |
| 26th | 13:02 | The nearly first quarter Moon is south of the Pleiades star cluster. (Taurus, evening sky.) |
| 27th | 08:06 | First Quarter Moon. (Evening sky.) |
| | 11:17 | The just-past first quarter Moon is north of the bright star Aldebaran. (Taurus, evening sky.) |
| 28th | 05:15 | The just-past first quarter Moon is north of Mars. (Evening sky.) |

# March 1st to 10th, 2023

## The Moon

    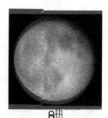

| 1st | 3rd | 5th | 7th | 9th |

| Date | Con | R.A. | Dec | Mag | Diam | Ill. | Elon. | Phase | Close To |
|------|-----|------|-----|-----|------|------|-------|-------|----------|
| 1st | Gem | 6h 20m | 27° 43' | -10.9 | 30' | 70% | 113° E | +G | |
| 2nd | Gem | 7h 14m | 27° 11' | -11.3 | 29' | 79% | 126° E | +G | |
| 3rd | Cnc | 8h 6m | 25° 25' | -11.6 | 29' | 86% | 138° E | +G | Praesepe |
| 4th | Cnc | 8h 56m | 22° 33' | -11.8 | 29' | 92% | 149° E | +G | Praesepe |
| 5th | Leo | 9h 44m | 18° 43' | -12.1 | 30' | 96% | 160° E | FM | Regulus |
| 6th | Leo | 10h 30m | 14° 7' | -12.4 | 30' | 99% | 171° E | FM | Regulus |
| 7th | Leo | 11h 15m | 8° 56' | -12.6 | 30' | 100% | 179° W | FM | |
| 8th | Vir | 11h 59m | 3° 20' | -12.4 | 30' | 99% | 169° W | FM | |
| 9th | Vir | 12h 43m | -2° 27' | -12.1 | 30' | 96% | 159° W | FM | Spica |
| 10th | Vir | 13h 28m | -8° 14' | -11.8 | 31' | 91% | 148° W | -G | Spica |

## Mercury and Venus

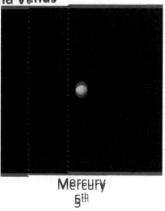 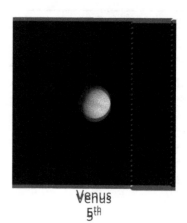

Mercury
5th

Venus
5th

### Mercury

| Date | Con. | R.A. | Dec. | Mag. | Diam. | Ill. | Elon. | Vis. | Rat. | Close To |
|------|------|------|------|------|-------|------|-------|------|------|----------|
| 1st | Aqr | 22h 1m | -14° 20' | -0.5 | 5" | 93% | 12° W | NV | N/A | Saturn |
| 3rd | Aqr | 22h 15m | -13° 10' | -0.6 | 5" | 94% | 10° W | NV | N/A | Saturn |
| 5th | Aqr | 22h 28m | -11° 54' | -0.7 | 5" | 95% | 9° W | NV | N/A | Saturn |
| 7th | Aqr | 22h 41m | -10° 34' | -0.8 | 5" | 96% | 7° W | NV | N/A | Saturn |
| 9th | Aqr | 22h 55m | -9° 9' | -0.9 | 5" | 97% | 6° W | NV | N/A | |

## Venus

| Date | Con. | R.A. | Dec. | Mag. | Diam. | Ill. | Elon. | Vis. | Rat. | Close To |
|------|------|------|------|------|-------|------|-------|------|------|----------|
| 1st | Psc | 0h 42m | 3° 51' | -3.4 | 12" | 86% | 28° E | PM | * | Jupiter |
| 3rd | Psc | 0h 51m | 4° 53' | -3.4 | 12" | 85% | 29° E | PM | * | Jupiter |
| 5th | Psc | 1h 0m | 5° 55' | -3.4 | 12" | 85% | 29° E | PM | * | Jupiter |
| 7th | Psc | 1h 9m | 6° 56' | -3.4 | 12" | 84% | 30° E | PM | * | Jupiter |
| 9th | Psc | 1h 18m | 7° 57' | -3.4 | 13" | 84% | 30° E | PM | * | Jupiter |

## Mars and the Outer Planets

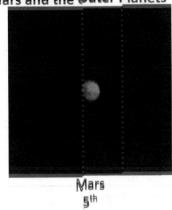

Mars
5th

Jupiter
5th

Saturn
5th

## Mars

| Date | Con. | R.A. | Dec. | Mag. | Diam. | Ill. | Elon. | Vis. | Rat. | Close To |
|------|------|------|------|------|-------|------|-------|------|------|----------|
| 1st | Tau | 5h 12m | 25° 24' | 0.6 | 8" | 90% | 96° E | PM | ** | Aldebaran |
| 5th | Tau | 5h 19m | 25° 29' | 0.7 | 8" | 90% | 94° E | PM | ** | |
| 10th | Tau | 5h 29m | 25° 33' | 0.8 | 8" | 90% | 92° E | PM | ** | |

## The Outer Planets

| Planet | Date | Con. | R.A. | Dec. | Mag. | Diam. | Elon. | Vis. | Rat. | Close To |
|--------|------|------|------|------|------|-------|-------|------|------|----------|
| Jupiter | 5th | Psc | 0h 49m | 4° 4' | -1.6 | 34" | 26° E | PM | * | Venus |
| Saturn | 5th | Aqr | 22h 9m | -12° 45' | 1.1 | 15" | 14° W | NV | N/A | Mercury |
| Uranus | 5th | Ari | 2h 53m | 16° 13' | 6.1 | 4" | 57° E | PM | * | |
| Neptune | 5th | Psc | 23h 42m | -3° 10' | 7.8 | 2" | 10° E | NV | N/A | |

## Highlights

| Date | Time (UT) | Event |
|------|-----------|-------|
| 2nd | 10:37 | Venus is 0.5° north of Jupiter. (Evening sky.) |
| 6th | 00:59 | The nearly full Moon is north of the bright star Regulus. (Leo, visible all night.) |
| 7th | 12:41 | Full Moon. (Visible all night.) |
| 10th | 11:24 | The waning gibbous Moon is north of the bright star Spica. (Virgo, morning sky.) |

# March 11<sup>th</sup> to 20<sup>th</sup>, 2023

## The Moon

| | 11<sup>th</sup> | | 13<sup>th</sup> | | 15<sup>th</sup> | | 17<sup>th</sup> | | 19<sup>th</sup> |

| Date | Con | R.A. | Dec | Mag | Diam | Ill. | Elon. | Phase | Close To |
|------|-----|------|-----|-----|------|------|-------|-------|----------|
| 11th | Vir | 14h 15m | -13° 47' | -11.5 | 31' | 85% | 137° W | -G | |
| 12th | Lib | 15h 6m | -18° 49' | -11.2 | 31' | 77% | 126° W | -G | |
| 13th | Sco | 16h 0m | -23° 3' | -10.8 | 31' | 67% | 113° W | -G | Antares |
| 14th | Oph | 16h 58m | -26° 7' | -10.4 | 32' | 57% | 100° W | LQ | Antares |
| 15th | Sgr | 17h 59m | -27° 41' | -9.9 | 32' | 45% | 85° W | LQ | |
| 16th | Sgr | 19h 3m | -27° 30' | -9.4 | 32' | 34% | 70° W | -Cr | |
| 17th | Sgr | 20h 7m | -25° 29' | -8.7 | 33' | 24% | 55° W | -Cr | |
| 18th | Cap | 21h 8m | -21° 45' | -7.9 | 33' | 14% | 41° W | -Cr | |
| 19th | Aqr | 22h 7m | -16° 37' | -6.9 | 33' | 7% | 27° W | NM | Saturn |
| 20th | Aqr | 23h 3m | -10° 29' | -5.7 | 33' | 2% | 14° W | NM | Saturn, Neptune |

## Mercury and Venus

Mercury
15<sup>th</sup>

Venus
15<sup>th</sup>

### Mercury

| Date | Con. | R.A. | Dec. | Mag. | Diam. | Ill. | Elon. | Vis. | Rat. | Close To |
|------|------|------|------|------|-------|------|-------|------|------|----------|
| 11th | Aqr | 23h 8m | -7° 38' | -1.1 | 5" | 98% | 4° W | NV | N/A | |
| 13th | Aqr | 23h 22m | -6° 3' | -1.2 | 5" | 99% | 3° W | NV | N/A | |
| 15th | Aqr | 23h 36m | -4° 24' | -1.4 | 5" | 100% | 1° W | NV | N/A | Neptune |
| 17th | Psc | 23h 50m | -2° 40' | -1.5 | 5" | 100% | 1° E | NV | N/A | Neptune |
| 19th | Psc | 0h 4m | 0° 52' | -1.5 | 5" | 100% | 2° E | NV | N/A | Neptune |

## Venus

| Date | Con. | R.A. | Dec. | Mag. | Diam. | Ill. | Elon. | Vis. | Rat. | Close To |
|------|------|------|------|------|-------|------|-------|------|------|----------|
| 11th | Psc | 1h 26m | 8° 56' | -3.4 | 13" | 83% | 30° E | PM | * | Jupiter |
| 13th | Psc | 1h 35m | 9° 55' | -3.4 | 13" | 83% | 31° E | PM | * | |
| 15th | Psc | 1h 44m | 10° 54' | -3.4 | 13" | 82% | 31° E | PM | * | |
| 17th | Ari | 1h 54m | 11° 51' | -3.5 | 13" | 82% | 32° E | PM | * | |
| 19th | Ari | 2h 3m | 12° 47' | -3.5 | 13" | 81% | 32° E | PM | * | |

## Mars and the Outer Planets

Mars
15th

Jupiter
15th

Saturn
15th

### Mars

| Date | Con. | R.A. | Dec. | Mag. | Diam. | Ill. | Elon. | Vis. | Rat. | Close To |
|------|------|------|------|------|-------|------|-------|------|------|----------|
| 11th | Tau | 5h 31m | 25° 34' | 0.8 | 7" | 90% | 91° E | PM | ** | |
| 15th | Tau | 5h 39m | 25° 36' | 0.9 | 7" | 90% | 90° E | PM | ** | |
| 20th | Tau | 5h 49m | 25° 36' | 1.0 | 7" | 90% | 88° E | PM | ** | |

### The Outer Planets

| Planet | Date | Con. | R.A. | Dec. | Mag. | Diam. | Elon. | Vis. | Rat. | Close To |
|--------|------|------|------|------|------|-------|-------|------|------|----------|
| Jupiter | 15th | Psc | 0h 57m | 4° 59' | -1.6 | 34" | 19° E | PM | * | |
| Saturn | 15th | Aqr | 22h 13m | -12° 22' | 1.1 | 16" | 22° W | AM | * | |
| Uranus | 15th | Ari | 2h 54m | 16° 20' | 6.2 | 3" | 49° E | PM | * | |
| Neptune | 15th | Psc | 23h 43m | -3° 1' | 7.8 | 2" | 1° E | NV | N/A | Mercury |

## Highlights

| Date | Time (UT) | Event |
|------|-----------|-------|
| 13th | 22:39 | The waning gibbous Moon is north of the bright star Antares. (Scorpius, morning sky.) |
| 15th | 02:09 | Last Quarter Moon. (Morning sky.) |
| 16th | 12:18 | Neptune is in conjunction with the Sun. (Not visible.) |
| 17th | 10:33 | Mercury is at superior conjunction with the Sun. (Not visible.) |
| 18th | N/A | Good opportunity to see Earthshine on the waning crescent Moon. (Morning sky.) |
| 19th | 17:08 | The waning crescent Moon is south of Saturn. (Morning sky.) |
| 20th | 21:25 | Spring equinox. |

# March 21st to 31st, 2023

## The Moon

| | 21st | | 23rd | | 25th | | 27th | | 29th | | 31st |
|---|---|---|---|---|---|---|---|---|---|---|---|

| Date | Con | R.A. | Dec | Mag | Diam | Ill. | Elon: | Phase | Close To |
|---|---|---|---|---|---|---|---|---|---|
| 21st | Aqr | 23h 55m | -3° 50' | -4.4 | 33' | 0% | 2° W | NM | Mercury, Neptune |
| 22nd | Psc | 0h 47m | 3° 56' | -5.0 | 32' | 1% | 10° E | NM | Mercury, Jupiter |
| 23rd | Psc | 1h 38m | 9° 34' | -6.3 | 32' | 4% | 22° E | NM | Venus, Jupiter |
| 24th | Ari | 2h 29m | 15° 15' | -7.3 | 32' | 10% | 34° E | NM | Venus, Uranus |
| 25th | Ari | 3h 21m | 20° 12' | -8.1 | 31' | 17% | 46° E | +Cr | Uranus, Pleiades |
| 26th | Tau | 4h 14m | 24° 1' | -8.8 | 31' | 25% | 59° E | +Cr | Pleiades, Hyades, Aldebaran |
| 27th | Tau | 5h 8m | 26° 35' | -9.4 | 30' | 35% | 71° E | +Cr | Hyades, Aldebaran |
| 28th | Gem | 6h 3m | 27° 48' | -9.9 | 30' | 44% | 84° E | FQ | Mars |
| 29th | Gem | 6h 58m | 27° 40' | -10.3 | 30' | 54% | 97° E | FQ | |
| 30th | Gem | 7h 51m | 26° 14' | -10.7 | 30' | 63% | 109° E | FQ | |
| 31st | Cnc | 8h 41m | 23° 39' | -11.0 | 30' | 72% | 121° E | +G | Praesepe |

## Mercury and Venus

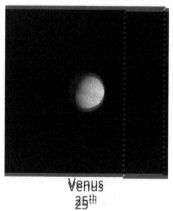

Mercury
25th

Venus
25th

### Mercury

| Date | Con. | R.A. | Dec. | Mag. | Diam. | Ill. | Elon. | Vis. | Rat. | Close To |
|---|---|---|---|---|---|---|---|---|---|---|
| 21st | Psc | 0h 18m | 0° 58' | -1.5 | 5" | 99% | 4° E | NV | N/A | Moon |
| 23rd | Psc | 0h 32m | 2° 51' | -1.5 | 5" | 97% | 6° E | NV | N/A | Jupiter |
| 25th | Psc | 0h 47m | 4° 44' | -1.4 | 5" | 95% | 8° E | NV | N/A | Jupiter |
| 27th | Psc | 1h 1m | 6° 37' | -1.3 | 5" | 91% | 9° E | NV | N/A | Jupiter |
| 29th | Psc | 1h 15m | 8° 28' | -1.2 | 6" | 87% | 11° E | NV | N/A | Jupiter |
| 31st | Psc | 1h 29m | 10° 14' | -1.0 | 6" | 81% | 13° E | NV | N/A | Jupiter |

## Venus

| Date | Con. | R.A. | Dec. | Mag. | Diam. | Ill. | Elon. | Vis. | Rat. | Close To |
|------|------|------|------|------|-------|------|-------|------|------|----------|
| 21st | Ari | 2h 12m | 13° 42' | -3.5 | 13" | 81% | 33° E | PM | * | |
| 23rd | Ari | 2h 21m | 14° 35' | -3.5 | 13" | 80% | 33° E | PM | * | Moon |
| 25th | Ari | 2h 30m | 15° 28' | -3.5 | 14" | 79% | 33° E | PM | * | |
| 27th | Ari | 2h 39m | 16° 19' | -3.5 | 14" | 79% | 34° E | PM | * | Uranus |
| 29th | Ari | 2h 49m | 17° 8' | -3.5 | 14" | 78% | 34° E | PM | * | Uranus |
| 31st | Ari | 2h 58m | 17° 55' | -3.5 | 14" | 78% | 35° E | PM | * | Uranus |

## Mars and the Outer Planets

Mars
25th

Jupiter
25th

Saturn
25th

### Mars

| Date | Con. | R.A. | Dec. | Mag. | Diam. | Ill. | Elon. | Vis. | Rat. | Close To |
|------|------|------|------|------|-------|------|-------|------|------|----------|
| 21st | Tau | 5h 51m | 25° 36' | 1.0 | 7" | 90% | 87° E | PM | ** | |
| 25th | Tau | 6h 0m | 25° 35' | 1.0 | 7" | 90% | 86° E | PM | ** | |
| 31st | Gem | 6h 13m | 25° 29' | 1.1 | 6" | 90% | 84° E | PM | ** | |

### The Outer Planets

| Planet | Date | Con. | R.A. | Dec. | Mag. | Diam. | Elon. | Vis. | Rat. | Close To |
|--------|------|------|------|------|------|-------|-------|------|------|----------|
| Jupiter | 25th | Psc | 1h 6m | 5° 54' | -1.6 | 33" | 12° E | NV | N/A | Mercury |
| Saturn | 25th | Aqr | 22h 18m | -11° 59' | 1.2 | 16" | 30° W | AM | * | |
| Uranus | 25th | Ari | 2h 56m | 16° 27' | 6.2 | 3" | 40° E | PM | * | Moon, Venus |
| Neptune | 25th | Psc | 23h 45m | -2° 52' | 7.8 | 2" | 8° W | NV | N/A | |

## Highlights

| Date | Time (UT) | Event |
|------|-----------|-------|
| 21st | 17:24 | New Moon. (Not visible.) |
| 24th | 08:26 | The waxing crescent Moon is south of Venus. (Evening sky.) |
| 25th | 01:22 | The waxing crescent Moon is north of Uranus. (Evening sky.) |
| | N/A | Good opportunity to see Earthshine on the waxing crescent Moon. (Evening sky.) |
| 26th | 00:54 | The waxing crescent Moon is south of the Pleaides star cluster. (Taurus, evening sky.) |
| | 23:18 | The waxing crescent Moon is north of the bright star Aldebaran. (Taurus, evening sky.) |
| 28th | 11:27 | The nearly first quarter Moon is north of Mars. (Evening sky.) |
| 29th | 02:33 | First Quarter Moon. (Evening sky.) |
| 31st | 06:04 | Venus is 1.3° north of Uranus. (Evening sky.) |

# April 1st to 10th, 2023

## The Moon

| 1st | 3rd | 5th | 7th | 9th |

| Date | Con | R.A. | Dec | Mag | Diam | Ill. | Elon. | Phase | Close To |
|------|-----|------|-----|-----|------|------|-------|-------|----------|
| 1st | Leo | 9h 30m | 20° 5' | -11.3 | 30' | 80% | 132° E | +G | Regulus |
| 2nd | Leo | 10h 17m | 15° 40' | -11.6 | 30' | 87% | 143° E | +G | Regulus |
| 3rd | Leo | 11h 2m | 10° 36' | -11.9 | 30' | 93% | 153° E | +G | |
| 4th | Vir | 11h 46m | 5° 4' | -12.2 | 30' | 97% | 163° E | FM | |
| 5th | Vir | 12h 30m | 0° 46' | -12.5 | 30' | 99% | 173° E | FM | |
| 6th | Vir | 13h 16m | -6° 40' | -12.6 | 31' | 100% | 176° W | FM | Spica |
| 7th | Vir | 14h 3m | -12° 25' | -12.3 | 31' | 98% | 165° W | FM | Spica |
| 8th | Lib | 14h 53m | -17° 43' | -12.0 | 31' | 94% | 154° W | -G | |
| 9th | Lib | 15h 46m | -22° 15' | -11.7 | 31' | 88% | 141° W | -G | Antares |
| 10th | Sco | 16h 44m | -25° 39' | -11.3 | 32' | 80% | 128° W | -G | Antares |

## Mercury and Venus

Mercury
5th

Venus
5th

### Mercury

| Date | Con. | R.A. | Dec. | Mag. | Diam. | Ill. | Elon. | Vis. | Rat. | Close To |
|------|------|------|------|------|-------|------|-------|------|------|----------|
| 1st | Psc | 1h 35m | 11° 5' | -0.9 | 6" | 78% | 13° E | NV | N/A | Jupiter |
| 3rd | Ari | 1h 48m | 12° 42' | -0.8 | 6" | 71% | 15° E | NV | N/A | Jupiter |
| 5th | Ari | 2h 0m | 14° 11' | -0.5 | 6" | 64% | 16° E | PM | ** | |
| 7th | Ari | 2h 11m | 15° 29' | -0.3 | 7" | 56% | 17° E | PM | ** | |
| 9th | Ari | 2h 20m | 16° 37' | 0.0 | 7" | 49% | 17° E | PM | ** | |

**Venus**

| Date | Con. | R.A. | Dec. | Mag. | Diam. | Ill. | Elon. | Vis. | Rat. | Close To |
|------|------|------|------|------|-------|------|-------|------|------|----------|
| 1st | Ari | 3h 3m | 18° 19' | -3.5 | 14" | 77% | 35° E | PM | * | Uranus |
| 3rd | Ari | 3h 12m | 19° 3' | -3.5 | 14" | 77% | 36° E | PM | * | Uranus, Pleiades |
| 5th | Ari | 3h 22m | 19° 47' | -3.5 | 14" | 76% | 36° E | PM | * | Pleiades |
| 7th | Tau | 3h 32m | 20° 28' | -3.5 | 15" | 75% | 37° E | PM | * | Pleiades |
| 9th | Tau | 3h 41m | 21° 7' | -3.5 | 15" | 75% | 38° E | PM | * | Pleiades |

## Mars and the Outer Planets

Mars
5<sup>th</sup>

Jupiter
5<sup>th</sup>

Saturn
5<sup>th</sup>

**Mars**

| Date | Con. | R.A. | Dec. | Mag. | Diam. | Ill. | Elon. | Vis. | Rat. | Close To |
|------|------|------|------|------|-------|------|-------|------|------|----------|
| 1st | Gem | 6h 15m | 25° 27' | 1.2 | 6" | 90% | 83° E | PM | ** | |
| 5th | Gem | 6h 24m | 25° 21' | 1.2 | 6" | 90% | 82° E | PM | ** | |
| 10th | Gem | 6h 35m | 25° 10' | 1.3 | 6" | 90% | 80° E | PM | * | |

**The Outer Planets**

| Planet | Date | Con. | R.A. | Dec. | Mag. | Diam. | Elon. | Vis. | Rat. | Close To |
|--------|------|------|------|------|------|-------|-------|------|------|----------|
| Jupiter | 5th | Psc | 1h 16m | 6° 54' | -1.6 | 33" | 5° E | NV | N/A | |
| Saturn | 5th | Aqr | 22h 22m | -11° 35' | 1.2 | 16" | 39° W | AM | * | |
| Uranus | 5th | Ari | 2h 58m | 16° 37' | 6.2 | 3" | 30° E | PM | * | Venus |
| Neptune | 5th | Psc | 23h 46m | -2° 43' | 7.8 | 2" | 18° W | AM | ** | |

## Highlights

| Date | Time (UT) | Event |
|------|-----------|-------|
| 2<sup>nd</sup> | 08:03 | The waxing gibbous Moon is north of the bright star Regulus. (Leo, evening sky.) |
| 6<sup>th</sup> | 04:35 | Full Moon. (Visible all night.) |
| | 15:22 | The full Moon is north of the bright star Spica. (Virgo, visible all night.) |
| 10<sup>th</sup> | 07:21 | The waning gibbous Moon is north of the bright star Antares. (Scorpius, morning sky.) |
| | 14:16 | Venus is 2.7° south of the Pleiades star cluster. (Taurus, evening sky.) |

# April 11th to 20th, 2023

## The Moon

| | 11th | | 13th | | 15th | | 17th | | 19th |
|---|---|---|---|---|---|---|---|---|---|

| Date | Con | R.A. | Dec | Mag | Diam | Ill. | Elon. | Phase | Close To |
|---|---|---|---|---|---|---|---|---|---|
| 11th | Sgr | 17h 45m | -27° 36' | -11.0 | 32' | 71% | 113° W | -G | |
| 12th | Sgr | 18h 48m | -27° 50' | -10.5 | 32' | 60% | 99° W | LQ | |
| 13th | Sgr | 19h 51m | -26° 16' | -10.1 | 32' | 49% | 84° W | LQ | |
| 14th | Cap | 20h 52m | -23° 2' | -9.5 | 32' | 37% | 69° W | -Cr | |
| 15th | Cap | 21h 49m | -18° 24' | -8.9 | 32' | 26% | 56° W | -Cr | Saturn |
| 16th | Aqr | 22h 44m | -12° 42' | -8.1 | 32' | 17% | 43° W | -Cr | Saturn |
| 17th | Aqr | 23h 36m | -6° 20' | -7.2 | 32' | 9% | 31° W | NM | Neptune |
| 18th | Psc | 0h 27m | 0° 18' | -6.1 | 32' | 4% | 19° W | NM | Neptune |
| 19th | Psc | 1h 17m | 6° 51' | -4.8 | 32' | 1% | 8° W | NM | Jupiter |
| 20th | Ari | 2h 8m | 12° 58' | -4.3 | 32' | 0% | 4° E | NM | Mercury, Jupiter |

## Mercury and Venus

Mercury
15th

Venus
15th

### Mercury

| Date | Con. | R.A. | Dec. | Mag. | Diam. | Ill. | Elon. | Vis. | Rat. | Close To |
|---|---|---|---|---|---|---|---|---|---|---|
| 11th | Ari | 2h 29m | 17° 33' | 0.2 | 8" | 41% | 18° E | PM | *** | |
| 13th | Ari | 2h 36m | 18° 18' | 0.5 | 8" | 34% | 18° E | PM | *** | |
| 15th | Ari | 2h 41m | 18° 50' | 0.8 | 9" | 28% | 17° E | PM | *** | Uranus |
| 17th | Ari | 2h 45m | 19° 10' | 1.1 | 9" | 23% | 16° E | PM | ** | Uranus |
| 19th | Ari | 2h 48m | 19° 17' | 1.3 | 10" | 17% | 15° E | NV | N/A | Uranus |

## Venus

| Date | Con. | R.A. | Dec. | Mag. | Diam. | Ill. | Elon. | Vis. | Rat. | Close To |
|------|------|------|------|------|-------|------|-------|------|------|----------|
| 11th | Tau | 3h 51m | 21° 44' | -3.6 | 15" | 74% | 38° E | PM | * | Pleiades, Hyades |
| 13th | Tau | 4h 1m | 22° 18' | -3.6 | 15" | 73% | 39° E | PM | ** | Pleiades, Hyades, Aldebaran |
| 15th | Tau | 4h 10m | 22° 51' | -3.6 | 15" | 72% | 39° E | PM | ** | Pleiades, Hyades, Aldebaran |
| 17th | Tau | 4h 20m | 23° 21' | -3.6 | 15" | 72% | 40° E | PM | ** | Pleiades, Hyades, Aldebaran |
| 19th | Tau | 4h 30m | 23° 49' | -3.6 | 16" | 71% | 41° E | PM | ** | Hyades, Aldebaran |

## Mars and the Outer Planets

Mars
15th

Jupiter
15th

Saturn
15th

## Mars

| Date | Con. | R.A. | Dec. | Mag. | Diam. | Ill. | Elon. | Vis. | Rat. | Close To |
|------|------|------|------|------|-------|------|-------|------|------|----------|
| 11th | Gem | 6h 37m | 25° 7' | 1.3 | 6" | 90% | 80° E | PM | * | |
| 15th | Gem | 6h 46m | 24° 55' | 1.3 | 6" | 91% | 78° E | PM | * | |
| 20th | Gem | 6h 58m | 24° 38' | 1.4 | 6" | 91% | 77° E | PM | * | |

## The Outer Planets

| Planet | Date | Con. | R.A. | Dec. | Mag. | Diam. | Elon. | Vis. | Rat. | Close To |
|--------|------|------|------|------|------|-------|-------|------|------|----------|
| Jupiter | 15th | Psc | 1h 25m | 7° 49' | -1.6 | 33" | 2° W | NV | N/A | |
| Saturn | 15th | Aqr | 22h 26m | -11° 16' | 1.2 | 16" | 47° W | AM | ** | Moon |
| Uranus | 15th | Ari | 3h 0m | 16° 46' | 6.2 | 3" | 22° E | PM | * | Mercury |
| Neptune | 15th | Psc | 23h 48m | -2° 35' | 7.8 | 2" | 26° W | AM | ** | |

## Highlights

| Date | Time (UT) | Event |
|------|-----------|-------|
| 11th | 22:06 | Mercury is at greatest eastern elongation from the Sun. (Evening sky.) |
| 12th | 12:02 | Jupiter is in conjunction with the Sun. (Not visible.) |
| 13th | 09:11 | Last Quarter Moon. (Morning sky.) |
| 16th | 02:06 | The waning crescent Moon is south of Saturn. (Morning sky.) |
| 17th | 18:52 | The waning crescent Moon is south of Neptune. (Morning sky.) |
| | N/A | Good opportunity to see Earthshine on the waning crescent Moon. |
| 20th | 04:13 | New Moon (Not visible.) |
| | 04:18 | Total solar eclipse. Visible from South-east Asia, Australia and the Indian and Pacific Oceans. |
| | 13:58 | Venus is 7.6° north of the bright star Aldebaran. (Taurus, evening sky.) |

# April 21<sup>st</sup> to 30<sup>th</sup>, 2023

## The Moon

21<sup>st</sup>

23<sup>rd</sup>

25<sup>th</sup>

27<sup>th</sup>

29<sup>th</sup>

| Date | Con | R.A. | Dec | Mag | Diam | Ill. | Elon. | Phase | Close To |
|------|-----|------|-----|-----|------|------|-------|-------|----------|
| 21st | Ari | 2h 59m | 18° 21' | -5.6 | 31' | 2% | 16° E | NM | Mercury, Uranus, Pleiades |
| 22nd | Tau | 3h 53m | 22° 42' | -6.7 | 31' | 6% | 28° E | NM | Pleiades, Hyades, Aldebaran |
| 23rd | Tau | 4h 48m | 25° 49' | -7.6 | 31' | 12% | 41° E | NM | Venus, Hyades, Aldebaran |
| 24th | Tau | 5h 43m | 27° 34' | -8.4 | 30' | 20% | 54° E | +Cr | |
| 25th | Aur | 6h 39m | 27° 55' | -9.0 | 30' | 28% | 67° E | +Cr | Mars |
| 26th | Gem | 7h 33m | 26° 55' | -9.5 | 30' | 37% | 80° E | +Cr | Mars |
| 27th | Cnc | 8h 25m | 24° 41' | -10.0 | 30' | 46% | 92° E | FQ | Praesepe |
| 28th | Cnc | 9h 14m | 21° 24' | -10.4 | 30' | 56% | 103° E | FQ | Praesepe |
| 29th | Leo | 10h 1m | 17° 15' | -10.7 | 30' | 65% | 114° E | FQ | Regulus |
| 30th | Leo | 10h 47m | 12° 25' | -11.1 | 30' | 74% | 124° E | +G | Regulus |

## Mercury and Venus

Mercury
25<sup>th</sup>

Venus
25<sup>th</sup>

### Mercury

| Date | Con. | R.A. | Dec. | Mag. | Diam. | Ill. | Elon. | Vis. | Rat. | Close To |
|------|------|------|------|------|-------|------|-------|------|------|----------|
| 21st | Ari | 2h 49m | 19° 12' | 1.6 | 10" | 12% | 13° E | NV | N/A | Moon, Uranus |
| 23rd | Ari | 2h 48m | 18° 55' | 2.0 | 11" | 8% | 11° E | NV | N/A | Uranus |
| 25th | Ari | 2h 46m | 18° 26' | 2.3 | 11" | 5% | 9° E | NV | N/A | Uranus |
| 27th | Ari | 2h 43m | 17° 48' | 2.7 | 11" | 2% | 6° E | NV | N/A | Uranus |
| 29th | Ari | 2h 39m | 17° 0' | 3.0 | 12" | 1% | 3° E | NV | N/A | |

## Venus

| Date | Con. | R.A. | Dec. | Mag. | Diam. | Ill. | Elon. | Vis. | Rat. | Close To |
|------|------|------|------|------|-------|------|-------|------|------|----------|
| 21st | Tau | 4h 40m | 24° 14' | -3.6 | 16" | 70% | 41° E | PM | ** | Hyades, Aldebaran |
| 23rd | Tau | 4h 50m | 24° 37' | -3.6 | 16" | 69% | 42° E | PM | ** | Moon, Hyades, Aldebaran |
| 25th | Tau | 5h 0m | 24° 57' | -3.6 | 16" | 69% | 42° E | PM | ** | Hyades, Aldebaran |
| 27th | Tau | 5h 10m | 25° 14' | -3.6 | 17" | 68% | 43° E | PM | ** | Aldebaran |
| 29th | Tau | 5h 20m | 25° 29' | -3.7 | 17" | 67% | 43° E | PM | ** | |

## Mars and the Outer Planets

Mars
25th

Jupiter
25th

Saturn
25th

### Mars

| Date | Con. | R.A. | Dec. | Mag. | Diam. | Ill. | Elon. | Vis. | Rat. | Close To |
|------|------|------|------|------|-------|------|-------|------|------|----------|
| 21st | Gem | 7h 0m | 24° 34' | 1.4 | 6" | 91% | 76° E | PM | * | |
| 25th | Gem | 7h 10m | 24° 17' | 1.5 | 6" | 91% | 75° E | PM | * | Moon |
| 30th | Gem | 7h 22m | 23° 53' | 1.5 | 5" | 91% | 73° E | PM | * | |

### The Outer Planets

| Planet | Date | Con. | R.A. | Dec. | Mag. | Diam. | Elon. | Vis. | Rat. | Close To |
|--------|------|------|------|------|------|-------|-------|------|------|----------|
| Jupiter | 25th | Psc | 1h 34m | 8° 42' | -1.6 | 33" | 9° W | NV | N/A | |
| Saturn | 25th | Aqr | 22h 29m | -10° 59' | 1.2 | 16" | 55° W | AM | ** | |
| Uranus | 25th | Ari | 3h 2m | 16° 55' | 6.2 | 3" | 13° E | NV | N/A | Mercury |
| Neptune | 25th | Psc | 23h 49m | -2° 27' | 7.8 | 2" | 35° W | AM | *** | |

## Highlights

| Date | Time (UT) | Event |
|------|-----------|-------|
| 21st | 15:49 | Mercury is stationary prior to beginning retrograde motion. (Not visible.) |
| 22nd | 07:26 | The waxing crescent Moon is south of the Pleiades star cluster. (Taurus, evening sky.) |
| | N/A | The Lyrid meteor shower is at its maximum. (ZHR: 18) |
| 23rd | 05:29 | The waxing crescent Moon is north of the bright star Aldebaran. (Taurus, evening sky.) |
| | 11:41 | The waxing crescent Moon is north of Venus. (Evening sky.) |
| | N/A | Good opportunity to see Earthshine on the waxing crescent Moon. (Evening sky.) |
| 26th | 03:12 | The waxing crescent Moon is north of Mars. (Evening sky.) |
| 27th | 21:21 | First Quarter Moon. (Evening sky.) |
| 29th | 13:21 | The waxing gibbous Moon is north of the bright star Regulus. (Leo, evening sky.) |

# May 1st to 10th, 2023

## The Moon

  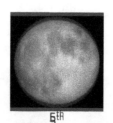  

| 1st | 3rd | 5th | 7th | 9th |

| Date | Con | R.A. | Dec | Mag | Diam | Ill. | Elon. | Phase | Close To |
|------|-----|------|-----|-----|------|------|-------|-------|----------|
| 1st | Leo | 11h 31m | 7° 2' | -11.4 | 30' | 82% | 134° E | +G | |
| 2nd | Vir | 12h 15m | 1° 17' | -11.7 | 30' | 89% | 144° E | +G | |
| 3rd | Vir | 13h 0m | -4° 38' | -12.0 | 31' | 94% | 155° E | +G | Spica |
| 4th | Vir | 13h 47m | -10° 31' | -12.3 | 31' | 98% | 165° E | FM | Spica |
| 5th | Lib | 14h 36m | -16° 4' | -12.7 | 31' | 100% | 177° E | FM | |
| 6th | Lib | 15h 30m | -20° 58' | -12.5 | 32' | 99% | 171° W | FM | |
| 7th | Oph | 16h 27m | -24° 50' | -12.1 | 32' | 96% | 157° W | FM | Antares |
| 8th | Oph | 17h 29m | -27° 15' | -11.8 | 32' | 91% | 143° W | -G | |
| 9th | Sgr | 18h 33m | -27° 56' | -11.4 | 32' | 83% | 128° W | -G | |
| 10th | Sgr | 19h 37m | -26° 46' | -11.1 | 32' | 73% | 113° W | -G | |

## Mercury and Venus

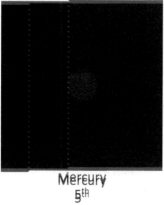

Mercury
5th

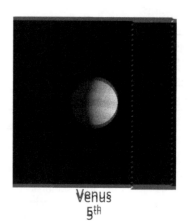

Venus
5th

### Mercury

| Date | Con. | R.A. | Dec. | Mag. | Diam. | Ill. | Elon. | Vis. | Rat. | Close To |
|------|------|------|------|------|-------|------|-------|------|------|----------|
| 1st | Ari | 2h 35m | 16° 7' | 3.4 | 12" | 0% | 0° E | NV | N/A | |
| 3rd | Ari | 2h 31m | 15° 11' | 3.3 | 12" | 0% | 3° W | NV | N/A | |
| 5th | Ari | 2h 27m | 14° 14' | 2.9 | 12" | 1% | 5° W | NV | N/A | |
| 7th | Ari | 2h 23m | 13° 20' | 2.6 | 12" | 3% | 8° W | NV | N/A | Jupiter |
| 9th | Ari | 2h 20m | 12° 31' | 2.3 | 12" | 5% | 11° W | NV | N/A | Jupiter |

## Venus

| Date | Con. | R.A. | Dec. | Mag. | Diam. | Ill. | Elon. | Vis. | Rat. | Close To |
|------|------|------|------|------|-------|------|-------|------|------|----------|
| 1st | Tau | 5h 29m | 25° 42' | -3.7 | 17" | 66% | 44° E | PM | ** | |
| 3rd | Tau | 5h 39m | 25° 52' | -3.7 | 17" | 65% | 45° E | PM | ** | |
| 5th | Tau | 5h 49m | 25° 59' | -3.7 | 18" | 65% | 45° E | PM | ** | |
| 7th | Tau | 5h 59m | 26° 3' | -3.7 | 18" | 64% | 46° E | PM | ** | |
| 9th | Gem | 6h 9m | 26° 5' | -3.7 | 18" | 63% | 46° E | PM | ** | |

## Mars and the Outer Planets

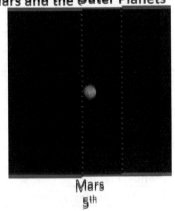

Mars
5th

Jupiter
5th

Saturn
5th

## Mars

| Date | Con. | R.A. | Dec. | Mag. | Diam. | Ill. | Elon. | Vis. | Rat. | Close To |
|------|------|------|------|------|-------|------|-------|------|------|----------|
| 1st | Gem | 7h 24m | 23° 48' | 1.5 | 5" | 91% | 73° E | PM | * | |
| 5th | Gem | 7h 34m | 23° 25' | 1.6 | 5" | 92% | 71° E | PM | * | |
| 10th | Gem | 7h 45m | 22° 54' | 1.6 | 5" | 92% | 69° E | PM | * | |

## The Outer Planets

| Planet | Date | Con. | R.A. | Dec. | Mag. | Diam. | Elon. | Vis. | Rat. | Close To |
|--------|------|------|------|------|------|-------|-------|------|------|----------|
| Jupiter | 5th | Psc | 1h 43m | 8° 33' | -1.6 | 33" | 16° W | AM | * | |
| Saturn | 5th | Aqr | 22h 32m | -10° 44' | 1.2 | 16" | 64° W | AM | ** | |
| Uranus | 5th | Ari | 3h 5m | 17° 5' | 6.2 | 3" | 4° E | NV | N/A | Mercury |
| Neptune | 5th | Psc | 23h 50m | -2° 20' | 7.8 | 2" | 45° W | AM | *** | |

## Highlights

| Date | Time (UT) | Event |
|------|-----------|-------|
| 1st | 23:21 | Mercury is in inferior conjunction with the Sun. (Not visible.) |
| 2nd | 09:24 | Dwarf planet Pluto is stationary prior to beginning retrograde motion. (Morning sky.) |
| 4th | 02:38 | The waxing gibbous Moon is north of the bright star Spica. (Virgo, evening sky.) |
| 5th | 17:22 | Penumbral lunar eclipse. Visible from Africa, Asia, the Atlantic, Australia, Europe and the Indian and Pacific oceans. |
| | 17:35 | Full Moon. (Visible all night.) |
| 7th | 12:52 | The waning gibbous Moon is north of the bright star Antares. (Scorpius, morning sky.) |
| | N/A | The Eta Aquariid meteor shower is at its maximum. (ZHR: 40) |
| 9th | 22:18 | Uranus is in conjunction with the Sun. (Not visible.) |
| | N/A | The Eta Lyrid meteor shower is at its maximum. (ZHR: 3) |

# May 11<sup>th</sup> to 20<sup>th</sup>, 2023

## The Moon

  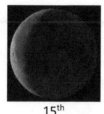  

| 11<sup>th</sup> | 13<sup>th</sup> | 15<sup>th</sup> | 17<sup>th</sup> | 19<sup>th</sup> |

| Date | Con | R.A. | Dec | Mag | Diam | Ill. | Elon. | Phase | Close To |
|------|-----|------|-----|-----|------|------|-------|-------|----------|
| 11th | Cap | 20h 38m | -23° 53' | -10.6 | 32' | 63% | 98° W | LQ | |
| 12th | Cap | 21h 36m | -19° 33' | -10.2 | 32' | 51% | 85° W | LQ | |
| 13th | Aqr | 22h 31m | -14° 8' | -9.7 | 32' | 40% | 72° W | LQ | Saturn |
| 14th | Aqr | 23h 23m | -8° 2' | -9.0 | 32' | 29% | 60° W | -Cr | Neptune |
| 15th | Psc | 0h 13m | -1° 36' | -8.3 | 32' | 19% | 49° W | -Cr | Neptune |
| 16th | Psc | 1h 1m | 4° 52' | -7.5 | 32' | 11% | 38° W | NM | |
| 17th | Ari | 1h 51m | 11° 1' | -6.5 | 32' | 5% | 26° W | NM | Mercury, Jupiter |
| 18th | Ari | 2h 41m | 16° 35' | -5.4 | 31' | 2% | 15° W | NM | Mercury, Jupiter, Uranus |
| 19th | Tau | 3h 34m | 21° 16' | -4.2 | 31' | 0% | 3° W | NM | Uranus, Pleiades |
| 20th | Tau | 4h 28m | 24° 49' | -5.0 | 31' | 1% | 10° E | NM | Pleiades, Hyades, Aldebaran |

## Mercury and Venus

Mercury
15<sup>th</sup>

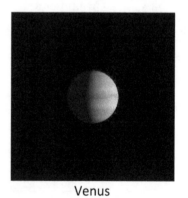

Venus
15<sup>th</sup>

### Mercury

| Date | Con. | R.A. | Dec. | Mag. | Diam. | Ill. | Elon. | Vis. | Rat. | Close To |
|------|------|------|------|------|-------|------|-------|------|------|----------|
| 11th | Ari | 2h 18m | 11° 50' | 2.1 | 11" | 8% | 13° W | NV | N/A | Jupiter |
| 13th | Ari | 2h 17m | 11° 18' | 1.9 | 11" | 11% | 16° W | AM | *** | Jupiter |
| 15th | Ari | 2h 17m | 10° 55' | 1.7 | 11" | 14% | 18° W | AM | *** | Jupiter |
| 17th | Ari | 2h 19m | 10° 42' | 1.5 | 10" | 18% | 19° W | AM | *** | Moon, Jupiter |
| 19th | Ari | 2h 21m | 10° 38' | 1.3 | 10" | 21% | 21° W | AM | *** | Jupiter |

**Venus**

| Date | Con. | R.A. | Dec. | Mag. | Diam. | Ill. | Elon. | Vis. | Rat. | Close To |
|------|------|------|------|------|-------|------|-------|------|------|----------|
| 11th | Gem | 6h 18m | 26° 4' | -3.7 | 19" | 62% | 47° E | PM | ** | |
| 13th | Gem | 6h 28m | 26° 1' | -3.8 | 19" | 61% | 47° E | PM | ** | |
| 15th | Gem | 6h 38m | 25° 55' | -3.8 | 19" | 60% | 47° E | PM | ** | |
| 17th | Gem | 6h 47m | 25° 47' | -3.8 | 20" | 59% | 48° E | PM | ** | |
| 19th | Gem | 6h 56m | 25° 36' | -3.8 | 20" | 58% | 48° E | PM | *** | |

## Mars and the Outer Planets

Mars
15<sup>th</sup>     Jupiter
15<sup>th</sup>     Saturn
15<sup>th</sup>

**Mars**

| Date | Con. | R.A. | Dec. | Mag. | Diam. | Ill. | Elon. | Vis. | Rat. | Close To |
|------|------|------|------|------|-------|------|-------|------|------|----------|
| 11th | Gem | 7h 48m | 22° 48' | 1.6 | 5" | 92% | 69° E | PM | * | |
| 15th | Gem | 7h 57m | 22° 20' | 1.6 | 5" | 92% | 67° E | PM | * | |
| 20th | Cnc | 8h 9m | 21° 43' | 1.7 | 5" | 92% | 65° E | PM | * | Praesepe |

**The Outer Planets**

| Planet | Date | Con. | R.A. | Dec. | Mag. | Diam. | Elon. | Vis. | Rat. | Close To |
|--------|------|------|------|------|------|-------|-------|------|------|----------|
| Jupiter | 15th | Psc | 1h 52m | 10° 22' | -1.6 | 34" | 24° W | AM | * | Mercury |
| Saturn | 15th | Aqr | 22h 34m | -10° 33' | 1.2 | 17" | 73° W | AM | ** | |
| Uranus | 15th | Ari | 3h 7m | 17° 14' | 6.2 | 3" | 5° W | NV | N/A | Pleiades |
| Neptune | 15th | Psc | 23h 51m | -2° 15' | 7.7 | 2" | 54° W | AM | *** | Moon |

## Highlights

| Date | Time (UT) | Event |
|------|-----------|-------|
| 12<sup>th</sup> | 06:25 | Dwarf planet Ceres is stationary prior to resuming prograde motion. (Evening sky.) |
| | 14:29 | Last Quarter Moon. (Morning sky.) |
| 13<sup>th</sup> | 14:52 | The just-past last quarter Moon is south of Saturn. (Morning sky.) |
| 14<sup>th</sup> | 06:31 | Mercury is stationary prior to resuming prograde motion. (Morning sky.) |
| | 23:58 | The waning crescent Moon is south of Neptune. (Morning sky.) |
| 16<sup>th</sup> | N/A | Good opportunity to see Earthshine on the waning crescent Moon. (Morning sky.) |
| 17<sup>th</sup> | 15:13 | The waning crescent Moon is north of Jupiter. (Morning sky.) |
| 18<sup>th</sup> | 00:39 | The waning crescent Moon is north of Mercury. (Morning sky.) |
| 19<sup>th</sup> | 15:54 | New Moon. (Not visible.) |

# May 21st to 31st, 2023

## The Moon

| | 21st | 23rd | 25th | 27th | 29th | 31st |

| Date | Con | R.A. | Dec | Mag | Diam | Ill. | Elon: | Phase | Close To |
|------|-----|------|-----|-----|------|------|-------|-------|----------|
| 21st | Tau | 5h 23m | 27° 4' | -6.1 | 30' | 4% | 23° E | NM | |
| 22nd | Gem | 6h 19m | 27° 54' | -7.1 | 30' | 8% | 36° E | NM | |
| 23rd | Gem | 7h 14m | 27° 20' | -7.9 | 30' | 14% | 49° E | +Cr | Venus |
| 24th | Cnc | 8h 8m | 25° 29' | -8.5 | 30' | 22% | 61° E | +Cr | Mars, Praesepe |
| 25th | Cnc | 8h 58m | 22° 32' | -9.1 | 30' | 30% | 73° E | +Cr | Mars, Praesepe |
| 26th | Leo | 9h 46m | 18° 39' | -9.6 | 30' | 39% | 84° E | FQ | Regulus |
| 27th | Leo | 10h 31m | 14° 3' | -10.1 | 30' | 49% | 94° E | FQ | Regulus |
| 28th | Leo | 11h 15m | 8° 53' | -10.5 | 30' | 58% | 104° E | FQ | |
| 29th | Vir | 11h 59m | 3° 10' | -10.8 | 30' | 68% | 114° E | +G | |
| 30th | Vir | 12h 43m | -2° 29' | -11.2 | 30' | 77% | 124° E | +G | Spica |
| 31st | Vir | 13h 28m | -8° 21' | -11.5 | 31' | 85% | 134° E | +G | Spica |

## Mercury and Venus

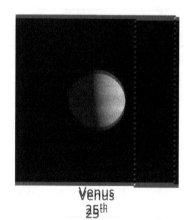

Mercury
25th

Venus
25th

### Mercury

| Date | Con: | R.A. | Dec. | Mag. | Diam. | Ill. | Elon. | Vis. | Rat. | Close To |
|------|------|------|------|------|-------|------|-------|------|------|----------|
| 21st | Ari | 2h 24m | 10° 44' | 1.2 | 10" | 25% | 23° W | AM | *** | Jupiter |
| 23rd | Ari | 2h 28m | 10° 50' | 1.1 | 9" | 28% | 23° W | AM | *** | Jupiter |
| 25th | Ari | 2h 34m | 11° 21' | 1.0 | 9" | 32% | 23° W | AM | *** | Jupiter |
| 27th | Ari | 2h 40m | 11° 51' | 0.8 | 9" | 35% | 24° W | AM | *** | Jupiter |
| 29th | Ari | 2h 47m | 12° 27' | 0.7 | 8" | 39% | 24° W | AM | *** | |
| 31st | Ari | 2h 55m | 13° 9' | 0.6 | 8" | 43% | 24° W | AM | *** | Uranus |

## Venus

| Date | Con. | R.A. | Dec. | Mag. | Diam. | Ill. | Elon. | Vis. | Rat. | Close To |
|------|------|------|------|------|-------|------|-------|------|------|----------|
| 21st | Gem | 7h 6m | 25° 23' | -3.8 | 20" | 57% | 49° E | PM | *** | |
| 23rd | Gem | 7h 15m | 25° 8' | -3.8 | 21" | 56% | 49° E | PM | *** | Moon |
| 25th | Gem | 7h 24m | 24° 50' | -3.9 | 21" | 55% | 49° E | PM | *** | |
| 27th | Gem | 7h 33m | 24° 30' | -3.9 | 22" | 54% | 49° E | PM | *** | |
| 29th | Gem | 7h 41m | 24° 8' | -3.9 | 22" | 53% | 49° E | PM | *** | |
| 31st | Gem | 7h 50m | 23° 45' | -3.9 | 23" | 52% | 49° E | PM | *** | |

## Mars and the Outer Planets

Mars
25th

Jupiter
25th

Saturn
25th

## Mars

| Date | Con. | R.A. | Dec. | Mag. | Diam. | Ill. | Elon. | Vis. | Rat. | Close To |
|------|------|------|------|------|-------|------|-------|------|------|----------|
| 21st | Cnc | 8h 12m | 21° 35' | 1.7 | 5" | 92% | 65° E | PM | * | Praesepe |
| 25th | Cnc | 8h 21m | 21° 2' | 1.7 | 5" | 93% | 63° E | PM | * | Moon, Praesepe |
| 31st | Cnc | 8h 36m | 20° 9' | 1.8 | 5" | 93% | 61° E | PM | * | Praesepe |

## The Outer Planets

| Planet | Date | Con. | R.A. | Dec. | Mag. | Diam. | Elon. | Vis. | Rat. | Close To |
|--------|------|------|------|------|------|-------|-------|------|------|----------|
| Jupiter | 25th | Ari | 2h 0m | 11° 9' | -1.6 | 34" | 32° W | AM | * | Mercury |
| Saturn | 25th | Aqr | 22h 36m | -10° 25' | 1.2 | 17" | 83° W | AM | ** | |
| Uranus | 25th | Ari | 3h 9m | 17° 24' | 6.2 | 3" | 15° W | NV | N/A | Mercury, Pleiades |
| Neptune | 25th | Psc | 23h 52m | -2° 10' | 7.7 | 2" | 64° W | AM | *** | |

## Highlights

| Date | Time (UT) | Event |
|------|-----------|-------|
| 23rd | 10:17 | The waxing crescent Moon is north of Venus. (Evening sky.) |
| | N/A | Good opportunity to see Earthshine on the waxing crescent Moon. (Evening sky.) |
| 24th | 18:49 | The waxing crescent Moon is north of Mars. (Evening sky.) |
| | N/A | The May Camelopardalid meteor shower is at its maximum. (ZHR: Variable.) |
| 25th | 03:32 | The waxing crescent Moon is north of the Praesepe star cluster. (Evening sky.) |
| 27th | 01:18 | The almost first quarter Moon is north of the bright star Regulus. (Leo, evening sky.) |
| | 15:23 | First Quarter Moon. (Evening sky.) |
| 29th | 05:27 | Mercury is at greatest western elongation from the Sun. (Morning sky.) |
| 31st | 09:40 | The waxing gibbous Moon is north of the bright star Spica. (Virgo, evening sky.) |

# June 1ˢᵗ to 10ᵗʰ, 2023

## The Moon

| | 1ˢᵗ | | 3ʳᵈ | | 5ᵗʰ | | 7ᵗʰ | | 9ᵗʰ |

| Date | Con | R.A. | Dec | Mag | Diam | Ill. | Elon. | Phase | Close To |
|------|-----|------|-----|-----|------|------|-------|-------|----------|
| 1st | Vir | 14h 17m | -14° 1' | -11.8 | 31' | 91% | 145° E | +G | |
| 2nd | Lib | 15h 9m | -19° 13' | -12.2 | 32' | 96% | 157° E | FM | |
| 3rd | Sco | 16h 6m | -23° 32' | -12.5 | 32' | 99% | 170° E | FM | Antares |
| 4th | Oph | 17h 7m | -26° 32' | -12.6 | 32' | 100% | 175° W | FM | Antares |
| 5th | Sgr | 18h 12m | -27° 50' | -12.2 | 33' | 97% | 160° W | FM | |
| 6th | Sgr | 19h 18m | -27° 13' | -11.9 | 33' | 92% | 145° W | -G | |
| 7th | Cap | 20h 22m | -24° 43' | -11.5 | 33' | 85% | 130° W | -G | |
| 8th | Cap | 21h 22m | -20° 37' | -11.1 | 33' | 76% | 116° W | -G | |
| 9th | Aqr | 22h 19m | -15° 21' | -10.7 | 32' | 65% | 103° W | LQ | Saturn |
| 10th | Aqr | 23h 11m | -9° 19' | -10.3 | 32' | 54% | 91° W | LQ | Saturn, Neptune |

## Mercury and Venus

Mercury
5ᵗʰ

Venus
5ᵗʰ

**Mercury**

| Date | Con. | R.A. | Dec. | Mag. | Diam. | Ill. | Elon. | Vis. | Rat. | Close To |
|------|------|------|------|------|-------|------|-------|------|------|----------|
| 1st | Ari | 2h 59m | 13° 32' | 0.5 | 8" | 45% | 24° W | AM | *** | Uranus |
| 3rd | Ari | 3h 8m | 14° 21' | 0.4 | 7" | 49% | 24° W | AM | *** | Uranus, Pleiades |
| 5th | Ari | 3h 18m | 15° 13' | 0.3 | 7" | 53% | 24° W | AM | *** | Uranus, Pleiades |
| 7th | Tau | 3h 29m | 16° 9' | 0.1 | 7" | 57% | 23° W | AM | *** | Uranus, Pleiades |
| 9th | Tau | 3h 41m | 17° 7' | 0.0 | 7" | 61% | 22° W | AM | *** | Pleiades |

## Venus

| Date | Con. | R.A. | Dec. | Mag. | Diam. | Ill. | Elon. | Vis. | Rat. | Close To |
|------|------|------|------|------|-------|------|-------|------|------|----------|
| 1st | Gem | 7h 54m | 23° 32' | -3.9 | 23" | 51% | 49° E | PM | *** | |
| 3rd | Cnc | 8h 2m | 23° 6' | -3.9 | 23" | 50% | 49° E | PM | *** | Praesepe |
| 5th | Cnc | 8h 10m | 22° 37' | -4.0 | 24" | 49% | 49° E | PM | *** | Mars, Praesepe |
| 7th | Cnc | 8h 18m | 22° 8' | -4.0 | 24" | 48% | 49° E | PM | *** | Mars, Praesepe |
| 9th | Cnc | 8h 26m | 21° 36' | -4.0 | 25" | 47% | 49° E | PM | *** | Mars, Praesepe |

## Mars and the Outer Planets

Mars
5th

Jupiter
5th

Saturn
5th

### Mars

| Date | Con. | R.A. | Dec. | Mag. | Diam. | Ill. | Elon. | Vis. | Rat. | Close To |
|------|------|------|------|------|-------|------|-------|------|------|----------|
| 1st | Cnc | 8h 38m | 19° 59' | 1.8 | 5" | 93% | 60° E | PM | * | Praesepe |
| 5th | Cnc | 8h 47m | 19° 21' | 1.8 | 5" | 93% | 59° E | PM | * | Venus, Praesepe |
| 10th | Cnc | 8h 59m | 18° 30' | 1.8 | 5" | 94% | 57° E | PM | * | Venus, Praesepe |

### The Outer Planets

| Planet | Date | Con. | R.A. | Dec. | Mag. | Diam. | Elon. | Vis. | Rat. | Close To |
|--------|------|------|------|------|------|-------|-------|------|------|----------|
| Jupiter | 5th | Ari | 2h 9m | 11° 57' | -1.7 | 35" | 41° W | AM | * | |
| Saturn | 5th | Aqr | 22h 37m | -10° 20' | 1.1 | 17" | 94° W | AM | ** | |
| Uranus | 5th | Ari | 3h 12m | 17° 34' | 6.2 | 3" | 25° W | AM | * | Mercury, Pleiades |
| Neptune | 5th | Psc | 23h 52m | -2° 6' | 7.7 | 2" | 75° W | AM | **** | |

## Highlights

| Date | Time (UT) | Event |
|------|-----------|-------|
| 2nd | 03:36 | Mars is 0.2° north of the Praesepe star cluster. (Cancer, evening sky.) |
| 3rd | 19:56 | The nearly full Moon is north of the bright star Antares. (Scorpius, evening sky.) |
| 4th | 03:42 | Full Moon. (Visible all night.) |
| | 04:39 | Mercury is 2.9° south of Uranus. (Morning sky.) |
| | 10:45 | Venus is at greatest eastern elongation from the Sun. (Evening sky.) |
| 9th | 19:29 | The waning gibbous Moon is south of Saturn. (Morning sky.) |
| 10th | 10:55 | Mercury is 6.5° south of the Pleiades star cluster. (Morning sky.) |
| | 19:32 | Last Quarter Moon. (Morning sky.) |

# June 11th to 20th, 2023

## The Moon

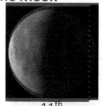    

| 11th | 13th | 15th | 17th | 19th |

| Date | Con | R.A. | Dec | Mag | Diam | Ill. | Elon. | Phase | Close To |
|------|-----|------|-----|-----|------|------|-------|-------|----------|
| 11th | Psc | 0h 1m | -2° 56' | -9.8 | 32' | 42% | 79° W | LQ | Neptune |
| 12th | Psc | 0h 50m | 3° 28' | -9.2 | 32' | 32% | 68° W | -Cr | |
| 13th | Psc | 1h 38m | 9° 37' | -8.6 | 31' | 22% | 57° W | -Cr | Jupiter |
| 14th | Ari | 2h 28m | 15° 15' | -7.8 | 31' | 14% | 46° W | -Cr | Jupiter, Uranus |
| 15th | Ari | 3h 19m | 20° 5' | -6.9 | 31' | 7% | 34° W | NM | Uranus, Pleiades |
| 16th | Tau | 4h 11m | 23° 55' | -5.9 | 31' | 3% | 22° W | NM | Mercury, Pleiades, Hyades, Aldeb |
| 17th | Tau | 5h 6m | 26° 30' | -4.8 | 30' | 1% | 9° W | NM | Mercury, Hyades, Aldebaran |
| 18th | Gem | 6h 2m | 27° 45' | -4.5 | 30' | 0% | 4° E | NM | |
| 19th | Gem | 6h 57m | 27° 35' | -5.5 | 30' | 2% | 17° E | NM | |
| 20th | Gem | 7h 51m | 26° 5' | -6.5 | 30' | 5% | 29° E | NM | |

## Mercury and Venus

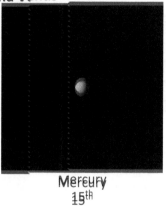

Mercury
15th

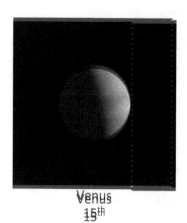

Venus
15th

### Mercury

| Date | Con. | R.A. | Dec. | Mag. | Diam. | Ill. | Elon. | Vis. | Rat. | Close To |
|------|------|------|------|------|-------|------|-------|------|------|----------|
| 11th | Tau | 3h 53m | 18° 6' | -0.2 | 6" | 65% | 21° W | AM | *** | Pleiades, Hyades |
| 13th | Tau | 4h 6m | 19° 5' | -0.4 | 6" | 70% | 20° W | AM | *** | Pleiades, Hyades, Aldebara |
| 15th | Tau | 4h 21m | 20° 3' | -0.5 | 6" | 75% | 18° W | AM | ** | Pleiades, Hyades, Aldebara |
| 17th | Tau | 4h 36m | 20° 59' | -0.7 | 6" | 80% | 17° W | AM | ** | Moon, Hyades, Aldebaran |
| 19th | Tau | 4h 52m | 21° 51' | -0.9 | 6" | 84% | 15° W | NV | N/A | Hyades, Aldebaran |

## Venus

| Date | Con. | R.A. | Dec. | Mag. | Diam. | Ill. | Elon. | Vis. | Rat. | Close To |
|------|------|------|------|------|-------|------|-------|------|------|----------|
| 11th | Cnc | 8h 33m | 21° 4' | -4.0 | 26" | 46% | 49° E | PM | *** | Mars, Praesepe |
| 13th | Cnc | 8h 40m | 20° 30' | -4.0 | 26" | 44% | 49° E | PM | *** | Mars, Praesepe |
| 15th | Cnc | 8h 47m | 19° 55' | -4.0 | 27" | 43% | 48° E | PM | *** | Mars, Praesepe |
| 17th | Cnc | 8h 54m | 19° 19' | -4.1 | 28" | 42% | 48° E | PM | *** | Mars, Praesepe |
| 19th | Cnc | 9h 1m | 18° 42' | -4.1 | 28" | 41% | 48° E | PM | *** | Mars, Praesepe |

## Mars and the Outer Planets

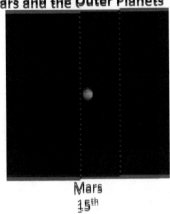

Mars
15th

Jupiter
15th

Saturn
15th

## Mars

| Date | Con. | R.A. | Dec. | Mag. | Diam. | Ill. | Elon. | Vis. | Rat. | Close To |
|------|------|------|------|------|-------|------|-------|------|------|----------|
| 11th | Cnc | 9h 2m | 18° 20' | 1.8 | 5" | 94% | 56° E | PM | * | Venus, Praesepe |
| 15th | Cnc | 9h 11m | 17° 37' | 1.8 | 4" | 94% | 54° E | PM | * | Venus, Praesepe |
| 20th | Leo | 9h 23m | 16° 40' | 1.9 | 4" | 94% | 52° E | PM | * | Venus |

## The Outer Planets

| Planet | Date | Con. | R.A. | Dec. | Mag. | Diam. | Elon. | Vis. | Rat. | Close To |
|--------|------|------|------|------|------|-------|-------|------|------|----------|
| Jupiter | 15th | Ari | 2h 17m | 12° 36' | -1.7 | 35" | 49° W | AM | * | |
| Saturn | 15th | Aqr | 22h 38m | -10° 20' | 1.1 | 18" | 104° W | AM | *** | |
| Uranus | 15th | Ari | 3h 14m | 17° 42' | 6.2 | 3" | 35° W | AM | * | Moon, Pleiades |
| Neptune | 15th | Psc | 23h 53m | -2° 4' | 7.7 | 2" | 85° W | AM | **** | |

## Highlights

| Date | Time (UT) | Event |
|------|-----------|-------|
| 11th | 08:52 | The just-past last quarter Moon is south of Neptune. (Morning sky.) |
| 13th | 05:16 | Venus is 0.9° north of the Praesepe open star cluster. (Cancer, evening sky.) |
| 14th | 05:00 | The waning crescent Moon is north of Jupiter. (Morning sky.) |
| 15th | 10:09 | The waning crescent Moon is north of Uranus. (Morning sky.) |
| | 23:58 | The waning crescent Moon is south of the Pleiades star cluster. (Morning sky.) |
| | N/A | Good opportunity to see Earthshine on the waning crescent Moon. (Morning sky.) |
| 16th | 21:26 | The waning crescent Moon is north of Mercury. (Morning sky.) |
| 18th | 04:28 | New Moon. (Not visible.) |
| | 13:13 | Saturn is stationary prior to beginning retrograde motion. (Morning sky.) |

# June 21ˢᵗ to 30ᵗʰ, 2023

## The Moon

| 21ˢᵗ | 23ʳᵈ | 25ᵗʰ | 27ᵗʰ | 29ᵗʰ |

| Date | Con | R.A. | Dec | Mag | Diam | Ill. | Elon. | Phase | Close To |
|------|-----|------|-----|-----|------|------|-------|-------|----------|
| 21st | Cnc | 8h 42m | 23° 25' | -7.4 | 30' | 10% | 41° E | NM | Venus, Mars, Praesepe |
| 22nd | Leo | 9h 31m | 19° 47' | -8.1 | 29' | 17% | 52° E | +Cr | Venus, Mars, Regulus |
| 23rd | Leo | 10h 17m | 15° 23' | -8.7 | 29' | 24% | 62° E | +Cr | Mars, Regulus |
| 24th | Leo | 11h 1m | 10° 25' | -9.3 | 30' | 33% | 72° E | +Cr | |
| 25th | Vir | 11h 44m | 5° 1' | -9.8 | 30' | 42% | 82° E | FQ | |
| 26th | Vir | 12h 27m | 0° 38' | -10.2 | 30' | 52% | 92° E | FQ | |
| 27th | Vir | 13h 11m | -6° 22' | -10.6 | 30' | 62% | 102° E | FQ | Spica |
| 28th | Vir | 13h 57m | -12° 2' | -11.0 | 31' | 71% | 112° E | +G | Spica |
| 29th | Lib | 14h 47m | -17° 20' | -11.3 | 31' | 80% | 124° E | +G | |
| 30th | Lib | 15h 41m | -21° 59' | -11.7 | 32' | 88% | 136° E | +G | Antares |

## Mercury and Venus

Mercury
25ᵗʰ

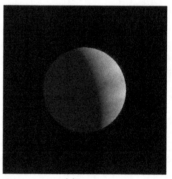

Venus
25ᵗʰ

**Mercury**

| Date | Con. | R.A. | Dec. | Mag. | Diam. | Ill. | Elon. | Vis. | Rat. | Close To |
|------|------|------|------|------|-------|------|-------|------|------|----------|
| 21st | Tau | 5h 9m | 22° 38' | -1.1 | 5" | 89% | 12° W | NV | N/A | Aldebaran |
| 23rd | Tau | 5h 27m | 23° 18' | -1.3 | 5" | 93% | 10° W | NV | N/A | |
| 25th | Tau | 5h 45m | 23° 49' | -1.5 | 5" | 96% | 8° W | NV | N/A | |
| 27th | Gem | 6h 4m | 24° 11' | -1.7 | 5" | 98% | 5° W | NV | N/A | |
| 29th | Gem | 6h 23m | 24° 23' | -1.8 | 5" | 100% | 2° W | NV | N/A | |

**Venus**

| Date | Con. | R.A. | Dec. | Mag. | Diam. | Ill. | Elon. | Vis. | Rat. | Close To |
|------|------|------|------|------|-------|------|-------|------|------|----------|
| 21st | Cnc | 9h 7m | 18° 4' | -4.1 | 29" | 39% | 47° E | PM | *** | Moon, Mars, Praesepe |
| 23rd | Cnc | 9h 13m | 17° 26' | -4.1 | 30" | 38% | 46° E | PM | *** | Mars, Praesepe |
| 25th | Cnc | 9h 19m | 16° 47' | -4.1 | 31" | 36% | 46° E | PM | *** | Mars, Praesepe |
| 27th | Leo | 9h 24m | 16° 8' | -4.1 | 32" | 35% | 45° E | PM | *** | Mars |
| 29th | Leo | 9h 29m | 15° 29' | -4.2 | 33" | 33% | 44° E | PM | *** | Mars, Regulus |

## Mars and the Outer Planets

Mars
25th

Jupiter
25th

Saturn
25th

**Mars**

| Date | Con. | R.A. | Dec. | Mag. | Diam. | Ill. | Elon. | Vis. | Rat. | Close To |
|------|------|------|------|------|-------|------|-------|------|------|----------|
| 21st | Leo | 9h 25m | 16° 29' | 1.9 | 4" | 94% | 52° E | PM | * | Moon, Venus |
| 25th | Leo | 9h 35m | 15° 41' | 1.9 | 4" | 95% | 50° E | PM | * | Venus, Regulus |
| 30th | Leo | 9h 46m | 14° 40' | 1.9 | 4" | 95% | 47° E | PM | * | Venus, Regulus |

**The Outer Planets**

| Planet | Date | Con. | R.A. | Dec. | Mag. | Diam. | Elon. | Vis. | Rat. | Close To |
|--------|------|------|------|------|------|-------|-------|------|------|----------|
| Jupiter | 25th | Ari | 2h 25m | 13° 12' | -1.8 | 36" | 58° W | AM | ** | |
| Saturn | 25th | Aqr | 22h 37m | -10° 23' | 1.0 | 18" | 115° W | AM | *** | |
| Uranus | 25th | Ari | 3h 16m | 17° 50' | 6.2 | 3" | 45° W | AM | * | Pleiades |
| Neptune | 25th | Psc | 23h 53m | -2° 3' | 7.7 | 2" | 96° W | AM | **** | |

## Highlights

| Date | Time (UT) | Event |
|------|-----------|-------|
| 21st | 08:43 | The waxing crescent Moon is north of the Praesepe star cluster. (Cancer, evening sky.) |
| | 14:58 | Summer solstice. |
| | N/A | Good opportunity to see Earthshine on the waxing crescent Moon. (Evening sky.) |
| 22nd | 01:36 | The waxing crescent Moon is north of Venus. (Evening sky.) |
| | 08:13 | The waxing crescent Moon is north of Mars. (Evening sky.) |
| 23rd | 06:05 | The waxing crescent Moon is north of the bright star Regulus. (Leo, evening sky.) |
| 26th | 07:50 | First Quarter Moon. (Evening sky.) |
| 27th | 19:09 | The just-past first quarter Moon is north of the bright star Spica. (Evening sky.) |
| | N/A | The Bootid meteor shower is at its maximum. (ZHR: Variable.) |

# July 1st to 10th, 2023

## The Moon

 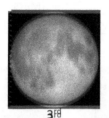 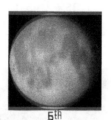  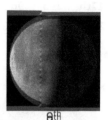

| 1st | 3rd | 5th | 7th | 9th |

| Date | Con | R.A. | Dec | Mag | Diam | Ill. | Elon. | Phase | Close To |
|------|-----|------|-----|-----|------|------|-------|-------|----------|
| 1st | Sco | 16h 41m | -25° 32' | -12.0 | 32' | 94% | 150° E | +G | Antares |
| 2nd | Sgr | 17h 45m | -27° 32' | -12.4 | 33' | 98% | 165° E | FM | |
| 3rd | Sgr | 18h 51m | -27° 39' | -12.6 | 33' | 100% | 179° W | FM | |
| 4th | Sgr | 19h 58m | -25° 46' | -12.3 | 33' | 98% | 164° W | FM | |
| 5th | Cap | 21h 2m | -22° 3' | -12.0 | 33' | 94% | 149° W | -G | |
| 6th | Aqr | 22h 1m | -16° 55' | -11.6 | 33' | 87% | 135° W | -G | Saturn |
| 7th | Aqr | 22h 56m | -10° 53' | -11.2 | 33' | 78% | 122° W | -G | Saturn |
| 8th | Aqr | 23h 48m | -4° 23' | -10.8 | 32' | 67% | 110° W | -G | Neptune |
| 9th | Cet | 0h 38m | 2° 10' | -10.4 | 32' | 56% | 99° W | LQ | Neptune |
| 10th | Psc | 1h 27m | 8° 28' | -9.9 | 32' | 45% | 88° W | LQ | |

## Mercury and Venus

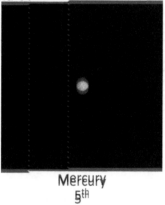

Mercury
5th

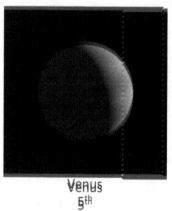

Venus
5th

### Mercury

| Date | Con. | R.A. | Dec. | Mag. | Diam. | Ill. | Elon. | Vis. | Rat. | Close To |
|------|------|------|------|------|-------|------|-------|------|------|----------|
| 1st | Gem | 6h 42m | 24° 23' | -1.8 | 5" | 100% | 0° E | NV | N/A | |
| 3rd | Gem | 7h 1m | 24° 12' | -1.7 | 5" | 99% | 3° E | NV | N/A | |
| 5th | Gem | 7h 20m | 23° 50' | -1.5 | 5" | 98% | 6° E | NV | N/A | |
| 7th | Gem | 7h 38m | 23° 18' | -1.3 | 5" | 96% | 8° E | NV | N/A | |
| 9th | Gem | 7h 56m | 22° 37' | -1.1 | 5" | 93% | 11° E | NV | N/A | |

## Venus

| Date | Con. | R.A. | Dec. | Mag. | Diam. | Ill. | Elon. | Vis. | Rat. | Close To |
|------|------|------|------|------|-------|------|-------|------|------|----------|
| 1st | Leo | 9h 34m | 14° 50' | -4.2 | 34" | 32% | 43° E | PM | *** | Mars, Regulus |
| 3rd | Leo | 9h 38m | 14° 11' | -4.2 | 35" | 30% | 42° E | PM | *** | Mars, Regulus |
| 5th | Leo | 9h 42m | 13° 32' | -4.2 | 36" | 29% | 41° E | PM | *** | Mars, Regulus |
| 7th | Leo | 9h 46m | 12° 54' | -4.2 | 37" | 27% | 40° E | PM | *** | Mars, Regulus |
| 9th | Leo | 9h 49m | 12° 17' | -4.2 | 38" | 25% | 39° E | PM | *** | Mars, Regulus |

## Mars and the Outer Planets

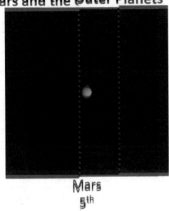

Mars
5th

Jupiter
5th

Saturn
5th

## Mars

| Date | Con. | R.A. | Dec. | Mag. | Diam. | Ill. | Elon. | Vis. | Rat. | Close To |
|------|------|------|------|------|-------|------|-------|------|------|----------|
| 1st | Leo | 9h 49m | 14° 28' | 1.9 | 4" | 95% | 47° E | PM | * | Venus, Regulus |
| 5th | Leo | 9h 58m | 13° 37' | 1.9 | 4" | 95% | 45° E | PM | * | Venus, Regulus |
| 10th | Leo | 10h 9m | 12° 31' | 1.9 | 4" | 96% | 43° E | PM | * | Venus, Regulus |

## The Outer Planets

| Planet | Date | Con. | R.A. | Dec. | Mag. | Diam. | Elon. | Vis. | Rat. | Close To |
|--------|------|------|------|------|------|-------|-------|------|------|----------|
| Jupiter | 5th | Ari | 2h 32m | 13° 44' | -1.8 | 37" | 66° W | AM | ** | |
| Saturn | 5th | Aqr | 22h 37m | -10° 29' | 1.0 | 18" | 125° W | AM | *** | |
| Uranus | 5th | Ari | 3h 18m | 17° 56' | 6.1 | 3" | 55° W | AM | * | Pleiades |
| Neptune | 5th | Psc | 23h 53m | -2° 3' | 7.7 | 2" | 106° W | AM | ***** | |

## Highlights

| Date | Time (UT) | Event |
|------|-----------|-------|
| 1st | 04:52 | Mercury is at superior conjunction with the Sun. (Not visible.) |
| | 07:10 | Neptune is stationary prior to beginning prograde motion. (Morning sky.) |
| | 08:06 | The waxing gibbous Moon is north of the bright star Antares. (Scorpius, evening sky.) |
| 3rd | 11:39 | Full Moon. (Visible all night.) |
| 7th | 02:55 | The waning gibbous Moon is south of Saturn. (Morning sky.) |
| 8th | 15:05 | The waning gibbous Moon is south of Neptune. (Morning sky.) |
| 9th | 18:54 | Mars is 0.7° north of the bright star Regulus. (Leo, evening sky.) |
| 10th | 01:48 | Last quarter Moon. (Morning sky.) |

# July 11<sup>th</sup> to 20<sup>th</sup>, 2023

## The Moon

  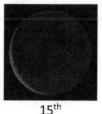 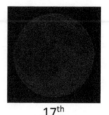 

|        | 11<sup>th</sup> | 13<sup>th</sup> | 15<sup>th</sup> | 17<sup>th</sup> | 19<sup>th</sup> |

| Date | Con | R.A. | Dec | Mag | Diam | Ill. | Elon. | Phase | Close To |
|------|-----|------|-----|-----|------|------|-------|-------|----------|
| 11th | Ari | 2h 16m | 14° 13' | -9.4 | 31' | 35% | 76° W | -Cr | Jupiter |
| 12th | Ari | 3h 6m | 19° 13' | -8.8 | 31' | 25% | 65° W | -Cr | Jupiter, Uranus, Pleiades |
| 13th | Tau | 3h 59m | 23° 14' | -8.1 | 31' | 17% | 53° W | -Cr | Uranus, Pleiades, Hyades, Aldeba |
| 14th | Tau | 4h 52m | 26° 5' | -7.3 | 30' | 10% | 40° W | NM | Hyades, Aldebaran |
| 15th | Tau | 5h 47m | 27° 36' | -6.5 | 30' | 5% | 28° W | NM | |
| 16th | Gem | 6h 43m | 27° 45' | -5.5 | 30' | 2% | 15° W | NM | |
| 17th | Gem | 7h 37m | 26° 34' | -4.5 | 30' | 0% | 2° W | NM | |
| 18th | Cnc | 8h 28m | 24° 10' | -4.9 | 29' | 1% | 10° E | NM | Mercury, Praesepe |
| 19th | Cnc | 9h 18m | 20° 45' | -5.9 | 29' | 3% | 21° E | NM | Mercury, Venus, Praesepe |
| 20th | Leo | 10h 4m | 16° 31' | -6.8 | 29' | 7% | 32° E | NM | Mercury, Venus, Mars, Regulus |

## Mercury and Venus

Mercury
15<sup>th</sup>

Venus
15<sup>th</sup>

**Mercury**

| Date | Con. | R.A. | Dec. | Mag. | Diam. | Ill. | Elon. | Vis. | Rat. | Close To |
|------|------|------|------|------|-------|------|-------|------|------|----------|
| 11th | Cnc | 8h 13m | 21° 49' | -0.9 | 5" | 91% | 13° E | NV | N/A | Praesepe |
| 13th | Cnc | 8h 29m | 20° 54' | -0.7 | 5" | 88% | 15° E | NV | N/A | Praesepe |
| 15th | Cnc | 8h 44m | 19° 53' | -0.6 | 5" | 85% | 17° E | PM | ** | Praesepe |
| 17th | Cnc | 8h 59m | 18° 48' | -0.4 | 6" | 82% | 18° E | PM | ** | Praesepe |
| 19th | Cnc | 9h 12m | 17° 39' | -0.3 | 6" | 79% | 20° E | PM | *** | Moon, Praesepe |

**Venus**

| Date | Con. | R.A. | Dec. | Mag. | Diam. | Ill. | Elon. | Vis. | Rat. | Close To |
|------|------|------|------|------|-------|------|-------|------|------|----------|
| 11th | Leo | 9h 52m | 11° 41' | -4.2 | 39" | 23% | 38° E | PM | *** | Mars, Regulus |
| 13th | Leo | 9h 54m | 11° 6' | -4.2 | 41" | 22% | 36° E | PM | *** | Mars, Regulus |
| 15th | Leo | 9h 56m | 10° 32' | -4.2 | 42" | 20% | 34° E | PM | *** | Mars, Regulus |
| 17th | Leo | 9h 57m | 9° 60' | -4.1 | 43" | 18% | 33° E | PM | *** | Mars, Regulus |
| 19th | Leo | 9h 58m | 9° 30' | -4.1 | 45" | 16% | 31° E | PM | *** | Moon, Mars, Regulus |

## Mars and the Outer Planets

Mars
15th

Jupiter
15th

Saturn
15th

**Mars**

| Date | Con. | R.A. | Dec. | Mag. | Diam. | Ill. | Elon. | Vis. | Rat. | Close To |
|------|------|------|------|------|-------|------|-------|------|------|----------|
| 11th | Leo | 10h 12m | 12° 17' | 1.9 | 4" | 96% | 43° E | PM | * | Venus, Regulus |
| 15th | Leo | 10h 21m | 11° 23' | 1.9 | 4" | 96% | 41° E | PM | * | Venus, Regulus |
| 20th | Leo | 10h 33m | 10° 13' | 2.0 | 4" | 96% | 39° E | PM | * | Moon, Venus, Regulus |

**The Outer Planets**

| Planet | Date | Con. | R.A. | Dec. | Mag. | Diam. | Elon. | Vis. | Rat. | Close To |
|--------|------|------|------|------|------|-------|-------|------|------|----------|
| Jupiter | 15th | Ari | 2h 38m | 14° 12' | -1.9 | 38" | 75° W | AM | ** | |
| Saturn | 15th | Aqr | 22h 35m | -10° 39' | 0.9 | 18" | 136° W | AM | *** | |
| Uranus | 15th | Ari | 3h 19m | 18° 2' | 6.1 | 4" | 65° W | AM | * | Pleiades |
| Neptune | 15th | Psc | 23h 53m | -2° 5' | 7.7 | 2" | 116° W | AM | ***** | |

## Highlights

| Date | Time (UT) | Event |
|------|-----------|-------|
| 11th | 20:42 | The waning crescent Moon is north of Jupiter. (Morning sky.) |
| 12th | 18:30 | The waning crescent Moon is north of Uranus. (Morning sky.) |
| 13th | 05:08 | The waning crescent Moon is south of the Pleiades star cluster. (Taurus, morning sky.) |
| 14th | 02:42 | The waning crescent Moon is north of the bright star Aldebaran. (Taurus, morning sky.) |
| | 21:28 | Mercury is 0.5° north of the Praesepe star cluster. (Cancer, evening sky.) |
| | N/A | Good opportunity to see Earthshine on the waning crescent Moon. (Morning sky.) |
| 17th | 18:32 | New Moon. (Not visible.) |
| 19th | 06:45 | The waxing crescent Moon is north of Mercury. (Evening sky.) |
| 20th | 06:46 | The waxing crescent Moon is north of Venus. (Evening sky.) |
| | 10:01 | Venus is stationary prior to beginning retrograde motion. (Evening sky.) |
| | 13:27 | The waxing crescent Moon is north of the bright star Regulus. (Leo, evening sky.) |

# July 21ˢᵗ to 31ˢᵗ, 2023

## The Moon

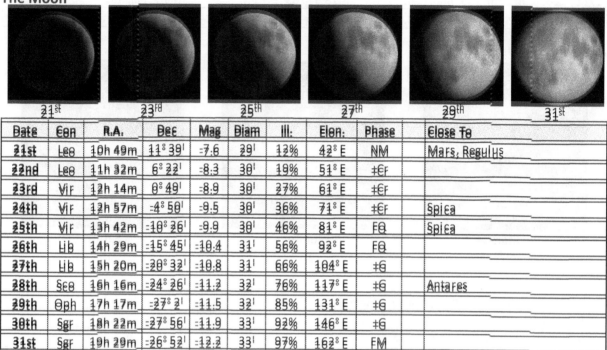

| | 21ˢᵗ | | 23ʳᵈ | | 25ᵗʰ | | 27ᵗʰ | | 29ᵗʰ | 31ˢᵗ |
|---|---|---|---|---|---|---|---|---|---|---|

| Date | Con | R.A. | Dec | Mag | Diam | Ill. | Elon. | Phase | Close To |
|---|---|---|---|---|---|---|---|---|---|
| 21st | Leo | 10h 49m | 11° 39' | -7.6 | 29' | 12% | 42° E | NM | Mars, Regulus |
| 22nd | Leo | 11h 32m | 6° 22' | -8.3 | 30' | 19% | 51° E | +Cr | |
| 23rd | Vir | 12h 14m | 0° 49' | -8.9 | 30' | 27% | 61° E | +Cr | |
| 24th | Vir | 12h 57m | -4° 50' | -9.5 | 30' | 36% | 71° E | +Cr | Spica |
| 25th | Vir | 13h 42m | -10° 26' | -9.9 | 30' | 46% | 81° E | FQ | Spica |
| 26th | Lib | 14h 29m | -15° 45' | -10.4 | 31' | 56% | 92° E | FQ | |
| 27th | Lib | 15h 20m | -20° 32' | -10.8 | 31' | 66% | 104° E | +G | |
| 28th | Sco | 16h 16m | -24° 26' | -11.2 | 32' | 76% | 117° E | +G | Antares |
| 29th | Oph | 17h 17m | -27° 3' | -11.5 | 32' | 85% | 131° E | +G | |
| 30th | Sgr | 18h 22m | -27° 56' | -11.9 | 33' | 92% | 146° E | +G | |
| 31st | Sgr | 19h 29m | -26° 52' | -12.2 | 33' | 97% | 162° E | FM | |

## Mercury and Venus

Mercury
25ᵗʰ

Venus
25ᵗʰ

### Mercury

| Date | Con. | R.A. | Dec. | Mag. | Diam. | Ill. | Elon. | Vis. | Rat. | Close To |
|---|---|---|---|---|---|---|---|---|---|---|
| 21st | Leo | 9h 26m | 16° 27' | -0.2 | 6" | 76% | 21° E | PM | *** | Venus |
| 23rd | Leo | 9h 38m | 15° 14' | -0.1 | 6" | 73% | 22° E | PM | *** | Venus, Regulus |
| 25th | Leo | 9h 50m | 13° 59' | 0.0 | 6" | 71% | 23° E | PM | *** | Venus, Regulus |
| 27th | Leo | 10h 1m | 12° 43' | 0.1 | 6" | 68% | 24° E | PM | *** | Venus, Regulus |
| 29th | Leo | 10h 12m | 11° 27' | 0.2 | 6" | 65% | 24° E | PM | *** | Venus, Regulus |
| 31st | Leo | 10h 21m | 10° 11' | 0.3 | 7" | 63% | 25° E | PM | *** | Venus, Mars, Regulus |

## Venus

| Date | Con. | R.A. | Dec. | Mag. | Diam. | Ill. | Elon. | Vis. | Rat. | Close To |
|------|------|------|------|------|-------|------|-------|------|------|----------|
| 21st | Leo | 9h 58m | 9° 2' | -4.1 | 46" | 14% | 29° E | PM | *** | Mercury, Mars, Regulus |
| 23rd | Leo | 9h 57m | 8° 36' | -4.0 | 48" | 13% | 27° E | PM | *** | Mercury, Regulus |
| 25th | Leo | 9h 56m | 8° 13' | -4.0 | 49" | 11% | 25° E | PM | ** | Mercury, Regulus |
| 27th | Leo | 9h 55m | 7° 53' | -3.9 | 51" | 9% | 22° E | PM | ** | Mercury, Regulus |
| 29th | Leo | 9h 53m | 7° 36' | -3.8 | 52" | 7% | 20° E | PM | ** | Mercury, Regulus |
| 31st | Leo | 9h 50m | 7° 22' | -3.7 | 53" | 6% | 17° E | PM | ** | Mercury, Regulus |

## Mars and the Outer Planets

Mars
25th

Jupiter
25th

Saturn
25th

## Mars

| Date | Con. | R.A. | Dec. | Mag. | Diam. | Ill. | Elon. | Vis. | Rat. | Close To |
|------|------|------|------|------|-------|------|-------|------|------|----------|
| 21st | Leo | 10h 35m | 9° 59' | 2.0 | 4" | 96% | 38° E | PM | * | Moon, Venus, Regulus |
| 25th | Leo | 10h 44m | 9° 3' | 2.0 | 4" | 96% | 37° E | PM | * | Regulus |
| 31st | Leo | 10h 58m | 7° 34' | 2.0 | 4" | 97% | 34° E | PM | * | Mercury |

## The Outer Planets

| Planet | Date | Con. | R.A. | Dec. | Mag. | Diam. | Elon. | Vis. | Rat. | Close To |
|--------|------|------|------|------|------|-------|-------|------|------|----------|
| Jupiter | 25th | Ari | 2h 43m | 14° 35' | -1.9 | 39" | 84° W | AM | ** | |
| Saturn | 25th | Aqr | 22h 34m | -10° 52' | 0.8 | 19" | 146° W | AM | **** | |
| Uranus | 25th | Ari | 3h 21m | 18° 7' | 6.1 | 4" | 74° W | AM | * | Jupiter, Pleiades |
| Neptune | 25th | Psc | 23h 52m | -2° 8' | 7.7 | 2" | 126° W | AM | ***** | |

## Highlights

| Date | Time (UT) | Event |
|------|-----------|-------|
| 21st | 03:31 | The waxing crescent Moon is north of Mars. (Evening sky.) |
| | N/A | Good opportunity to see Earthshine on the waxing crescent Moon. (Evening sky.) |
| 22nd | 10:18 | Dwarf planet Pluto is at opposition. (Visible all night.) |
| 24th | N/A | The June Boötid meteor shower is at its maximum. (ZHR: Var) |
| 25th | 03:45 | The almost first quarter Moon is north of the bright star Spica. (Virgo, evening sky.) |
| | 22:07 | First Quarter Moon. (Evening sky.) |
| 26th | 12:33 | Mercury is 5.3° north of Venus. (Evening sky.) |
| 28th | 15:32 | The waxing gibbous Moon is north of the bright star Antares. (Scorpius, evening sky.) |
| | 18:56 | Mercury is 0.1° south of the bright star Regulus. (Leo, evening sky.) |
| 30th | N/A | The Southern Delta Aquariid meteor shower is at its maximum. (ZHR: 16) |

# August 1ˢᵗ to 10ᵗʰ, 2023

## The Moon

| | 1ˢᵗ | | 3ʳᵈ | | 5ᵗʰ | | 7ᵗʰ | | 9ᵗʰ |

| Date | Con | R.A. | Dec | Mag | Diam | Ill. | Elon. | Phase | Close To |
|------|-----|------|-----|-----|------|------|-------|-------|----------|
| 1st | Cap | 20h 35m | -23° 49' | -12.6 | 33' | 100% | 177° E | FM | |
| 2nd | Cap | 21h 37m | -19° 5' | -12.4 | 33' | 99% | 168° W | FM | |
| 3rd | Aqr | 22h 35m | -13° 9' | -12.1 | 33' | 95% | 154° W | FM | Saturn |
| 4th | Aqr | 23h 30m | -6° 32' | -11.7 | 33' | 89% | 142° W | -G | Neptune |
| 5th | Psc | 0h 22m | 0° 17' | -11.3 | 33' | 80% | 130° W | -G | Neptune |
| 6th | Psc | 1h 12m | 6° 55' | -10.9 | 32' | 71% | 118° W | -G | |
| 7th | Ari | 2h 3m | 13° 0' | -10.5 | 32' | 60% | 106° W | LQ | Jupiter |
| 8th | Ari | 2h 54m | 18° 19' | -10.1 | 31' | 49% | 95° W | LQ | Jupiter, Uranus |
| 9th | Tau | 3h 46m | 22° 37' | -9.6 | 31' | 39% | 83° W | LQ | Uranus, Pleiades, Hyades |
| 10th | Tau | 4h 40m | 25° 44' | -9.1 | 30' | 29% | 70° W | -Cr | Hyades, Aldebaran |

## Mercury and Venus

Mercury
5ᵗʰ

Venus
5ᵗʰ

### Mercury

| Date | Con. | R.A. | Dec. | Mag. | Diam. | Ill. | Elon. | Vis. | Rat. | Close To |
|------|------|------|------|------|-------|------|-------|------|------|----------|
| 1st | Leo | 10h 26m | 9° 33' | 0.3 | 7" | 61% | 25° E | PM | *** | Venus, Mars, Regulus |
| 3rd | Leo | 10h 35m | 8° 19' | 0.4 | 7" | 59% | 26° E | PM | *** | Mars, Regulus |
| 5th | Leo | 10h 43m | 7° 7' | 0.5 | 7" | 56% | 26° E | PM | *** | Mars, Regulus |
| 7th | Sex | 10h 51m | 5° 57' | 0.5 | 7" | 53% | 26° E | PM | *** | Mars |
| 9th | Leo | 10h 58m | 4° 50' | 0.6 | 7" | 50% | 26° E | PM | *** | Mars |

**Venus**

| Date | Con. | R.A. | Dec. | Mag. | Diam. | Ill. | Elon. | Vis. | Rat. | Close To |
|------|------|------|------|------|-------|------|-------|------|------|----------|
| 1st | Leo | 9h 48m | 7° 17' | -3.7 | 54" | 5% | 16° E | PM | ** | Mercury, Regulus |
| 3rd | Leo | 9h 44m | 7° 9' | -3.6 | 55" | 4% | 13° E | NV | N/A | Regulus |
| 5th | Leo | 9h 40m | 7° 4' | -3.4 | 56" | 3% | 10° E | NV | N/A | Regulus |
| 7th | Leo | 9h 36m | 7° 4' | -3.3 | 57" | 2% | 7° E | NV | N/A | Regulus |
| 9th | Leo | 9h 31m | 7° 6' | -3.2 | 57" | 1% | 4° E | NV | N/A | Regulus |

## Mars and the Outer Planets

Mars
5th

Jupiter
5th

Saturn
5th

**Mars**

| Date | Con. | R.A. | Dec. | Mag. | Diam. | Ill. | Elon. | Vis. | Rat. | Close To |
|------|------|------|------|------|-------|------|-------|------|------|----------|
| 1st | Leo | 11h 0m | 7° 19' | 2.0 | 4" | 97% | 34° E | PM | * | Mercury |
| 5th | Leo | 11h 9m | 6° 19' | 2.0 | 4" | 97% | 32° E | PM | * | Mercury |
| 10th | Leo | 11h 21m | 5° 3' | 2.0 | 4" | 97% | 30° E | PM | * | Mercury |

**The Outer Planets**

| Planet | Date | Con. | R.A. | Dec. | Mag. | Diam. | Elon. | Vis. | Rat. | Close To |
|--------|------|------|------|------|------|-------|-------|------|------|----------|
| Jupiter | 5th | Ari | 2h 48m | 14° 54' | -2.0 | 40" | 93° W | AM | *** | |
| Saturn | 5th | Aqr | 22h 31m | -11° 9' | 0.8 | 19" | 157° W | AM | **** | |
| Uranus | 5th | Ari | 3h 22m | 18° 11' | 6.1 | 4" | 85° W | AM | * | Jupiter, Pleiades |
| Neptune | 5th | Psc | 23h 52m | -2° 12' | 7.7 | 2" | 137° W | AM | ***** | Moon |

## Highlights

| Date | Time (UT) | Event |
|------|-----------|-------|
| 1st | 18:32 | Full Moon. (Visible all night.) |
| 3rd | 11:30 | The waning gibbous Moon is south of Saturn. (Morning sky.) |
| 4th | 20:18 | The waning gibbous Moon is south of Neptune. (Morning sky.) |
| 8th | 10:29 | Last Quarter Moon. (Morning sky.) |
| | 11:37 | The just-past last quarter Moon is north of Jupiter. (Morning sky.) |
| | 23:15 | The just-past last quarter Moon is north of Uranus. (Morning sky.) |
| 9th | 13:59 | The just-past last quarter Moon is south of the Pleiades star cluster. (Taurus, morning sky.) |
| 10th | 01:39 | Mercury is at greatest eastern elongation from the Sun. (Evening sky.) |
| | 11:47 | The waning crescent Moon is north of the bright star Aldebaran. (Taurus, morning sky.) |

# August 11<sup>th</sup> to 20<sup>th</sup>, 2023

## The Moon

| 11<sup>th</sup> | 13<sup>th</sup> | 15<sup>th</sup> | 17<sup>th</sup> | 19<sup>th</sup> |

| Date | Con | R.A. | Dec | Mag | Diam | Ill. | Elon. | Phase | Close To |
|------|-----|------|-----|-----|------|------|-------|-------|----------|
| 11th | Tau | 5h 35m | 27° 32' | -8.5 | 30' | 21% | 57° W | -Cr | |
| 12th | Aur | 6h 30m | 27° 58' | -7.8 | 30' | 13% | 44° W | -Cr | |
| 13th | Gem | 7h 24m | 27° 3' | -7.0 | 30' | 8% | 32° W | NM | |
| 14th | Cnc | 8h 16m | 24° 54' | -6.1 | 29' | 3% | 20° W | NM | Praesepe |
| 15th | Cnc | 9h 6m | 21° 41' | -5.1 | 29' | 1% | 8° W | NM | Venus, Praesepe |
| 16th | Leo | 9h 53m | 17° 35' | -4.4 | 29' | 0% | 3° E | NM | Venus, Regulus |
| 17th | Leo | 10h 38m | 12° 49' | -5.2 | 29' | 1% | 13° E | NM | Mercury, Regulus |
| 18th | Leo | 11h 21m | 7° 35' | -6.2 | 29' | 4% | 23° E | NM | Mercury, Mars |
| 19th | Vir | 12h 3m | 2° 3' | -7.1 | 30' | 8% | 32° E | NM | Mercury, Mars |
| 20th | Vir | 12h 46m | -3° 36' | -7.9 | 30' | 15% | 42° E | +Cr | Spica |

## Mercury and Venus

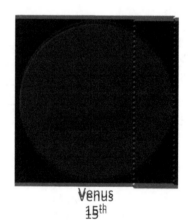

Mercury
15<sup>th</sup>

Venus
15<sup>th</sup>

### Mercury

| Date | Con. | R.A. | Dec. | Mag. | Diam. | Ill. | Elon. | Vis. | Rat. | Close To |
|------|------|------|------|------|-------|------|-------|------|------|----------|
| 11th | Leo | 11h 4m | 3° 47' | 0.6 | 8" | 47% | 25° E | PM | *** | Mars |
| 13th | Leo | 11h 10m | 3° 48' | 0.7 | 8" | 44% | 25° E | PM | *** | Mars |
| 15th | Leo | 11h 15m | 1° 54' | 0.8 | 8" | 40% | 24° E | PM | *** | Mars |
| 17th | Leo | 11h 18m | 1° 7' | 0.9 | 9" | 36% | 23° E | PM | *** | Moon, Mars |
| 19th | Leo | 11h 21m | 0° 27' | 0.9 | 9" | 32% | 22° E | PM | *** | Moon, Mars |

## Venus

| Date | Con. | R.A. | Dec. | Mag. | Diam. | Ill. | Elon. | Vis. | Rat. | Close To |
|------|------|------|------|------|-------|------|-------|------|------|----------|
| 11th | Leo | 9h 26m | 7° 13' | -3.1 | 58" | 1% | 1° E | NV | N/A | |
| 13th | Cnc | 9h 21m | 7° 22' | -3.1 | 58" | 1% | 3° W | NV | N/A | |
| 15th | Cnc | 9h 16m | 7° 34' | -3.1 | 58" | 1% | 6° W | NV | N/A | Moon, Praesepe |
| 17th | Cnc | 9h 11m | 7° 48' | -3.2 | 57" | 2% | 9° W | NV | N/A | Praesepe |
| 19th | Cnc | 9h 7m | 8° 4' | -3.4 | 57" | 2% | 12° W | NV | N/A | Praesepe |

## Mars and the Outer Planets

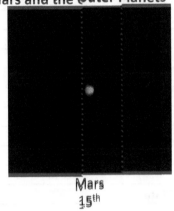

Mars
15th

Jupiter
15th

Saturn
15th

### Mars

| Date | Con. | R.A. | Dec. | Mag. | Diam. | Ill. | Elon. | Vis. | Rat. | Close To |
|------|------|------|------|------|-------|------|-------|------|------|----------|
| 11th | Leo | 11h 23m | 4° 48' | 2.0 | 4" | 97% | 30° E | PM | * | Mercury |
| 15th | Leo | 11h 33m | 3° 46' | 2.0 | 4" | 98% | 28° E | PM | * | Mercury |
| 20th | Vir | 11h 44m | 2° 28' | 2.0 | 4" | 98% | 27° E | PM | * | Mercury |

### The Outer Planets

| Planet | Date | Con. | R.A. | Dec. | Mag. | Diam. | Elon. | Vis. | Rat. | Close To |
|--------|------|------|------|------|------|-------|-------|------|------|----------|
| Jupiter | 15th | Ari | 2h 51m | 15° 6' | -2.1 | 42" | 102° W | AM | *** | |
| Saturn | 15th | Aqr | 22h 28m | -11° 25' | 0.7 | 19" | 168° W | AM | **** | |
| Uranus | 15th | Ari | 3h 22m | 18° 13' | 6.1 | 4" | 94° W | AM | * | Jupiter, Pleiades |
| Neptune | 15th | Psc | 23h 51m | -2° 17' | 7.6 | 2" | 147° W | AM | ***** | |

## Highlights

| Date | Time (UT) | Event |
|------|-----------|-------|
| 12th | N/A | The Perseid meteor shower is at its maximum: (ZHR: 100) |
| 13th | 11:18 | Venus is at inferior conjunction with the Sun. (Not visible.) |
| | N/A | Good opportunity to see Earthshine on the waning crescent Moon. (Morning sky.) |
| 14th | 23:11 | The waning crescent Moon is north of the Praesepe star cluster. (Cancer, morning sky.) |
| 16th | 09:38 | New Moon (Not visible.) |
| 18th | 09:36 | The waxing crescent Moon is north of Mercury. (Evening sky.) |
| 19th | 00:05 | The waxing crescent Moon is north of Mars. (Evening sky.) |
| | N/A | Good opportunity to see Earthshine on the waxing crescent Moon. (Evening sky.) |

# August 21<sup>st</sup> to 31<sup>st</sup>, 2023

## The Moon

21<sup>st</sup>

23<sup>rd</sup>

25<sup>th</sup>

27<sup>th</sup>

29<sup>th</sup>

31<sup>st</sup>

| Date | Con | R.A. | Dec | Mag | Diam | Ill. | Elon. | Phase | Close To |
|------|-----|------|-----|-----|------|------|-------|-------|----------|
| 21st | Vir | 13h 29m | -9° 12' | -8.6 | 30' | 22% | 52° E | +Cr | Spica |
| 22nd | Vir | 14h 15m | -14° 33' | -9.2 | 30' | 31% | 63° E | +Cr | |
| 23rd | Lib | 15h 4m | -19° 25' | -9.7 | 31' | 40% | 74° E | FQ | |
| 24th | Sco | 15h 57m | -23° 31' | -10.2 | 31' | 51% | 86° E | FQ | Antares |
| 25th | Oph | 16h 55m | -26° 30' | -10.6 | 32' | 62% | 100° E | FQ | Antares |
| 26th | Sgr | 17h 57m | -27° 60' | -11.0 | 32' | 72% | 114° E | +G | |
| 27th | Sgr | 19h 1m | -27° 42' | -11.4 | 33' | 82% | 130° E | +G | |
| 28th | Sgr | 20h 6m | -25° 28' | -11.8 | 33' | 90% | 145° E | +G | |
| 29th | Cap | 21h 9m | -21° 26' | -12.1 | 33' | 96% | 160° E | FM | |
| 30th | Aqr | 22h 9m | -15° 56' | -12.5 | 33' | 99% | 174° E | FM | Saturn |
| 31st | Aqr | 23h 6m | -9° 26' | -12.5 | 33' | 100% | 173° W | FM | Saturn, Neptune |

## Mercury and Venus

Mercury
25<sup>th</sup>

Venus
25<sup>th</sup>

### Mercury

| Date | Con. | R.A. | Dec. | Mag. | Diam. | Ill. | Elon. | Vis. | Rat. | Close To |
|------|------|------|------|------|-------|------|-------|------|------|----------|
| 21st | Leo | 11h 23m | 0° 5' | 1.1 | 9" | 28% | 20° E | PM | *** | Mars |
| 23rd | Leo | 11h 23m | 0° 27' | 1.2 | 10" | 24% | 19° E | PM | *** | Mars |
| 25th | Leo | 11h 22m | 0° 37' | 1.3 | 10" | 20% | 17° E | PM | *** | Mars |
| 27th | Leo | 11h 20m | 0° 34' | 1.5 | 10" | 15% | 14° E | NV | N/A | |
| 29th | Leo | 11h 17m | 0° 18' | 1.8 | 10" | 11% | 12° E | NV | N/A | |
| 31st | Leo | 11h 12m | 0° 13' | 2.1 | 11" | 7% | 9° E | NV | N/A | |

## Venus

| Date | Con. | R.A. | Dec. | Mag. | Diam. | Ill. | Elon. | Vis. | Rat. | Close To |
|------|------|------|------|------|-------|------|-------|------|------|----------|
| 21st | Cnc | 9h 3m | 8° 22' | -3.5 | 56" | 3% | 15° W | NV | N/A | Praesepe |
| 23rd | Cnc | 8h 59m | 8° 40' | -3.6 | 55" | 4% | 17° W | AM | ** | Praesepe |
| 25th | Cnc | 8h 56m | 8° 59' | -3.7 | 54" | 6% | 20° W | AM | ** | Praesepe |
| 27th | Cnc | 8h 54m | 9° 18' | -3.8 | 53" | 7% | 22° W | AM | ** | Praesepe |
| 29th | Cnc | 8h 52m | 9° 37' | -3.9 | 52" | 9% | 25° W | AM | ** | Praesepe |
| 31st | Cnc | 8h 51m | 9° 55' | -4.0 | 50" | 11% | 27° W | AM | *** | Praesepe |

## Mars and the Outer Planets

Mars
25th

Jupiter
25th

Saturn
25th

### Mars

| Date | Con. | R.A. | Dec. | Mag. | Diam. | Ill. | Elon. | Vis. | Rat. | Close To |
|------|------|------|------|------|-------|------|-------|------|------|----------|
| 21st | Vir | 11h 46m | 2° 13' | 2.0 | 4" | 98% | 26° E | PM | * | Mercury |
| 25th | Vir | 11h 56m | 1° 10' | 2.0 | 4" | 98% | 25° E | PM | * | Mercury |
| 31st | Vir | 12h 10m | 0° 25' | 2.0 | 4" | 98% | 23° E | PM | * | |

### The Outer Planets

| Planet | Date | Con. | R.A. | Dec. | Mag. | Diam. | Elon. | Vis. | Rat. | Close To |
|--------|------|------|------|------|------|-------|-------|------|------|----------|
| Jupiter | 25th | Ari | 2h 53m | 15° 13' | -2.1 | 43" | 111° W | AM | *** | |
| Saturn | 25th | Aqr | 22h 26m | -11° 42' | 0.6 | 19" | 178° W | AN | **** | |
| Uranus | 25th | Ari | 3h 23m | 18° 14' | 6.1 | 4" | 103° W | AM | ** | Jupiter, Pleiades |
| Neptune | 25th | Psc | 23h 50m | -2° 23' | 7.6 | 2" | 156° W | AM | ***** | |

## Highlights

| Date | Time (UT) | Event |
|------|-----------|-------|
| 21st | 07:40 | The waxing crescent Moon is north of the bright star Spica. (Virgo, evening sky.) |
| 23rd | 04:40 | Mercury is stationary prior to beginning retrograde motion. (Evening sky.) |
| 24th | 09:58 | First Quarter Moon. (Evening sky.) |
| 25th | 02:52 | The just-past first quarter Moon is north of the bright star Antares. (Scorpius, evening sky.) |
| 28th | 00:09 | Saturn is at opposition. (Visible all night.) |
| 30th | 16:30 | The nearly full Moon is south of Saturn. (Evening sky.) |
| 31st | 01:36 | Full Moon. (Visible all night.) |

# September 1ˢᵗ to 10ᵗʰ, 2023

## The Moon

| 1ˢᵗ | | 3ʳᵈ | | 5ᵗʰ | | 7ᵗʰ | | 9ᵗʰ |

| Date | Con | R.A. | Dec | Mag | Diam | Ill. | Elon. | Phase | Close To |
|------|-----|------|-----|-----|------|------|-------|-------|----------|
| 1st | Psc | 0h 0m | -2° 28' | -12.2 | 33' | 97% | 160° W | FM | Neptune |
| 2nd | Psc | 0h 52m | 4° 30' | -11.8 | 33' | 91% | 148° W | -G | |
| 3rd | Psc | 1h 44m | 11° 4' | -11.5 | 32' | 84% | 136° W | -G | |
| 4th | Ari | 2h 36m | 16° 52' | -11.1 | 32' | 75% | 124° W | -G | Jupiter, Uranus |
| 5th | Ari | 3h 30m | 21° 39' | -10.7 | 31' | 65% | 111° W | LQ | Jupiter, Uranus, Pleiades |
| 6th | Tau | 4h 24m | 25° 12' | -10.3 | 31' | 54% | 99° W | LQ | Pleiades, Hyades, Aldebaran |
| 7th | Tau | 5h 20m | 27° 24' | -9.9 | 30' | 44% | 86° W | LQ | Aldebaran |
| 8th | Aur | 6h 16m | 28° 11' | -9.4 | 30' | 35% | 73° W | -Cr | |
| 9th | Gem | 7h 10m | 27° 34' | -8.8 | 30' | 26% | 60° W | -Cr | |
| 10th | Cnc | 8h 3m | 25° 41' | -8.2 | 30' | 18% | 48° W | -Cr | Praesepe |

## Mercury and Venus

 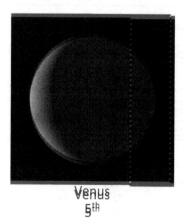

Mercury
5ᵗʰ

Venus
5ᵗʰ

### Mercury

| Date | Con. | R.A. | Dec. | Mag. | Diam. | Ill. | Elon. | Vis. | Rat. | Close To |
|------|------|------|------|------|-------|------|-------|------|------|----------|
| 1st | Leo | 11h 9m | 0° 33' | 2.2 | 11" | 5% | 7° E | NV | N/A | |
| 3rd | Leo | 11h 3m | 1° 24' | 2.6 | 11" | 3% | 4° E | NV | N/A | |
| 5th | Leo | 10h 56m | 2° 26' | 2.9 | 11" | 1% | 0° E | NV | N/A | |
| 7th | Sex | 10h 50m | 3° 34' | 2.9 | 10" | 1% | 3° W | NV | N/A | |
| 9th | Sex | 10h 44m | 4° 44' | 2.5 | 10" | 3% | 6° W | NV | N/A | Regulus |

## Venus

| Date | Con. | R.A. | Dec. | Mag. | Diam. | Ill. | Elon. | Vis. | Rat. | Close To |
|------|------|------|------|------|-------|------|-------|------|------|----------|
| 1st | Cnc | 8h 50m | 18° 3' | -4.0 | 50" | 12% | 28° W | AM | *** | Praesepe |
| 3rd | Cnc | 8h 50m | 18° 19' | -4.1 | 48" | 13% | 30° W | AM | *** | Praesepe |
| 5th | Cnc | 8h 51m | 18° 34' | -4.1 | 47" | 15% | 31° W | AM | *** | Praesepe |
| 7th | Cnc | 8h 52m | 18° 47' | -4.2 | 45" | 17% | 33° W | AM | *** | Praesepe |
| 9th | Cnc | 8h 53m | 18° 59' | -4.2 | 44" | 19% | 34° W | AM | *** | Praesepe |

## Mars and the Outer Planets

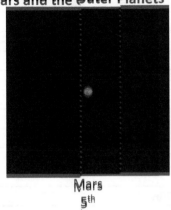

Mars
5th

Jupiter
5th

Saturn
5th

## Mars

| Date | Con. | R.A. | Dec. | Mag. | Diam. | Ill. | Elon. | Vis. | Rat. | Close To |
|------|------|------|------|------|-------|------|-------|------|------|----------|
| 1st | Vir | 12h 12m | 0° 41' | 2.0 | 4" | 98% | 23° E | PM | * | |
| 5th | Vir | 12h 22m | -1° 44' | 1.9 | 4" | 98% | 21° E | PM | * | |
| 10th | Vir | 12h 33m | -3° 4' | 1.9 | 4" | 99% | 20° E | PM | * | |

## The Outer Planets

| Planet | Date | Con. | R.A. | Dec. | Mag. | Diam. | Elon. | Vis. | Rat. | Close To |
|--------|------|------|------|------|------|-------|-------|------|------|----------|
| Jupiter | 5th | Ari | 2h 54m | 15° 14' | -2.2 | 45" | 121° W | AM | **** | Moon |
| Saturn | 5th | Aqr | 22h 22m | -12° 1' | 0.6 | 19" | 172° E | AN | **** | |
| Uranus | 5th | Ari | 3h 22m | 18° 14' | 6.0 | 4" | 113° W | AM | ** | Moon, Jupiter, Pleiades |
| Neptune | 5th | Psc | 23h 49m | -2° 30' | 7.6 | 2" | 167° W | AM | ***** | |

## Highlights

| Date | Time (UT) | Event |
|------|-----------|-------|
| 1st | 09:02 | The just-past full Moon is south of Neptune. (Visible all night.) |
| 2nd | 17:29 | Venus is stationary prior to beginning retrograde motion. (Evening sky.) |
| 4th | 18:38 | The waning gibbous Moon is north of Jupiter. (Morning sky.) |
| | 20:26 | Jupiter is stationary prior to beginning retrograde motion. (Morning sky.) |
| 5th | 10:35 | The nearly last quarter Moon is north of Uranus. (Morning sky.) |
| | 18:35 | The nearly last quarter Moon is south of the Pleiades star cluster. (Taurus, morning sky.) |
| 6th | 11:03 | Mercury is at inferior conjunction with the Sun. (Not visible.) |
| | 17:01 | The almost last quarter Moon is north of the bright star Aldebaran. (Taurus, morning sky.) |
| | 22:22 | Last Quarter Moon. (Morning sky.) |

# September 11<sup>th</sup> to 20<sup>th</sup>, 2023

## The Moon

 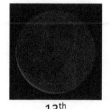   

| 11<sup>th</sup> | 13<sup>th</sup> | 15<sup>th</sup> | 17<sup>th</sup> | 19<sup>th</sup> |

| Date | Con | R.A. | Dec | Mag | Diam | Ill. | Elon. | Phase | Close To |
|------|-----|------|-----|-----|------|------|-------|-------|----------|
| 11th | Cnc | 8h 53m | 22° 40' | -7.5 | 29' | 11% | 36° W | NM | Venus, Praesepe |
| 12th | Leo | 9h 41m | 18° 45' | -6.7 | 29' | 6% | 25° W | NM | Venus, Regulus |
| 13th | Leo | 10h 27m | 14° 6' | -5.7 | 29' | 2% | 14° W | NM | Mercury, Regulus |
| 14th | Leo | 11h 10m | 8° 55' | -4.7 | 30' | 0% | 4° W | NM | Mercury |
| 15th | Vir | 11h 53m | 3° 23' | -4.5 | 30' | 0% | 5° E | NM | |
| 16th | Vir | 12h 35m | -2° 19' | -5.6 | 30' | 2% | 15° E | NM | Mars |
| 17th | Vir | 13h 19m | -7° 60' | -6.6 | 30' | 5% | 25° E | NM | Mars, Spica |
| 18th | Vir | 14h 4m | -13° 28' | -7.4 | 30' | 11% | 35° E | NM | Spica |
| 19th | Lib | 14h 52m | -18° 28' | -8.2 | 30' | 18% | 46° E | +Cr | |
| 20th | Lib | 15h 43m | -22° 45' | -8.9 | 31' | 26% | 58° E | +Cr | Antares |

## Mercury and Venus

Mercury
15<sup>th</sup>

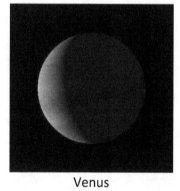

Venus
15<sup>th</sup>

### Mercury

| Date | Con. | R.A. | Dec. | Mag. | Diam. | Ill. | Elon. | Vis. | Rat. | Close To |
|------|------|------|------|------|-------|------|-------|------|------|----------|
| 11th | Sex | 10h 40m | 5° 51' | 2.0 | 10" | 6% | 9° W | NV | N/A | Regulus |
| 13th | Leo | 10h 37m | 6° 49' | 1.5 | 9" | 11% | 12° W | NV | N/A | Moon, Regulus |
| 15th | Leo | 10h 37m | 7° 34' | 1.1 | 9" | 18% | 14° W | NV | N/A | Regulus |
| 17th | Leo | 10h 38m | 8° 4' | 0.7 | 8" | 26% | 15° W | AM | ** | Regulus |
| 19th | Leo | 10h 42m | 8° 17' | 0.3 | 8" | 35% | 16° W | AM | ** | Regulus |

### Venus

| Date | Con. | R.A. | Dec. | Mag. | Diam. | Ill. | Elon. | Vis. | Rat. | Close To |
|------|------|------|------|------|-------|------|-------|------|------|----------|
| 11th | Cnc | 8h 55m | 11° 8' | -4.2 | 43" | 21% | 35° W | AM | *** | Moon, Praesepe |
| 13th | Cnc | 8h 58m | 11° 16' | -4.2 | 41" | 22% | 37° W | AM | *** | Praesepe |
| 15th | Cnc | 9h 1m | 11° 21' | -4.2 | 40" | 24% | 38° W | AM | *** | Praesepe |
| 17th | Cnc | 9h 5m | 11° 25' | -4.3 | 39" | 26% | 38° W | AM | *** | Praesepe |
| 19th | Cnc | 9h 9m | 11° 26' | -4.3 | 38" | 28% | 39° W | AM | *** | Praesepe |

## Mars and the Outer Planets

Mars
15th

Jupiter
15th

Saturn
15th

### Mars

| Date | Con. | R.A. | Dec. | Mag. | Diam. | Ill. | Elon. | Vis. | Rat. | Close To |
|------|------|------|------|------|-------|------|-------|------|------|----------|
| 11th | Vir | 12h 36m | -3° 20' | 1.9 | 4" | 99% | 20° E | PM | * | |
| 15th | Vir | 12h 45m | -4° 23' | 1.9 | 4" | 99% | 18° E | PM | * | Spica |
| 20th | Vir | 12h 57m | -5° 42' | 1.9 | 4" | 99% | 17° E | PM | * | Spica |

### The Outer Planets

| Planet | Date | Con. | R.A. | Dec. | Mag. | Diam. | Elon. | Vis. | Rat. | Close To |
|--------|------|------|------|------|------|-------|-------|------|------|----------|
| Jupiter | 15th | Ari | 2h 53m | 15° 9' | -2.3 | 46" | 130° W | AM | **** | |
| Saturn | 15th | Aqr | 22h 20m | -12° 17' | 0.7 | 19" | 162° E | PM | **** | |
| Uranus | 15th | Ari | 3h 22m | 18° 12' | 6.0 | 4" | 122° W | AM | ** | Jupiter, Pleiades |
| Neptune | 15th | Psc | 23h 48m | -2° 37' | 7.6 | 2" | 176° W | AN | **** | |

## Highlights

| Date | Time (UT) | Event |
|------|-----------|-------|
| 11th | 03:18 | The waning crescent Moon is north of the Praesepe star cluster. (Cancer, morning sky.) |
| | N/A | Good opportunity to see Earthshine on the waning crescent Moon. (Morning sky.) |
| 13th | 00:54 | The waning crescent Moon is north of the bright star Regulus. (Leo, morning sky.) |
| | 19:20 | The waning crescent Moon is north of Mercury. (Morning sky.) |
| 15th | 00:09 | Mercury is stationary prior to beginning prograde motion. (Morning sky.) |
| | 01:40 | New Moon. (Not visible) |
| 17th | 16:33 | The waxing crescent Moon is north of the bright star Spica. (Virgo, evening sky.) |
| 18th | N/A | Good opportunity to see Earthshine on the waxing crescent Moon. (Evening sky.) |
| 20th | 00:30 | Neptune is at opposition. (Visible all night.) |

# September 21st to 30th, 2023

## The Moon

21st

23rd

25th

27th

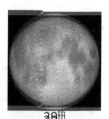
29th

| Date | Con | R.A. | Dec | Mag | Diam | Ill. | Elon. | Phase | Close To |
|------|-----|------|-----|-----|------|------|-------|-------|----------|
| 21st | Sco | 16h 39m | -26° 0' | -9.5 | 31' | 36% | 71° E | +Cr | Antares |
| 22nd | Oph | 17h 38m | -27° 54' | -10.0 | 32' | 47% | 85° E | FQ | |
| 23rd | Sgr | 18h 40m | -28° 9' | -10.4 | 32' | 58% | 100° E | FQ | |
| 24th | Sgr | 19h 43m | -26° 37' | -10.9 | 32' | 69% | 115° E | +G | |
| 25th | Cap | 20h 45m | -23° 19' | -11.3 | 33' | 79% | 129° E | +G | |
| 26th | Cap | 21h 45m | -18° 28' | -11.7 | 33' | 88% | 143° E | +G | Saturn |
| 27th | Aqr | 22h 41m | -12° 26' | -12.0 | 33' | 95% | 157° E | +G | Saturn |
| 28th | Aqr | 23h 36m | -5° 39' | -12.4 | 33' | 99% | 169° E | FM | Neptune |
| 29th | Cet | 0h 29m | 1° 25' | -12.7 | 33' | 100% | 178° W | FM | Neptune |
| 30th | Psc | 1h 21m | 8° 19' | -12.3 | 33' | 98% | 166° W | FM | |

## Mercury and Venus

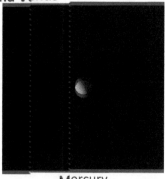
Mercury
25th

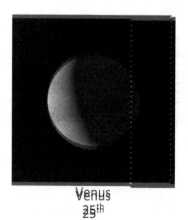
Venus
25th

### Mercury

| Date | Con. | R.A. | Dec. | Mag. | Diam. | Ill. | Elon. | Vis. | Rat. | Close To |
|------|------|------|------|------|-------|------|-------|------|------|----------|
| 21st | Leo | 10h 48m | 8° 12' | 0.0 | 7" | 44% | 16° W | AM | ** | |
| 23rd | Leo | 10h 56m | 7° 50' | -0.3 | 7" | 53% | 16° W | AM | ** | |
| 25th | Leo | 11h 6m | 7° 13' | -0.5 | 7" | 62% | 15° W | AM | ** | |
| 27th | Leo | 11h 16m | 6° 22' | -0.7 | 6" | 70% | 15° W | NV | N/A | |
| 29th | Leo | 11h 28m | 5° 19' | -0.8 | 6" | 77% | 13° W | NV | N/A | |

## Venus

| Date | Con. | R.A. | Dec. | Mag. | Diam. | Ill. | Elon. | Vis. | Rat. | Close To |
|------|------|------|------|------|-------|------|-------|------|------|----------|
| 21st | Cnc | 9h 14m | 11° 25' | -4.3 | 37" | 29% | 40° W | AM | *** | Praesepe |
| 23rd | Cnc | 9h 18m | 11° 22' | -4.2 | 36" | 31% | 40° W | AM | *** | Praesepe |
| 25th | Leo | 9h 23m | 11° 17' | -4.2 | 35" | 32% | 41° W | AM | *** | |
| 27th | Leo | 9h 29m | 11° 9' | -4.2 | 34" | 34% | 41° W | AM | *** | Regulus |
| 29th | Leo | 9h 35m | 10° 60' | -4.2 | 33" | 35% | 42° W | AM | *** | Regulus |

## Mars and the Outer Planets

Mars
25th

Jupiter
25th

Saturn
25th

## Mars

| Date | Con. | R.A. | Dec. | Mag. | Diam. | Ill. | Elon. | Vis. | Rat. | Close To |
|------|------|------|------|------|-------|------|-------|------|------|----------|
| 21st | Vir | 13h 0m | -5° 58' | 1.9 | 4" | 99% | 17° E | PM | * | Spica |
| 25th | Vir | 13h 10m | -7° 0' | 1.9 | 4" | 99% | 16° E | PM | * | Spica |
| 30th | Vir | 13h 22m | -8° 18' | 1.9 | 4" | 99% | 14° E | NV | N/A | Spica |

## The Outer Planets

| Planet | Date | Con. | R.A. | Dec. | Mag. | Diam. | Elon. | Vis. | Rat. | Close To |
|--------|------|------|------|------|------|-------|-------|------|------|----------|
| Jupiter | 25th | Ari | 2h 51m | 14° 58' | -2.3 | 47" | 139° W | AM | **** | |
| Saturn | 25th | Aqr | 22h 17m | -12° 31' | 0.7 | 19" | 152° E | PM | **** | |
| Uranus | 25th | Ari | 3h 21m | 18° 9' | 6.0 | 4" | 132° W | AM | ** | Jupiter, Pleiades |
| Neptune | 25th | Psc | 23h 47m | -2° 43' | 7.6 | 2" | 175° E | AN | ***** | |

## Highlights

| Date | Time (UT) | Event |
|------|-----------|-------|
| 21st | 06:50 | The nearly first quarter Moon is north of the bright star Antares. (Scorpius, evening sky.) |
| 22nd | 13:08 | Mercury is at greatest western elongation from the Sun. (Morning sky.) |
| | 19:33 | First Quarter Moon. (Evening sky.) |
| 23rd | 06:51 | Autumnal Equinox. |
| 27th | 03:14 | The waxing gibbous Moon is south of Saturn. (Evening sky.) |
| 28th | 15:23 | The waxing gibbous Moon is south of Neptune. (Evening sky.) |
| 29th | 09:58 | Full Moon. (Visible all night.) |

# October 1st to 10th, 2023

## The Moon

1st     3rd     5th     7th     9th

| Date | Con | R.A. | Dec | Mag | Diam | Ill. | Elon. | Phase | Close To |
|------|-----|------|-----|-----|------|------|-------|-------|----------|
| 1st | Ari | 2h 14m | 14° 38' | -12.0 | 32' | 94% | 154° W | -G | Jupiter |
| 2nd | Ari | 3h 9m | 20° 1' | -11.6 | 32' | 88% | 141° W | -G | Jupiter, Uranus, Pleiades |
| 3rd | Tau | 4h 4m | 24° 11' | -11.3 | 31' | 79% | 128° W | -G | Uranus, Pleiades, Hyades, Aldeba |
| 4th | Tau | 5h 1m | 26° 57' | -10.9 | 31' | 70% | 115° W | -G | Hyades, Aldebaran |
| 5th | Tau | 5h 58m | 28° 13' | -10.6 | 30' | 61% | 101° W | LQ | |
| 6th | Gem | 6h 54m | 28° 0' | -10.2 | 30' | 51% | 88° W | LQ | |
| 7th | Gem | 7h 48m | 26° 27' | -9.7 | 30' | 41% | 76° W | LQ | |
| 8th | Cnc | 8h 40m | 23° 43' | -9.2 | 30' | 32% | 64° W | -Cr | Praesepe |
| 9th | Leo | 9h 28m | 20° 0' | -8.7 | 29' | 23% | 53° W | -Cr | Venus, Regulus |
| 10th | Leo | 10h 14m | 15° 31' | -8.0 | 29' | 16% | 42° W | -Cr | Venus, Regulus |

## Mercury and Venus

Mercury
5th

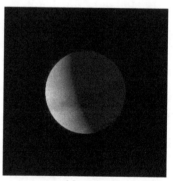

Venus
5th

### Mercury

| Date | Con. | R.A. | Dec. | Mag. | Diam. | Ill. | Elon. | Vis. | Rat. | Close To |
|------|------|------|------|------|-------|------|-------|------|------|----------|
| 1st | Vir | 11h 40m | 4° 7' | -0.9 | 6" | 83% | 12° W | NV | N/A | |
| 3rd | Vir | 11h 53m | 2° 47' | -1.0 | 5" | 87% | 11° W | NV | N/A | |
| 5th | Vir | 12h 6m | 1° 23' | -1.1 | 5" | 91% | 9° W | NV | N/A | |
| 7th | Vir | 12h 19m | 0° 4' | -1.1 | 5" | 94% | 8° W | NV | N/A | |
| 9th | Vir | 12h 31m | -1° 34' | -1.1 | 5" | 96% | 7° W | NV | N/A | |

**Venus**

| Date | Con. | R.A. | Dec. | Mag. | Diam. | Ill. | Elon. | Vis. | Rat. | Close To |
|------|------|------|------|------|-------|------|-------|------|------|----------|
| 1st | Leo | 9h 41m | 10° 48' | -4.2 | 32" | 37% | 42° W | AM | *** | Regulus |
| 3rd | Leo | 9h 47m | 10° 34' | -4.2 | 31" | 38% | 42° W | AM | *** | Regulus |
| 5th | Leo | 9h 53m | 10° 17' | -4.2 | 30" | 40% | 43° W | AM | *** | Regulus |
| 7th | Leo | 10h 0m | 9° 59' | -4.2 | 29" | 41% | 43° W | AM | *** | Regulus |
| 9th | Leo | 10h 7m | 9° 38' | -4.1 | 29" | 42% | 43° W | AM | *** | Moon, Regulus |

## Mars and the Outer Planets

| Mars | Jupiter | Saturn |
|------|---------|--------|
| 5th | 5th | 5th |

**Mars**

| Date | Con. | R.A. | Dec. | Mag. | Diam. | Ill. | Elon. | Vis. | Rat. | Close To |
|------|------|------|------|------|-------|------|-------|------|------|----------|
| 1st | Vir | 13h 25m | -8° 33' | 1.9 | 4" | 99% | 14° E | NV | N/A | Spica |
| 5th | Vir | 13h 35m | -9° 34' | 1.9 | 4" | 99% | 13° E | NV | N/A | Spica |
| 10th | Vir | 13h 47m | -10° 49' | 1.8 | 4" | 100% | 11° E | NV | N/A | Spica |

**The Outer Planets**

| Planet | Date | Con. | R.A. | Dec. | Mag. | Diam. | Elon. | Vis. | Rat. | Close To |
|--------|------|------|------|------|------|-------|-------|------|------|----------|
| Jupiter | 5th | Ari | 2h 48m | 14° 42' | -2.4 | 48" | 149° W | AM | **** | |
| Saturn | 5th | Aqr | 22h 15m | -12° 42' | 0.8 | 19" | 143° E | PM | **** | |
| Uranus | 5th | Ari | 3h 20m | 18° 5' | 6.0 | 4" | 141° W | AM | ** | Jupiter, Pleiades |
| Neptune | 5th | Psc | 23h 46m | -2° 50' | 7.6 | 2" | 166° E | PM | ***** | |

## Highlights

| Date | Time (UT) | Event |
|------|-----------|-------|
| 2nd | 04:20 | The waning gibbous Moon is north of Jupiter. (Morning sky.) |
| | 16:26 | The waning gibbous Moon is north of Uranus. (Morning sky.) |
| 3rd | 05:31 | The waning gibbous Moons is south of the Pleiades star cluster. (Taurus, morning sky.) |
| | 23:36 | The waning gibbous Moon is north of the bright star Aldebaran. (Taurus, morning sky.) |
| 6th | 13:48 | Last Quarter Moon. (Morning sky.) |
| 8th | 13:49 | The waning crescent Moon is north of the Praesepe star cluster. (Cancer, morning sky.) |
| | N/A | The Draconid meteor shower is at its maximum. (ZHR: Variable.) |
| 9th | 19:50 | Venus is 2.4° south of the bright star Regulus. (Leo, morning sky.) |
| 10th | 08:26 | The waning crescent Moon is north of the bright star Regulus. (Leo, morning sky.) |
| | 10:45 | The waning crescent Moon is north of Venus. (Leo, morning sky.) |
| | 11:08 | Pluto is stationary prior to resuming prograde motion. (Evening sky.) |

# October 11<sup>th</sup> to 20<sup>th</sup>, 2023

## The Moon

11<sup>th</sup>  13<sup>th</sup>  15<sup>th</sup>  17<sup>th</sup>  19<sup>th</sup>

| Date | Con | R.A. | Dec | Mag | Diam | Ill. | Elon. | Phase | Close To |
|------|-----|------|-----|-----|------|------|-------|-------|----------|
| 11th | Leo | 10h 58m | 18° 27' | -7.3 | 30' | 10% | 32° W | NM | Venus |
| 12th | Vir | 11h 41m | 4° 58' | -6.4 | 30' | 5% | 22° W | NM | |
| 13th | Vir | 12h 24m | 0° 45' | -5.4 | 30' | 2% | 12° W | NM | Mercury |
| 14th | Vir | 13h 7m | -6° 31' | -4.2 | 30' | 0% | 2° W | NM | Mercury, Spica |
| 15th | Vir | 13h 52m | -12° 8' | -4.8 | 30' | 1% | 8° E | NM | Mercury, Mars, Spica |
| 16th | Lib | 14h 40m | -17° 21' | -6.0 | 31' | 3% | 19° E | NM | Mars |
| 17th | Lib | 15h 31m | -21° 53' | -7.0 | 31' | 8% | 31° E | NM | |
| 18th | Sco | 16h 26m | -25° 25' | -7.8 | 31' | 14% | 44° E | +Cr | Antares |
| 19th | Oph | 17h 24m | -27° 39' | -8.6 | 31' | 22% | 57° E | +Cr | |
| 20th | Sgr | 18h 25m | -28° 18' | -9.2 | 32' | 32% | 72° E | +Cr | |

## Mercury and Venus

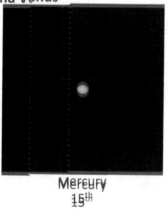
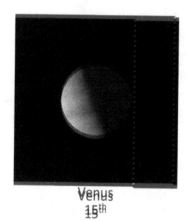

Mercury
15<sup>th</sup>

Venus
15<sup>th</sup>

### Mercury

| Date | Con. | R.A. | Dec. | Mag. | Diam. | Ill. | Elon. | Vis. | Rat. | Close To |
|------|------|------|------|------|-------|------|-------|------|------|----------|
| 11th | Vir | 12h 44m | -3° 5' | -1.1 | 5" | 98% | 5° W | NV | N/A | |
| 13th | Vir | 12h 57m | -4° 35' | -1.1 | 5" | 99% | 4° W | NV | N/A | Moon, Spica |
| 15th | Vir | 13h 10m | -6° 5' | -1.1 | 5" | 99% | 3° W | NV | N/A | Moon, Spica |
| 17th | Vir | 13h 22m | -7° 33' | -1.1 | 5" | 100% | 1° W | NV | N/A | Spica |
| 19th | Vir | 13h 35m | -8° 59' | -1.1 | 5" | 100% | 0° W | NV | N/A | Mars, Spica |

## Venus

| Date | Con. | R.A. | Dec. | Mag. | Diam. | Ill. | Elon. | Vis. | Rat. | Close To |
|------|------|------|------|------|-------|------|-------|------|------|----------|
| 11th | Leo | 10h 14m | 9° 16' | -4.1 | 28" | 43% | 43° W | AM | *** | Moon, Regulus |
| 13th | Leo | 10h 21m | 8° 51' | -4.1 | 27" | 45% | 43° W | AM | *** | Regulus |
| 15th | Leo | 10h 28m | 8° 25' | -4.1 | 26" | 46% | 43° W | AM | *** | Regulus |
| 17th | Leo | 10h 35m | 7° 56' | -4.1 | 26" | 47% | 43° W | AM | *** | Regulus |
| 19th | Leo | 10h 43m | 7° 26' | -4.1 | 25" | 48% | 43° W | AM | *** | Regulus |

## Mars and the Outer Planets

Mars
15th

Jupiter
15th

Saturn
15th

## Mars

| Date | Con. | R.A. | Dec. | Mag. | Diam. | Ill. | Elon. | Vis. | Rat. | Close To |
|------|------|------|------|------|-------|------|-------|------|------|----------|
| 11th | Vir | 13h 50m | -11° 4' | 1.8 | 4" | 100% | 11° E | NV | N/A | Spica |
| 15th | Vir | 14h 0m | -12° 3' | 1.8 | 4" | 100% | 10° E | NV | N/A | Moon, Spica |
| 20th | Vir | 14h 13m | -13° 15' | 1.8 | 4" | 100% | 9° E | NV | N/A | Mercury |

## The Outer Planets

| Planet | Date | Con. | R.A. | Dec. | Mag. | Diam. | Elon. | Vis. | Rat. | Close To |
|--------|------|------|------|------|------|-------|-------|------|------|----------|
| Jupiter | 15th | Ari | 2h 43m | 14° 22' | -2.4 | 49" | 159° W | AM | ***** | |
| Saturn | 15th | Aqr | 22h 14m | -12° 50' | 0.8 | 18" | 133° E | PM | *** | |
| Uranus | 15th | Ari | 3h 19m | 18° 1' | 6.0 | 4" | 150° W | AM | *** | Jupiter, Pleiades |
| Neptune | 15th | Psc | 23h 45m | -3° 56' | 7.6 | 2" | 156° E | PM | ***** | |

## Highlights

| Date | Time (UT) | Event |
|------|-----------|-------|
| 11th | N/A | Good opportunity to see Earthshine on the waning crescent Moon. (Morning sky.) |
| 14th | 17:56 | New Moon |
| | 18:01 | Annular solar eclipse. Visible from Western Africa, the Atlantic, North America, the Pacific and South America. |
| 17th | N/A | Good opportunity to see Earthshine on the waxing crescent Moon. (Evening sky.) |
| 18th | 12:09 | The waxing crescent Moon is north of the bright star Antares. (Scorpius, evening sky.) |
| 20th | 05:22 | Mercury is at superior conjunction with the Sun. (Not visible.) |

# October 21ˢᵗ to 31ˢᵗ, 2023

## The Moon

| 21ˢᵗ | 23ʳᵈ | 25ᵗʰ | 27ᵗʰ | 29ᵗʰ | 31ˢᵗ |

| Date | Con | R.A. | Dec | Mag | Diam | Ill. | Elon. | Phase | Close To |
|------|-----|------|-----|-----|------|------|-------|-------|----------|
| 21st | Sgr | 19h 27m | -27° 14' | -9.8 | 32' | 43% | 86° E | FQ | |
| 22nd | Cap | 20h 28m | -24° 28' | -10.3 | 32' | 54% | 100° E | FQ | |
| 23rd | Cap | 21h 26m | -20° 10' | -10.7 | 32' | 65% | 114° E | FQ | Saturn |
| 24th | Aqr | 22h 22m | -14° 40' | -11.2 | 33' | 76% | 127° E | +G | Saturn |
| 25th | Aqr | 23h 15m | -8° 18' | -11.5 | 33' | 85% | 139° E | +G | Neptune |
| 26th | Psc | 0h 7m | -1° 29' | -11.9 | 33' | 93% | 151° E | +G | Neptune |
| 27th | Psc | 0h 59m | 5° 25' | -12.3 | 33' | 98% | 163° E | FM | |
| 28th | Ari | 1h 51m | 11° 59' | -12.6 | 32' | 100% | 175° E | FM | Jupiter |
| 29th | Ari | 2h 45m | 17° 49' | -12.5 | 32' | 99% | 172° W | FM | Jupiter, Uranus |
| 30th | Tau | 3h 41m | 22° 35' | -12.2 | 32' | 96% | 159° W | FM | Uranus, Pleiades, Hyades |
| 31st | Tau | 4h 38m | 26° 0' | -11.8 | 31' | 91% | 146° W | -G | Hyades, Aldebaran |

## Mercury and Venus

Mercury
25ᵗʰ

Venus
25ᵗʰ

**Mercury**

| Date | Con. | R.A. | Dec. | Mag. | Diam. | Ill. | Elon. | Vis. | Rat. | Close To |
|------|------|------|------|------|-------|------|-------|------|------|----------|
| 21st | Vir | 13h 47m | -10° 24' | -1.0 | 5" | 100% | 1° E | NV | N/A | Mars, Spica |
| 23rd | Vir | 13h 59m | -11° 46' | -0.9 | 5" | 100% | 2° E | NV | N/A | Mars, Spica |
| 25th | Vir | 14h 12m | -13° 5' | -0.8 | 5" | 100% | 3° E | NV | N/A | Mars |
| 27th | Lib | 14h 24m | -14° 22' | -0.7 | 5" | 99% | 5° E | NV | N/A | Mars |
| 29th | Lib | 14h 36m | -15° 35' | -0.7 | 5" | 99% | 6° E | NV | N/A | Mars |
| 31st | Lib | 14h 49m | -16° 45' | -0.6 | 5" | 98% | 7° E | NV | N/A | Mars |

## Venus

| Date | Con. | R.A. | Dec. | Mag. | Diam. | Ill. | Elon. | Vis. | Rat. | Close To |
|------|------|------|------|------|-------|------|-------|------|------|----------|
| 21st | Leo | 10h 50m | 6° 54' | -4.0 | 25" | 49% | 43° W | AM | *** | |
| 23rd | Leo | 10h 58m | 6° 21' | -4.0 | 24" | 50% | 43° W | AM | *** | |
| 25th | Leo | 11h 6m | 5° 46' | -4.0 | 24" | 51% | 43° W | AM | *** | |
| 27th | Leo | 11h 14m | 5° 9' | -4.0 | 23" | 52% | 43° W | AM | *** | |
| 29th | Leo | 11h 21m | 4° 31' | -4.0 | 23" | 53% | 43° W | AM | *** | |
| 31st | Leo | 11h 29m | 3° 51' | -4.0 | 22" | 54% | 43° W | AM | *** | |

## Mars and the Outer Planets

Mars
25th

Jupiter
25th

Saturn
25th

### Mars

| Date | Con. | R.A. | Dec. | Mag. | Diam. | Ill. | Elon. | Vis. | Rat. | Close To |
|------|------|------|------|------|-------|------|-------|------|------|----------|
| 21st | Vir | 14h 16m | -13° 29' | 1.8 | 4" | 100% | 8° E | NV | N/A | Mercury |
| 25th | Lib | 14h 26m | -14° 24' | 1.8 | 4" | 100% | 7° E | NV | N/A | Mercury |
| 31st | Lib | 14h 43m | -15° 45' | 1.7 | 4" | 100% | 5° E | NV | N/A | Mercury |

### The Outer Planets

| Planet | Date | Con. | R.A. | Dec. | Mag. | Diam. | Elon. | Vis. | Rat. | Close To |
|--------|------|------|------|------|------|-------|-------|------|------|----------|
| Jupiter | 25th | Ari | 2h 38m | 13° 59' | -2.4 | 49" | 170° W | AM | ***** | |
| Saturn | 25th | Aqr | 22h 13m | -12° 54' | 0.9 | 18" | 124° E | PM | *** | |
| Uranus | 25th | Ari | 3h 18m | 17° 55' | 6.0 | 4" | 160° W | AM | *** | Jupiter, Pleiades |
| Neptune | 25th | Psc | 23h 44m | -3° 1' | 7.6 | 2" | 147° E | PM | ***** | Moon |

## Highlights

| Date | Time (UT) | Event |
|------|-----------|-------|
| 22nd | 03:30 | First Quarter Moon. (Evening sky.) |
| | N/A | The Orionid meteor shower is at its maximum. (ZHR: 15) |
| 23rd | 22:59 | Venus is at greatest western elongation from the Sun. (Morning sky.) |
| 24th | 07:59 | The waxing gibbous Moon is south of Saturn. (Evening sky.) |
| 26th | 03:17 | The waxing gibbous Moon is south of Neptune. (Evening sky.) |
| 28th | 20:14 | Partial lunar eclipse. Visible from Africa, Asia, the Atlantic, Australia, Europe, north-eastern North America, the Pacific and north-eastern South America |
| | 20:25 | Full Moon. (Visible all night.) |
| 29th | 09:25 | The just-past full Moon is north of Jupiter. (Visible all night.) |
| 30th | 02:47 | The waning gibbous Moon is north of Uranus. (Morning sky.) |
| 31st | 11:49 | The waning gibbous Moon is north of the bright star Aldebaran. (Taurus, morning sky.) |

# November 1st to 10th, 2023

## The Moon

    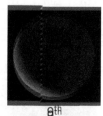

| 1st | 3rd | 5th | 7th | 9th |

| Date | Con | R.A. | Dec | Mag | Diam | Ill. | Elon. | Phase | Close To |
|------|-----|------|-----|-----|------|------|-------|-------|----------|
| 1st | Tau | 5h 36m | 27° 54' | -11.5 | 31' | 85% | 132° W | -G | |
| 2nd | Aur | 6h 34m | 28° 14' | -11.2 | 30' | 76% | 119° W | -G | |
| 3rd | Gem | 7h 30m | 27° 7' | -10.8 | 30' | 68% | 106° W | -G | |
| 4th | Cnc | 8h 23m | 24° 43' | -10.5 | 30' | 58% | 93° W | LQ | Praesepe |
| 5th | Cnc | 9h 13m | 21° 16' | -10.1 | 30' | 49% | 82° W | LQ | Praesepe |
| 6th | Leo | 10h 0m | 16° 60' | -9.6 | 30' | 39% | 71° W | LQ | Regulus |
| 7th | Leo | 10h 44m | 12° 6' | -9.1 | 30' | 30% | 61° W | -Cr | Regulus |
| 8th | Leo | 11h 27m | 6° 45' | -8.6 | 30' | 22% | 51° W | -Cr | Venus |
| 9th | Vir | 12h 10m | 1° 6' | -7.9 | 30' | 15% | 43° W | -Cr | Venus |
| 10th | Vir | 12h 53m | -4° 40' | -7.1 | 30' | 8% | 32° W | NM | Venus, Spica |

## Mercury and Venus

Mercury
5th

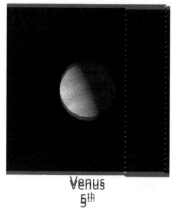

Venus
5th

### Mercury

| Date | Con. | R.A. | Dec. | Mag. | Diam. | Ill. | Elon. | Vis. | Rat. | Close To |
|------|------|------|------|------|-------|------|-------|------|------|----------|
| 1st | Lib | 14h 55m | -17° 19' | -0.6 | 5" | 98% | 7° E | NV | N/A | Mars |
| 3rd | Lib | 15h 7m | -18° 24' | -0.5 | 5" | 97% | 9° E | NV | N/A | Mars |
| 5th | Lib | 15h 20m | -19° 25' | -0.5 | 5" | 96% | 10° E | NV | N/A | Mars |
| 7th | Lib | 15h 32m | -20° 23' | -0.4 | 5" | 96% | 11° E | NV | N/A | Mars |
| 9th | Lib | 15h 45m | -21° 16' | -0.4 | 5" | 95% | 12° E | NV | N/A | Mars |

## Venus

| Date | Con. | R.A. | Dec. | Mag. | Diam. | Ill. | Elon. | Vis. | Rat. | Close To |
|------|------|------|------|------|-------|------|-------|------|------|----------|
| 1st | Leo | 11h 33m | 3° 31' | -3.9 | 22" | 55% | 43° W | AM | *** | |
| 3rd | Vir | 11h 41m | 2° 50' | -3.9 | 22" | 56% | 43° W | AM | ** | |
| 5th | Vir | 11h 50m | 2° 7' | -3.9 | 21" | 57% | 43° W | AM | ** | |
| 7th | Vir | 11h 58m | 1° 24' | -3.9 | 21" | 58% | 43° W | AM | ** | |
| 9th | Vir | 12h 6m | 0° 39' | -3.9 | 20" | 59% | 43° W | AM | ** | Moon |

## Mars and the Outer Planets

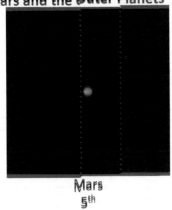

Mars
5th

Jupiter
5th

Saturn
5th

## Mars

| Date | Con. | R.A. | Dec. | Mag. | Diam. | Ill. | Elon. | Vis. | Rat. | Close To |
|------|------|------|------|------|-------|------|-------|------|------|----------|
| 1st | Lib | 14h 45m | -15° 58' | 1.7 | 4" | 100% | 5° E | NV | N/A | Mercury |
| 5th | Lib | 14h 56m | -16° 49' | 1.7 | 4" | 100% | 4° E | NV | N/A | Mercury |
| 10th | Lib | 15h 10m | -17° 49' | 1.7 | 4" | 100% | 2° E | NV | N/A | |

## The Outer Planets

| Planet | Date | Con. | R.A. | Dec. | Mag. | Diam. | Elon. | Vis. | Rat. | Close To |
|--------|------|------|------|------|------|-------|-------|------|------|----------|
| Jupiter | 5th | Ari | 2h 32m | 13° 32' | -2.4 | 49" | 178° E | AN | ***** | |
| Saturn | 5th | Aqr | 22h 12m | -12° 55' | 0.9 | 18" | 113° E | PM | *** | |
| Uranus | 5th | Ari | 3h 16m | 17° 48' | 6.0 | 4" | 171° W | AN | *** | Pleiades |
| Neptune | 5th | Psc | 23h 44m | -3° 5' | 7.7 | 2" | 136° E | PM | ***** | |

## Highlights

| Date | Time (UT) | Event |
|------|-----------|-------|
| 3rd | 14:34 | Jupiter is at opposition. (Visible all night.) |
| 4th | 15:49 | Saturn is stationary prior to resuming prograde motion. (Evening sky.) |
| | 19:05 | The nearly last quarter Moon is north of the Praesepe star cluster. (Cancer, morning sky.) |
| 5th | 08:37 | Last Quarter Moon. (Morning sky.) |
| 6th | 17:14 | The just-past last quarter Moon is north of the bright star Regulus. (Leo, morning sky.) |
| 8th | N/A | The Andromedid meteor shower is at its maximum. (ZHR: Var) |
| 9th | 10:17 | The waning crescent Moon is north of Venus. (Morning sky.) |
| | N/A | Good opportunity to see Earthshine on the waning crescent Moon. (Morning sky.) |

# November 11<sup>th</sup> to 20<sup>th</sup>, 2023

## The Moon

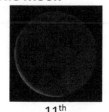
11<sup>th</sup>

13<sup>th</sup>

15<sup>th</sup>

17<sup>th</sup>

19<sup>th</sup>

| Date | Con | R.A. | Dec | Mag | Diam | Ill. | Elon. | Phase | Close To |
|------|-----|------|-----|-----|------|------|-------|-------|----------|
| 11th | Vir | 13h 38m | -10° 22' | -6.2 | 30' | 4% | 22° W | NM | Spica |
| 12th | Lib | 14h 25m | -15° 47' | -5.0 | 31' | 1% | 11° W | NM | |
| 13th | Lib | 15h 16m | -20° 37' | -4.2 | 31' | 0% | 1° E | NM | Mars |
| 14th | Sco | 16h 10m | -24° 32' | -5.4 | 31' | 1% | 13° E | NM | Mercury, Antares |
| 15th | Oph | 17h 9m | -27° 10' | -6.5 | 32' | 5% | 27° E | NM | Mercury, Antares |
| 16th | Sgr | 18h 11m | -28° 14' | -7.5 | 32' | 11% | 41° E | NM | |
| 17th | Sgr | 19h 13m | -27° 33' | -8.3 | 32' | 19% | 56° E | +Cr | |
| 18th | Cap | 20h 14m | -25° 8' | -9.0 | 32' | 29% | 70° E | +Cr | |
| 19th | Cap | 21h 13m | -21° 11' | -9.6 | 32' | 39% | 84° E | FQ | |
| 20th | Aqr | 22h 8m | -16° 1' | -10.2 | 32' | 51% | 97° E | FQ | Saturn |

## Mercury and Venus

Mercury
15<sup>th</sup>

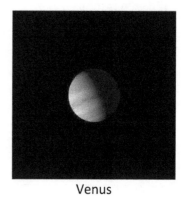
Venus
15<sup>th</sup>

**Mercury**

| Date | Con. | R.A. | Dec. | Mag. | Diam. | Ill. | Elon. | Vis. | Rat. | Close To |
|------|------|------|------|------|-------|------|-------|------|------|----------|
| 11th | Sco | 15h 57m | -22° 6' | -0.4 | 5" | 94% | 13° E | NV | N/A | Antares |
| 13th | Sco | 16h 10m | -22° 51' | -0.4 | 5" | 93% | 14° E | NV | N/A | Antares |
| 15th | Sco | 16h 22m | -23° 31' | -0.3 | 5" | 91% | 15° E | PM | ** | Moon, Antares |
| 17th | Oph | 16h 35m | -24° 7' | -0.3 | 5" | 90% | 16° E | PM | ** | Antares |
| 19th | Oph | 16h 48m | -24° 38' | -0.3 | 5" | 88% | 17° E | PM | ** | Antares |

## Venus

| Date | Con. | R.A. | Dec. | Mag. | Diam. | Ill. | Elon. | Vis. | Rat. | Close To |
|------|------|------|------|------|-------|------|-------|------|------|----------|
| 11th | Vir | 12h 14m | 0° 6' | -3.9 | 20" | 60% | 43° W | AM | ** | |
| 13th | Vir | 12h 22m | 0° 52' | -3.8 | 20" | 61% | 43° W | AM | ** | |
| 15th | Vir | 12h 31m | -1° 39' | -3.8 | 19" | 61% | 43° W | AM | ** | |
| 17th | Vir | 12h 39m | -2° 27' | -3.8 | 19" | 62% | 43° W | AM | ** | |
| 19th | Vir | 12h 48m | -3° 14' | -3.8 | 19" | 63% | 43° W | AM | ** | Spica |

## Mars and the Outer Planets

Mars
15th

Jupiter
15th

Saturn
15th

### Mars

| Date | Con. | R.A. | Dec. | Mag. | Diam. | Ill. | Elon. | Vis. | Rat. | Close To |
|------|------|------|------|------|-------|------|-------|------|------|----------|
| 11th | Lib | 15h 13m | -18° 1' | 1.7 | 4" | 100% | 2° E | NV | N/A | |
| 15th | Lib | 15h 25m | -18° 47' | 1.7 | 4" | 100% | 1° E | NV | N/A | |
| 20th | Lib | 15h 39m | -19° 40' | 1.6 | 4" | 100% | 1° W | NV | N/A | |

### The Outer Planets

| Planet | Date | Con. | R.A. | Dec. | Mag. | Diam. | Elon. | Vis. | Rat. | Close To |
|--------|------|------|------|------|------|-------|-------|------|------|----------|
| Jupiter | 15th | Ari | 2h 27m | 13° 8' | -2.4 | 49" | 166° E | PM | ***** | |
| Saturn | 15th | Aqr | 22h 13m | -12° 51' | 1.0 | 17" | 103° E | PM | *** | |
| Uranus | 15th | Ari | 3h 14m | 17° 42' | 6.0 | 4" | 178° E | AN | *** | Pleiades |
| Neptune | 15th | Psc | 23h 43m | -3° 9' | 7.7 | 2" | 125° E | PM | ***** | |

## Highlights

| Date | Time (UT) | Event |
|------|-----------|-------|
| 11th | 03:11 | The waning crescent Moon is north of the bright star Spica. (Virgo, morning sky.) |
| 13th | 09:28 | New Moon. (Not visible.) |
| | 16:58 | Uranus is at opposition. (Visible all night.) |
| 14th | 16:14 | The waxing crescent Moon is south of Mercury. (Evening sky.) |
| 16th | N/A | Good opportunity to see Earthshine on the waxing crescent Moon. (Evening sky.) |
| 18th | 03:40 | Mars is in conjunction with the Sun. (Not visible.) |
| | N/A | The Leonid meteor shower is at its maximum. (ZHR: 15) |
| 20th | 10:51 | First Quarter Moon. (Evening sky.) |
| | 12:16 | The first quarter Moon is south of Saturn. (Evening sky.) |
| | 18:19 | Dwarf planet Ceres is in conjunction with the Sun. (Not visible.) |

# November 21ˢᵗ to 30ᵗʰ, 2023

## The Moon

| 21ˢᵗ | 23ʳᵈ | 25ᵗʰ | 27ᵗʰ | 29ᵗʰ |

| Date | Con | R.A. | Dec | Mag | Diam | Ill. | Elon. | Phase | Close To |
|------|-----|------|-----|-----|------|------|-------|-------|----------|
| 21st | Aqr | 23h 1m | -9° 60' | -10.6 | 32' | 62% | 109° E | FQ | Saturn, Neptune |
| 22nd | Aqr | 23h 52m | -3° 28' | -11.0 | 32' | 73% | 120° E | +G | Neptune |
| 23rd | Psc | 0h 42m | 3° 14' | -11.4 | 32' | 82% | 132° E | +G | |
| 24th | Psc | 1h 32m | 9° 45' | -11.8 | 32' | 90% | 143° E | +G | |
| 25th | Ari | 2h 24m | 15° 44' | -12.1 | 32' | 96% | 155° E | FM | Jupiter |
| 26th | Ari | 3h 18m | 20° 51' | -12.4 | 32' | 99% | 168° E | FM | Uranus, Pleiades |
| 27th | Tau | 4h 15m | 24° 47' | -12.6 | 31' | 100% | 179° W | FM | Pleiades, Hyades, Aldebaran |
| 28th | Tau | 5h 13m | 27° 16' | -12.4 | 31' | 98% | 166° W | FM | Hyades, Aldebaran |
| 29th | Aur | 6h 12m | 28° 12' | -12.0 | 31' | 95% | 152° W | -G | |
| 30th | Gem | 7h 9m | 27° 35' | -11.7 | 30' | 90% | 139° W | -G | |

## Mercury and Venus

 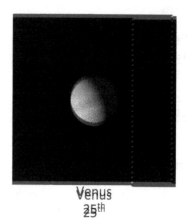

Mercury
25ᵗʰ

Venus
25ᵗʰ

### Mercury

| Date | Con. | R.A. | Dec. | Mag. | Diam. | Ill. | Elon. | Vis. | Rat. | Close To |
|------|------|------|------|------|-------|------|-------|------|------|----------|
| 21st | Oph | 17h 0m | -25° 3' | -0.3 | 5" | 86% | 18° E | PM | ** | Antares |
| 23rd | Oph | 17h 12m | -25° 24' | -0.3 | 5" | 84% | 19° E | PM | ** | |
| 25th | Oph | 17h 25m | -25° 39' | -0.3 | 6" | 81% | 20° E | PM | *** | |
| 27th | Oph | 17h 37m | -25° 48' | -0.3 | 6" | 78% | 21° E | PM | *** | |
| 29th | Sgr | 17h 48m | -25° 52' | -0.3 | 6" | 75% | 22° E | PM | *** | |

## Venus

| Date | Con. | R.A. | Dec. | Mag. | Diam. | Ill. | Elon. | Vis. | Rat. | Close To |
|------|------|------|------|------|-------|------|-------|------|------|----------|
| 21st | Vir | 12h 56m | -4° 2' | -3.8 | 18" | 64% | 43° W | AM | ** | Spica |
| 23rd | Vir | 13h 5m | -4° 51' | -3.8 | 18" | 65% | 43° W | AM | ** | Spica |
| 25th | Vir | 13h 13m | -5° 39' | -3.8 | 18" | 65% | 43° W | AM | ** | Spica |
| 27th | Vir | 13h 22m | -6° 28' | -3.7 | 18" | 66% | 43° W | AM | ** | Spica |
| 29th | Vir | 13h 31m | -7° 17' | -3.7 | 17" | 67% | 43° W | AM | ** | Spica |

## Mars and the Outer Planets

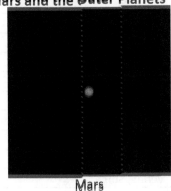

Mars
25th

Jupiter
25th

Saturn
25th

## Mars

| Date | Con. | R.A. | Dec. | Mag. | Diam. | Ill. | Elon. | Vis. | Rat. | Close To |
|------|------|------|------|------|-------|------|-------|------|------|----------|
| 21st | Lib | 15h 42m | -19° 50' | 1.6 | 4" | 100% | 1° W | NV | N/A | |
| 25th | Sco | 15h 54m | -20° 30' | 1.6 | 4" | 100% | 2° W | NV | N/A | Antares |
| 30th | Sco | 16h 9m | -21° 15' | 1.6 | 4" | 100% | 4° W | NV | N/A | Antares |

## The Outer Planets

| Planet | Date | Con. | R.A. | Dec. | Mag. | Diam. | Elon. | Vis. | Rat. | Close To |
|--------|------|------|------|------|------|-------|-------|------|------|----------|
| Jupiter | 25th | Ari | 2h 23m | 12° 47' | -2.4 | 48" | 155° E | PM | **** | Moon |
| Saturn | 25th | Aqr | 22h 14m | -12° 44' | 1.0 | 17" | 93° E | PM | ** | |
| Uranus | 25th | Ari | 3h 12m | 17° 35' | 6.0 | 4" | 167° E | PM | *** | Pleiades |
| Neptune | 25th | Psc | 23h 43m | -3° 10' | 7.7 | 2" | 115° E | PM | ***** | |

## Highlights

| Date | Time (UT) | Event |
|------|-----------|-------|
| 21st | N/A | The Alpha Monocerotid meteor shower is at its maximum. (ZHR: Var) |
| 22nd | 07:46 | The waxing gibbous Moon is south of Neptune. (Evening sky.) |
| 25th | 10:55 | The waxing gibbous Moon is north of Jupiter. (Evening sky.) |
| 26th | 09:50 | The nearly full Moon is north of Uranus. (Evening sky.) |
| 27th | 09:17 | Full Moon. (Visible all night.) |
| | 18:44 | The full Moon is north of the bright star Aldebaran. (Taurus, visible all night.) |
| 28th | 02:01 | Venus is 4.5° north of the bright star Spica. (Virgo, morning sky.) |

# December 1<sup>st</sup> to 10<sup>th</sup>, 2023

## The Moon

| | 1<sup>st</sup> | | 3<sup>rd</sup> | | 5<sup>th</sup> | | 7<sup>th</sup> | | 9<sup>th</sup> | |

| Date | Con | R.A. | Dec | Mag | Diam | Ill. | Elon. | Phase | | Close To |
|------|-----|------|-----|-----|------|------|-------|-------|--|----------|
| 1st | Cnc | 8h 4m | 25° 35' | -11.4 | 30' | 83% | 126° W | -G | | Praesepe |
| 2nd | Cnc | 8h 56m | 22° 27' | -11.1 | 30' | 75% | 114° W | -G | | Praesepe |
| 3rd | Leo | 9h 44m | 18° 24' | -10.8 | 30' | 66% | 103° W | -G | | Regulus |
| 4th | Leo | 10h 29m | 13° 41' | -10.4 | 30' | 57% | 93° W | LQ | | Regulus |
| 5th | Leo | 11h 12m | 8° 29' | -10.0 | 30' | 48% | 83° W | LQ | | |
| 6th | Vir | 11h 55m | 2° 58' | -9.6 | 30' | 38% | 74° W | LQ | | |
| 7th | Vir | 12h 37m | -2° 43' | -9.1 | 30' | 29% | 64° W | -Cr | | Spica |
| 8th | Vir | 13h 21m | -8° 24' | -8.5 | 30' | 21% | 55° W | -Cr | | Spica |
| 9th | Vir | 14h 7m | -13° 54' | -7.7 | 31' | 13% | 44° W | -Cr | | Venus, Spica |
| 10th | Lib | 14h 56m | -18° 57' | -6.9 | 31' | 7% | 33° W | NM | | Venus |

## Mercury and Venus

Mercury
5<sup>th</sup>

Venus
5<sup>th</sup>

**Mercury**

| Date | Con. | R.A. | Dec. | Mag. | Diam. | Ill. | Elon. | Vis. | Rat. | Close To |
|------|------|------|------|------|-------|------|-------|------|------|----------|
| 1st | Sgr | 17h 59m | -25° 50' | -0.3 | 6" | 71% | 23° E | PM | *** | |
| 3rd | Sgr | 18h 9m | -25° 43' | -0.2 | 6" | 66% | 23° E | PM | *** | |
| 5th | Sgr | 18h 18m | -25° 30' | -0.2 | 7" | 60% | 23° E | PM | *** | |
| 7th | Sgr | 18h 26m | -25° 12' | -0.1 | 7" | 54% | 23° E | PM | *** | |
| 9th | Sgr | 18h 32m | -24° 50' | 0.1 | 7" | 46% | 22° E | PM | *** | |

**Venus**

| Date | Con. | R.A. | Dec. | Mag. | Diam. | Ill. | Elon. | Vis. | Rat. | Close To |
|------|------|------|------|------|-------|------|-------|------|------|----------|
| 1st | Vir | 13h 40m | -8° 5' | -3.7 | 17" | 68% | 42° W | AM | ** | Spica |
| 3rd | Vir | 13h 49m | -8° 53' | -3.7 | 17" | 69% | 42° W | AM | ** | Spica |
| 5th | Vir | 13h 58m | -9° 41' | -3.7 | 17" | 69% | 42° W | AM | ** | Spica |
| 7th | Vir | 14h 7m | -10° 28' | -3.7 | 16" | 70% | 42° W | AM | ** | |
| 9th | Vir | 14h 16m | -11° 15' | -3.7 | 16" | 71% | 42° W | AM | ** | Moon |

## Mars and the Outer Planets

Mars
5th

Jupiter
5th

Saturn
5th

**Mars**

| Date | Con. | R.A. | Dec. | Mag. | Diam. | Ill. | Elon. | Vis. | Rat. | Close To |
|------|------|------|------|------|-------|------|-------|------|------|----------|
| 1st | Sco | 16h 12m | -21° 23' | 1.6 | 4" | 100% | 4° W | NV | N/A | Antares |
| 5th | Sco | 16h 24m | -21° 55' | 1.6 | 4" | 100% | 6° W | NV | N/A | Antares |
| 10th | Oph | 16h 39m | -22° 30' | 1.6 | 4" | 100% | 7° W | NV | N/A | Antares |

**The Outer Planets**

| Planet | Date | Con. | R.A. | Dec. | Mag. | Diam. | Elon. | Vis. | Rat. | Close To |
|--------|------|------|------|------|------|-------|-------|------|------|----------|
| Jupiter | 5th | Ari | 2h 19m | 12° 30' | -2.4 | 47" | 143° E | PM | **** | |
| Saturn | 5th | Aqr | 22h 15m | -12° 34' | 1.1 | 17" | 82° E | PM | ** | |
| Uranus | 5th | Ari | 3h 11m | 17° 29' | 6.0 | 4" | 156° E | PM | *** | Pleiades |
| Neptune | 5th | Aqr | 23h 43m | -3° 11' | 7.7 | 2" | 104° E | PM | ***** | |

## Highlights

| Date | Time (UT) | Event |
|------|-----------|-------|
| 3rd | 22:39 | The waning gibbous Moon is north of the bright star Regulus. (Leo, morning sky.) |
| 4th | 14:22 | Mercury is at greatest eastern elongation from the Sun. (Evening sky.) |
| 5th | 05:50 | Last Quarter Moon. (Morning sky.) |
| 6th | 19:03 | Neptune is stationary prior to resuming prograde motion. (Evening sky.) |
| 8th | 15:52 | The waning crescent Moon is north of the bright star Spica. (Virgo, morning sky.) |
| 9th | 18:18 | The waning crescent Moon is south of Venus. (Morning sky.) |
| | N/A | Good opportunity to see Earthshine on the waning crescent Moon. (Morning sky.) |

# December 11ᵗʰ to 20ᵗʰ, 2023

## The Moon

|  |  |  |  |  |
|---|---|---|---|---|
| 11ᵗʰ | 13ᵗʰ | 15ᵗʰ | 17ᵗʰ | 19ᵗʰ |

| Date | Con | R.A. | Dec | Mag | Diam | Ill. | Elon. | Phase | Close To |
|---|---|---|---|---|---|---|---|---|---|
| 11th | Sco | 15h 50m | -23° 14' | -5.9 | 31' | 3% | 21° W | NM | Antares |
| 12th | Sco | 16h 48m | -26° 23' | -4.7 | 32' | 0% | 7° W | NM | Mars, Antares |
| 13th | Sgr | 17h 50m | -28° 0' | -4.8 | 32' | 1% | 7° E | NM | Mercury |
| 14th | Sgr | 18h 54m | -27° 50' | -6.0 | 32' | 3% | 22° E | NM | Mercury |
| 15th | Sgr | 19h 58m | -25° 49' | -7.1 | 32' | 8% | 37° E | NM | |
| 16th | Cap | 20h 58m | -22° 8' | -8.0 | 32' | 16% | 51° E | +Cr | |
| 17th | Cap | 21h 56m | -17° 8' | -8.8 | 32' | 25% | 64° E | +Cr | Saturn |
| 18th | Aqr | 22h 49m | -11° 12' | -9.4 | 32' | 36% | 76° E | +Cr | Saturn |
| 19th | Aqr | 23h 40m | -4° 46' | -10.0 | 32' | 47% | 88° E | FQ | Neptune |
| 20th | Cet | 0h 30m | 1° 51' | -10.5 | 32' | 58% | 99° E | FQ | Neptune |

## Mercury and Venus

Mercury
15ᵗʰ

Venus
15ᵗʰ

### Mercury

| Date | Con. | R.A. | Dec. | Mag. | Diam. | Ill. | Elon. | Vis. | Rat. | Close To |
|---|---|---|---|---|---|---|---|---|---|---|
| 11th | Sgr | 18h 36m | -24° 24' | 0.2 | 8" | 38% | 21° E | PM | *** | |
| 13th | Sgr | 18h 37m | -23° 54' | 0.5 | 8" | 29% | 19° E | PM | *** | Moon |
| 15th | Sgr | 18h 35m | -23° 22' | 0.9 | 9" | 20% | 16° E | PM | ** | |
| 17th | Sgr | 18h 29m | -22° 48' | 1.3 | 9" | 11% | 13° E | NV | N/A | |
| 19th | Sgr | 18h 21m | -22° 13' | 1.9 | 10" | 5% | 8° E | NV | N/A | |

## Venus

| Date | Con. | R.A. | Dec. | Mag. | Diam. | Ill. | Elon. | Vis. | Rat. | Close To |
|------|------|------|------|------|-------|------|-------|------|------|----------|
| 11th | Lib | 14h 25m | -12° 1' | -3.6 | 16" | 71% | 43° W | AM | ** | |
| 13th | Lib | 14h 34m | -12° 46' | -3.6 | 16" | 72% | 42° W | AM | ** | |
| 15th | Lib | 14h 43m | -13° 30' | -3.6 | 16" | 73% | 42° W | AM | ** | |
| 17th | Lib | 14h 53m | -14° 13' | -3.6 | 15" | 73% | 42° W | AM | ** | |
| 19th | Lib | 15h 2m | -14° 56' | -3.6 | 15" | 74% | 41° W | AM | ** | |

## Mars and the Outer Planets

Mars
15th

Jupiter
15th

Saturn
15th

### Mars

| Date | Con. | R.A. | Dec. | Mag. | Diam. | Ill. | Elon. | Vis. | Rat. | Close To |
|------|------|------|------|------|-------|------|-------|------|------|----------|
| 11th | Oph | 16h 42m | -22° 37' | 1.6 | 4" | 100% | 7° W | NV | N/A | Antares |
| 15th | Oph | 16h 55m | -22° 60' | 1.6 | 4" | 100% | 9° W | NV | N/A | Antares |
| 20th | Oph | 17h 11m | -23° 24' | 1.6 | 4" | 100% | 10° W | NV | N/A | |

### The Outer Planets

| Planet | Date | Con. | R.A. | Dec. | Mag. | Diam. | Elon. | Vis. | Rat. | Close To |
|--------|------|------|------|------|------|-------|-------|------|------|----------|
| Jupiter | 15th | Ari | 2h 16m | 12° 19' | -2.3 | 46" | 131° E | PM | **** | |
| Saturn | 15th | Aqr | 22h 18m | -12° 20' | 1.1 | 17" | 73° E | PM | ** | |
| Uranus | 15th | Ari | 3h 9m | 17° 34' | 6.0 | 4" | 145° E | PM | *** | Pleiades |
| Neptune | 15th | Psc | 23h 43m | -3° 10' | 7.7 | 2" | 93° E | PM | **** | |

## Highlights

| Date | Time (UT) | Event |
|------|-----------|-------|
| 12th | 23:33 | New Moon. (Not visible.) |
| 13th | 04:47 | Mercury is stationary prior to beginning retrograde motion. (Evening sky.) |
| | N/A | The Geminid meteor shower is at its maximum. (ZHR: 120) |
| 14th | 04:09 | The waxing crescent Moon is south of Mercury. (Evening sky.) |
| 15th | N/A | Good opportunity to see Earthshine on the waxing crescent Moon. (Evening sky.) |
| 17th | 23:45 | The waxing crescent Moon is south of Saturn. (Evening sky.) |
| 19th | 11:23 | The almost first quarter Moon is south of Neptune. (Evening sky.) |
| | 18:40 | First Quarter Moon. (Evening sky.) |

# December 21ˢᵗ to 31ˢᵗ, 2023

## The Moon

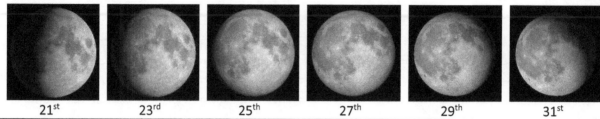

| 21ˢᵗ | 23ʳᵈ | 25ᵗʰ | 27ᵗʰ | 29ᵗʰ | 31ˢᵗ |

| Date | Con | R.A. | Dec | Mag | Diam | Ill. | Elon. | Phase | Close To |
|------|-----|------|-----|-----|------|------|-------|-------|----------|
| 21st | Psc | 1h 19m | 8° 18' | -10.9 | 32' | 69% | 111° E | +G | |
| 22nd | Ari | 2h 10m | 14° 18' | -11.3 | 32' | 79% | 122° E | +G | Jupiter |
| 23rd | Ari | 3h 2m | 19° 32' | -11.6 | 31' | 87% | 134° E | +G | Jupiter, Uranus, Pleiades |
| 24th | Tau | 3h 57m | 23° 44' | -11.9 | 31' | 93% | 147° E | +G | Pleiades, Hyades, Aldebaran |
| 25th | Tau | 4h 54m | 26° 37' | -12.2 | 31' | 97% | 160° E | FM | Hyades, Aldebaran |
| 26th | Tau | 5h 52m | 28° 1' | -12.5 | 31' | 100% | 173° E | FM | |
| 27th | Gem | 6h 49m | 27° 53' | -12.5 | 30' | 100% | 173° W | FM | |
| 28th | Gem | 7h 45m | 26° 18' | -12.3 | 30' | 98% | 161° W | FM | |
| 29th | Cnc | 8h 38m | 23° 29' | -12.0 | 30' | 94% | 149° W | -G | Praesepe |
| 30th | Leo | 9h 27m | 19° 40' | -11.7 | 30' | 88% | 137° W | -G | Praesepe, Regulus |
| 31st | Leo | 10h 14m | 15° 7' | -11.4 | 30' | 82% | 127° W | -G | Regulus |

## Mercury and Venus

Mercury
25ᵗʰ

Venus
25ᵗʰ

**Mercury**

| Date | Con. | R.A. | Dec. | Mag. | Diam. | Ill. | Elon. | Vis. | Rat. | Close To |
|------|------|------|------|------|-------|------|-------|------|------|----------|
| 21st | Sgr | 18h 10m | -21° 39' | 2.6 | 10" | 1% | 3° E | NV | N/A | |
| 23rd | Sgr | 17h 58m | -21° 6' | 2.7 | 10" | 1% | 2° W | NV | N/A | Mars |
| 25th | Sgr | 17h 47m | -20° 39' | 2.1 | 10" | 4% | 7° W | NV | N/A | Mars |
| 27th | Oph | 17h 38m | -20° 20' | 1.5 | 10" | 9% | 11° W | NV | N/A | Mars |
| 29th | Oph | 17h 31m | -20° 9' | 1.1 | 9" | 17% | 15° W | AM | ** | Mars |
| 31st | Oph | 17h 27m | -20° 8' | 0.7 | 9" | 25% | 18° W | AM | *** | Mars |

**Venus**

| Date | Con. | R.A. | Dec. | Mag. | Diam. | Ill. | Elon. | Vis. | Rat. | Close To |
|------|------|------|------|------|-------|------|-------|------|------|----------|
| 21st | Lib | 15h 12m | -15° 36' | -3.6 | 15" | 75% | 41° W | AM | ** | |
| 23rd | Lib | 15h 21m | -16° 16' | -3.6 | 15" | 75% | 41° W | AM | ** | |
| 25th | Lib | 15h 31m | -16° 54' | -3.6 | 15" | 76% | 41° W | AM | ** | |
| 27th | Lib | 15h 41m | -17° 30' | -3.6 | 14" | 77% | 41° W | AM | ** | |
| 29th | Lib | 15h 51m | -18° 5' | -3.5 | 14" | 77% | 40° W | AM | ** | Antares |
| 31st | Lib | 16h 1m | -18° 38' | -3.5 | 14" | 78% | 40° W | AM | ** | Antares |

## Mars and the Outer Planets

Mars
25th

Jupiter
25th

Saturn
25th

**Mars**

| Date | Con. | R.A. | Dec. | Mag. | Diam. | Ill. | Elon. | Vis. | Rat. | Close To |
|------|------|------|------|------|-------|------|-------|------|------|----------|
| 21st | Oph | 17h 14m | -23° 28' | 1.6 | 4" | 100% | 11° W | NV | N/A | |
| 25th | Oph | 17h 27m | -23° 43' | 1.6 | 4" | 100% | 12° W | NV | N/A | Mercury |
| 31st | Sgr | 17h 46m | -23° 57' | 1.6 | 4" | 99% | 14° W | NV | N/A | Mercury |

**The Outer Planets**

| Planet | Date | Con. | R.A. | Dec. | Mag. | Diam. | Elon. | Vis. | Rat. | Close To |
|--------|------|------|------|------|------|-------|-------|------|------|----------|
| Jupiter | 25th | Ari | 2h 14m | 12° 15' | -2.2 | 45" | 120° E | PM | **** | |
| Saturn | 25th | Aqr | 22h 21m | -12° 3' | 1.1 | 16" | 62° E | PM | ** | |
| Uranus | 25th | Ari | 3h 8m | 17° 19' | 6.0 | 4" | 133° E | PM | ** | Pleiades |
| Neptune | 25th | Psc | 23h 43m | -3° 8' | 7.7 | 2" | 82° E | PM | **** | |

## Highlights

| Date | Time (UT) | Event |
|------|-----------|-------|
| 21st | 19:56 | Asteroid Vesta is at opposition. (Visible all night.) |
| 22nd | 03:28 | Winter solstice. |
| | 12:45 | The waxing gibbous Moon is north of Jupiter. (Evening sky.) |
| | 18:47 | Mercury is at inferior conjunction with the Sun. (Not visible.) |
| | N/A | The Ursid meteor shower is at its maximum. (ZHR: 10) |
| 23rd | 13:19 | The waxing gibbous Moon is north of Uranus. (Evening sky.) |
| 25th | 06:04 | The waxing gibbous Moon is north of the bright star Aldebaran. (Taurus, evening sky.) |
| 27th | 00:34 | Full Moon. (Visible all night.) |
| 31st | 10:44 | The waning gibbous Moon is north of the bright star Regulus. (Leo, morning sky.) |
| | 14:32 | Jupiter is stationary prior to resuming prograde motion. (Evening sky.) |

# Planet Visibility Ratings

| | | Morning Sky | | | | | | | Evening Sky | | | | | | |
|---|---|---|---|---|---|---|---|---|---|---|---|---|---|---|---|
| | | Me | Ve | Ma | Ju | Sa | Ur | Ne | Me | Ve | Ma | Ju | Sa | Ur | Ne |
| Jan | 5th | *** | * | * | | | | | | | | *** | ** | ** | **** |
| | 15th | *** | * | * | | | | | | | | *** | * | ** | *** |
| | 25th | *** | * | * | | | | | | | | *** | * | ** | *** |
| Feb | 5th | ** | * | * | | | | | | | | ** | * | * | *** |
| | 15th | | * | * | | | | | | | | ** | | * | *** |
| | 25th | | * | * | | | | | | | | ** | | * | ** |
| Mar | 5th | | * | * | | | | | | | | ** | | * | |
| | 15th | | * | * | | | | | | | | * | | * | |
| | 25th | | * | * | | * | | | ** | | | * | | * | |
| Apr | 5th | | | * | | * | | | | | | * | | * | |
| | 15th | | | * | | * | | ** | | | | * | | * | |
| | 25th | *** | | * | | * | | *** | | | | * | | * | |
| May | 5th | *** | | * | | ** | | *** | | | | | | | |
| | 15th | *** | | * | | ** | | *** | | | | | | | |
| | 25th | *** | | * | | ** | | *** | | | | | | | |
| Jun | 5th | | | * | | ** | * | **** | | | | | | | |
| | 15th | | | * | * | ** | * | **** | | | | | | | |
| | 25th | | | * | * | *** | * | **** | | | | | | | |
| Jul | 5th | | | * | * | *** | * | ***** | *** | | | | | | |
| | 15th | | | * | * | *** | * | ***** | **** | | | | | | |
| | 25th | | | * | * | *** | * | ***** | **** | | | | | | |
| Aug | 5th | | | * | ** | **** | * | ***** | *** | * | | | | | |
| | 15th | | | * | ** | **** | * | ***** | | * | | | | | |
| | 25th | | | ** | ** | **** | ** | ***** | | * | | | | | |
| Sep | 5th | ** | | ** | ** | **** | ** | ***** | | * | | | | | |
| | 15th | | | ** | *** | | ** | ***** | | * | | | **** | | |
| | 25th | | | ** | *** | | ** | | | * | | | **** | | ***** |
| Oct | 5th | | | ** | *** | | ** | | | * | | | **** | | ***** |
| | 15th | | | ** | *** | | *** | | | * | | | **** | | ***** |
| | 25th | | | ** | **** | | *** | | | * | | | **** | | ***** |
| Nov | 5th | | | ** | **** | | *** | | *** | * | | | *** | | ***** |
| | 15th | | | ** | **** | | *** | | *** | ** | | | *** | | ***** |
| | 25th | | | *** | **** | | | | *** | ** | | | *** | *** | ***** |
| Dec | 5th | | | *** | ***** | | | | | ** | | | ** | *** | ***** |
| | 15th | *** | | *** | | | | | | *** | | **** | ** | *** | **** |
| | 25th | *** | | **** | | | | | | *** | | **** | ** | ** | **** |

# Solar and Lunar Eclipses

| Date | Time (UT) | Type | Visible From |
|---|---|---|---|
| Mar 25th | 07:12 | Penumbral Lunar | Central America, North America and South America |
| Apr 8th | 18:18 | Total Solar | Central America and North America |
| Sep 18th | 02:44 | Partial Lunar | Western Africa, Central America, western Europe, eastern North America and South America |
| Oct 2nd | 18:46 | Annular Solar | Southern South America. |

# Planetary Highlights

| Date | Time (UT) | Elon. | Vis. | Description |
|------|-----------|-------|------|-------------|
| Jan 12th | 14:31 | 25° W | AM | Mercury is at greatest western elongation from the Sun. (Sagittarius) |
| Jan 27th | 16:07 | 21° W | AM | Mercury is 0.2° north of Mars. (Sagittarius) |
| Feb 22nd | 15:39 | 26° W | AM | Venus is 0.6° north of Mars. (Capricornus) |
| Mar 22nd | 01:51 | 18° W | AM | Venus is 0.3° north of Saturn. (Aquarius) |
| Apr 11th | 03:02 | 33° W | AM | Mars is 0.5° north of Saturn. (Aquarius) |
| Apr 20th | 02:43 | 25° E | PM | Jupiter is 0.5° south of Uranus. (Aries) |
| Apr 29th | 03:55 | 38° W | AM | Mars is 0.0° south of Neptune. (Pisces) |
| May 9th | 21:25 | 24° W | AM | Mercury is at greatest western elongation from the Sun. (Pisces) |
| Jul 6th | 21:28 | 23° E | PM | Mercury is 0.2° north of the Praesepe open star cluster. (Cancer) |
| Jul 11th | 15:31 | 45° W | AM | Jupiter is 4.8° north of the bright star Aldebaran. (Taurus) |
| Jul 15th | 09:13 | 61° W | AM | Mars is 0.6° south of Uranus. (Taurus) |
| Jul 22nd | 06:34 | 27° E | PM | Mercury is at greatest eastern elongation from the Sun. (Leo) |
| Aug 5th | 06:34 | 67° W | AM | Mars is 5.0° north of the bright star Aldebaran. (Taurus) |
| Aug 14th | 16:51 | 69° W | AM | Mars is 0.3° north of Jupiter. (Taurus) |
| Sep 9th | 01:41 | 180° | AN | Saturn is at opposition from the Sun. (Aquarius) |
| Sep 17th | 05:52 | 26° E | PM | Venus is 2.6° north of the bright star Spica. (Virgo) |
| Sep 21st | 12:57 | 180° | AN | Neptune is at opposition from the Sun. (Pisces) |
| Nov 16th | 08:01 | 23° E | PM | Mercury is at greatest eastern elongation from the Sun. (Ophiuchus) |
| Nov 17th | 05:56 | 180° | AN | Uranus is at opposition from the Sun. (Taurus) |
| Dec 7th | 22:05 | 180° | AN | Jupiter is at opposition from the Sun. (Taurus) |
| Dec 25th | 02:14 | 24° W | AM | Mercury is at greatest western elongation from the Sun. (Ophiuchus |

# Major Meteor Showers

| Shower Name | Start Date | End Date | Peak | ZHR | Speed | Brightness | Moon |
|-------------|-----------|----------|------|-----|-------|-----------|------|
| Quadrantids | Dec 28th | Jan 12th | Jan 3rd | 120 | *** | ***** | ◑ |
| Lyrids | Apr 18th | Apr 25th | Apr 22nd | 18 | *** | ***** | ○ |
| Eta Aquariids | Apr 24th | May 19th | May 7th | 40 | * | **** | ● |
| June Bootids | Jun 23rd | Jun 25th | Jun 24th | Var | ***** | ***** | ○ |
| Alpha Capricornids | Jul 8th | Aug 10th | Jul 27th | 5 | ***** | **** | ◑ |
| Southern Delta Aquariids | Jul 21st | Aug 23rd | Jul 30th | 16 | *** | * | ◐ |
| Perseids | Jul 13th | Aug 26th | Aug 12th | 100 | * | ***** | ◐ |
| Kappa Cygnids | Aug 6th | Aug 31st | Aug 17th | 3 | ***** | ** | ○ |
| Aurigids | Aug 29th | Sep 4th | Sep 1st | 6 | * | **** | ● |
| September Epsilon Perseids | Sep 5th | Sep 28th | Sep 9th | 5 | * | ** | ◕ |
| Draconids | Oct 6th | Oct 10th | Oct 8th | Var | ***** | *** | ◕ |
| Southern Taurids | Sep 7th | Nov 19th | Oct 10th | 5 | **** | **** | ◑ |
| Orionids | Aug 25th | Nov 19th | Oct 22nd | 15 | * | **** | ○ |
| Andromedids | Oct 26th | Nov 20th | Nov 8th | Var | ***** | **** | ● |
| Northern Taurids | Oct 25th | Dec 4th | Nov 11th | 5 | **** | **** | ○ |
| Leonids | Nov 5th | Dec 3rd | Nov 18th | 15 | * | **** | ○ |
| Alpha Monocerotids | Nov 21st | Nov 23rd | Nov 21st | Var | * | **** | ◑ |
| Geminids | Nov 30th | Dec 17th | Dec 13th | 120 | **** | *** | ○ |
| December Leonis Minorids | Dec 6th | Jan 18th | Dec 20th | 5 | * | ** | ◑ |
| Ursids | Dec 17th | Dec 24th | Dec 22nd | 10 | **** | ** | ◑ |
| Coma Berenicids | Dec 24th | Jan 3rd | Dec 31st | 5 | * | ** | ● |

# January 1st to 10th, 2024

## The Moon

| 1st | 3rd | 5th | 7th | 9th |

| Date | Con | R.A. | Dec | Mag | Diam | Ill. | Elon. | Phase | Close To |
|------|-----|------|-----|-----|------|------|-------|-------|----------|
| 1st | Leo | 10h 36m | 12° 38' | -11.2 | 30' | 78% | 122° W | -G | Regulus |
| 2nd | Leo | 11h 19m | 7° 22' | -10.9 | 30' | 70% | 112° W | -G | |
| 3rd | Vir | 12h 1m | 1° 51' | -10.6 | 30' | 61% | 103° W | LQ | |
| 4th | Vir | 12h 43m | -3° 47' | -10.2 | 30' | 51% | 93° W | LQ | Spica |
| 5th | Vir | 13h 26m | -9° 21' | -9.8 | 30' | 42% | 84° W | LQ | Spica |
| 6th | Vir | 14h 12m | -14° 42' | -9.3 | 30' | 32% | 73° W | -Cr | Spica |
| 7th | Lib | 15h 1m | -19° 35' | -8.7 | 31' | 23% | 62° W | -Cr | |
| 8th | Sco | 15h 54m | -23° 41' | -7.9 | 31' | 15% | 50° W | -Cr | Venus, Antares |
| 9th | Sco | 16h 53m | -26° 40' | -7.1 | 32' | 8% | 36° W | NM | Mercury, Venus, Antares |
| 10th | Sgr | 17h 55m | -28° 6' | -6.0 | 32' | 3% | 22° W | NM | Mercury, Mars |

## Mercury and Venus

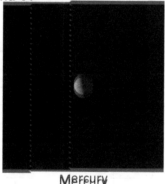

Mercury
5th

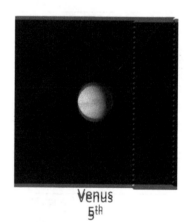

Venus
5th

### Mercury

| Date | Con. | R.A. | Dec. | Mag. | Diam. | Ill. | Elon. | Vis. | Rat. | Close To |
|------|------|------|------|------|-------|------|-------|------|------|----------|
| 1st | Oph | 17h 27m | -20° 9' | 0.7 | 9" | 27% | 19° W | AM | *** | Mars |
| 3rd | Oph | 17h 26m | -20° 18' | 0.4 | 8" | 35% | 21° W | AM | *** | Mars |
| 5th | Oph | 17h 29m | -20° 33' | 0.3 | 8" | 42% | 23° W | AM | *** | Mars |
| 7th | Oph | 17h 33m | -20° 51' | 0.2 | 7" | 49% | 24° W | AM | *** | Mars |
| 9th | Oph | 17h 39m | -21° 12' | 0.1 | 7" | 55% | 25° W | AM | *** | Moon, Mars |

## Venus

| Date | Con. | R.A. | Dec. | Mag. | Diam. | Ill. | Elon. | Vis. | Rat. | Close To |
|------|------|------|------|------|-------|------|-------|------|------|----------|
| 1st | Sco | 16h 3m | -18° 46' | -3.5 | 14" | 78% | 40° W | AM | ** | Antares |
| 3rd | Sco | 16h 13m | -19° 17' | -3.5 | 14" | 78% | 40° W | AM | * | Antares |
| 5th | Sco | 16h 24m | -19° 46' | -3.5 | 14" | 79% | 39° W | AM | * | Antares |
| 7th | Oph | 16h 34m | -20° 12' | -3.5 | 14" | 80% | 39° W | AM | * | Antares |
| 9th | Oph | 16h 44m | -20° 37' | -3.5 | 14" | 80% | 39° W | AM | * | Moon, Antares |

## Mars and the Outer Planets

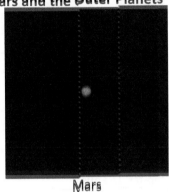

Mars
5th

Jupiter
5th

Saturn
5th

## Mars

| Date | Con. | R.A. | Dec. | Mag. | Diam. | Ill. | Elon. | Vis. | Rat. | Close To |
|------|------|------|------|------|-------|------|-------|------|------|----------|
| 1st | Sgr | 17h 48m | -23° 58' | 1.6 | 4" | 99% | 14° W | NV | N/A | Mercury |
| 5th | Sgr | 18h 1m | -24° 2' | 1.6 | 4" | 99% | 15° W | AM | * | Mercury |
| 10th | Sgr | 18h 17m | -24° 1' | 1.6 | 4" | 99% | 16° W | AM | * | Moon, Mercury |

## The Outer Planets

| Planet | Date | Con. | R.A. | Dec. | Mag. | Diam. | Elon. | Vis. | Rat. | Close To |
|--------|------|------|------|------|------|-------|-------|------|------|----------|
| Jupiter | 5th | Ari | 2h 14m | 12° 18' | -2.2 | 43" | 108° E | PM | *** | |
| Saturn | 5th | Aqr | 22h 34m | -11° 42' | 1.2 | 16" | 51° E | PM | ** | |
| Uranus | 5th | Ari | 3h 7m | 17° 16' | 6.0 | 4" | 122° E | PM | ** | Pleiades |
| Neptune | 5th | Psc | 23h 44m | -3° 4' | 7.7 | 2" | 71° E | PM | **** | |

## Highlights

| Date | Time (UT) | Event |
|------|-----------|-------|
| 2nd | 03:48 | Mercury is stationary prior to resuming prograde motion. (Morning sky.) |
| 3rd | N/A | The Quadrantid meteor shower is at its maximum. (ZHR: 120) |
| 4th | 03:31 | Last Quarter Moon. (Morning sky.) |
| | 21:36 | The just-past last quarter Moon is north of the bright star Spica. (Morning sky.) |
| 6th | 14:09 | Venus is 6.4° north of the bright star Antares. (Morning sky.) |
| 7th | N/A | Good opportunity to see Earthshine on the waning crescent Moon. (Morning sky.) |
| 8th | 16:07 | The waning crescent Moon is north of the bright star Antares. (Morning sky.) |
| | 20:38 | The waning crescent Moon is south of Venus. (Morning sky.) |
| 9th | 19:54 | The waning crescent Moon is south of Mercury. (Morning sky.) |
| 10th | 07:03 | The nearly new Moon is south of Mars. (Morning sky.) |

# January 11th to 20th, 2024

## The Moon

  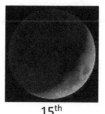  

| 11th | 13th | 15th | 17th | 19th |

| Date | Con | R.A. | Dec | Mag | Diam | Ill. | Elon. | Phase | Close To |
|------|-----|------|-----|-----|------|------|-------|-------|----------|
| 11th | Sgr | 19h 0m | -27° 42' | -4.8 | 33' | 1% | 7° W | NM | Mars |
| 12th | Sgr | 20h 5m | -25° 23' | -4.8 | 33' | 1% | 8° E | NM | |
| 13th | Cap | 21h 7m | -21° 20' | -6.1 | 33' | 3% | 23° E | NM | |
| 14th | Aqr | 22h 5m | -15° 55' | -7.2 | 33' | 9% | 36° E | NM | Saturn |
| 15th | Aqr | 23h 0m | -9° 38' | -8.1 | 33' | 17% | 49° E | +Cr | Saturn, Neptune |
| 16th | Psc | 23h 52m | -2° 55' | -8.9 | 33' | 26% | 61° E | +Cr | Neptune |
| 17th | Psc | 0h 42m | 3° 50' | -9.5 | 32' | 37% | 72° E | +Cr | |
| 18th | Psc | 1h 33m | 10° 16' | -10.0 | 32' | 48% | 84° E | FQ | Jupiter |
| 19th | Ari | 2h 24m | 16° 4' | -10.5 | 32' | 59% | 96° E | FQ | Jupiter, Uranus |
| 20th | Ari | 3h 16m | 20° 59' | -10.9 | 31' | 69% | 108° E | +G | Uranus, Pleiades |

## Mercury and Venus

Mercury
15th

Venus
15th

**Mercury**

| Date | Con. | R.A. | Dec. | Mag. | Diam. | Ill. | Elon. | Vis. | Rat. | Close To |
|------|------|------|------|------|-------|------|-------|------|------|----------|
| 11th | Sgr | 17h 46m | -21° 33' | 0.0 | 7" | 60% | 25° W | AM | *** | Mars |
| 13th | Sgr | 17h 54m | -21° 54' | 0.0 | 7" | 65% | 25° W | AM | *** | Mars |
| 15th | Sgr | 18h 4m | -22° 13' | 0.0 | 6" | 68% | 25° W | AM | *** | Mars |
| 17th | Sgr | 18h 14m | -22° 29' | -0.1 | 6" | 72% | 25° W | AM | *** | Mars |
| 19th | Sgr | 18h 24m | -22° 42' | -0.1 | 6" | 75% | 24° W | AM | *** | Mars |

**Venus**

| Date | Con. | R.A. | Dec. | Mag. | Diam. | Ill. | Elon. | Vis. | Rat. | Close To |
|------|------|------|------|------|-------|------|-------|------|------|----------|
| 11th | Oph | 16h 54m | -20° 59' | -3.5 | 13" | 81% | 38° W | AM | * | Antares |
| 13th | Oph | 17h 5m | -21° 19' | -3.5 | 13" | 81% | 38° W | AM | * | Antares |
| 15th | Oph | 17h 15m | -21° 37' | -3.5 | 13" | 82% | 37° W | AM | * | |
| 17th | Oph | 17h 26m | -21° 52' | -3.5 | 13" | 82% | 37° W | AM | * | |
| 19th | Oph | 17h 36m | -22° 4' | -3.5 | 13" | 83% | 36° W | AM | * | |

## Mars and the Outer Planets

Mars
15th

Jupiter
15th

Saturn
15th

**Mars**

| Date | Con. | R.A. | Dec. | Mag. | Diam. | Ill. | Elon. | Vis. | Rat. | Close To |
|------|------|------|------|------|-------|------|-------|------|------|----------|
| 11th | Sgr | 18h 20m | -23° 60' | 1.6 | 4" | 99% | 17° W | AM | * | Moon, Mercury |
| 15th | Sgr | 18h 33m | -23° 54' | 1.6 | 4" | 99% | 18° W | AM | * | Mercury |
| 20th | Sgr | 18h 50m | -23° 40' | 1.6 | 4" | 99% | 19° W | AM | * | Mercury |

**The Outer Planets**

| Planet | Date | Con. | R.A. | Dec. | Mag. | Diam. | Elon. | Vis. | Rat. | Close To |
|--------|------|------|------|------|------|-------|-------|------|------|----------|
| Jupiter | 15th | Ari | 2h 16m | 12° 27' | -2.1 | 42" | 98° E | PM | *** | |
| Saturn | 15th | Aqr | 22h 28m | -11° 20' | 1.2 | 16" | 41° E | PM | * | Moon |
| Uranus | 15th | Ari | 3h 7m | 17° 13' | 6.0 | 4" | 111° E | PM | ** | Pleiades |
| Neptune | 15th | Psc | 23h 44m | -2° 59' | 7.7 | 2" | 60° E | PM | *** | Moon |

## Highlights

| Date | Time (UT) | Event |
|------|-----------|-------|
| 11th | 11:58 | New Moon. (Not visible.) |
| 12th | 14:31 | Mercury is at greatest western elongation from the Sun. (Morning sky.) |
| 14th | 07:50 | The waxing crescent Moon is south of Saturn. (Evening sky.) |
| | N/A | Good opportunity to see Earthshine on the waxing crescent Moon. (Evening sky.) |
| 15th | 22:18 | The waxing crescent Moon is south of Neptune. (Evening sky.) |
| 18th | 03:53 | First Quarter Moon. (Evening sky.) |
| | 22:19 | The first quarter Moon is north of Jupiter. (Evening sky.) |
| 19th | 20:06 | The just-past first quarter Moon is north of Uranus. (Evening sky.) |
| 20th | 20:27 | Dwarf planet Pluto is in conjunction with the Sun. (Not visible.) |

# January 21st to 31st, 2024

## The Moon

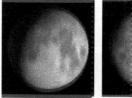  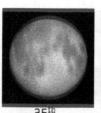  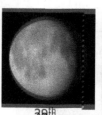 

21st    23rd    25th    27th    29th    31st

| Date | Con | R.A. | Dec | Mag | Diam | Ill. | Elon. | Phase | Close To |
|------|-----|------|-----|-----|------|------|-------|-------|----------|
| 21st | Tau | 4h 11m | 24° 47' | -11.3 | 31' | 79% | 120° E | +G | Pleiades, Hyades, Aldebaran |
| 22nd | Tau | 5h 7m | 27° 13' | -11.6 | 31' | 86% | 133° E | +G | Hyades, Aldebaran |
| 23rd | Aur | 6h 5m | 28° 12' | -11.9 | 30' | 93% | 147° E | +G | |
| 24th | Gem | 7h 1m | 27° 42' | -12.2 | 30' | 97% | 160° E | FM | |
| 25th | Gem | 7h 56m | 25° 48' | -12.5 | 30' | 99% | 172° E | FM | Praesepe |
| 26th | Cnc | 8h 48m | 22° 44' | -12.6 | 30' | 100% | 176° W | FM | Praesepe |
| 27th | Leo | 9h 37m | 18° 43' | -12.4 | 30' | 98% | 165° W | FM | Regulus |
| 28th | Leo | 10h 22m | 13° 59' | -12.1 | 30' | 95% | 154° W | FM | Regulus |
| 29th | Leo | 11h 6m | 8° 48' | -11.8 | 29' | 91% | 144° W | -G | |
| 30th | Vir | 11h 48m | 3° 19' | -11.5 | 29' | 84% | 135° W | -G | |
| 31st | Vir | 12h 30m | -2° 18' | -11.2 | 30' | 77% | 125° W | -G | |

## Mercury and Venus

Mercury
25th

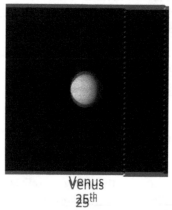

Venus
25th

### Mercury

| Date | Con. | R.A. | Dec. | Mag. | Diam. | Ill. | Elon. | Vis. | Rat. | Close To |
|------|------|------|------|------|-------|------|-------|------|------|----------|
| 21st | Sgr | 18h 35m | -22° 51' | -0.1 | 6" | 78% | 24° W | AM | *** | Mars |
| 23rd | Sgr | 18h 47m | -22° 57' | -0.1 | 6" | 80% | 23° W | AM | *** | Mars |
| 25th | Sgr | 18h 59m | -22° 58' | -0.1 | 6" | 82% | 22° W | AM | *** | Mars |
| 27th | Sgr | 19h 11m | -22° 54' | -0.1 | 5" | 84% | 21° W | AM | *** | Mars |
| 29th | Sgr | 19h 23m | -22° 46' | -0.2 | 5" | 86% | 20° W | AM | ** | Mars |
| 31st | Sgr | 19h 36m | -22° 33' | -0.2 | 5" | 87% | 19° W | AM | ** | Mars |

## Venus

| Date | Con. | R.A. | Dec. | Mag. | Diam. | Ill. | Elon. | Vis. | Rat. | Close To |
|------|------|------|------|------|-------|------|-------|------|------|----------|
| 21st | Sgr | 17h 47m | -22° 14' | -3.4 | 13" | 83% | 36° W | AM | * | |
| 23rd | Sgr | 17h 58m | -22° 22' | -3.4 | 13" | 84% | 35° W | AM | * | |
| 25th | Sgr | 18h 8m | -22° 27' | -3.4 | 13" | 84% | 35° W | AM | * | |
| 27th | Sgr | 18h 19m | -22° 29' | -3.4 | 12" | 85% | 34° W | AM | * | |
| 29th | Sgr | 18h 30m | -22° 28' | -3.4 | 12" | 85% | 33° W | AM | * | |
| 31st | Sgr | 18h 40m | -22° 25' | -3.4 | 12" | 86% | 33° W | AM | * | |

## Mars and the Outer Planets

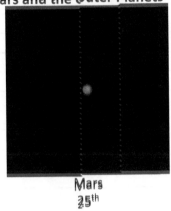

Mars
25th

Jupiter
25th

Saturn
25th

## Mars

| Date | Con. | R.A. | Dec. | Mag. | Diam. | Ill. | Elon. | Vis. | Rat. | Close To |
|------|------|------|------|------|-------|------|-------|------|------|----------|
| 21st | Sgr | 18h 53m | -23° 36' | 1.6 | 4" | 99% | 19° W | AM | * | Mercury |
| 25th | Sgr | 19h 6m | -23° 20' | 1.6 | 4" | 99% | 20° W | AM | * | Mercury |
| 31st | Sgr | 19h 26m | -22° 47' | 1.5 | 4" | 99% | 21° W | AM | * | Mercury |

## The Outer Planets

| Planet | Date | Con. | R.A. | Dec. | Mag. | Diam. | Elon. | Vis. | Rat. | Close To |
|--------|------|------|------|------|------|-------|-------|------|------|----------|
| Jupiter | 25th | Ari | 2h 18m | 12° 43' | -2.0 | 41" | 88° E | PM | *** | |
| Saturn | 25th | Aqr | 22h 32m | -10° 56' | 1.2 | 16" | 31° E | PM | * | |
| Uranus | 25th | Ari | 3h 6m | 17° 13' | 6.0 | 4" | 100° E | PM | ** | |
| Neptune | 25th | Psc | 23h 45m | -2° 53' | 7.8 | 2" | 50° E | PM | *** | |

## Highlights

| Date | Time (UT) | Event |
|------|-----------|-------|
| 21st | 09:55 | The waxing gibbous Moon is north of the bright star Aldebaran. (Evening sky.) |
| 25th | 17:55 | Full Moon. (Visible all night.) |
| 27th | 07:51 | Uranus is stationary prior to resuming prograde motion. (Evening sky.) |
| | 15:25 | The just-past full Moon is north of the bright star Regulus. (Morning sky.) |
| | 16:07 | Mercury is 0.2° north of Mars. (Morning sky.) |

# February 1st to 10th, 2024

## The Moon

 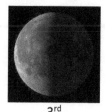   

| | 1st | 3rd | 5th | 7th | 9th |

| Date | Con | R.A. | Dec | Mag | Diam | Ill. | Elon. | Phase | Close To |
|------|-----|------|-----|-----|------|------|-------|-------|----------|
| 1st | Vir | 13h 12m | -7° 51' | -10.9 | 30' | 69% | 116° W | -G | Spica |
| 2nd | Vir | 13h 56m | -13° 13' | -10.5 | 30' | 59% | 106° W | LQ | Spica |
| 3rd | Lib | 14h 43m | -18° 10' | -10.1 | 30' | 50% | 95° W | LQ | |
| 4th | Lib | 15h 33m | -22° 29' | -9.7 | 31' | 40% | 84° W | LQ | |
| 5th | Sco | 16h 28m | -25° 50' | -9.1 | 31' | 30% | 71° W | -Cr | Antares |
| 6th | Oph | 17h 28m | -27° 52' | -8.4 | 32' | 20% | 57° W | -Cr | |
| 7th | Sgr | 18h 31m | -28° 14' | -7.6 | 32' | 12% | 42° W | NM | Venus |
| 8th | Sgr | 19h 36m | -26° 44' | -6.6 | 33' | 6% | 27° W | NM | Venus, Mars |
| 9th | Cap | 20h 39m | -23° 22' | -5.4 | 33' | 2% | 12° W | NM | Mercury, Mars |
| 10th | Cap | 21h 40m | -18° 23' | -4.3 | 33' | 0% | 2° E | NM | |

## Mercury and Venus

Mercury
5th

Venus
5th

**Mercury**

| Date | Con. | R.A. | Dec. | Mag. | Diam. | Ill. | Elon. | Vis. | Rat. | Close To |
|------|------|------|------|------|-------|------|-------|------|------|----------|
| 1st | Sgr | 19h 42m | -22° 24' | -0.2 | 5" | 88% | 18° W | AM | ** | Mars |
| 3rd | Sgr | 19h 55m | -22° 3' | -0.2 | 5" | 89% | 17° W | AM | ** | Mars |
| 5th | Sgr | 20h 8m | -21° 37' | -0.3 | 5" | 91% | 16° W | AM | ** | Mars |
| 7th | Cap | 20h 21m | -21° 5' | -0.3 | 5" | 92% | 15° W | NV | N/A | Mars |
| 9th | Cap | 20h 34m | -20° 28' | -0.4 | 5" | 93% | 13° W | NV | N/A | Moon, Mars |

## Venus

| Date | Con. | R.A. | Dec. | Mag. | Diam. | Ill. | Elon. | Vis. | Rat. | Close To |
|------|------|------|------|------|-------|------|-------|------|------|----------|
| 1st | Sgr | 18h 46m | -22° 23' | -3.4 | 12" | 86% | 33° W | AM | * | |
| 3rd | Sgr | 18h 56m | -22° 15' | -3.4 | 12" | 86% | 32° W | AM | * | Mars |
| 5th | Sgr | 19h 7m | -22° 6' | -3.4 | 12" | 87% | 31° W | AM | * | Mars |
| 7th | Sgr | 19h 18m | -21° 53' | -3.4 | 12" | 87% | 31° W | AM | * | Moon, Mars |
| 9th | Sgr | 19h 28m | -21° 38' | -3.4 | 12" | 87% | 30° W | AM | * | Mars |

## Mars and the Outer Planets

Mars
5th

Jupiter
5th

Saturn
5th

## Mars

| Date | Con. | R.A. | Dec. | Mag. | Diam. | Ill. | Elon. | Vis. | Rat. | Close To |
|------|------|------|------|------|-------|------|-------|------|------|----------|
| 1st | Sgr | 19h 29m | -22° 41' | 1.5 | 4" | 98% | 22° W | AM | * | Mercury |
| 5th | Sgr | 19h 42m | -22° 13' | 1.5 | 4" | 98% | 22° W | AM | * | Mercury, Venus |
| 10th | Sgr | 19h 59m | -21° 33' | 1.5 | 4" | 98% | 23° W | AM | * | Venus |

## The Outer Planets

| Planet | Date | Con. | R.A. | Dec. | Mag. | Diam. | Elon. | Vis. | Rat. | Close To |
|--------|------|------|------|------|------|-------|-------|------|------|----------|
| Jupiter | 5th | Ari | 2h 22m | 13° 7' | -1.9 | 39" | 78° E | PM | ** | |
| Saturn | 5th | Aqr | 22h 36m | -10° 28' | 1.2 | 16" | 21° E | PM | * | |
| Uranus | 5th | Ari | 3h 7m | 17° 13' | 6.1 | 4" | 89° E | PM | * | Pleiades |
| Neptune | 5th | Psc | 23h 46m | -2° 45' | 7.8 | 2" | 39° E | PM | *** | |

## Highlights

| Date | Time (UT) | Event |
|------|-----------|-------|
| 1st | 08:51 | The nearly last quarter Moon is north of the bright star Spica. (Morning sky.) |
| 2nd | 23:19 | Last Quarter Moon. (Morning sky.) |
| 4th | 22:55 | The waning crescent Moon appears north of the bright star Antares. (Morning sky.) |
| 6th | N/A | Good opportunity to see Earthshine on the waning crescent Moon. (Morning sky.) |
| 8th | 04:44 | The nearly new Moon is south of Mars. (Morning sky.) |
| 9th | 23:00 | New Moon. (Not visible.) |

# February 11th to 20th, 2024

## The Moon

    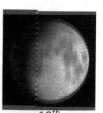

| 11th | 13th | 15th | 17th | 19th |

| Date | Con | R.A. | Dec | Mag | Diam | Ill. | Elon. | Phase | Close To |
|------|-----|------|-----|-----|------|------|-------|-------|----------|
| 11th | Aqr | 22h 38m | -12° 12' | -5.5 | 33' | 2% | 15° E | NM | Saturn |
| 12th | Aqr | 23h 32m | -5° 20' | -6.7 | 33' | 6% | 28° E | NM | Neptune |
| 13th | Psc | 0h 25m | 1° 44' | -7.7 | 33' | 13% | 40° E | +Cr | Neptune |
| 14th | Psc | 1h 17m | 8° 34' | -8.6 | 33' | 22% | 52° E | +Cr | |
| 15th | Ari | 2h 9m | 14° 48' | -9.3 | 32' | 32% | 64° E | +Cr | Jupiter |
| 16th | Ari | 3h 3m | 20° 6' | -9.8 | 32' | 43% | 77° E | FQ | Jupiter, Uranus, Pleiades |
| 17th | Tau | 3h 58m | 24° 14' | -10.3 | 31' | 54% | 90° E | FQ | Pleiades, Hyades, Aldebaran |
| 18th | Tau | 4h 54m | 27° 0' | -10.7 | 31' | 64% | 103° E | FQ | Hyades, Aldebaran |
| 19th | Tau | 5h 51m | 28° 18' | -11.1 | 30' | 74% | 116° E | +G | |
| 20th | Gem | 6h 48m | 28° 6' | -11.4 | 30' | 83% | 129° E | +G | |

## Mercury and Venus

 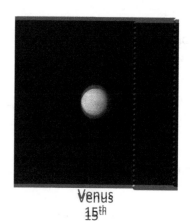

Mercury
15th

Venus
15th

## Mercury

| Date | Con. | R.A. | Dec. | Mag. | Diam. | Ill. | Elon. | Vis. | Rat. | Close To |
|------|------|------|------|------|-------|------|-------|------|------|----------|
| 11th | Cap | 20h 48m | -19° 46' | -0.4 | 5" | 94% | 13° W | NV | N/A | |
| 13th | Cap | 21h 1m | -18° 58' | -0.5 | 5" | 95% | 11° W | NV | N/A | |
| 15th | Cap | 21h 14m | -18° 5' | -0.6 | 5" | 96% | 9° W | NV | N/A | |
| 17th | Cap | 21h 28m | -17° 6' | -0.7 | 5" | 97% | 8° W | NV | N/A | |
| 19th | Cap | 21h 41m | -16° 2' | -0.8 | 5" | 98% | 6° W | NV | N/A | |

## Venus

| Date | Con. | R.A. | Dec. | Mag. | Diam. | Ill. | Elon. | Vis. | Rat. | Close To |
|------|------|------|------|------|-------|------|-------|------|------|----------|
| 11th | Sgr | 19h 39m | -21° 20' | -3.4 | 12" | 88% | 29° W | AM | * | Mars |
| 13th | Sgr | 19h 49m | -21° 0' | -3.4 | 12" | 88% | 29° W | AM | * | Mars |
| 15th | Sgr | 20h 0m | -20° 37' | -3.4 | 12" | 89% | 28° W | AM | * | Mars |
| 17th | Cap | 20h 10m | -20° 12' | -3.4 | 12" | 89% | 27° W | AM | * | Mars |
| 19th | Cap | 20h 20m | -19° 45' | -3.4 | 11" | 89% | 27° W | AM | * | Mars |

## Mars and the Outer Planets

Mars
15th

Jupiter
15th

Saturn
15th

## Mars

| Date | Con. | R.A. | Dec. | Mag. | Diam. | Ill. | Elon. | Vis. | Rat. | Close To |
|------|------|------|------|------|-------|------|-------|------|------|----------|
| 11th | Sgr | 20h 2m | -21° 25' | 1.5 | 4" | 98% | 24° W | AM | * | Venus |
| 15th | Cap | 20h 15m | -20° 47' | 1.5 | 4" | 98% | 24° W | AM | * | Venus |
| 20th | Cap | 20h 31m | -19° 56' | 1.5 | 4" | 98% | 25° W | AM | * | Venus |

## The Outer Planets

| Planet | Date | Con. | R.A. | Dec. | Mag. | Diam. | Elon. | Vis. | Rat. | Close To |
|--------|------|------|------|------|------|-------|-------|------|------|----------|
| Jupiter | 15th | Ari | 2h 27m | 13° 34' | -1.9 | 38" | 69° E | PM | ** | Moon |
| Saturn | 15th | Aqr | 22h 41m | -10° 2' | 1.2 | 16" | 12° E | NV | N/A | |
| Uranus | 15th | Ari | 3h 7m | 17° 15' | 6.1 | 4" | 79° E | PM | * | Jupiter, Pleiades |
| Neptune | 15th | Psc | 23h 48m | -2° 37' | 7.8 | 2" | 29° E | PM | *** | |

## Highlights

| Date | Time (UT) | Event |
|------|-----------|-------|
| 13th | 05:26 | The waxing crescent Moon is south of Neptune. (Evening sky.) |
| | N/A | Good opportunity to see Earthshine on the waxing crescent Moon. (Evening sky.) |
| 15th | 07:12 | The nearly first quarter Moon is north of Jupiter. (Evening sky.) |
| 16th | 03:04 | The almost first quarter Moon is north of Uranus. (Evening sky.) |
| | 15:02 | First Quarter Moon. (Evening sky.) |
| | 20:08 | The first quarter Moon is south of the Pleiades star cluster. (Evening sky.) |

# February 21st to 29th, 2024

## The Moon

21st

23rd

25th

27th

29th

| Date | Con | R.A. | Dec | Mag | Diam | Ill. | Elon. | Phase | Close To |
|------|-----|------|-----|-----|------|------|-------|-------|----------|
| 21st | Gem | 7h 43m | 26° 30' | -11.7 | 30' | 89% | 142° E | +G | |
| 22nd | Cnc | 8h 35m | 23° 41' | -12.0 | 30' | 94% | 154° E | +G | Praesepe |
| 23rd | Leo | 9h 24m | 19° 52' | -12.3 | 30' | 98% | 165° E | FM | Praesepe, Regulus |
| 24th | Leo | 10h 10m | 15° 17' | -12.6 | 29' | 100% | 176° E | FM | Regulus |
| 25th | Leo | 10h 54m | 10° 10' | -12.6 | 29' | 100% | 174° W | FM | Regulus |
| 26th | Leo | 11h 37m | 4° 42' | -12.3 | 29' | 98% | 164° W | FM | |
| 27th | Vir | 12h 18m | 0° 55' | -12.0 | 29' | 95% | 155° W | -G | |
| 28th | Vir | 13h 0m | -6° 32' | -11.7 | 30' | 90% | 145° W | -G | Spica |
| 29th | Vir | 13h 44m | -11° 58' | -11.4 | 30' | 83% | 135° W | -G | Spica |

## Mercury and Venus

Mercury
25th

Venus
25th

### Mercury

| Date | Con. | R.A. | Dec. | Mag. | Diam. | Ill. | Elon. | Vis. | Rat. | Close To |
|------|------|------|------|------|-------|------|-------|------|------|----------|
| 21st | Cap | 21h 55m | -14° 52' | -0.9 | 5" | 98% | 5° W | NV | N/A | |
| 23rd | Aqr | 22h 9m | -13° 36' | -1.0 | 5" | 99% | 3° W | NV | N/A | Saturn |
| 25th | Aqr | 22h 22m | -12° 16' | -1.1 | 5" | 99% | 2° W | NV | N/A | Saturn |
| 27th | Aqr | 22h 36m | -10° 49' | -1.2 | 5" | 100% | 0° W | NV | N/A | Saturn |
| 29th | Aqr | 22h 50m | -9° 18' | -1.3 | 5" | 100% | 1° E | NV | N/A | Saturn |

**Venus**

| Date | Con. | R.A. | Dec. | Mag. | Diam. | Ill. | Elon. | Vis. | Rat. | Close To |
|------|------|------|------|------|-------|------|-------|------|------|----------|
| 21st | Cap | 20h 31m | -19° 15' | -3.4 | 11" | 90% | 26° W | AM | * | Mars |
| 23rd | Cap | 20h 41m | -18° 43' | -3.3 | 11" | 90% | 25° W | AM | * | Mars |
| 25th | Cap | 20h 51m | -18° 8' | -3.3 | 11" | 90% | 25° W | AM | * | Mars |
| 27th | Cap | 21h 1m | -17° 32' | -3.3 | 11" | 91% | 24° W | AM | * | Mars |
| 29th | Cap | 21h 11m | -16° 54' | -3.3 | 11" | 91% | 24° W | AM | * | Mars |

## Mars and the Outer Planets

Mars
25th

Jupiter
25th

Saturn
25th

**Mars**

| Date | Con. | R.A. | Dec. | Mag. | Diam. | Ill. | Elon. | Vis. | Rat. | Close To |
|------|------|------|------|------|-------|------|-------|------|------|----------|
| 21st | Cap | 20h 34m | -19° 45' | 1.5 | 4" | 98% | 25° W | AM | * | Venus |
| 25th | Cap | 20h 47m | -18° 59' | 1.5 | 4" | 97% | 26° W | AM | * | Venus |
| 29th | Cap | 20h 59m | -18° 10' | 1.5 | 4" | 97% | 27° W | AM | * | Venus |

**The Outer Planets**

| Planet | Date | Con. | R.A. | Dec. | Mag. | Diam. | Elon. | Vis. | Rat. | Close To |
|--------|------|------|------|------|------|-------|-------|------|------|----------|
| Jupiter | 25th | Ari | 2h 33m | 14° 5' | -1.8 | 37" | 61° E | PM | ** | |
| Saturn | 25th | Aqr | 22h 45m | -9° 35' | 1.2 | 16" | 4° E | NV | N/A | Mercury |
| Uranus | 25th | Ari | 3h 8m | 17° 19' | 6.1 | 4" | 69° E | PM | * | Jupiter, Pleiades |
| Neptune | 25th | Psc | 23h 49m | -2° 29' | 7.8 | 2" | 20° E | PM | ** | |

## Highlights

| Date | Time (UT) | Event |
|------|-----------|-------|
| 22nd | 15:39 | Venus is 0.6° north of Mars. (Morning sky.) |
| 23rd | 21:37 | The nearly full Moon is north of the bright star Regulus. (Evening sky.) |
| 24th | 12:31 | Full Moon. (Visible all night.) |
| 28th | 08:31 | Mercury is at superior conjunction with the Sun. (Not visible.) |
| | 14:04 | The waning gibbous Moon is north of the bright star Spica. (Morning sky.) |
| 29th | 15:29 | Saturn is in conjunction with the Sun. (Not visible.) |

# March 1st to 10th, 2024

## The Moon

1st

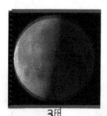
3rd

5th

7th

9th

| Date | Con | R.A. | Dec | Mag | Diam | Ill. | Elon. | Phase | Close To |
|------|-----|------|-----|-----|------|------|-------|-------|----------|
| 1st | Lib | 14h 29m | -17° 1' | -11.1 | 30' | 75% | 125° W | -G | |
| 2nd | Lib | 15h 17m | -21° 29' | -10.8 | 30' | 66% | 114° W | -G | |
| 3rd | Sco | 16h 10m | -25° 5' | -10.4 | 31' | 57% | 102° W | LQ | Antares |
| 4th | Oph | 17h 6m | -27° 31' | -10.0 | 31' | 46% | 88° W | LQ | Antares |
| 5th | Sgr | 18h 7m | -28° 29' | -9.4 | 32' | 36% | 74° W | -Cr | |
| 6th | Sgr | 19h 9m | -27° 44' | -8.8 | 32' | 26% | 60° W | -Cr | |
| 7th | Cap | 20h 11m | -25° 11' | -8.1 | 33' | 16% | 45° W | -Cr | |
| 8th | Cap | 21h 12m | -20° 56' | -7.1 | 33' | 8% | 31° W | NM | Venus, Mars |
| 9th | Aqr | 22h 11m | -15° 15' | -5.9 | 33' | 3% | 17° W | NM | Venus, Mars, Saturn |
| 10th | Aqr | 23h 7m | -8° 35' | -4.5 | 33' | 0% | 4° W | NM | Saturn, Neptune |

## Mercury and Venus

Mercury
5th

Venus
5th

### Mercury

| Date | Con. | R.A. | Dec. | Mag. | Diam. | Ill. | Elon. | Vis. | Rat. | Close To |
|------|------|------|------|------|-------|------|-------|------|------|----------|
| 1st | Aqr | 22h 57m | -8° 31' | -1.4 | 5" | 100% | 3° E | NV | N/A | Saturn |
| 3rd | Aqr | 23h 11m | -6° 52' | -1.4 | 5" | 99% | 4° E | NV | N/A | Saturn |
| 5th | Aqr | 23h 25m | -5° 9' | -1.4 | 5" | 98% | 5° E | NV | N/A | Saturn |
| 7th | Aqr | 23h 39m | -3° 23' | -1.4 | 5" | 97% | 7° E | NV | N/A | Neptune |
| 9th | Psc | 23h 52m | -1° 33' | -1.3 | 5" | 94% | 8° E | NV | N/A | Neptune |

## Venus

| Date | Con. | R.A. | Dec. | Mag. | Diam. | Ill. | Elon. | Vis. | Rat. | Close To |
|------|------|------|------|------|-------|------|-------|------|------|----------|
| 1st | Cap | 21h 16m | -16° 34' | -3.3 | 11" | 91% | 23° W | AM | * | Mars |
| 3rd | Cap | 21h 26m | -15° 53' | -3.3 | 11" | 92% | 23° W | AM | * | Mars |
| 5th | Cap | 21h 36m | -15° 10' | -3.3 | 11" | 92% | 22° W | AM | * | Mars |
| 7th | Cap | 21h 46m | -14° 25' | -3.3 | 11" | 92% | 21° W | AM | * | Mars |
| 9th | Cap | 21h 55m | -13° 39' | -3.3 | 11" | 93% | 21° W | AM | * | Moon, Mars |

## Mars and the Outer Planets

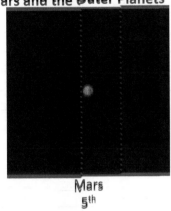

Mars
5th

Jupiter
5th

Saturn
5th

## Mars

| Date | Con. | R.A. | Dec. | Mag. | Diam. | Ill. | Elon. | Vis. | Rat. | Close To |
|------|------|------|------|------|-------|------|-------|------|------|----------|
| 1st | Cap | 21h 2m | -17° 57' | 1.5 | 4" | 97% | 27° W | AM | * | Venus |
| 5th | Cap | 21h 15m | -17° 4' | 1.4 | 4" | 97% | 27° W | AM | * | Venus |
| 10th | Cap | 21h 30m | -15° 54' | 1.4 | 4" | 97% | 28° W | AM | * | Venus |

## The Outer Planets

| Planet | Date | Con. | R.A. | Dec. | Mag. | Diam. | Elon. | Vis. | Rat. | Close To |
|--------|------|------|------|------|------|-------|-------|------|------|----------|
| Jupiter | 5th | Ari | 2h 39m | 14° 35' | -1.7 | 36" | 54° E | PM | ** | |
| Saturn | 5th | Aqr | 22h 50m | -9° 10' | 1.2 | 16" | 4° W | NV | N/A | Mercury |
| Uranus | 5th | Ari | 3h 9m | 17° 23' | 6.1 | 4" | 61° E | PM | * | Jupiter, Pleiades |
| Neptune | 5th | Psc | 23h 50m | -2° 21' | 7.8 | 2" | 12° E | NV | N/A | Mercury |

## Highlights

| Date | Time (UT) | Event |
|------|-----------|-------|
| 3rd | 05:28 | The almost last quarter Moon is north of the bright star Aldebaran. (Morning sky.) |
| | 15:24 | Last Quarter Moon. (Morning sky.) |
| 6th | N/A | Good opportunity to see Earthshine on the waning crescent Moon. (Morning sky.) |
| 8th | 03:12 | The waning crescent Moon is south of Mars. (Morning sky.) |
| | 18:40 | The waning crescent Moon is south of Venus. (Morning sky.) |
| 10th | 09:01 | New Moon. (Not visible.) |

# March 11th to 20th, 2024

## The Moon

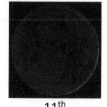    

| 11th | 13th | 15th | 17th | 19th |

| Date | Con | R.A. | Dec | Mag | Diam | Ill. | Elon. | Phase | Close To |
|------|-----|------|-----|-----|------|------|-------|-------|----------|
| 11th | Psc | 0h 1m | -1° 24' | -4.9 | 33' | 1% | 9° E | NM | Mercury, Neptune |
| 12th | Psc | 0h 54m | 5° 48' | -6.2 | 33' | 4% | 21° E | NM | Mercury |
| 13th | Ari | 1h 48m | 12° 34' | -7.3 | 33' | 10% | 34° E | NM | |
| 14th | Ari | 2h 43m | 18° 29' | -8.2 | 32' | 18% | 46° E | +Cr | Jupiter, Uranus |
| 15th | Tau | 3h 40m | 23° 12' | -9.0 | 32' | 28% | 60° E | +Cr | Uranus, Pleiades, Hyades |
| 16th | Tau | 4h 38m | 26° 30' | -9.6 | 31' | 38% | 73° E | FQ | Hyades, Aldebaran |
| 17th | Tau | 5h 36m | 28° 15' | -10.0 | 31' | 48% | 87° E | FQ | |
| 18th | Aur | 6h 34m | 28° 25' | -10.5 | 30' | 58% | 101° E | FQ | |
| 19th | Gem | 7h 30m | 27° 8' | -10.8 | 30' | 68% | 114° E | +G | |
| 20th | Cnc | 8h 23m | 24° 34' | -11.2 | 30' | 77% | 126° E | +G | Praesepe |

## Mercury and Venus

Mercury
15th

Venus
15th

### Mercury

| Date | Con. | R.A. | Dec. | Mag. | Diam. | Ill. | Elon. | Vis. | Rat. | Close To |
|------|------|------|------|------|-------|------|-------|------|------|----------|
| 11th | Psc | 0h 6m | 0° 17' | -1.2 | 5" | 91% | 10° E | NV | N/A | Moon, Neptune |
| 13th | Psc | 0h 19m | 2° 8' | -1.1 | 6" | 87% | 11° E | NV | N/A | |
| 15th | Psc | 0h 32m | 3° 57' | -1.0 | 6" | 81% | 13° E | NV | N/A | |
| 17th | Psc | 0h 44m | 5° 42' | -0.9 | 6" | 74% | 14° E | NV | N/A | |
| 19th | Psc | 0h 56m | 7° 20' | -0.7 | 6" | 67% | 15° E | PM | ** | |

## Venus

| Date | Con. | R.A. | Dec. | Mag. | Diam. | Ill. | Elon. | Vis. | Rat. | Close To |
|------|------|------|------|------|-------|------|-------|------|------|----------|
| 11th | Aqr | 22h 5m | -12° 51' | -3.3 | 11" | 93% | 20° W | AM | * | Mars |
| 13th | Aqr | 22h 15m | -12° 2' | -3.3 | 11" | 93% | 20° W | AM | * | Mars, Saturn |
| 15th | Aqr | 22h 24m | -11° 12' | -3.3 | 11" | 94% | 19° W | AM | * | Mars, Saturn |
| 17th | Aqr | 22h 34m | -10° 20' | -3.3 | 11" | 94% | 19° W | AM | * | Saturn |
| 19th | Aqr | 22h 43m | -9° 27' | -3.3 | 11" | 94% | 18° W | AM | * | Saturn |

## Mars and the Outer Planets

Mars
15th

Jupiter
15th

Saturn
15th

**Mars**

| Date | Con. | R.A. | Dec. | Mag. | Diam. | Ill. | Elon. | Vis. | Rat. | Close To |
|------|------|------|------|------|-------|------|-------|------|------|----------|
| 11th | Cap | 21h 33m | -15° 40' | 1.4 | 4" | 97% | 28° W | AM | * | Venus |
| 15th | Cap | 21h 45m | -14° 41' | 1.4 | 4" | 97% | 29° W | AM | * | Venus |
| 20th | Aqr | 22h 1m | -13° 23' | 1.4 | 4" | 96% | 30° W | AM | * | |

**The Outer Planets**

| Planet | Date | Con. | R.A. | Dec. | Mag. | Diam. | Elon. | Vis. | Rat. | Close To |
|--------|------|------|------|------|------|-------|-------|------|------|----------|
| Jupiter | 15th | Ari | 2h 46m | 15° 10' | -1.7 | 35" | 46° E | PM | * | |
| Saturn | 15th | Aqr | 22h 54m | -8° 43' | 1.2 | 16" | 12° W | NV | N/A | Venus |
| Uranus | 15th | Ari | 3h 10m | 17° 29' | 6.1 | 3" | 52° E | PM | * | Moon, Jupiter, Pleiades |
| Neptune | 15th | Psc | 23h 51m | -2° 12' | 7.8 | 2" | 3° E | NV | N/A | |

## Highlights

| Date | Time (UT) | Event |
|------|-----------|-------|
| 13th | N/A | Good opportunity to see Earthshine on the waxing crescent Moon. (Evening sky.) |
| 14th | 01:47 | The waxing crescent Moon is north of Jupiter. (Evening sky.) |
| | 09:49 | The waxing crescent Moon is north of Uranus. (Evening sky.) |
| 15th | 03:19 | The waxing crescent Moon is south of the Pleiades open star cluster. (Evening sky.) |
| 17th | 04:11 | First Quarter Moon. (Evening sky.) |
| 18th | 00:34 | Neptune is in conjunction with the Sun. (Not visible.) |
| 20th | 03:07 | Vernal equinox. |

# March 21st to 31st, 2024

## The Moon

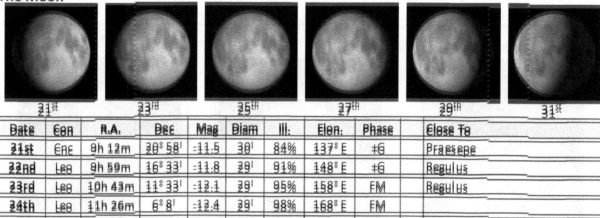

21st　　23rd　　25th　　27th　　29th　　31st

| Date | Con | R.A. | Dec | Mag | Diam | Ill. | Elon. | Phase | Close To |
|------|-----|------|-----|-----|------|------|-------|-------|----------|
| 21st | Cnc | 9h 12m | 20° 58' | -11.5 | 30' | 84% | 137° E | +G | Praesepe |
| 22nd | Leo | 9h 59m | 16° 33' | -11.8 | 29' | 91% | 148° E | +G | Regulus |
| 23rd | Leo | 10h 43m | 11° 33' | -12.1 | 29' | 95% | 158° E | FM | Regulus |
| 24th | Leo | 11h 26m | 6° 8' | -12.4 | 29' | 98% | 168° E | FM | |
| 25th | Vir | 12h 8m | 0° 30' | -12.6 | 29' | 100% | 178° E | FM | |
| 26th | Vir | 12h 50m | -5° 10' | -12.5 | 30' | 100% | 173° W | FM | Spica |
| 27th | Vir | 13h 33m | -10° 42' | -12.3 | 30' | 97% | 163° W | FM | Spica |
| 28th | Vir | 14h 18m | -15° 54' | -12.0 | 30' | 93% | 153° W | -G | |
| 29th | Lib | 15h 5m | -20° 32' | -11.7 | 30' | 88% | 142° W | -G | |
| 30th | Sco | 15h 56m | -24° 22' | -11.3 | 30' | 80% | 130° W | -G | Antares |
| 31st | Sco | 16h 51m | -27° 6' | -11.0 | 31' | 72% | 117° W | -G | Antares |

## Mercury and Venus

Mercury
25th

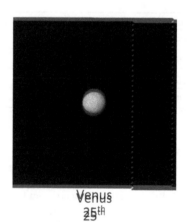

Venus
25th

### Mercury

| Date | Con. | R.A. | Dec. | Mag. | Diam. | Ill. | Elon. | Vis. | Rat. | Close To |
|------|------|------|------|------|-------|------|-------|------|------|----------|
| 21st | Psc | 1h 6m | 8° 51' | -0.4 | 7" | 59% | 16° E | PM | ** | |
| 23rd | Psc | 1h 15m | 10° 12' | -0.2 | 7" | 51% | 16° E | PM | ** | |
| 25th | Psc | 1h 22m | 11° 20' | 0.1 | 8" | 43% | 16° E | PM | ** | |
| 27th | Psc | 1h 28m | 12° 16' | 0.4 | 8" | 34% | 16° E | PM | ** | |
| 29th | Psc | 1h 33m | 12° 57' | 0.7 | 8" | 27% | 15° E | PM | ** | |
| 31st | Psc | 1h 35m | 13° 22' | 1.0 | 9" | 20% | 14° E | NV | N/A | |

## Venus

| Date | Con. | R.A. | Dec. | Mag. | Diam. | Ill. | Elon. | Vis. | Rat. | Close To |
|------|------|------|------|------|-------|------|-------|------|------|----------|
| 21st | Aqr | 22h 52m | -8° 33' | -3.3 | 11" | 94% | 18° W | AM | * | Saturn |
| 23rd | Aqr | 23h 2m | -7° 39' | -3.3 | 10" | 95% | 17° W | AM | * | Saturn |
| 25th | Aqr | 23h 11m | -6° 43' | -3.3 | 10" | 95% | 17° W | AM | * | Saturn |
| 27th | Aqr | 23h 20m | -5° 47' | -3.3 | 10" | 95% | 16° W | AM | * | Saturn |
| 29th | Aqr | 23h 29m | -4° 50' | -3.3 | 10" | 95% | 16° W | AM | * | Saturn |
| 31st | Aqr | 23h 38m | -3° 52' | -3.3 | 10" | 96% | 15° W | AM | * | Saturn, Neptune |

## Mars and the Outer Planets

Mars
25th

Jupiter
25th

Saturn
25th

## Mars

| Date | Con. | R.A. | Dec. | Mag. | Diam. | Ill. | Elon. | Vis. | Rat. | Close To |
|------|------|------|------|------|-------|------|-------|------|------|----------|
| 21st | Aqr | 22h 4m | -13° 7' | 1.4 | 4" | 96% | 30° W | AM | * | |
| 25th | Aqr | 22h 16m | -12° 3' | 1.4 | 4" | 96% | 30° W | AM | * | |
| 31st | Aqr | 22h 33m | -10° 21' | 1.4 | 4" | 96% | 31° W | AM | * | Saturn |

## The Outer Planets

| Planet | Date | Con. | R.A. | Dec. | Mag. | Diam. | Elon. | Vis. | Rat. | Close To |
|--------|------|------|------|------|------|-------|-------|------|------|----------|
| Jupiter | 25th | Ari | 2h 54m | 15° 47' | -1.7 | 35" | 39° E | PM | * | Uranus |
| Saturn | 25th | Aqr | 22h 59m | -8° 17' | 1.3 | 16" | 20° W | AM | * | Venus |
| Uranus | 25th | Ari | 3h 12m | 17° 36' | 6.1 | 3" | 44° E | PM | * | Jupiter, Pleiades |
| Neptune | 25th | Psc | 23h 53m | -2° 3' | 7.8 | 2" | 6° W | NV | N/A | |

## Highlights

| Date | Time (UT) | Event |
|------|-----------|-------|
| 22nd | 01:51 | Venus is 0.3° north of Saturn. (Morning sky.) |
| | 06:24 | The waxing gibbous Moon is north of the bright star Regulus. (Evening sky.) |
| 24th | 22:26 | Mercury is at greatest eastern elongation from the Sun. (Evening sky.) |
| 25th | 07:01 | Full Moon. (Visible all night.) |
| | 07:12 | Penumbral eclipse of the Moon. Visible from Antarctica, Central America, North America, the Pacific and South America |
| 26th | 17:44 | The just-past full Moon is north of the bright star Spica. (Visible all night.) |
| 30th | 14:56 | The waning gibbous Moon is north of the bright star Antares. (Morning sky.) |

# April 1st to 10th, 2024

## The Moon

| 1st | 3rd | 5th | 7th | 9th |

| Date | Con | R.A. | Dec | Mag | Diam | Ill. | Elon. | Phase | Close To |
|------|-----|------|-----|-----|------|------|-------|-------|----------|
| 1st | Sgr | 17h 49m | -28° 28' | -10.6 | 31' | 62% | 103° W | LQ | |
| 2nd | Sgr | 18h 50m | -28° 14' | -10.2 | 31' | 51% | 89° W | LQ | |
| 3rd | Sgr | 19h 50m | -26° 20' | -9.7 | 32' | 41% | 75° W | LQ | |
| 4th | Cap | 20h 50m | -22° 46' | -9.1 | 32' | 30% | 61° W | -Cr | |
| 5th | Cap | 21h 47m | -17° 46' | -8.4 | 33' | 20% | 48° W | -Cr | |
| 6th | Aqr | 22h 43m | -11° 37' | -7.5 | 33' | 11% | 35° W | NM | Mars, Saturn |
| 7th | Aqr | 23h 36m | -4° 44' | -6.4 | 33' | 5% | 22° W | NM | Venus, Mars, Saturn, Neptune |
| 8th | Cet | 0h 30m | 2° 29' | -5.0 | 33' | 1% | 10° W | NM | Venus, Neptune |
| 9th | Psc | 1h 23m | 9° 32' | -4.2 | 33' | 0% | 3° E | NM | Mercury |
| 10th | Ari | 2h 18m | 15° 59' | -5.7 | 33' | 2% | 16° E | NM | |

## Mercury and Venus

Mercury
5th

Venus
5th

**Mercury**

| Date | Con. | R.A. | Dec. | Mag. | Diam. | Ill. | Elon. | Vis. | Rat. | Close To |
|------|------|------|------|------|-------|------|-------|------|------|----------|
| 1st | Psc | 1h 36m | 13° 29' | 1.2 | 9" | 17% | 13° E | NV | N/A | |
| 3rd | Psc | 1h 35m | 13° 32' | 1.6 | 10" | 12% | 11° E | NV | N/A | |
| 5th | Psc | 1h 34m | 13° 18' | 1.9 | 10" | 7% | 9° E | NV | N/A | |
| 7th | Psc | 1h 31m | 12° 50' | 2.3 | 11" | 4% | 7° E | NV | N/A | |
| 9th | Psc | 1h 27m | 12° 9' | 2.8 | 11" | 2% | 4° E | NV | N/A | Moon |

**Venus**

| Date | Con. | R.A. | Dec. | Mag. | Diam. | Ill. | Elon. | Vis. | Rat. | Close To |
|------|------|------|------|------|-------|------|-------|------|------|----------|
| 1st | Aqr | 23h 43m | -3° 24' | -3.3 | 10" | 96% | 15° W | NV | N/A | Neptune |
| 3rd | Psc | 23h 52m | -2° 25' | -3.3 | 10" | 96% | 14° W | NV | N/A | Neptune |
| 5th | Psc | 0h 1m | -1° 27' | -3.3 | 10" | 96% | 14° W | NV | N/A | Neptune |
| 7th | Psc | 0h 10m | 0° 28' | -3.3 | 10" | 97% | 14° W | NV | N/A | Moon, Neptune |
| 9th | Psc | 0h 19m | 0° 31' | -3.3 | 10" | 97% | 13° W | NV | N/A | |

## Mars and the Outer Planets

Mars
5th

Jupiter
5th

Saturn
5th

**Mars**

| Date | Con. | R.A. | Dec. | Mag. | Diam. | Ill. | Elon. | Vis. | Rat. | Close To |
|------|------|------|------|------|-------|------|-------|------|------|----------|
| 1st | Aqr | 22h 36m | -10° 4' | 1.4 | 4" | 96% | 32° W | AM | * | Saturn |
| 5th | Aqr | 22h 48m | -8° 55' | 1.4 | 5" | 95% | 32° W | AM | * | Saturn |
| 10th | Aqr | 23h 3m | -7° 26' | 1.4 | 5" | 95% | 33° W | AM | * | Saturn |

**The Outer Planets**

| Planet | Date | Con. | R.A. | Dec. | Mag. | Diam. | Elon. | Vis. | Rat. | Close To |
|--------|------|------|------|------|------|-------|-------|------|------|----------|
| Jupiter | 5th | Ari | 3h 3m | 16° 29' | -1.6 | 34" | 32° E | PM | * | Uranus |
| Saturn | 5th | Aqr | 23h 3m | -7° 49' | 1.3 | 16" | 29° W | AM | * | Mars |
| Uranus | 5th | Ari | 3h 14m | 17° 44' | 6.2 | 3" | 34° E | PM | * | Jupiter, Pleiades |
| Neptune | 5th | Psc | 23h 54m | -1° 54' | 7.8 | 2" | 16° W | NV | N/A | Venus |

## Highlights

| Date | Time (UT) | Event |
|------|-----------|-------|
| 1st | 20:04 | Mercury is stationary prior to beginning retrograde motion. (Not visible.) |
| 2nd | 03:15 | Last Quarter Moon. (Morning sky.) |
| 5th | N/A | Good opportunity to see Earthshine on the waning crescent Moon. (Morning sky.) |
| 6th | 02:07 | The waning crescent Moon is south of Mars. (Morning sky.) |
| | 08:49 | The waning crescent Moon is south of Saturn. (Morning sky.) |
| 7th | 06:30 | The nearly new Moon is south of Neptune. (Morning sky.) |
| 8th | 18:18 | Total solar eclipse. Visible from the Atlantic, Central America, North America and the Pacific. |
| | 18:22 | New Moon. (Not visible.) |
| 10th | 22:32 | The waxing crescent Moon is north of Jupiter. (Evening sky.) |

# April 11th to 20th, 2024

## The Moon

| 11th | 13th | 15th | 17th | 19th |

| Date | Con | R.A. | Dec | Mag | Diam | Ill. | Elon. | Phase | Close To |
|------|-----|------|-----|-----|------|------|-------|-------|----------|
| 11th | Ari | 3h 16m | 21° 24' | -6.9 | 32' | 7% | 29° E | NM | Jupiter, Uranus, Pleiades |
| 12th | Tau | 4h 15m | 25° 26' | -7.8 | 32' | 14% | 43° E | +Cr | Pleiades, Hyades, Aldebaran |
| 13th | Tau | 5h 15m | 27° 50' | -8.6 | 31' | 23% | 57° E | +Cr | Aldebaran |
| 14th | Aur | 6h 15m | 28° 34' | -9.2 | 31' | 32% | 71° E | +Cr | |
| 15th | Gem | 7h 13m | 27° 42' | -9.8 | 31' | 42% | 85° E | FQ | |
| 16th | Cnc | 8h 8m | 25° 26' | -10.2 | 30' | 52% | 97° E | FQ | Praesepe |
| 17th | Cnc | 8h 59m | 22° 4' | -10.6 | 30' | 62% | 109° E | FQ | Praesepe |
| 18th | Leo | 9h 47m | 17° 49' | -11.0 | 30' | 71% | 120° E | +G | Regulus |
| 19th | Leo | 10h 31m | 12° 57' | -11.3 | 30' | 79% | 131° E | +G | Regulus |
| 20th | Leo | 11h 14m | 7° 38' | -11.6 | 29' | 86% | 140° E | +G | |

## Mercury and Venus

 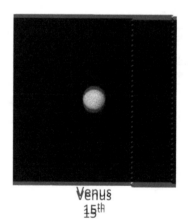

Mercury
15th

Venus
15th

### Mercury

| Date | Con. | R.A. | Dec. | Mag. | Diam. | Ill. | Elon. | Vis. | Rat. | Close To |
|------|------|------|------|------|-------|------|-------|------|------|----------|
| 11th | Psc | 1h 22m | 11° 17' | 3.1 | 11" | 0% | 1° E | NV | N/A | |
| 13th | Psc | 1h 17m | 10° 19' | 3.2 | 12" | 0% | 2° W | NV | N/A | Venus |
| 15th | Psc | 1h 12m | 9° 17' | 2.9 | 12" | 1% | 5° W | NV | N/A | Venus |
| 17th | Psc | 1h 8m | 8° 15' | 2.6 | 12" | 3% | 8° W | NV | N/A | Venus |
| 19th | Psc | 1h 4m | 7° 18' | 2.3 | 12" | 5% | 11° W | NV | N/A | Venus |

## Venus

| Date | Con. | R.A. | Dec. | Mag. | Diam. | Ill. | Elon. | Vis. | Rat. | Close To |
|------|------|------|------|------|-------|------|-------|------|------|----------|
| 11th | Cet | 0h 28m | 1° 29' | -3.3 | 10" | 97% | 13° W | NV | N/A | |
| 13th | Cet | 0h 37m | 2° 28' | -3.3 | 10" | 97% | 12° W | NV | N/A | Mercury |
| 15th | Psc | 0h 46m | 3° 27' | -3.3 | 10" | 97% | 12° W | NV | N/A | Mercury |
| 17th | Psc | 0h 56m | 4° 25' | -3.3 | 10" | 98% | 11° W | NV | N/A | Mercury |
| 19th | Psc | 1h 5m | 5° 23' | -3.3 | 10" | 98% | 11° W | NV | N/A | Mercury |

## Mars and the Outer Planets

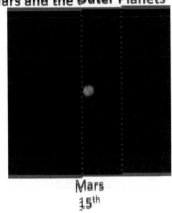

Mars
15th

Jupiter
15th

Saturn
15th

## Mars

| Date | Con. | R.A. | Dec. | Mag. | Diam. | Ill. | Elon. | Vis. | Rat. | Close To |
|------|------|------|------|------|-------|------|-------|------|------|----------|
| 11th | Aqr | 23h 5m | -7° 9' | 1.4 | 5" | 95% | 33° W | AM | * | Saturn |
| 15th | Aqr | 23h 17m | -5° 57' | 1.3 | 5" | 95% | 34° W | AM | * | Saturn |
| 20th | Aqr | 23h 31m | -4° 25' | 1.3 | 5" | 95% | 35° W | AM | * | Saturn |

## The Outer Planets

| Planet | Date | Con. | R.A. | Dec. | Mag. | Diam. | Elon. | Vis. | Rat. | Close To |
|--------|------|------|------|------|------|-------|-------|------|------|----------|
| Jupiter | 15th | Ari | 3h 12m | 17° 6' | -1.6 | 33" | 25° E | PM | * | Uranus, Pleiades |
| Saturn | 15th | Aqr | 23h 7m | -7° 26' | 1.4 | 16" | 37° W | AM | * | Mars |
| Uranus | 15th | Ari | 3h 16m | 17° 53' | 6.2 | 3" | 25° E | PM | * | Jupiter, Pleiades |
| Neptune | 15th | Psc | 23h 56m | -1° 45' | 7.8 | 2" | 25° W | AM | ** | Mars |

## Highlights

| Date | Time (UT) | Event |
|------|-----------|-------|
| 11th | 00:31 | The waxing crescent Moon is north of Uranus (Evening sky.) |
| | 03:02 | Mars is 0.5° north of Saturn. (Morning sky.) |
| | 11:24 | The waxing crescent Moon is south of the Pleiades open star cluster. (Evening sky.) |
| | 22:56 | Mercury is at inferior conjunction with the Sun. (Not visible.) |
| 13th | 06:51 | The waxing crescent Moon is north of the bright star Aldebaran. (Evening sky.) |
| | N/A | Good opportunity to see Earthshine on the waxing crescent Moon. (Evening sky.) |
| 15th | 19:14 | First Quarter Moon. (Evening sky.) |
| 18th | 10:16 | The waxing gibbous Moon is north of the bright star Regulus. (Evening sky.) |
| 20th | 02:43 | Jupiter is 0.5° south of Uranus. (Evening sky.) |

# April 21st to 30th, 2024

## The Moon

21st

23rd

25th

27th

29th

| Date | Con | R.A. | Dec | Mag | Diam | Ill. | Elon. | Phase | Close To |
|------|-----|------|-----|-----|------|------|-------|-------|----------|
| 21st | Vir | 11h 56m | 2° 3' | -11.9 | 29' | 92% | 150° E | +G | |
| 22nd | Vir | 12h 38m | -3° 38' | -12.2 | 30' | 96% | 160° E | FM | Spica |
| 23rd | Vir | 13h 21m | -9° 14' | -12.4 | 30' | 99% | 169° E | FM | Spica |
| 24th | Vir | 14h 6m | -14° 35' | -12.7 | 30' | 100% | 179° E | FM | Spica |
| 25th | Lib | 14h 53m | -19° 25' | -12.4 | 30' | 99% | 170° W | FM | |
| 26th | Lib | 15h 44m | -23° 30' | -12.1 | 30' | 96% | 158° W | FM | Antares |
| 27th | Sco | 16h 38m | -26° 32' | -11.8 | 31' | 91% | 145° W | -G | Antares |
| 28th | Oph | 17h 36m | -28° 14' | -11.5 | 31' | 84% | 132° W | -G | |
| 29th | Sgr | 18h 35m | -28° 23' | -11.2 | 31' | 76% | 118° W | -G | |
| 30th | Sgr | 19h 35m | -26° 54' | -10.8 | 32' | 66% | 104° W | -G | |

## Mercury and Venus

Mercury
25th

Venus
25th

### Mercury

| Date | Con. | R.A. | Dec. | Mag. | Diam. | Ill. | Elon. | Vis. | Rat. | Close To |
|------|------|------|------|------|-------|------|-------|------|------|----------|
| 21st | Psc | 1h 2m | 6° 26' | 2.0 | 11" | 8% | 14° W | NV | N/A | Venus |
| 23rd | Psc | 1h 0m | 5° 44' | 1.8 | 11" | 12% | 16° W | AM | *** | Venus |
| 25th | Psc | 1h 0m | 5° 11' | 1.6 | 11" | 15% | 18° W | AM | *** | Venus |
| 27th | Psc | 1h 1m | 4° 49' | 1.4 | 10" | 19% | 19° W | AM | *** | |
| 29th | Psc | 1h 3m | 4° 37' | 1.3 | 10" | 23% | 21° W | AM | *** | |

**Venus**

| Date | Con. | R.A. | Dec. | Mag. | Diam. | Ill. | Elon. | Vis. | Rat. | Close To |
|------|------|------|------|------|-------|------|-------|------|------|----------|
| 21st | Psc | 1h 14m | 6° 21' | -3.3 | 10" | 98% | 11° W | NV | N/A | Mercury |
| 23rd | Psc | 1h 23m | 7° 18' | -3.4 | 10" | 98% | 10° W | NV | N/A | Mercury |
| 25th | Psc | 1h 32m | 8° 14' | -3.4 | 10" | 98% | 10° W | NV | N/A | Mercury |
| 27th | Psc | 1h 41m | 9° 10' | -3.4 | 10" | 98% | 9° W | NV | N/A | |
| 29th | Psc | 1h 51m | 10° 4' | -3.4 | 10" | 99% | 9° W | NV | N/A | |

## Mars and the Outer Planets

Mars
25th

Jupiter
25th

Saturn
25th

**Mars**

| Date | Con. | R.A. | Dec. | Mag. | Diam. | Ill. | Elon. | Vis. | Rat. | Close To |
|------|------|------|------|------|-------|------|-------|------|------|----------|
| 21st | Aqr | 23h 34m | -4° 7' | 1.3 | 5" | 95% | 36° W | AM | * | Saturn |
| 25th | Psc | 23h 46m | -2° 54' | 1.3 | 5" | 94% | 36° W | AM | * | Saturn, Neptune |
| 30th | Psc | 0h 0m | -1° 22' | 1.3 | 5" | 94% | 38° W | AM | * | Neptune |

**The Outer Planets**

| Planet | Date | Con. | R.A. | Dec. | Mag. | Diam. | Elon. | Vis. | Rat. | Close To |
|--------|------|------|------|------|------|-------|-------|------|------|----------|
| Jupiter | 25th | Ari | 3h 22m | 17° 43' | -1.6 | 33" | 18° E | PM | * | Uranus, Pleiades |
| Saturn | 25th | Aqr | 23h 11m | -7° 5' | 1.4 | 16" | 45° W | AM | * | Mars |
| Uranus | 25th | Ari | 3h 18m | 18° 1' | 6.2 | 3" | 17° E | PM | * | Jupiter, Pleiades |
| Neptune | 25th | Psc | 23h 57m | -1° 38' | 7.8 | 2" | 34° W | AM | *** | Mars |

## Highlights

| Date | Time (UT) | Event |
|------|-----------|-------|
| 22nd | N/A | The Lyrid meteor shower is at its maximum. (ZHR: 18) |
| 23rd | 03:55 | The almost full Moon is north of the bright star Spica. (Visible all night.) |
| | 23:50 | Full Moon. (Visible all night.) |
| 24th | 08:27 | Mercury is stationary prior to resuming prograde motion. (Morning sky.) |
| 26th | 18:25 | The waning gibbous Moon is north of the bright star Antares. (Morning sky.) |
| 29th | 03:55 | Mars is 0.0° south of Neptune. (Morning sky.) |

# May 1st to 10th, 2024

## The Moon

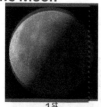 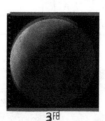   

| 1st | 3rd | 5th | 7th | 9th |

| Date | Con | R.A. | Dec | Mag | Diam | Ill. | Elon. | Phase | Close To |
|------|-----|------|-----|-----|------|------|-------|-------|----------|
| 1st | Cap | 20h 34m | -23° 49' | -10.4 | 32' | 55% | 90° W | LQ | |
| 2nd | Cap | 21h 30m | -19° 19' | -9.9 | 32' | 44% | 77° W | LQ | |
| 3rd | Aqr | 22h 25m | -13° 40' | -9.3 | 32' | 33% | 64° W | -Cr | |
| 4th | Aqr | 23h 17m | -7° 13' | -8.6 | 33' | 22% | 52° W | -Cr | Saturn, Neptune |
| 5th | Psc | 0h 9m | 0° 18' | -7.8 | 33' | 13% | 40° W | -Cr | Mars, Neptune |
| 6th | Psc | 1h 1m | 6° 41' | -6.7 | 33' | 6% | 28° W | NM | Mercury, Mars |
| 7th | Ari | 1h 54m | 13° 19' | -5.5 | 33' | 2% | 16° W | NM | Mercury, Venus |
| 8th | Ari | 2h 50m | 19° 10' | -4.2 | 33' | 0% | 3° W | NM | Venus, Jupiter, Uranus |
| 9th | Tau | 3h 49m | 23° 50' | -5.2 | 32' | 1% | 11° E | NM | Jupiter, Uranus, Pleiades, Hyades |
| 10th | Tau | 4h 50m | 26° 58' | -6.4 | 32' | 5% | 25° E | NM | Hyades, Aldebaran |

## Mercury and Venus

 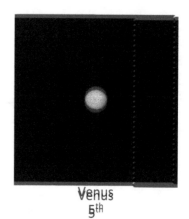

Mercury
5th

Venus
5th

### Mercury

| Date | Con | R.A. | Dec. | Mag. | Diam. | Ill. | Elon. | Vis. | Rat. | Close To |
|------|-----|------|------|------|-------|------|-------|------|------|----------|
| 1st | Psc | 1h 6m | 4° 35' | 1.1 | 10" | 26% | 22° W | AM | *** | |
| 3rd | Psc | 1h 11m | 4° 42' | 1.0 | 9" | 30% | 23° W | AM | *** | |
| 5th | Psc | 1h 16m | 4° 59' | 0.9 | 9" | 33% | 24° W | AM | *** | |
| 7th | Psc | 1h 21m | 5° 24' | 0.8 | 9" | 37% | 24° W | AM | *** | Moon |
| 9th | Psc | 1h 28m | 5° 56' | 0.7 | 8" | 40% | 24° W | AM | *** | |

## Venus

| Date | Con. | R.A. | Dec. | Mag. | Diam. | Ill. | Elon. | Vis. | Rat. | Close To |
|------|------|------|------|------|-------|------|-------|------|------|----------|
| 1st | Ari | 2h 0m | 10° 58' | -3.4 | 10" | 99% | 9° W | NV | N/A | |
| 3rd | Ari | 2h 9m | 11° 51' | -3.4 | 10" | 99% | 8° W | NV | N/A | |
| 5th | Ari | 2h 19m | 12° 43' | -3.4 | 10" | 99% | 8° W | NV | N/A | |
| 7th | Ari | 2h 28m | 13° 34' | -3.4 | 10" | 99% | 7° W | NV | N/A | Moon |
| 9th | Ari | 2h 38m | 14° 23' | -3.4 | 10" | 99% | 7° W | NV | N/A | |

## Mars and the Outer Planets

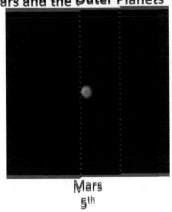

Mars
5th

Jupiter
5th

Saturn
5th

## Mars

| Date | Con. | R.A. | Dec. | Mag. | Diam. | Ill. | Elon. | Vis. | Rat. | Close To |
|------|------|------|------|------|-------|------|-------|------|------|----------|
| 1st | Psc | 0h 3m | -1° 3' | 1.3 | 5" | 94% | 38° W | AM | * | Neptune |
| 5th | Psc | 0h 14m | 0° 11' | 1.3 | 5" | 94% | 39° W | AM | * | Moon, Neptune |
| 10th | Cet | 0h 28m | 1° 42' | 1.3 | 5" | 93% | 40° W | AM | * | |

## The Outer Planets

| Planet | Date | Con. | R.A. | Dec. | Mag. | Diam. | Elon. | Vis. | Rat. | Close To |
|--------|------|------|------|------|------|-------|-------|------|------|----------|
| Jupiter | 5th | Tau | 3h 31m | 18° 18' | -1.5 | 33" | 10° E | NV | N/A | Uranus, Pleiades |
| Saturn | 5th | Aqr | 23h 14m | -6° 46' | 1.4 | 16" | 54° W | AM | ** | |
| Uranus | 5th | Ari | 3h 21m | 18° 10' | 6.2 | 3" | 8° E | NV | N/A | Jupiter, Pleiades |
| Neptune | 5th | Psc | 23h 58m | -1° 31' | 7.8 | 2" | 43° W | AM | *** | Moon, Mars |

## Highlights

| Date | Time (UT) | Event |
|------|-----------|-------|
| 1st | 11:28 | Last Quarter Moon. (Morning sky.) |
| 3rd | 13:15 | Dwarf planet Pluto is stationary prior to beginning retrograde motion. (Morning sky.) |
| | 21:53 | The waning crescent Moon is south of Saturn. (Morning sky.) |
| 4th | 19:38 | The waning crescent Moon is south of Neptune. (Morning sky.) |
| 5th | 00:53 | The waning crescent Moon is south of Mars. (Morning sky.) |
| | N/A | Good opportunity to see Earthshine on the waning crescent Moon. (Morning sky.) |
| 6th | 06:54 | The waning crescent Moon is north of Mercury. (Morning sky.) |
| 7th | N/A | The Eta Aquariid meteor shower is at its maximum. (ZHR: 40) |
| 8th | 03:23 | New Moon. (Not visible.) |
| 9th | 21:25 | Mercury is at greatest western elongation from the Sun. (Morning sky.) |
| 10th | N/A | The Eta Lyrid meteor shower is at its maximum. (ZHR: 3) |

# May 11<sup>th</sup> to 20<sup>th</sup>, 2024

## The Moon

11<sup>th</sup>

13<sup>th</sup>

15<sup>th</sup>

17<sup>th</sup>

19<sup>th</sup>

| Date | Con | R.A. | Dec | Mag | Diam | Ill. | Elon. | Phase | Close To |
|------|-----|------|-----|-----|------|------|-------|-------|----------|
| 11th | Tau | 5h 51m | 28° 23' | -7.4 | 31' | 10% | 40° E | NM | |
| 12th | Gem | 6h 52m | 28° 6' | -8.2 | 31' | 18% | 54° E | +Cr | |
| 13th | Gem | 7h 49m | 26° 16' | -8.9 | 31' | 26% | 67° E | +Cr | |
| 14th | Cnc | 8h 42m | 23° 11' | -9.4 | 30' | 36% | 79° E | +Cr | Praesepe |
| 15th | Leo | 9h 32m | 19° 8' | -9.9 | 30' | 45% | 91° E | FQ | Regulus |
| 16th | Leo | 10h 18m | 14° 23' | -10.3 | 30' | 55% | 101° E | FQ | Regulus |
| 17th | Leo | 11h 1m | 9° 10' | -10.7 | 30' | 64% | 111° E | FQ | |
| 18th | Vir | 11h 44m | 3° 40' | -11.0 | 30' | 73% | 121° E | +G | |
| 19th | Vir | 12h 25m | -1° 59' | -11.4 | 30' | 81% | 130° E | +G | |
| 20th | Vir | 13h 8m | -7° 37' | -11.7 | 30' | 88% | 140° E | +G | Spica |

## Mercury and Venus

Mercury
15<sup>th</sup>

Venus
15<sup>th</sup>

**Mercury**

| Date | Con. | R.A. | Dec. | Mag. | Diam. | Ill. | Elon. | Vis. | Rat. | Close To |
|------|------|------|------|------|-------|------|-------|------|------|----------|
| 11th | Psc | 1h 36m | 6° 35' | 0.7 | 8" | 43% | 24° W | AM | *** | |
| 13th | Psc | 1h 44m | 7° 21' | 0.6 | 8" | 47% | 24° W | AM | *** | |
| 15th | Psc | 1h 52m | 8° 12' | 0.5 | 7" | 50% | 24° W | AM | *** | |
| 17th | Psc | 2h 2m | 9° 8' | 0.4 | 7" | 53% | 24° W | AM | *** | |
| 19th | Cet | 2h 12m | 10° 8' | 0.3 | 7" | 57% | 23° W | AM | *** | |

## Venus

| Date | Con. | R.A. | Dec. | Mag. | Diam. | Ill. | Elon. | Vis. | Rat. | Close To |
|------|------|------|------|------|-------|------|-------|------|------|----------|
| 11th | Ari | 2h 48m | 15° 11' | -3.4 | 10" | 99% | 6° W | NV | N/A | |
| 13th | Ari | 2h 57m | 15° 57' | -3.4 | 10" | 99% | 6° W | NV | N/A | |
| 15th | Ari | 3h 7m | 16° 42' | -3.4 | 10" | 100% | 5° W | NV | N/A | Jupiter, Uranus, Pleiades |
| 17th | Ari | 3h 17m | 17° 25' | -3.4 | 10" | 100% | 5° W | NV | N/A | Jupiter, Uranus, Pleiades |
| 19th | Tau | 3h 27m | 18° 6' | -3.4 | 10" | 100% | 4° W | NV | N/A | Jupiter, Uranus, Pleiades |

## Mars and the Outer Planets

Mars
15th

Jupiter
15th

Saturn
15th

### Mars

| Date | Con. | R.A. | Dec. | Mag. | Diam. | Ill. | Elon. | Vis. | Rat. | Close To |
|------|------|------|------|------|-------|------|-------|------|------|----------|
| 11th | Cet | 0h 31m | 2° 0' | 1.3 | 5" | 93% | 41° W | AM | * | |
| 15th | Psc | 0h 42m | 3° 13' | 1.3 | 5" | 93% | 42° W | AM | * | |
| 20th | Psc | 0h 56m | 4° 42' | 1.3 | 5" | 93% | 43° W | AM | * | |

### The Outer Planets

| Planet | Date | Con. | R.A. | Dec. | Mag. | Diam. | Elon. | Vis. | Rat. | Close To |
|--------|------|------|------|------|------|-------|-------|------|------|----------|
| Jupiter | 15th | Tau | 3h 41m | 18° 52' | -1.5 | 33" | 3° E | NV | N/A | Venus, Uranus, Pleiades |
| Saturn | 15th | Aqr | 23h 17m | -6° 30' | 1.4 | 17" | 63° W | AM | ** | |
| Uranus | 15th | Ari | 3h 23m | 18° 19' | 6.2 | 3" | 1° W | NV | N/A | Venus, Jupiter, Pleiades |
| Neptune | 15th | Psc | 23h 59m | -1° 25' | 7.7 | 2" | 52° W | AM | *** | |

## Highlights

| Date | Time (UT) | Event |
|------|-----------|-------|
| 11th | N/A | Good opportunity to see Earthshine on the waxing crescent Moon. (Evening sky.) |
| 13th | 10:59 | Uranus is in conjunction with the Sun. (Not visible.) |
| 14th | 00:32 | The waxing crescent Moon is north of the Praesepe open star cluster. (Evening sky.) |
| 15th | 11:49 | First Quarter Moon. (Evening sky.) |
| | 19:00 | The first quarter Moon is north of the bright star Regulus. (Evening sky.) |
| 19th | 00:22 | Jupiter is in conjunction with the Sun. (Not visible.) |
| 20th | 09:24 | The just-past first quarter Moon is north of the bright star Spica. (Evening sky.) |

# May 21ˢᵗ to 31ˢᵗ, 2024

## The Moon

| 21ˢᵗ | 23ʳᵈ | 25ᵗʰ | 27ᵗʰ | 29ᵗʰ | 31ˢᵗ |
|------|------|------|------|------|------|

| Date | Con | R.A. | Dec | Mag | Diam | Ill. | Elon. | Phase | Close To |
|------|-----|------|-----|-----|------|------|-------|-------|----------|
| 21st | Vir | 13h 52m | -13° 3' | -12.0 | 30' | 93% | 150° E | +G | Spica |
| 22nd | Lib | 14h 39m | -18° 4' | -12.2 | 30' | 97% | 161° E | FM | |
| 23rd | Lib | 15h 29m | -22° 24' | -12.5 | 30' | 100% | 172° E | FM | |
| 24th | Sco | 16h 23m | -25° 46' | -12.6 | 31' | 100% | 175° W | FM | Antares |
| 25th | Oph | 17h 21m | -27° 51' | -12.3 | 31' | 98% | 162° W | FM | |
| 26th | Sgr | 18h 21m | -28° 23' | -12.0 | 31' | 93% | 148° W | -G | |
| 27th | Sgr | 19h 21m | -27° 15' | -11.6 | 32' | 87% | 134° W | -G | |
| 28th | Cap | 20h 21m | -24° 29' | -11.3 | 32' | 79% | 120° E | -G | |
| 29th | Cap | 21h 18m | -20° 17' | -10.9 | 32' | 69% | 107° W | -G | |
| 30th | Aqr | 22h 12m | -14° 56' | -10.5 | 32' | 58% | 94° W | LQ | |
| 31st | Aqr | 23h 4m | -8° 46' | -10.0 | 32' | 47% | 82° W | LQ | Saturn |

## Mercury and Venus

Mercury
25ᵗʰ

Venus
25ᵗʰ

### Mercury

| Date | Con. | R.A. | Dec. | Mag. | Diam. | Ill. | Elon. | Vis. | Rat. | Close To |
|------|------|------|------|------|-------|------|-------|------|------|----------|
| 21st | Ari | 2h 22m | 11° 13' | 0.1 | 7" | 60% | 23° W | AM | *** | |
| 23rd | Ari | 2h 34m | 12° 20' | 0.0 | 6" | 64% | 22° W | AM | *** | |
| 25th | Ari | 2h 46m | 13° 31' | -0.1 | 6" | 68% | 21° W | AM | *** | |
| 27th | Ari | 2h 59m | 14° 43' | -0.3 | 6" | 72% | 20° W | AM | *** | |
| 29th | Ari | 3h 12m | 15° 56' | -0.4 | 6" | 76% | 18° W | AM | ** | Uranus, Pleiades |
| 31st | Tau | 3h 26m | 17° 9' | -0.6 | 6" | 80% | 17° W | AM | ** | Jupiter, Uranus, Pleiades |

## Venus

| Date | Con. | R.A. | Dec. | Mag. | Diam. | Ill. | Elon. | Vis. | Rat. | Close To |
|------|------|------|------|------|-------|------|-------|------|------|----------|
| 21st | Tau | 3h 37m | 18° 46' | -3.4 | 10" | 100% | 4° W | NV | N/A | Jupiter, Uranus, Pleiades |
| 23rd | Tau | 3h 47m | 19° 23' | -3.5 | 10" | 100% | 3° W | NV | N/A | Jupiter, Pleiades, Hyades |
| 25th | Tau | 3h 57m | 19° 59' | -3.5 | 10" | 100% | 3° W | NV | N/A | Jupiter, Pleiades, Hyades |
| 27th | Tau | 4h 7m | 20° 32' | -3.5 | 10" | 100% | 2° W | NV | N/A | Jupiter, Pleiades, Hyades |
| 29th | Tau | 4h 18m | 21° 3' | -3.5 | 10" | 100% | 2° W | NV | N/A | Jupiter, Hyades, Aldebaran |
| 31st | Tau | 4h 28m | 21° 32' | -3.5 | 10" | 100% | 1° W | NV | N/A | Jupiter, Hyades, Aldebaran |

## Mars and the Outer Planets

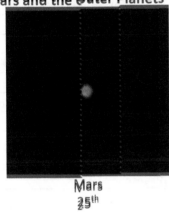

Mars
25th

Jupiter
25th

Saturn
25th

## Mars

| Date | Con. | R.A. | Dec. | Mag. | Diam. | Ill. | Elon. | Vis. | Rat. | Close To |
|------|------|------|------|------|-------|------|-------|------|------|----------|
| 21st | Psc | 0h 59m | 5° 0' | 1.3 | 5" | 93% | 43° W | AM | * | |
| 25th | Psc | 1h 10m | 6° 11' | 1.3 | 5" | 93% | 45° W | AM | * | |
| 31st | Psc | 1h 27m | 7° 54' | 1.2 | 5" | 92% | 47° W | AM | * | |

## The Outer Planets

| Planet | Date | Con. | R.A. | Dec. | Mag. | Diam. | Elon. | Vis. | Rat. | Close To |
|--------|------|------|------|------|------|-------|-------|------|------|----------|
| Jupiter | 25th | Tau | 3h 50m | 19° 34' | -1.5 | 33" | 5° W | NV | N/A | Venus, Pleiades, Hyades |
| Saturn | 25th | Aqr | 23h 20m | -6° 17' | 1.4 | 17" | 72° W | AM | ** | |
| Uranus | 25th | Tau | 3h 25m | 18° 28' | 6.2 | 3" | 11° W | NV | N/A | Mercury, Venus, Jupiter, Pleiades |
| Neptune | 25th | Psc | 0h 0m | -1° 20' | 7.7 | 2" | 62° W | AM | *** | |

## Highlights

| Date | Time (UT) | Event |
|------|-----------|-------|
| 21st | 13:54 | Full Moon. (Visible all night.) |
| 24th | 04:04 | The just-past full Moon is south of Antares. (Visible all night.) |
| 30th | 17:13 | Last Quarter Moon. (Morning sky.) |
| 31st | 09:27 | The just-past last quarter Moon is south of Saturn. (Morning sky.) |

# June 1ˢᵗ to 10ᵗʰ, 2024

## The Moon

| 1ˢᵗ | 3ʳᵈ | 5ᵗʰ | 7ᵗʰ | 9ᵗʰ |

| Date | Con | R.A. | Dec | Mag | Diam | Ill. | Elon. | Phase | Close To |
|------|-----|------|-----|-----|------|------|-------|-------|----------|
| **1st** | Psc | 23h 54m | -2° 7' | -9.4 | 32' | 35% | 71° W | -Cr | Saturn, Neptune |
| **2nd** | Psc | 0h 45m | 4° 40' | -8.8 | 32' | 25% | 59° W | -Cr | Mars, Neptune |
| **3rd** | Psc | 1h 36m | 11° 14' | -8.0 | 32' | 16% | 47° W | -Cr | Mars |
| **4th** | Ari | 2h 30m | 17° 14' | -7.1 | 32' | 8% | 35° W | NM | |
| **5th** | Ari | 3h 26m | 22° 14' | -6.0 | 32' | 3% | 22° W | NM | Jupiter, Uranus, Pleiades |
| **6th** | Tau | 4h 26m | 25° 55' | -4.8 | 32' | 0% | 8° W | NM | Mercury, Venus, Jupiter, Hyades |
| **7th** | Tau | 5h 27m | 27° 58' | -4.7 | 32' | 0% | 6° E | NM | Venus |
| **8th** | Aur | 6h 28m | 28° 18' | -5.9 | 31' | 3% | 21° E | NM | |
| **9th** | Gem | 7h 27m | 26° 59' | -6.9 | 31' | 7% | 34° E | NM | |
| **10th** | Cnc | 8h 23m | 24° 16' | -7.8 | 30' | 13% | 47° E | +Cr | Praesepe |

## Mercury and Venus

Mercury
5ᵗʰ

Venus
5ᵗʰ

**Mercury**

| Date | Con. | R.A. | Dec. | Mag. | Diam. | Ill. | Elon. | Vis. | Rat. | Close To |
|------|------|------|------|------|-------|------|-------|------|------|----------|
| **1st** | Tau | 3h 34m | 17° 46' | -0.7 | 6" | 82% | 16° W | AM | ** | Jupiter, Uranus, Pleiades |
| **3rd** | Tau | 3h 49m | 18° 58' | -0.9 | 5" | 86% | 14° W | NV | N/A | Jupiter, Pleiades, Hyades |
| **5th** | Tau | 4h 5m | 20° 7' | -1.1 | 5" | 89% | 12° W | NV | N/A | Moon, Jupiter, Pleiades |
| **7th** | Tau | 4h 22m | 21° 12' | -1.3 | 5" | 93% | 10° W | NV | N/A | Jupiter, Hyades, Aldebaran |
| **9th** | Tau | 4h 40m | 22° 11' | -1.5 | 5" | 96% | 7° W | NV | N/A | Venus, Jupiter, Aldebaran |

## Venus

| Date | Con. | R.A. | Dec. | Mag. | Diam. | Ill. | Elon. | Vis. | Rat. | Close To |
|------|------|------|------|------|-------|------|-------|------|------|----------|
| 1st | Tau | 4h 33m | 21° 45' | -3.5 | 10" | 100% | 1° W | NV | N/A | Jupiter, Hyades, Aldebaran |
| 3rd | Tau | 4h 44m | 22° 10' | -3.5 | 10" | 100% | 0° W | NV | N/A | Hyades, Aldebaran |
| 5th | Tau | 4h 54m | 22° 33' | -3.5 | 10" | 100% | 0° E | NV | N/A | Hyades, Aldebaran |
| 7th | Tau | 5h 5m | 22° 52' | -3.5 | 10" | 100% | 1° E | NV | N/A | Moon, Hyades, Aldebaran |
| 9th | Tau | 5h 15m | 23° 10' | -3.5 | 10" | 100% | 1° E | NV | N/A | Mercury |

## Mars and the Outer Planets

Mars
5<sup>th</sup>

Jupiter
5<sup>th</sup>

Saturn
5<sup>th</sup>

### Mars

| Date | Con. | R.A. | Dec. | Mag. | Diam. | Ill. | Elon. | Vis. | Rat. | Close To |
|------|------|------|------|------|-------|------|-------|------|------|----------|
| 1st | Psc | 1h 30m | 8° 11' | 1.2 | 5" | 92% | 47° W | AM | * | |
| 5th | Psc | 1h 41m | 9° 17' | 1.2 | 5" | 92% | 48° W | AM | * | |
| 10th | Psc | 1h 55m | 10° 38' | 1.2 | 5" | 92% | 50° W | AM | * | |

### The Outer Planets

| Planet | Date | Con. | R.A. | Dec. | Mag. | Diam. | Elon. | Vis. | Rat. | Close To |
|--------|------|------|------|------|------|-------|-------|------|------|----------|
| Jupiter | 5th | Tau | 4h 1m | 19° 56' | -1.5 | 33" | 13° W | NV | N/A | Moon, Mercury, Pleiades, Hyades |
| Saturn | 5th | Aqr | 23h 22m | -6° 6' | 1.4 | 17" | 83° W | AM | ** | |
| Uranus | 5th | Tau | 3h 28m | 18° 37' | 6.2 | 3" | 21° W | AM | * | Moon, Mercury, Jupiter, Pleiades |
| Neptune | 5th | Psc | 0h 1m | -1° 16' | 7.7 | 2" | 73° W | AM | **** | Saturn |

## Highlights

| Date | Time (UT) | Event |
|------|-----------|-------|
| 1st | 01:05 | The waning crescent Moon is south of Neptune. (Morning sky.) |
| 2nd | 22:44 | The waning crescent Moon is north of Mars. (Morning sky.) |
| 3rd | N/A | Good opportunity to see Earthshine on the waning crescent Moon. (Morning sky.) |
| 4th | 14:59 | Venus is at superior conjunction with the Sun. (Not visible.) |
| | 23:59 | The waning crescent Moon is north of Uranus. (Morning sky.) |
| 6th | 12:38 | New Moon. (Not visible.) |
| 10th | 06:17 | The waxing crescent Moon is north of the Praesepe open star cluster. (Evening sky.) |
| | N/A | Good opportunity to see Earthshine on the waxing crescent Moon. (Evening sky.) |

# June 11<sup>th</sup> to 20<sup>th</sup>, 2024

## The Moon

11<sup>th</sup>

13<sup>th</sup>

15<sup>th</sup>

17<sup>th</sup>

19<sup>th</sup>

| Date | Con | R.A. | Dec | Mag | Diam | Ill. | Elon. | Phase | Close To |
|------|-----|------|-----|-----|------|------|-------|-------|----------|
| 11th | Cnc | 9h 14m | 20° 28' | -8.5 | 30' | 21% | 59° E | +Cr | Praesepe |
| 12th | Leo | 10h 2m | 15° 52' | -9.1 | 30' | 29% | 70° E | +Cr | Regulus |
| 13th | Leo | 10h 47m | 10° 45' | -9.6 | 30' | 38% | 80° E | FQ | Regulus |
| 14th | Leo | 11h 30m | 5° 17' | -10.0 | 30' | 48% | 90° E | FQ | |
| 15th | Vir | 12h 11m | 0° 19' | -10.4 | 30' | 57% | 99° E | FQ | |
| 16th | Vir | 12h 53m | -5° 57' | -10.8 | 30' | 67% | 109° E | +G | Spica |
| 17th | Vir | 13h 37m | -11° 25' | -11.1 | 30' | 75% | 118° E | +G | Spica |
| 18th | Vir | 14h 22m | -16° 33' | -11.4 | 30' | 83% | 129° E | +G | |
| 19th | Lib | 15h 11m | -21° 6' | -11.8 | 30' | 90% | 140° E | +G | |
| 20th | Sco | 16h 4m | -24° 49' | -12.1 | 31' | 95% | 152° E | FM | Antares |

## Mercury and Venus

Mercury
15<sup>th</sup>

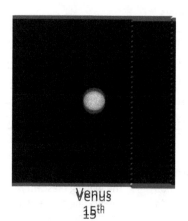
Venus
15<sup>th</sup>

### Mercury

| Date | Con. | R.A. | Dec. | Mag. | Diam. | Ill. | Elon. | Vis. | Rat. | Close To |
|------|------|------|------|------|-------|------|-------|------|------|----------|
| 11th | Tau | 4h 59m | 23° 3' | -1.7 | 5" | 98% | 5° W | NV | N/A | Venus, Hyades, Aldebaran |
| 13th | Tau | 5h 17m | 23° 47' | -1.8 | 5" | 100% | 2° W | NV | N/A | Venus |
| 15th | Tau | 5h 37m | 24° 20' | -1.9 | 5" | 100% | 0° E | NV | N/A | Venus |
| 17th | Tau | 5h 56m | 24° 42' | -1.7 | 5" | 99% | 3° E | NV | N/A | Venus |
| 19th | Gem | 6h 15m | 24° 53' | -1.6 | 5" | 98% | 6° E | NV | N/A | Venus |

## Venus

| Date | Con. | R.A. | Dec. | Mag. | Diam. | Ill. | Elon. | Vis. | Rat. | Close To |
|------|------|------|------|------|-------|------|-------|------|------|----------|
| 11th | Tau | 5h 26m | 23° 24' | -3.5 | 10" | 100% | 2° E | NV | N/A | Mercury |
| 13th | Tau | 5h 37m | 23° 36' | -3.5 | 10" | 100% | 2° E | NV | N/A | Mercury |
| 15th | Tau | 5h 47m | 23° 45' | -3.5 | 10" | 100% | 3° E | NV | N/A | Mercury |
| 17th | Tau | 5h 58m | 23° 53' | -3.5 | 10" | 100% | 4° E | NV | N/A | Mercury |
| 19th | Gem | 6h 9m | 23° 55' | -3.5 | 10" | 100% | 4° E | NV | N/A | Mercury |

## Mars and the Outer Planets

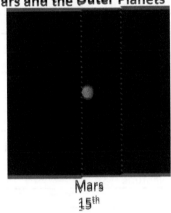

Mars
15th

Jupiter
15th

Saturn
15th

## Mars

| Date | Con. | R.A. | Dec. | Mag. | Diam. | Ill. | Elon. | Vis. | Rat. | Close To |
|------|------|------|------|------|-------|------|-------|------|------|----------|
| 11th | Ari | 1h 58m | 10° 54' | 1.2 | 5" | 92% | 50° W | AM | * | |
| 15th | Ari | 2h 9m | 11° 56' | 1.2 | 5" | 91% | 51° W | AM | * | |
| 20th | Ari | 2h 23m | 13° 11' | 1.2 | 5" | 91% | 53° W | AM | * | |

## The Outer Planets

| Planet | Date | Con. | R.A. | Dec. | Mag. | Diam. | Ill. | Elon. | Vis. | Rat. | Close To |
|--------|------|------|------|------|------|-------|------|-------|------|------|----------|
| Jupiter | 15th | Tau | 4h 11m | 20° 22' | -1.5 | 33" | | 21° W | AM | * | Pleiades, Hyades, Aldebaran |
| Saturn | 15th | Aqr | 23h 23m | -6° 1' | 1.3 | 17" | | 93° W | AM | ** | |
| Uranus | 15th | Tau | 3h 30m | 18° 45' | 6.2 | 3" | | 31° W | AM | * | Pleiades |
| Neptune | 15th | Psc | 0h 1m | -1° 13' | 7.7 | 2" | | 84° W | AM | **** | Saturn |

## Highlights

| Date | Time (UT) | Event |
|------|-----------|-------|
| 12th | 03:37 | The waxing crescent Moon is north of the bright star Regulus. (Evening sky.) |
| 14th | 05:19 | First Quarter Moon. (Evening sky.) |
| | 16:20 | Mercury is at superior conjunction with the Sun. (Not visible.) |
| 16th | 15:40 | The waxing gibbous Moon is north of the bright star Spica. (Evening sky.) |
| 20th | 10:35 | The waxing gibbous Moon is north of the bright star Antares. (Evening sky.) |
| | 20:52 | Estival (summer) solstice. |

# June 21st to 30th, 2024

## The Moon

21st

23rd

25th

27th

29th

| Date | Con | R.A. | Dec | Mag | Diam | Ill. | Elon. | Phase | | Close To |
|------|-----|------|-----|-----|------|------|-------|-------|--|----------|
| 21st | Oph | 17h 1m | -27° 20' | -12.4 | 31' | 99% | 165° E | FM | | Antares |
| 22nd | Sgr | 18h 2m | -28° 21' | -12.6 | 31' | 100% | 179° E | FM | | |
| 23rd | Sgr | 19h 3m | -27° 40' | -12.4 | 32' | 99% | 166° W | FM | | |
| 24th | Sgr | 20h 5m | -25° 15' | -12.1 | 32' | 95% | 152° W | FM | | |
| 25th | Cap | 21h 3m | -21° 18' | -11.7 | 32' | 89% | 138° W | -G | | |
| 26th | Cap | 21h 59m | -16° 5' | -11.4 | 32' | 81% | 125° W | -G | | |
| 27th | Aqr | 22h 52m | -10° 0' | -11.0 | 32' | 71% | 113° W | -G | | Saturn |
| 28th | Aqr | 23h 43m | -3° 25' | -10.6 | 32' | 60% | 102° W | LQ | | Saturn, Neptune |
| 29th | Psc | 0h 33m | 3° 19' | -10.1 | 32' | 49% | 90° W | LQ | | Neptune |
| 30th | Psc | 1h 23m | 9° 52' | -9.5 | 32' | 38% | 79° W | LQ | | |

## Mercury and Venus

Mercury
25th

Venus
25th

### Mercury

| Date | Con. | R.A. | Dec. | Mag. | Diam. | Ill. | Elon. | Vis. | Rat. | | Close To |
|------|------|------|------|------|-------|------|-------|------|------|--|----------|
| 21st | Gem | 6h 34m | 24° 53' | -1.4 | 5" | 95% | 8° E | NV | N/A | | Venus |
| 23rd | Gem | 6h 52m | 24° 42' | -1.2 | 5" | 92% | 11° E | NV | N/A | | Venus |
| 25th | Gem | 7h 10m | 24° 22' | -1.0 | 5" | 89% | 13° E | NV | N/A | | Venus |
| 27th | Gem | 7h 27m | 23° 52' | -0.8 | 5" | 86% | 15° E | PM | ** | | Venus |
| 29th | Gem | 7h 43m | 23° 15' | -0.6 | 6" | 82% | 17° E | PM | ** | | |

## Venus

| Date | Con. | R.A. | Dec. | Mag. | Diam. | Ill. | Elon. | Vis. | Rat. | Close To |
|------|------|------|------|------|-------|------|-------|------|------|----------|
| 21st | Gem | 6h 20m | 23° 56' | -3.4 | 10" | 100% | 5° E | NV | N/A | Mercury |
| 23rd | Gem | 6h 30m | 23° 54' | -3.4 | 10" | 100% | 5° E | NV | N/A | Mercury |
| 25th | Gem | 6h 41m | 23° 49' | -3.4 | 10" | 100% | 6° E | NV | N/A | Mercury |
| 27th | Gem | 6h 52m | 23° 41' | -3.4 | 10" | 99% | 7° E | NV | N/A | Mercury |
| 29th | Gem | 7h 3m | 23° 30' | -3.4 | 10" | 99% | 7° E | NV | N/A | |

## Mars and the Outer Planets

Mars
25th

Jupiter
25th

Saturn
25th

### Mars

| Date | Con. | R.A. | Dec. | Mag. | Diam. | Ill. | Elon. | Vis. | Rat. | Close To |
|------|------|------|------|------|-------|------|-------|------|------|----------|
| 21st | Ari | 2h 26m | 13° 25' | 1.2 | 5" | 91% | 53° W | AM | * | |
| 25th | Ari | 2h 38m | 14° 22' | 1.2 | 5" | 91% | 55° W | AM | * | |
| 30th | Ari | 2h 52m | 15° 29' | 1.2 | 5" | 91% | 56° W | AM | * | |

### The Outer Planets

| Planet | Date | Con. | R.A. | Dec. | Mag. | Diam. | Elon. | Vis. | Rat. | Close To |
|--------|------|------|------|------|------|-------|-------|------|------|----------|
| Jupiter | 25th | Tau | 4h 20m | 20° 46' | -1.6 | 33" | 29° W | AM | * | Pleiades, Hyades, Aldebaran |
| Saturn | 25th | Aqr | 23h 24m | -5° 60' | 1.3 | 18" | 103° W | AM | *** | |
| Uranus | 25th | Tau | 3h 32m | 18° 52' | 6.2 | 3" | 41° W | AM | * | Pleiades |
| Neptune | 25th | Psc | 0h 1m | -1° 12' | 7.7 | 2" | 94° W | AM | **** | Saturn |

## Highlights

| Date | Time (UT) | Event |
|------|-----------|-------|
| 22nd | 01:09 | Full Moon. (Visible all night.) |
| 24th | N/A | The June Bootid meteor shower is at its maximum. (ZHR: Var.) |
| 27th | 15:39 | The nearly last quarter Moon is north of Saturn. (Morning sky.) |
| 28th | 10:50 | The almost last quarter Moon is north of Neptune. (Morning sky.) |
| | 21:54 | Last Quarter Moon. (Morning sky.) |
| 30th | 19:54 | Saturn is stationary prior to beginning retrograde motion. (Morning sky.) |

# July 1ˢᵗ to 10ᵗʰ, 2024

## The Moon

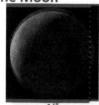    

| 1ˢᵗ | 3ʳᵈ | 5ᵗʰ | 7ᵗʰ | 9ᵗʰ |

| Date | Con | R.A. | Dec | Mag | Diam | Ill. | Elon. | Phase | Close To |
|------|-----|------|-----|-----|------|------|-------|-------|----------|
| 1st | Ari | 2h 15m | 15° 54' | -8.9 | 32' | 27% | 67° W | CF | Mars |
| 2nd | Ari | 3h 10m | 21° 4' | -8.2 | 32' | 18% | 54° W | CF | Mars, Uranus, Pleiades |
| 3rd | Tau | 4h 7m | 25° 2' | -7.3 | 32' | 10% | 41° W | NM | Jupiter, Uranus, Pleiades, Hyades |
| 4th | Tau | 5h 7m | 27° 31' | -6.4 | 31' | 5% | 27° W | NM | Jupiter, Hyades, Aldebaran |
| 5th | Aur | 6h 7m | 28° 22' | -5.2 | 31' | 1% | 13° W | NM | |
| 6th | Gem | 7h 7m | 27° 34' | -4.4 | 31' | 0% | 1° E | NM | Venus |
| 7th | Cnc | 8h 4m | 25° 16' | -5.3 | 31' | 1% | 14° E | NM | Mercury, Venus, Praesepe |
| 8th | Cnc | 8h 57m | 21° 45' | -6.4 | 30' | 5% | 27° E | NM | Mercury, Praesepe |
| 9th | Leo | 9h 46m | 17° 21' | -7.3 | 30' | 9% | 38° E | NM | Regulus |
| 10th | Leo | 10h 32m | 12° 19' | -8.0 | 30' | 16% | 48° E | CF | Regulus |

## Mercury and Venus

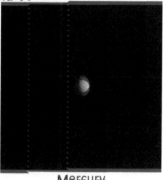 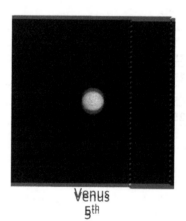

Mercury
5ᵗʰ

Venus
5ᵗʰ

### Mercury

| Date | Con. | R.A. | Dec. | Mag. | Diam. | Ill. | Elon. | Vis. | Rat. | Close To |
|------|------|------|------|------|-------|------|-------|------|------|----------|
| 1st | Gem | 7h 59m | 22° 31' | -0.4 | 6" | 78% | 19° E | PM | *** | |
| 3rd | Cnc | 8h 13m | 21° 41' | -0.3 | 6" | 75% | 21° E | PM | *** | Praesepe |
| 5th | Cnc | 8h 27m | 20° 47' | -0.2 | 6" | 73% | 22° E | PM | *** | Praesepe |
| 7th | Cnc | 8h 40m | 19° 48' | 0.0 | 6" | 68% | 23° E | PM | *** | Moon, Praesepe |
| 9th | Cnc | 8h 52m | 18° 47' | 0.1 | 6" | 65% | 24° E | PM | *** | Praesepe |

## Venus

| Date | Con. | R.A. | Dec. | Mag. | Diam. | Ill. | Elon. | Vis. | Rat. | Close To |
|------|------|------|------|------|-------|------|-------|------|------|----------|
| 1st | Gem | 7h 13m | 23° 17' | -3.4 | 10" | 99% | 8° E | NV | N/A | |
| 3rd | Gem | 7h 24m | 23° 1' | -3.4 | 10" | 99% | 9° E | NV | N/A | |
| 5th | Gem | 7h 34m | 22° 42' | -3.4 | 10" | 99% | 9° E | NV | N/A | |
| 7th | Gem | 7h 45m | 22° 21' | -3.4 | 10" | 99% | 10° E | NV | N/A | Moon |
| 9th | Gem | 7h 55m | 21° 56' | -3.4 | 10" | 99% | 10° E | NV | N/A | |

## Mars and the Outer Planets

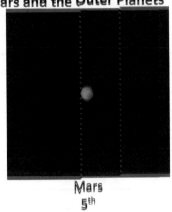

Mars
5th

Jupiter
5th

Saturn
5th

## Mars

| Date | Con. | R.A. | Dec. | Mag. | Diam. | Ill. | Elon. | Vis. | Rat. | Close To |
|------|------|------|------|------|-------|------|-------|------|------|----------|
| 1st | Ari | 2h 55m | 15° 42' | 1.2 | 5" | 91% | 57° W | AM | * | Moon |
| 5th | Ari | 3h 6m | 16° 33' | 1.2 | 5" | 90% | 58° W | AM | * | |
| 10th | Ari | 3h 20m | 17° 32' | 1.1 | 6" | 90% | 60° W | AM | * | Uranus, Pleiades |

## The Outer Planets

| Planet | Date | Con. | R.A. | Dec. | Mag. | Diam. | Elon. | Vis. | Rat. | Close To |
|--------|------|------|------|------|------|-------|-------|------|------|----------|
| Jupiter | 5th | Tau | 4h 29m | 21° 7' | -1.6 | 34" | 37° W | AM | * | Hyades, Aldebaran |
| Saturn | 5th | Aqr | 23h 24m | -6° 3' | 1.2 | 18" | 114° W | AM | *** | |
| Uranus | 5th | Tau | 3h 34m | 18° 59' | 6.1 | 3" | 51° W | AM | * | Mars, Pleiades |
| Neptune | 5th | Psc | 0h 1m | -1° 13' | 7.7 | 2" | 104° W | AM | ***** | Saturn |

## Highlights

| Date | Time (UT) | Event |
|------|-----------|-------|
| 1st | 19:05 | The waning crescent Moon is north of Mars. (Morning sky.) |
| 2nd | 11:13 | The waning crescent Moon is north of Uranus. (Morning sky.) |
| | 17:04 | The waning crescent Moon is south of the Pleiades. (Morning sky.) |
| | 21:17 | Neptune is stationary prior to beginning retrograde motion. (Morning sky.) |
| 3rd | 07:34 | The waning crescent Moon is north of Jupiter. (Morning sky.) |
| | N/A | Good opportunity to see Earthshine on the waning crescent Moon. (Morning sky.) |
| 5th | 14:30 | Dwarf planet Ceres is at opposition. (Visible all night.) |
| | 22:58 | New Moon. (Not visible.) |
| 6th | 21:28 | Mercury is 0.2° north of the Praesepe open star cluster. (Evening sky.) |
| 7th | 17:36 | The waxing crescent Moon is north of the Praesepe open star cluster. (Evening sky.) |
| | 20:35 | The waxing crescent Moon is north of Mercury. (Evening sky.) |
| 9th | 09:23 | The waxing crescent Moon is north of the bright star Regulus. (Evening sky.) |
| | N/A | Good opportunity to see Earthshine on the waxing crescent Moon. (Evening sky.) |

# July 11<sup>th</sup> to 20<sup>th</sup>, 2024

## The Moon

11<sup>th</sup>

13<sup>th</sup>

15<sup>th</sup>

17<sup>th</sup>

19<sup>th</sup>

| Date | Con | R.A. | Dec | Mag | Diam | Ill. | Elon. | Phase | Close To |
|------|-----|------|-----|-----|------|------|-------|-------|----------|
| 11th | Leo | 11h 15m | 6° 54' | -8.7 | 30' | 23% | 58° E | +Cr | |
| 12th | Vir | 11h 57m | 1° 18' | -9.2 | 30' | 32% | 68° E | +Cr | |
| 13th | Vir | 12h 39m | -4° 19' | -9.7 | 30' | 41% | 77° E | FQ | Spica |
| 14th | Vir | 13h 21m | -9° 49' | -10.1 | 30' | 50% | 87° E | FQ | Spica |
| 15th | Vir | 14h 6m | -15° 2' | -10.5 | 30' | 60% | 97° E | FQ | Spica |
| 16th | Lib | 14h 53m | -19° 45' | -10.9 | 30' | 69% | 108° E | +G | |
| 17th | Lib | 15h 44m | -23° 44' | -11.2 | 31' | 78% | 119° E | +G | Antares |
| 18th | Sco | 16h 39m | -26° 40' | -11.6 | 31' | 86% | 132° E | +G | Antares |
| 19th | Oph | 17h 39m | -28° 14' | -11.9 | 31' | 92% | 146° E | +G | |
| 20th | Sgr | 18h 40m | -28° 9' | -12.2 | 32' | 97% | 160° E | FM | |

## Mercury and Venus

Mercury
15<sup>th</sup>

Venus
15<sup>th</sup>

**Mercury**

| Date | Con. | R.A. | Dec. | Mag. | Diam. | Ill. | Elon. | Vis. | Rat. | Close To |
|------|------|------|------|------|-------|------|-------|------|------|----------|
| 11th | Cnc | 9h 4m | 17° 44' | 0.2 | 7" | 62% | 25° E | PM | *** | Praesepe |
| 13th | Cnc | 9h 15m | 16° 39' | 0.3 | 7" | 59% | 26° E | PM | *** | Praesepe |
| 15th | Leo | 9h 25m | 15° 34' | 0.4 | 7" | 56% | 26° E | PM | **** | |
| 17th | Leo | 9h 34m | 14° 28' | 0.5 | 7" | 53% | 27° E | PM | **** | Regulus |
| 19th | Leo | 9h 42m | 13° 23' | 0.5 | 7" | 50% | 27° E | PM | **** | Regulus |

**Venus**

| Date | Con. | R.A. | Dec. | Mag. | Diam. | Ill. | Elon. | Vis. | Rat. | Close To |
|------|------|------|------|------|-------|------|-------|------|------|----------|
| 11th | Cnc | 8h 6m | 21° 30' | -3.4 | 10" | 98% | 11° E | NV | N/A | Praesepe |
| 13th | Cnc | 8h 16m | 21° 1' | -3.4 | 10" | 98% | 11° E | NV | N/A | Praesepe |
| 15th | Cnc | 8h 26m | 20° 29' | -3.4 | 10" | 98% | 12° E | NV | N/A | Praesepe |
| 17th | Cnc | 8h 37m | 19° 55' | -3.4 | 10" | 98% | 12° E | NV | N/A | Praesepe |
| 19th | Cnc | 8h 47m | 19° 19' | -3.4 | 10" | 98% | 13° E | NV | N/A | Praesepe |

## Mars and the Outer Planets

Mars
15th

Jupiter
15th

Saturn
15th

**Mars**

| Date | Con. | R.A. | Dec. | Mag. | Diam. | Ill. | Elon. | Vis. | Rat. | Close To |
|------|------|------|------|------|-------|------|-------|------|------|----------|
| 11th | Ari | 3h 23m | 17° 44' | 1.1 | 6" | 90% | 60° W | AM | * | Uranus, Pleiades |
| 15th | Tau | 3h 35m | 18° 27' | 1.1 | 6" | 90% | 61° W | AM | * | Uranus, Pleiades |
| 20th | Tau | 3h 49m | 19° 18' | 1.1 | 6" | 90% | 63° W | AM | * | Uranus, Pleiades, Hyades |

**The Outer Planets**

| Planet | Date | Con. | R.A. | Dec. | Mag. | Diam. | Elon. | Vis. | Rat. | Close To |
|--------|------|------|------|------|------|-------|-------|------|------|----------|
| Jupiter | 15th | Tau | 4h 38m | 21° 26' | -1.6 | 34" | 45° W | AM | * | Hyades, Aldebaran |
| Saturn | 15th | Aqr | 23h 23m | -6° 8' | 1.2 | 18" | 124° W | AM | *** | |
| Uranus | 15th | Tau | 3h 36m | 19° 4' | 6.1 | 4" | 61° W | AM | * | Mars, Pleiades |
| Neptune | 15th | Psc | 0h 1m | -1° 14' | 7.7 | 2" | 114° W | AM | ***** | Saturn |

## Highlights

| Date | Time (UT) | Event |
|------|-----------|-------|
| 11th | 15:31 | Jupiter is 4.8° north of the bright star Aldebaran. (Morning sky.) |
| 13th | 22:49 | First Quarter Moon. (Evening sky.) |
| 14th | 03:07 | The just-past first quarter Moon is north of the bright star Spica. (Evening sky.) |
| 15th | 09:13 | Mars is 0.6° south of Uranus. (Morning sky.) |
| 17th | 18:44 | The waxing gibbous Moon is south of the bright star Antares. (Evening sky.) |
| 19th | 03:02 | Mars is 5.0° south of the Pleiades open star cluster. (Morning sky.) |

# July 21st to 31st, 2024

## The Moon

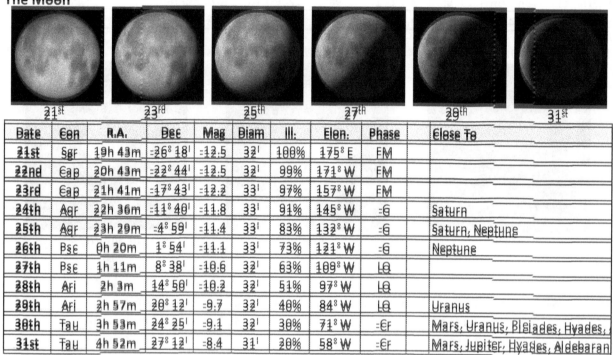

| 21st | 23rd | 25th | 27th | 29th | 31st |

| Date | Con | R.A. | Dec | Mag | Diam | Ill. | Elon. | Phase | Close To |
|------|-----|------|-----|-----|------|------|-------|-------|----------|
| 21st | Sgr | 19h 43m | -26° 18' | -12.5 | 32' | 100% | 175° E | FM | |
| 22nd | Cap | 20h 43m | -23° 44' | -12.5 | 32' | 99% | 171° W | FM | |
| 23rd | Cap | 21h 41m | -17° 43' | -12.3 | 33' | 97% | 157° W | FM | |
| 24th | Aqr | 22h 36m | -11° 40' | -11.8 | 33' | 91% | 145° W | =G | Saturn |
| 25th | Aqr | 23h 29m | -4° 59' | -11.4 | 33' | 83% | 132° W | =G | Saturn, Neptune |
| 26th | Psc | 0h 20m | 1° 54' | -11.1 | 33' | 73% | 121° W | =G | Neptune |
| 27th | Psc | 1h 11m | 8° 38' | -10.6 | 32' | 63% | 109° W | LQ | |
| 28th | Ari | 2h 3m | 14° 50' | -10.3 | 32' | 51% | 97° W | LQ | |
| 29th | Ari | 2h 57m | 20° 12' | -9.7 | 32' | 40% | 84° W | LQ | Uranus |
| 30th | Tau | 3h 53m | 24° 25' | -9.1 | 32' | 30% | 71° W | =Cr | Mars, Uranus, Pleiades, Hyades, |
| 31st | Tau | 4h 52m | 27° 12' | -8.4 | 31' | 20% | 58° W | =Cr | Mars, Jupiter, Hyades, Aldebaran |

## Mercury and Venus

Mercury
25th

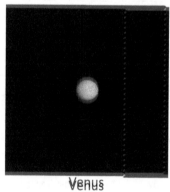

Venus
25th

### Mercury

| Date | Con. | R.A. | Dec. | Mag. | Diam. | Ill. | Elon. | Vis. | Rat. | Close To |
|------|------|------|------|------|-------|------|-------|------|------|----------|
| 21st | Leo | 9h 50m | 12° 20' | 0.6 | 8" | 47% | 27° E | PM | **** | Regulus |
| 23rd | Leo | 9h 57m | 11° 19' | 0.7 | 8" | 44% | 27° E | PM | **** | Regulus |
| 25th | Leo | 10h 3m | 10° 20' | 0.8 | 8" | 40% | 26° E | PM | **** | Regulus |
| 27th | Leo | 10h 8m | 9° 25' | 0.9 | 9" | 37% | 25° E | PM | **** | Regulus |
| 29th | Leo | 10h 12m | 8° 35' | 0.9 | 9" | 34% | 24° E | PM | **** | Venus, Regulus |
| 31st | Leo | 10h 15m | 7° 51' | 1.0 | 9" | 30% | 23° E | PM | *** | Venus, Regulus |

## Venus

| Date | Con. | R.A. | Dec. | Mag. | Diam. | Ill. | Elon. | Vis. | Rat. | Close To |
|---|---|---|---|---|---|---|---|---|---|---|
| 21st | Cnc | 8h 57m | 18° 41' | -3.4 | 10" | 97% | 13° E | NV | N/A | Praesepe |
| 23rd | Cnc | 9h 7m | 18° 1' | -3.4 | 10" | 97% | 14° E | NV | N/A | Praesepe |
| 25th | Cnc | 9h 17m | 17° 18' | -3.3 | 10" | 97% | 14° E | NV | N/A | Praesepe |
| 27th | Leo | 9h 26m | 16° 34' | -3.3 | 10" | 97% | 15° E | NV | N/A | |
| 29th | Leo | 9h 36m | 15° 48' | -3.3 | 10" | 97% | 15° E | PM | * | Mercury, Regulus |
| 31st | Leo | 9h 46m | 15° 1' | -3.3 | 10" | 96% | 16° E | PM | * | Mercury, Regulus |

## Mars and the Outer Planets

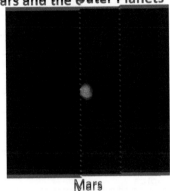

Mars
25th

Jupiter
25th

Saturn
25th

## Mars

| Date | Con. | R.A. | Dec. | Mag. | Diam. | Ill. | Elon. | Vis. | Rat. | Close To |
|---|---|---|---|---|---|---|---|---|---|---|
| 21st | Tau | 3h 52m | 19° 27' | 1.1 | 6" | 90% | 63° W | AM | * | Uranus, Pleiades, Hyades |
| 25th | Tau | 4h 3m | 20° 4' | 1.1 | 6" | 89% | 64° W | AM | * | Pleiades, Hyades, Aldebaran |
| 31st | Tau | 4h 20m | 20° 53' | 1.1 | 6" | 89% | 65° W | AM | * | Moon, Jupiter, Hyades, Aldebaran |

## The Outer Planets

| Planet | Date | Con. | R.A. | Dec. | Mag. | Diam. | Elon. | Vis. | Rat. | Close To |
|---|---|---|---|---|---|---|---|---|---|---|
| Jupiter | 25th | Tau | 4h 47m | 21° 41' | -1.7 | 35" | 53° W | AM | * | Hyades, Aldebaran |
| Saturn | 25th | Aqr | 23h 22m | -6° 18' | 1.1 | 19" | 134° W | AM | *** | Moon |
| Uranus | 25th | Tau | 3h 37m | 19° 9' | 6.1 | 4" | 70° W | AM | * | Mars, Pleiades |
| Neptune | 25th | Psc | 0h 1m | -1° 16' | 7.7 | 2" | 125° W | AM | ***** | Moon, Saturn |

## Highlights

| Date | Time (UT) | Event |
|---|---|---|
| 21st | 10:18 | Full Moon. (Visible all night.) |
| 22nd | 06:34 | Mercury is at greatest eastern elongation from the Sun. (Evening sky.) |
| 23rd | 13:14 | Dwarf planet Pluto is at opposition. (Visible all night.) |
| 24th | 19:11 | The waning gibbous Moon is south of Saturn. (Morning sky.) |
| 25th | 15:11 | The waning gibbous Moon is north of Neptune. (Morning sky.) |
| 26th | 22:24 | Mercury is 2.5° south of the bright star Regulus. (Evening sky.) |
| 28th | 02:52 | Last Quarter Moon. (Morning sky.) |
| 29th | 18:01 | The just-past last quarter Moon is north of Uranus. (Morning sky.) |
| | 20:36 | The just-past last quarter Moon is south of the Pleiades open star cluster. (Morning sky.) |
| 30th | 12:22 | The waning crescent Moon is north of Mars. (Morning sky.) |
| | 22:49 | The waning crescent Moon is north of Jupiter. (Morning sky.) |
| | N/A | The Southern Delta Aquariid meteor shower is at its maximum. (ZHR: 16) |

# August 1st to 10th, 2024

## The Moon

| | 1st | | 3rd | | 5th | | 7th | | 9th | |

| Date | Con | R.A. | Dec | Mag | Diam | Ill. | Elon. | Phase | Close To |
|------|-----|------|-----|-----|------|------|-------|-------|----------|
| 1st | Tau | 5h 51m | 28° 25' | -7.6 | 31' | 12% | 44° W | NM | |
| 2nd | Gem | 6h 50m | 28° 1' | -6.8 | 31' | 6% | 30° W | NM | |
| 3rd | Gem | 7h 47m | 26° 6' | -5.7 | 30' | 2% | 17° W | NM | |
| 4th | Cnc | 8h 41m | 22° 54' | -4.6 | 30' | 0% | 4° W | NM | Praesepe |
| 5th | Leo | 9h 31m | 18° 43' | -4.7 | 30' | 0% | 7° E | NM | Mercury, Venus, Regulus |
| 6th | Leo | 10h 18m | 13° 49' | -5.7 | 30' | 2% | 18° E | NM | Mercury, Venus, Regulus |
| 7th | Leo | 11h 2m | 8° 28' | -6.7 | 30' | 6% | 28° E | NM | Mercury, Venus |
| 8th | Vir | 11h 44m | 2° 52' | -7.5 | 30' | 11% | 38° E | NM | |
| 9th | Vir | 12h 26m | -2° 47' | -8.2 | 29' | 18% | 47° E | +Cr | |
| 10th | Vir | 13h 8m | -8° 20' | -8.8 | 30' | 26% | 57° E | +Cr | Spica |

## Mercury and Venus

Mercury
5th

Venus
5th

### Mercury

| Date | Con. | R.A. | Dec. | Mag. | Diam. | Ill. | Elon. | Vis. | Rat. | Close To |
|------|------|------|------|------|-------|------|-------|------|------|----------|
| 1st | Leo | 10h 16m | 7° 31' | 1.1 | 9" | 28% | 23° E | PM | *** | Venus, Regulus |
| 3rd | Leo | 10h 18m | 6° 56' | 1.2 | 10" | 24% | 21° E | PM | *** | Venus, Regulus |
| 5th | Leo | 10h 18m | 6° 30' | 1.4 | 10" | 21% | 19° E | PM | *** | Moon, Venus, Regulus |
| 7th | Sex | 10h 17m | 6° 14' | 1.5 | 10" | 17% | 17° E | PM | *** | Moon, Venus, Regulus |
| 9th | Sex | 10h 15m | 6° 7' | 1.7 | 11" | 13% | 14° E | NV | N/A | Venus, Regulus |

## Venus

| Date | Con. | R.A. | Dec. | Mag. | Diam. | Ill. | Elon. | Vis. | Rat. | Close To |
|------|------|------|------|------|-------|------|-------|------|------|----------|
| 1st | Leo | 9h 51m | 14° 36' | -3.3 | 10" | 96% | 16° E | PM | * | Mercury, Regulus |
| 3rd | Leo | 10h 0m | 13° 46' | -3.3 | 10" | 96% | 17° E | PM | * | Mercury, Regulus |
| 5th | Leo | 10h 10m | 12° 54' | -3.3 | 10" | 96% | 17° E | PM | * | Moon, Mercury, Regulus |
| 7th | Leo | 10h 19m | 12° 2' | -3.3 | 10" | 95% | 17° E | PM | * | Moon, Mercury, Regulus |
| 9th | Leo | 10h 28m | 11° 7' | -3.3 | 10" | 95% | 18° E | PM | * | Mercury, Regulus |

## Mars and the Outer Planets

Mars
5th

Jupiter
5th

Saturn
5th

### Mars

| Date | Con. | R.A. | Dec. | Mag. | Diam. | Ill. | Elon. | Vis. | Rat. | Close To |
|------|------|------|------|------|-------|------|-------|------|------|----------|
| 1st | Tau | 4h 23m | 21° 0' | 1.1 | 6" | 89% | 66° W | AM | * | Jupiter, Hyades, Aldebaran |
| 5th | Tau | 4h 35m | 21° 28' | 1.1 | 6" | 89% | 67° W | AM | * | Jupiter, Hyades, Aldebaran |
| 10th | Tau | 4h 49m | 21° 59' | 1.0 | 6" | 89% | 68° W | AM | * | Jupiter, Hyades, Aldebaran |

### The Outer Planets

| Planet | Date | Con. | R.A. | Dec. | Mag. | Diam. | Elon. | Vis. | Rat. | Close To |
|--------|------|------|------|------|------|-------|-------|------|------|----------|
| Jupiter | 5th | Tau | 4h 55m | 21° 55' | -1.7 | 36" | 62° W | AM | ** | Mars, Hyades, Aldebaran |
| Saturn | 5th | Aqr | 23h 20m | -6° 32' | 1.0 | 19" | 145° W | AM | **** | |
| Uranus | 5th | Tau | 3h 38m | 19° 13' | 6.1 | 4" | 81° W | AM | * | Pleiades |
| Neptune | 5th | Psc | 0h 0m | -1° 21' | 7.7 | 2" | 135° W | AM | ***** | |

## Highlights

| Date | Time (UT) | Event |
|------|-----------|-------|
| 1st | N/A | Good opportunity to see Earthshine on the waning crescent Moon. (Morning sky.) |
| 4th | 08:25 | Mercury is stationary prior to beginning retrograde motion. (Evening sky.) |
| | 11:14 | New Moon. (Not visible.) |
| | 15:20 | Venus is 1.1° north of the bright star Regulus. (Evening sky.) |
| 5th | 06:34 | Mars is 5.0° north of the bright star Aldebaran. (Morning sky.) |
| | 20:42 | The just-past new Moon is north of the bright star Regulus. (Evening sky.) |
| | 23:12 | The just-past new Moon is north of Venus. (Evening sky.) |
| 6th | 00:28 | The waxing crescent Moon is north of Mercury. (Evening sky.) |
| | 15:04 | Mercury is 5.9° south of Venus. (Evening sky.) |
| 8th | N/A | Good opportunity to see Earthshine on the waxing crescent Moon. (Evening sky.) |
| 10th | 07:52 | The waxing crescent Moon is north of the bright star Spica. (Evening sky.) |

# August 11th to 20th, 2024

## The Moon

   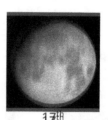 

| 11th | 13th | 15th | 17th | 19th |

| Date | Con | R.A. | Dec | Mag | Diam | Ill. | Elon. | Phase | Close To |
|------|-----|------|-----|-----|------|------|-------|-------|----------|
| 11th | Vir | 13h 51m | -13° 37' | -9.4 | 30' | 34% | 67° E | +Cr | Spica |
| 12th | Lib | 14h 37m | -18° 28' | -9.8 | 30' | 44% | 77° E | FQ | |
| 13th | Lib | 15h 25m | -22° 39' | -10.3 | 30' | 54% | 88° E | FQ | |
| 14th | Sco | 16h 18m | -25° 55' | -10.7 | 31' | 63% | 101° E | FQ | Antares |
| 15th | Oph | 17h 15m | -27° 59' | -11.0 | 31' | 73% | 114° E | +G | Antares |
| 16th | Sgr | 18h 15m | -28° 32' | -11.4 | 31' | 82% | 128° E | +G | |
| 17th | Sgr | 19h 17m | -27° 23' | -11.7 | 32' | 90% | 142° E | +G | |
| 18th | Cap | 20h 18m | -24° 27' | -12.1 | 32' | 95% | 157° E | FM | |
| 19th | Cap | 21h 18m | -19° 55' | -12.4 | 33' | 99% | 171° E | FM | |
| 20th | Aqr | 22h 15m | -14° 6' | -12.6 | 33' | 100% | 176° W | FM | |

## Mercury and Venus

Mercury
15th

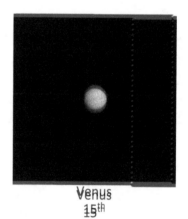

Venus
15th

### Mercury

| Date | Con. | R.A. | Dec. | Mag. | Diam. | Ill. | Elon. | Vis. | Rat. | Close To |
|------|------|------|------|------|-------|------|-------|------|------|----------|
| 11th | Sex | 10h 11m | 6° 13' | 1.9 | 11" | 9% | 12° E | NV | N/A | Venus, Regulus |
| 13th | Leo | 10h 7m | 6° 39' | 2.2 | 11" | 6% | 9° E | NV | N/A | Regulus |
| 15th | Leo | 10h 1m | 6° 58' | 2.5 | 11" | 3% | 5° E | NV | N/A | Regulus |
| 17th | Leo | 9h 55m | 7° 36' | 2.8 | 11" | 2% | 3° E | NV | N/A | Regulus |
| 19th | Leo | 9h 49m | 8° 23' | 2.9 | 11" | 1% | 1° W | NV | N/A | Regulus |

## Venus

| Date | Con. | R.A. | Dec. | Mag. | Diam. | Ill. | Elon. | Vis. | Rat. | Close To |
|------|------|------|------|------|-------|------|-------|------|------|----------|
| 11th | Leo | 10h 38m | 10° 12' | -3.3 | 10" | 95% | 18° E | PM | * | Mercury, Regulus |
| 13th | Leo | 10h 47m | 9° 16' | -3.3 | 10" | 94% | 19° E | PM | * | Regulus |
| 15th | Leo | 10h 56m | 8° 18' | -3.3 | 10" | 94% | 19° E | PM | * | |
| 17th | Leo | 11h 5m | 7° 20' | -3.3 | 11" | 94% | 19° E | PM | * | |
| 19th | Leo | 11h 14m | 6° 21' | -3.3 | 11" | 93% | 20° E | PM | * | |

## Mars and the Outer Planets

Mars
15th

Jupiter
15th

Saturn
15th

## Mars

| Date | Con. | R.A. | Dec. | Mag. | Diam. | Ill. | Elon. | Vis. | Rat. | Close To |
|------|------|------|------|------|-------|------|-------|------|------|----------|
| 11th | Tau | 4h 52m | 22° 4' | 1.0 | 6" | 89% | 68° W | AM | * | Jupiter, Hyades, Aldebaran |
| 15th | Tau | 5h 3m | 22° 25' | 1.0 | 6" | 88% | 69° W | AM | * | Jupiter, Hyades, Aldebaran |
| 20th | Tau | 5h 17m | 22° 46' | 1.0 | 6" | 88% | 70° W | AM | * | Jupiter |

## The Outer Planets

| Planet | Date | Con. | R.A. | Dec. | Mag. | Diam. | Elon. | Vis. | Rat. | Close To |
|--------|------|------|------|------|------|-------|-------|------|------|----------|
| Jupiter | 15th | Tau | 5h 2m | 22° 5' | -1.8 | 37" | 69° W | AM | ** | Mars, Hyades, Aldebaran |
| Saturn | 15th | Aqr | 23h 18m | -6° 48' | 0.9 | 19" | 155° W | AM | **** | |
| Uranus | 15th | Tau | 3h 39m | 19° 16' | 6.1 | 4" | 90° W | AM | * | Pleiades |
| Neptune | 15th | Psc | 0h 0m | -1° 25' | 7.6 | 2" | 145° W | AM | ***** | |

## Highlights

| Date | Time (UT) | Event |
|------|-----------|-------|
| 12th | 15:19 | First Quarter Moon. (Evening sky.) |
| | N/A | The Perseid meteor shower is at its maximum. (ZHR: 100) |
| 14th | 05:21 | The waxing gibbous Moon is north of the bright star Antares. (Evening sky.) |
| | 16:51 | Mars is 0.3° north of Jupiter. (Morning sky.) |
| 17th | N/A | The Kappa Cygnid meteor shower is at its maximum. (ZHR: 3) |
| 19th | 01:53 | Mercury is at inferior conjunction with the Sun. (Not visible.) |
| | 18:26 | Full Moon. (Visible all night.) |

# August 21ˢᵗ to 31ˢᵗ, 2024

## The Moon

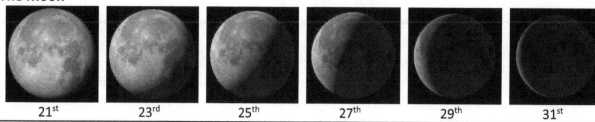

21ˢᵗ     23ʳᵈ     25ᵗʰ     27ᵗʰ     29ᵗʰ     31ˢᵗ

| Date | Con | R.A. | Dec | Mag | Diam | Ill. | Elon. | Phase | Close To |
|------|-----|------|-----|-----|------|------|-------|-------|----------|
| 21st | Aqr | 23h 10m | -7° 24' | -12.3 | 33' | 98% | 163° W | FM | Saturn |
| 22nd | Psc | 0h 3m | 0° 18' | -11.9 | 33' | 93% | 151° W | -G | Saturn, Neptune |
| 23rd | Psc | 0h 55m | 6° 46' | -11.5 | 33' | 85% | 139° W | -G | |
| 24th | Ari | 1h 48m | 13° 22' | -11.2 | 33' | 76% | 126° W | -G | |
| 25th | Ari | 2h 43m | 19° 8' | -10.8 | 32' | 65% | 113° W | LQ | |
| 26th | Tau | 3h 40m | 23° 44' | -10.3 | 32' | 54% | 100° W | LQ | Uranus, Pleiades, Hyades |
| 27th | Tau | 4h 38m | 26° 53' | -9.8 | 31' | 43% | 86° W | LQ | Jupiter, Hyades, Aldebaran |
| 28th | Tau | 5h 38m | 28° 27' | -9.3 | 31' | 33% | 72° W | -Cr | Mars, Jupiter |
| 29th | Aur | 6h 37m | 28° 23' | -8.7 | 31' | 24% | 59° W | -Cr | |
| 30th | Gem | 7h 34m | 26° 48' | -8.0 | 30' | 15% | 45° W | -Cr | |
| 31st | Cnc | 8h 28m | 23° 53' | -7.2 | 30' | 9% | 33° W | NM | Praesepe |

## Mercury and Venus

Mercury
25ᵗʰ

Venus
25ᵗʰ

**Mercury**

| Date | Con. | R.A. | Dec. | Mag. | Diam. | Ill. | Elon. | Vis. | Rat. | Close To |
|------|------|------|------|------|-------|------|-------|------|------|----------|
| 21st | Leo | 9h 43m | 9° 14' | 2.8 | 11" | 2% | 5° W | NV | N/A | Regulus |
| 23rd | Leo | 9h 38m | 10° 8' | 2.4 | 10" | 3% | 8° W | NV | N/A | Regulus |
| 25th | Leo | 9h 34m | 11° 0' | 2.0 | 10" | 7% | 11° W | NV | N/A | Regulus |
| 27th | Leo | 9h 32m | 11° 48' | 1.6 | 9" | 11% | 13° W | NV | N/A | Regulus |
| 29th | Leo | 9h 32m | 12° 27' | 1.2 | 9" | 17% | 15° W | NV | N/A | Regulus |
| 31st | Leo | 9h 34m | 12° 56' | 0.8 | 8" | 24% | 16° W | AM | ** | Regulus |

## Venus

| Date | Con. | R.A. | Dec. | Mag. | Diam. | Ill. | Elon. | Vis. | Rat. | Close To |
|------|------|------|------|------|-------|------|-------|------|------|----------|
| 21st | Leo | 11h 23m | 5° 22' | -3.3 | 11" | 93% | 20° E | PM | * | |
| 23rd | Leo | 11h 32m | 4° 21' | -3.3 | 11" | 93% | 21° E | PM | * | |
| 25th | Vir | 11h 41m | 3° 20' | -3.3 | 11" | 92% | 21° E | PM | * | |
| 27th | Vir | 11h 50m | 2° 19' | -3.3 | 11" | 92% | 21° E | PM | * | |
| 29th | Vir | 11h 59m | 1° 18' | -3.3 | 11" | 92% | 22° E | PM | * | |
| 31st | Vir | 12h 8m | 0° 16' | -3.3 | 11" | 91% | 22° E | PM | * | |

## Mars and the Outer Planets

Mars
25th

Jupiter
25th

Saturn
25th

### Mars

| Date | Con. | R.A. | Dec. | Mag. | Diam. | Ill. | Elon. | Vis. | Rat. | Close To |
|------|------|------|------|------|-------|------|-------|------|------|----------|
| 21st | Tau | 5h 19m | 22° 50' | 1.0 | 6" | 88% | 71° W | AM | ** | Jupiter |
| 25th | Tau | 5h 30m | 23° 3' | 1.0 | 6" | 88% | 72° W | AM | ** | Jupiter |
| 31st | Tau | 5h 47m | 23° 18' | 0.9 | 7" | 88% | 73° W | AM | ** | Jupiter |

### The Outer Planets

| Planet | Date | Con. | R.A. | Dec. | Mag. | Diam. | Elon. | Vis. | Rat. | Close To |
|--------|------|------|------|------|------|-------|-------|------|------|----------|
| Jupiter | 25th | Tau | 5h 8m | 22° 13' | -1.8 | 38" | 77° W | AM | ** | Mars, Aldebaran |
| Saturn | 25th | Aqr | 23h 15m | -7° 6' | 0.9 | 19" | 165° W | AM | **** | |
| Uranus | 25th | Tau | 3h 40m | 19° 17' | 6.1 | 4" | 99° W | AM | ** | Pleiades |
| Neptune | 25th | Psc | 23h 59m | -1° 31' | 7.6 | 2" | 154° W | AM | ***** | |

## Highlights

| Date | Time (UT) | Event |
|------|-----------|-------|
| 21st | 04:27 | The waning gibbous Moon is north of Saturn. (Morning sky.) |
| | 20:35 | The waning gibbous Moon is north of Neptune. (Morning sky.) |
| 22nd | 20:13 | Asteroid Vesta is in conjunction with the Sun. (Not visible.) |
| 25th | 03:35 | Dwarf planet Ceres is stationary prior to resuming prograde motion. (Evening sky.) |
| | 22:18 | The nearly last quarter Moon is north of Uranus. (Morning sky.) |
| 26th | 01:19 | The almost last quarter Moon is north of the Pleiades open star cluster. (Morning sky.) |
| | 09:27 | Last Quarter Moon. (Morning sky.) |
| 27th | 14:19 | The just-past last quarter Moon is north of Jupiter. (Morning sky.) |
| | 22:51 | The just-past last quarter Moon is north of Mars. (Morning sky.) |
| 28th | 02:35 | Mercury is stationary prior to resuming prograde motion. (Morning sky.) |
| 31st | 03:34 | The waning crescent Moon is north of the Praesepe open star cluster. (Morning sky.) |
| | N/A | Good opportunity to see Earthshine on the waning crescent Moon. (Morning sky.) |

# September 1st to 10th, 2024

## The Moon

1st

3rd

5th

7th

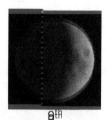
9th

| Date | Con | R.A. | Dec | Mag | Diam | Ill. | Elon. | Phase | Close To |
|------|-----|------|-----|-----|------|------|-------|-------|----------|
| 1st | Cnc | 9h 18m | 19° 56' | -6.3 | 30' | 4% | 21° W | NM | Mercury, Praesepe |
| 2nd | Leo | 10h 5m | 15° 12' | -5.2 | 30' | 1% | 10° W | NM | Mercury, Regulus |
| 3rd | Leo | 10h 50m | 9° 57' | -4.1 | 30' | 0% | 0° E | NM | Regulus |
| 4th | Leo | 11h 32m | 4° 23' | -5.0 | 29' | 1% | 10° E | NM | |
| 5th | Vir | 12h 14m | -1° 17' | -6.0 | 29' | 3% | 19° E | NM | Venus |
| 6th | Vir | 12h 56m | -6° 54' | -6.9 | 29' | 7% | 29° E | NM | Venus, Spica |
| 7th | Vir | 13h 38m | -12° 17' | -7.7 | 29' | 13% | 39° E | +Cr | Spica |
| 8th | Lib | 14h 23m | -17° 16' | -8.4 | 30' | 20% | 49° E | +Cr | |
| 9th | Lib | 15h 10m | -21° 37' | -9.0 | 30' | 28% | 60° E | +Cr | |
| 10th | Sco | 16h 1m | -25° 9' | -9.5 | 30' | 38% | 72° E | FQ | Antares |

## Mercury and Venus

Mercury
5th

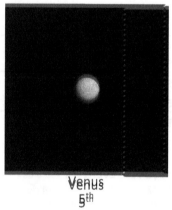
Venus
5th

### Mercury

| Date | Con. | R.A. | Dec. | Mag. | Diam. | Ill. | Elon. | Vis. | Rat. | Close To |
|------|------|------|------|------|-------|------|-------|------|------|----------|
| 1st | Leo | 9h 36m | 13° 7' | 0.6 | 8" | 28% | 17° W | AM | ** | Moon, Regulus |
| 3rd | Leo | 9h 41m | 13° 17' | 0.3 | 8" | 36% | 17° W | AM | ** | Regulus |
| 5th | Leo | 9h 48m | 13° 14' | 0.0 | 7" | 45% | 17° W | AM | ** | Regulus |
| 7th | Leo | 9h 57m | 12° 57' | -0.3 | 7" | 54% | 17° W | AM | ** | Regulus |
| 9th | Leo | 10h 8m | 12° 26' | -0.5 | 6" | 62% | 16° W | AM | ** | Regulus |

## Venus

| Date | Con. | R.A. | Dec. | Mag. | Diam. | Ill. | Elon. | Vis. | Rat. | Close To |
|------|------|------|------|------|-------|------|-------|------|------|----------|
| 1st | Vir | 12h 12m | 0° 15' | -3.3 | 11" | 91% | 23° E | PM | * | |
| 3rd | Vir | 12h 21m | -1° 17' | -3.3 | 11" | 91% | 23° E | PM | * | |
| 5th | Vir | 12h 30m | -2° 19' | -3.3 | 11" | 90% | 23° E | PM | * | Moon |
| 7th | Vir | 12h 39m | -3° 20' | -3.3 | 11" | 90% | 24° E | PM | * | |
| 9th | Vir | 12h 48m | -4° 22' | -3.3 | 11" | 90% | 24° E | PM | * | Spica |

## Mars and the Outer Planets

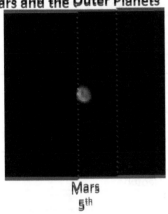

Mars
5th

Jupiter
5th

Saturn
5th

## Mars

| Date | Con. | R.A. | Dec. | Mag. | Diam. | Ill. | Elon. | Vis. | Rat. | Close To |
|------|------|------|------|------|-------|------|-------|------|------|----------|
| 1st | Tau | 5h 49m | 23° 19' | 0.9 | 7" | 88% | 73° W | AM | ** | Jupiter |
| 5th | Tau | 6h 0m | 23° 35' | 0.9 | 7" | 88% | 74° W | AM | ** | |
| 10th | Gem | 6h 13m | 23° 29' | 0.9 | 7" | 88% | 75° W | AM | ** | |

## The Outer Planets

| Planet | Date | Con. | R.A. | Dec. | Mag. | Diam. | Elon. | Vis. | Rat. | Close To |
|--------|------|------|------|------|------|-------|-------|------|------|----------|
| Jupiter | 5th | Tau | 5h 14m | 22° 19' | -1.9 | 39" | 86° W | AM | ** | Aldebaran |
| Saturn | 5th | Aqr | 23h 12m | -7° 26' | 0.8 | 19" | 176° W | AN | **** | |
| Uranus | 5th | Tau | 3h 40m | 19° 17' | 6.0 | 4" | 109° W | AM | ** | Pleiades |
| Neptune | 5th | Psc | 23h 58m | -1° 38' | 7.6 | 2" | 165° W | AM | ***** | |

## Highlights

| Date | Time (UT) | Event |
|------|-----------|-------|
| 1st | 07:46 | The waning crescent Moon is north of Mercury. (Morning sky.) |
| | 13:36 | Uranus is stationary prior to beginning retrograde motion. (Morning sky.) |
| | N/A | The Aurigid meteor shower is at its maximum. (ZHR: 6) |
| 3rd | 01:56 | New Moon. (Not visible.) |
| 5th | 02:25 | Mercury is at greatest western elongation from the Sun. (Morning sky.) |
| | 08:00 | The waxing crescent Moon is south of Venus. (Evening sky.) |
| 6th | 18:05 | The waxing crescent Moon is north of the bright star Spica. (Evening sky.) |
| | N/A | Good opportunity to see Earthshine on the waxing crescent Moon. (Evening sky.) |
| 9th | 01:03 | Mercury is 0.5° north of the bright star Regulus. (Morning sky.) |
| | 01:41 | Saturn is at opposition from the Sun. (Visible all night.) |
| | N/A | The September Epsilon Perseid meteor shower is at its maximum. (ZHR: 5) |
| 10th | 10:42 | The nearly first quarter Moon is north of the bright star Antares. (Evening sky.) |

# September 11<sup>th</sup> to 20<sup>th</sup>, 2024

## The Moon

11<sup>th</sup>

13<sup>th</sup>

15<sup>th</sup>

17<sup>th</sup>

19<sup>th</sup>

| Date | Con | R.A. | Dec | Mag | Diam | Ill. | Elon. | Phase | Close To |
|------|-----|------|-----|-----|------|------|-------|-------|----------|
| 11th | Oph | 16h 55m | -27° 35' | -10.0 | 30' | 47% | 84° E | FQ | Antares |
| 12th | Sgr | 17h 53m | -28° 39' | -10.4 | 31' | 58% | 98° E | FQ | |
| 13th | Sgr | 18h 53m | -28° 9' | -10.8 | 31' | 68% | 112° E | +G | |
| 14th | Sgr | 19h 53m | -25° 58' | -11.2 | 32' | 78% | 126° E | +G | |
| 15th | Cap | 20h 52m | -22° 9' | -11.6 | 32' | 86% | 140° E | +G | |
| 16th | Cap | 21h 50m | -16° 53' | -12.0 | 33' | 93% | 153° E | +G | |
| 17th | Aqr | 22h 45m | -10° 30' | -12.3 | 33' | 98% | 166° E | FM | Saturn |
| 18th | Aqr | 23h 39m | -3° 25' | -12.7 | 33' | 100% | 179° E | FM | Saturn, Neptune |
| 19th | Psc | 0h 33m | 3° 53' | -12.4 | 33' | 99% | 168° W | FM | Neptune |
| 20th | Psc | 1h 27m | 10° 57' | -12.0 | 33' | 95% | 156° W | -G | |

## Mercury and Venus

Mercury
15<sup>th</sup>

Venus
15<sup>th</sup>

### Mercury

| Date | Con. | R.A. | Dec. | Mag. | Diam. | Ill. | Elon. | Vis. | Rat. | Close To |
|------|------|------|------|------|-------|------|-------|------|------|----------|
| 11th | Leo | 10h 19m | 11° 42' | -0.7 | 6" | 70% | 15° W | NV | N/A | Regulus |
| 13th | Leo | 10h 32m | 10° 46' | -0.9 | 6" | 77% | 13° W | NV | N/A | Regulus |
| 15th | Leo | 10h 45m | 9° 39' | -1.0 | 6" | 83% | 12° W | NV | N/A | Regulus |
| 17th | Leo | 10h 59m | 8° 24' | -1.1 | 5" | 88% | 10° W | NV | N/A | |
| 19th | Leo | 11h 12m | 7° 3' | -1.2 | 5" | 92% | 9° W | NV | N/A | |

**Venus**

| Date | Con. | R.A. | Dec. | Mag. | Diam. | Ill. | Elon. | Vis. | Rat. | Close To |
|------|------|------|------|------|-------|------|-------|------|------|----------|
| 11th | Vir | 12h 57m | -5° 23' | -3.3 | 11" | 89% | 25° E | PM | * | Spica |
| 13th | Vir | 13h 6m | -6° 24' | -3.3 | 11" | 89% | 25° E | PM | * | Spica |
| 15th | Vir | 13h 15m | -7° 24' | -3.3 | 12" | 88% | 26° E | PM | * | Spica |
| 17th | Vir | 13h 24m | -8° 24' | -3.3 | 12" | 88% | 26° E | PM | * | Spica |
| 19th | Vir | 13h 33m | -9° 23' | -3.3 | 12" | 88% | 26° E | PM | * | Spica |

## Mars and the Outer Planets

Mars
15th

Jupiter
15th

Saturn
15th

**Mars**

| Date | Con. | R.A. | Dec. | Mag. | Diam. | Ill. | Elon. | Vis. | Rat. | Close To |
|------|------|------|------|------|-------|------|-------|------|------|----------|
| 11th | Gem | 6h 16m | 23° 29' | 0.8 | 7" | 88% | 76° W | AM | ** | |
| 15th | Gem | 6h 26m | 23° 29' | 0.8 | 7" | 88% | 77° W | AM | ** | |
| 20th | Gem | 6h 38m | 23° 25' | 0.8 | 7" | 87% | 78° W | AM | ** | |

**The Outer Planets**

| Planet | Date | Con. | R.A. | Dec. | Mag. | Diam. | Elon. | Vis. | Rat. | Close To |
|--------|------|------|------|------|------|-------|-------|------|------|----------|
| Jupiter | 15th | Tau | 5h 18m | 22° 23' | -2.0 | 40" | 94° W | AM | *** | |
| Saturn | 15th | Aqr | 23h 10m | -7° 44' | 0.8 | 19" | 174° E | AN | **** | |
| Uranus | 15th | Tau | 3h 39m | 19° 16' | 6.0 | 4" | 118° W | AM | ** | Pleiades |
| Neptune | 15th | Psc | 23h 57m | -1° 44' | 7.6 | 2" | 174° W | AN | ***** | |

## Highlights

| Date | Time (UT) | Event |
|------|-----------|-------|
| 11th | 06:06 | First Quarter Moon. (Evening sky.) |
| 17th | 05:52 | Venus is 2.6° north of the bright star Spica. (Evening sky.) |
| | 10:42 | The nearly full Moon is north of Saturn. (Visible all night.) |
| 18th | 02:35 | Full Moon. (Visible all night.) |
| | 02:44 | Partial lunar eclipse. Visible from Antarctica, the Atlantic, western Africa, Central America, western Europe, eastern North America, eastern Pacific and South America |
| | 08:58 | The just-past full Moon is north of Neptune. (Visible all night.) |

# September 21st to 30th, 2024

## The Moon

| 21st | 23rd | 25th | 27th | 29th |

| Date | Con | R.A. | Dec | Mag | Diam | Ill. | Elon. | Phase | Close To |
|------|-----|------|-----|-----|------|------|-------|-------|----------|
| 21st | Ari | 2h 23m | 17° 17' | -11.7 | 33' | 88% | 143° W | =G | |
| 22nd | Ari | 3h 21m | 22° 29' | -11.3 | 33' | 79% | 129° W | =G | Uranus, Pleiades |
| 23rd | Tau | 4h 21m | 26° 13' | -10.9 | 32' | 69% | 115° W | =G | Uranus, Pleiades, Hyades, Aldeba |
| 24th | Tau | 5h 22m | 28° 17' | -10.5 | 32' | 58% | 101° W | LQ | Jupiter, Aldebaran |
| 25th | Aur | 6h 22m | 28° 38' | -10.0 | 31' | 48% | 86° W | LQ | Mars |
| 26th | Gem | 7h 21m | 27° 22' | -9.5 | 31' | 37% | 73° W | =Cr | Mars |
| 27th | Cnc | 8h 15m | 24° 44' | -9.0 | 30' | 28% | 60° W | =Cr | Praesepe |
| 28th | Cnc | 9h 7m | 21° 0' | -8.4 | 30' | 20% | 48° W | =Cr | Praesepe |
| 29th | Leo | 9h 54m | 16° 27' | -7.6 | 30' | 13% | 37° W | NM | Regulus |
| 30th | Leo | 10h 39m | 11° 19' | -6.8 | 30' | 7% | 27° W | NM | Regulus |

## Mercury and Venus

 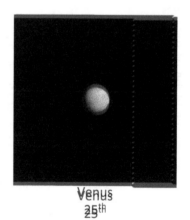

Mercury
25th

Venus
25th

## Mercury

| Date | Con. | R.A. | Dec. | Mag. | Diam. | Ill. | Elon. | Vis. | Rat. | Close To |
|------|------|------|------|------|-------|------|-------|------|------|----------|
| 21st | Leo | 11h 26m | 5° 36' | -1.2 | 5" | 95% | 7° W | NV | N/A | |
| 23rd | Vir | 11h 40m | 4° 5' | -1.2 | 5" | 97% | 5° W | NV | N/A | |
| 25th | Vir | 11h 53m | 2° 33' | -1.3 | 5" | 98% | 4° W | NV | N/A | |
| 27th | Vir | 12h 6m | 0° 59' | -1.3 | 5" | 99% | 3° W | NV | N/A | |
| 29th | Vir | 12h 20m | 0° 35' | -1.2 | 5" | 100% | 1° W | NV | N/A | |

## Venus

| Date | Con. | R.A. | Dec. | Mag. | Diam. | Ill. | Elon. | Vis. | Rat. | Close To |
|------|------|------|------|------|-------|------|-------|------|------|----------|
| 21st | Vir | 13h 42m | -10° 21' | -3.4 | 12" | 87% | 27° E | PM | * | Spica |
| 23rd | Vir | 13h 51m | -11° 18' | -3.4 | 12" | 87% | 27° E | PM | * | Spica |
| 25th | Vir | 14h 0m | -12° 14' | -3.4 | 12" | 86% | 28° E | PM | * | Spica |
| 27th | Vir | 14h 9m | -13° 10' | -3.4 | 12" | 86% | 28° E | PM | * | |
| 29th | Vir | 14h 19m | -14° 4' | -3.4 | 12" | 85% | 29° E | PM | * | |

## Mars and the Outer Planets

Mars
25th

Jupiter
25th

Saturn
25th

## Mars

| Date | Con. | R.A. | Dec. | Mag. | Diam. | Ill. | Elon. | Vis. | Rat. | Close To |
|------|------|------|------|------|-------|------|-------|------|------|----------|
| 21st | Gem | 6h 41m | 23° 24' | 0.8 | 7" | 87% | 78° W | AM | ** | |
| 25th | Gem | 6h 50m | 23° 19' | 0.7 | 7" | 87% | 80° W | AM | ** | Moon |
| 30th | Gem | 7h 2m | 23° 10' | 0.7 | 8" | 87% | 81° W | AM | ** | |

## The Outer Planets

| Planet | Date | Con. | R.A. | Dec. | Mag. | Diam. | Ill. | Elon. | Vis. | Rat. | Close To |
|--------|------|------|------|------|------|-------|------|-------|------|------|----------|
| Jupiter | 25th | Tau | 5h 21m | 22° 25' | -2.0 | 41" | | 102° W | AM | *** | |
| Saturn | 25th | Aqr | 23h 7m | -8° 1' | 0.8 | 19" | | 165° E | PM | **** | |
| Uranus | 25th | Tau | 3h 39m | 19° 14' | 6.0 | 4" | | 127° W | AM | ** | Pleiades |
| Neptune | 25th | Psc | 23h 56m | -1° 51' | 7.6 | 2" | | 177° E | AN | ***** | |

## Highlights

| Date | Time (UT) | Event |
|------|-----------|-------|
| 21st | 12:57 | Neptune is at opposition from the Sun. (Visible all night.) |
| 22nd | 08:58 | The waning gibbous Moon is north of Uranus. (Morning sky.) |
| | 12:44 | Autumnal equinox |
| 23rd | 21:42 | The nearly last quarter Moon is north of Jupiter. (Morning sky.) |
| 24th | 18:51 | Last Quarter Moon. (Morning sky.) |
| 25th | 13:31 | The just-past last quarter Moon is north of Mars. (Morning sky.) |
| 27th | 12:48 | The waning crescent Moon is north of the Praesepe open star cluster. (Morning sky.) |
| 29th | 05:30 | The waning crescent Moon is north of the bright star Regulus. (Morning sky.) |
| | N/A | Good opportunity to see Earthshine on the waning crescent Moon. (Morning sky.) |
| 30th | 20:52 | Mercury is at superior conjunction with the Sun. (Not visible.) |

# October 1st to 10th, 2024

## The Moon

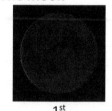
1st

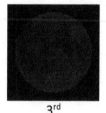
3rd

5th

7th

9th

| Date | Con | R.A. | Dec | Mag | Diam | Ill. | Elon. | Phase | Close To |
|------|-----|------|-----|-----|------|------|-------|-------|----------|
| 1st | Leo | 11h 22m | 5° 50' | -5.9 | 29' | 3% | 17° W | NM | |
| 2nd | Vir | 12h 3m | 0° 10' | -4.8 | 29' | 1% | 8° W | NM | Mercury |
| 3rd | Vir | 12h 45m | -5° 29' | -4.1 | 29' | 0% | 2° E | NM | Mercury, Spica |
| 4th | Vir | 13h 27m | -10° 58' | -5.3 | 29' | 1% | 12° E | NM | Mercury, Spica |
| 5th | Vir | 14h 12m | -16° 4' | -6.3 | 30' | 4% | 22° E | NM | Venus, Spica |
| 6th | Lib | 14h 58m | -20° 37' | -7.2 | 30' | 9% | 32° E | NM | Venus |
| 7th | Lib | 15h 48m | -24° 21' | -8.0 | 30' | 15% | 44° E | +Cr | Antares |
| 8th | Sco | 16h 41m | -27° 4' | -8.6 | 30' | 23% | 56° E | +Cr | Antares |
| 9th | Oph | 17h 36m | -28° 31' | -9.2 | 30' | 32% | 69° E | +Cr | |
| 10th | Sgr | 18h 34m | -28° 30' | -9.8 | 31' | 42% | 83° E | FQ | |

## Mercury and Venus

Mercury
5th

Venus
5th

### Mercury

| Date | Con. | R.A. | Dec. | Mag. | Diam. | Ill. | Elon. | Vis. | Rat. | Close To |
|------|------|------|------|------|-------|------|-------|------|------|----------|
| 1st | Vir | 12h 32m | -2° 9' | -1.2 | 5" | 100% | 1° E | NV | N/A | |
| 3rd | Vir | 12h 45m | -3° 42' | -1.1 | 5" | 100% | 2° E | NV | N/A | Moon, Spica |
| 5th | Vir | 12h 58m | -5° 13' | -1.0 | 5" | 100% | 3° E | NV | N/A | Spica |
| 7th | Vir | 13h 10m | -6° 43' | -0.9 | 5" | 99% | 5° E | NV | N/A | Spica |
| 9th | Vir | 13h 22m | -8° 11' | -0.8 | 5" | 99% | 6° E | NV | N/A | Spica |

**Venus**

| Date | Con. | R.A. | Dec. | Mag. | Diam. | Ill. | Elon. | Vis. | Rat. | Close To |
|------|------|------|------|------|-------|------|-------|------|------|----------|
| 1st | Lib | 14h 28m | -14° 56' | -3.4 | 12" | 85% | 29° E | PM | * | |
| 3rd | Lib | 14h 37m | -15° 48' | -3.4 | 12" | 84% | 30° E | PM | * | |
| 5th | Lib | 14h 47m | -16° 38' | -3.4 | 12" | 84% | 31° E | PM | * | Moon |
| 7th | Lib | 14h 57m | -17° 26' | -3.4 | 13" | 84% | 31° E | PM | * | |
| 9th | Lib | 15h 6m | -18° 13' | -3.4 | 13" | 83% | 32° E | PM | * | |

## Mars and the Outer Planets

Mars
5th

Jupiter
5th

Saturn
5th

**Mars**

| Date | Con. | R.A. | Dec. | Mag. | Diam. | Ill. | Elon. | Vis. | Rat. | Close To |
|------|------|------|------|------|-------|------|-------|------|------|----------|
| 1st | Gem | 7h 4m | 23° 8' | 0.7 | 8" | 87% | 81° W | AM | ** | |
| 5th | Gem | 7h 13m | 22° 59' | 0.6 | 8" | 88% | 83° W | AM | ** | |
| 10th | Gem | 7h 24m | 22° 47' | 0.6 | 8" | 88% | 85° W | AM | ** | |

**The Outer Planets**

| Planet | Date | Con. | R.A. | Dec. | Mag. | Diam. | Elon. | Vis. | Rat. | Close To |
|--------|------|------|------|------|------|-------|-------|------|------|----------|
| Jupiter | 5th | Tau | 5h 22m | 22° 26' | -2.1 | 43" | 111° W | AM | *** | |
| Saturn | 5th | Aqr | 23h 4m | -8° 16' | 0.9 | 19" | 155° E | PM | **** | |
| Uranus | 5th | Tau | 3h 38m | 19° 11' | 6.0 | 4" | 137° W | AM | ** | Pleiades |
| Neptune | 5th | Psc | 23h 55m | -1° 58' | 7.6 | 2" | 168° E | PM | ***** | |

## Highlights

| Date | Time (UT) | Event |
|------|-----------|-------|
| 2nd | 18:46 | Annular solar eclipse. Visible from Antarctica, the Pacific and southern South America. |
| | 18:50 | New Moon. (Not visible.) |
| 5th | 22:14 | The waxing crescent Moon is south of Venus. (Evening sky.) |
| 6th | N/A | Good opportunity to see Earthshine on the waxing crescent Moon. (Evening sky.) |
| 7th | 20:39 | The waxing crescent Moon is south of the bright star Antares. (Evening sky.) |
| 8th | N/A | The Draconid meteor shower is at its maximum. (ZHR: Var) |
| 9th | 06:27 | Jupiter is stationary prior to beginning retrograde motion. (Morning sky.) |
| 10th | 18:56 | First Quarter Moon. (Evening sky.) |
| | N/A | The Southern Taurid meteor shower is at its maximum. (ZHR: 5) |

# October 11th to 20th, 2024

## The Moon

    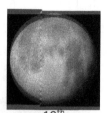

11th        13th        15th        17th        19th

| Date | Con | R.A. | Dec | Mag | Diam | Ill. | Elon. | Phase | Close To |
|------|-----|------|-----|-----|------|------|-------|-------|----------|
| 11th | Sgr | 19h 33m | -26°54' | -10.2 | 31' | 52% | 97° E | FQ | |
| 12th | Cap | 20h 31m | -23°45' | -10.7 | 32' | 63% | 110° E | FQ | |
| 13th | Cap | 21h 27m | -19°9' | -11.1 | 32' | 73% | 123° E | +G | |
| 14th | Aqr | 22h 22m | -13°22' | -11.4 | 33' | 83% | 136° E | +G | Saturn |
| 15th | Aqr | 23h 15m | -6°42' | -11.8 | 33' | 91% | 148° E | +G | Saturn, Neptune |
| 16th | Psc | 0h 8m | 0°29' | -12.2 | 33' | 97% | 161° E | FM | Neptune |
| 17th | Psc | 1h 2m | 7°43' | -12.6 | 33' | 100% | 173° E | FM | |
| 18th | Ari | 1h 58m | 14°31' | -12.5 | 33' | 100% | 174° W | FM | |
| 19th | Ari | 2h 56m | 20°24' | -12.2 | 33' | 96% | 160° W | FM | Uranus |
| 20th | Tau | 3h 57m | 24°54' | -11.8 | 33' | 91% | 146° W | -G | Uranus, Pleiades, Hyades, Aldeba |

## Mercury and Venus

            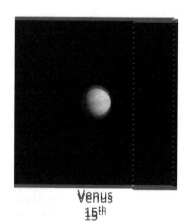

Mercury                      Venus
15th                         15th

### Mercury

| Date | Con. | R.A. | Dec. | Mag. | Diam. | Ill. | Elon. | Vis. | Rat. | Close To |
|------|------|------|------|------|-------|------|-------|------|------|----------|
| 11th | Vir | 13h 34m | -9°36' | -0.7 | 5" | 98% | 7° E | NV | N/A | Spica |
| 13th | Vir | 13h 46m | -10°59' | -0.6 | 5" | 97% | 8° E | NV | N/A | Spica |
| 15th | Vir | 13h 58m | -12°19' | -0.5 | 5" | 96% | 9° E | NV | N/A | Spica |
| 17th | Vir | 14h 10m | -13°37' | -0.5 | 5" | 96% | 10° E | NV | N/A | |
| 19th | Vir | 14h 22m | -14°52' | -0.4 | 5" | 95% | 11° E | NV | N/A | |

## Venus

| Date | Con. | R.A. | Dec. | Mag. | Diam. | Ill. | Elon. | Vis. | Rat. | Close To |
|------|------|------|------|------|-------|------|-------|------|------|----------|
| 11th | Lib | 15h 16m | -18° 57' | -3.4 | 13" | 83% | 32° E | PM | * | |
| 13th | Lib | 15h 26m | -19° 40' | -3.4 | 13" | 82% | 33° E | PM | * | |
| 15th | Lib | 15h 36m | -20° 21' | -3.4 | 13" | 82% | 34° E | PM | * | |
| 17th | Lib | 15h 46m | -20° 60' | -3.4 | 13" | 81% | 34° E | PM | * | |
| 19th | Sco | 15h 56m | -21° 36' | -3.4 | 13" | 81% | 35° E | PM | * | Antares |

## Mars and the Outer Planets

Mars
15th

Jupiter
15th

Saturn
15th

## Mars

| Date | Con. | R.A. | Dec. | Mag. | Diam. | Ill. | Elon. | Vis. | Rat. | Close To |
|------|------|------|------|------|-------|------|-------|------|------|----------|
| 11th | Gem | 7h 26m | 22° 44' | 0.6 | 8" | 88% | 85° W | AM | ** | |
| 15th | Gem | 7h 34m | 22° 33' | 0.5 | 8" | 88% | 87° W | AM | ** | |
| 20th | Gem | 7h 44m | 22° 19' | 0.4 | 8" | 88% | 89° W | AM | ** | |

## The Outer Planets

| Planet | Date | Con. | R.A. | Dec. | Mag. | Diam. | Elon. | Vis. | Rat. | Close To |
|--------|------|------|------|------|------|-------|-------|------|------|----------|
| Jupiter | 15th | Tau | 5h 22m | 22° 26' | -2.2 | 44" | 120° W | AM | *** | |
| Saturn | 15th | Aqr | 23h 2m | -8° 29' | 0.9 | 19" | 145° E | PM | **** | Moon |
| Uranus | 15th | Tau | 3h 37m | 19° 7' | 6.0 | 4" | 146° W | AM | *** | Pleiades |
| Neptune | 15th | Psc | 23h 54m | -2° 4' | 7.6 | 2" | 158° E | PM | ***** | Moon |

## Highlights

| Date | Time (UT) | Event |
|------|-----------|-------|
| 11th | 14:47 | Dwarf planet Pluto is stationary prior to resuming prograde motion. (Evening sky.) |
| 14th | 16:27 | The waxing gibbous Moon is south of Saturn. (Evening sky.) |
| 15th | 15:43 | The waxing gibbous Moon is north of Neptune. (Evening sky.) |
| 17th | 11:27 | Full Moon. (Visible all night.) |
| 18th | N/A | The Epsilon Geminid meteor shower is at its maximum. (ZHR: 3) |
| 19th | 15:13 | The waning gibbous Moon is north of Uranus. (Morning sky.) |

# October 21st to 31st, 2024

## The Moon

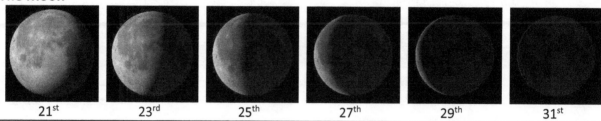

| 21st | 23rd | 25th | 27th | 29th | 31st |

| Date | Con | R.A. | Dec | Mag | Diam | Ill. | Elon. | Phase | Close To |
|------|-----|------|-----|-----|------|------|-------|-------|----------|
| 21st | Tau | 5h 0m | 27° 42' | -11.4 | 32' | 83% | 131° W | -G | Jupiter, Hyades, Aldebaran |
| 22nd | Aur | 6h 3m | 28° 40' | -11.1 | 32' | 74% | 116° W | -G | Jupiter |
| 23rd | Gem | 7h 3m | 27° 52' | -10.7 | 31' | 64% | 102° W | LQ | Mars |
| 24th | Gem | 8h 0m | 25° 33' | -10.3 | 31' | 53% | 89° W | LQ | Mars, Praesepe |
| 25th | Cnc | 8h 53m | 22° 2' | -9.8 | 30' | 43% | 76° W | LQ | Praesepe |
| 26th | Leo | 9h 42m | 17° 38' | -9.3 | 30' | 34% | 65° W | -Cr | Regulus |
| 27th | Leo | 10h 28m | 12° 37' | -8.8 | 30' | 25% | 55° W | -Cr | Regulus |
| 28th | Leo | 11h 11m | 7° 13' | -8.2 | 30' | 17% | 45° W | -Cr | |
| 29th | Vir | 11h 53m | 1° 36' | -7.4 | 29' | 11% | 35° W | NM | |
| 30th | Vir | 12h 34m | -4° 3' | -6.6 | 29' | 6% | 26° W | NM | |
| 31st | Vir | 13h 16m | -9° 35' | -5.6 | 29' | 2% | 16° W | NM | Spica |

## Mercury and Venus

Mercury
25th

Venus
25th

**Mercury**

| Date | Con. | R.A. | Dec. | Mag. | Diam. | Ill. | Elon. | Vis. | Rat. | Close To |
|------|------|------|------|------|-------|------|-------|------|------|----------|
| 21st | Lib | 14h 34m | -16° 3' | -0.4 | 5" | 94% | 12° E | NV | N/A | |
| 23rd | Lib | 14h 45m | -17° 11' | -0.3 | 5" | 93% | 14° E | NV | N/A | |
| 25th | Lib | 14h 57m | -18° 16' | -0.3 | 5" | 91% | 15° E | NV | N/A | |
| 27th | Lib | 15h 9m | -19° 17' | -0.2 | 5" | 90% | 16° E | PM | ** | |
| 29th | Lib | 15h 21m | -20° 14' | -0.2 | 5" | 89% | 17° E | PM | ** | |
| 31st | Lib | 15h 32m | -21° 7' | -0.2 | 5" | 87% | 18° E | PM | ** | |

**Venus**

| Date | Con. | R.A. | Dec. | Mag. | Diam. | Ill. | Elon. | Vis. | Rat. | Close To |
|------|------|------|------|------|-------|------|-------|------|------|----------|
| 21st | Sco | 16h 6m | -22° 11' | -3.4 | 13" | 80% | 36° E | PM | * | Antares |
| 23rd | Sco | 16h 16m | -22° 42' | -3.5 | 14" | 79% | 36° E | PM | * | Antares |
| 25th | Oph | 16h 26m | -23° 12' | -3.5 | 14" | 79% | 37° E | PM | * | Antares |
| 27th | Oph | 16h 37m | -23° 39' | -3.5 | 14" | 78% | 37° E | PM | * | Antares |
| 29th | Oph | 16h 47m | -24° 3' | -3.5 | 14" | 78% | 38° E | PM | * | Antares |
| 31st | Oph | 16h 57m | -24° 25' | -3.5 | 14" | 77% | 39° E | PM | * | Antares |

## Mars and the Outer Planets

Mars
25th

Jupiter
25th

Saturn
25th

**Mars**

| Date | Con. | R.A. | Dec. | Mag. | Diam. | Ill. | Elon. | Vis. | Rat. | Close To |
|------|------|------|------|------|-------|------|-------|------|------|----------|
| 21st | Gem | 7h 46m | 22° 16' | 0.4 | 9" | 88% | 89° W | AM | ** | |
| 25th | Gem | 7h 53m | 22° 4' | 0.4 | 9" | 88% | 91° W | AM | ** | |
| 31st | Cnc | 8h 3m | 21° 47' | 0.3 | 9" | 89% | 95° W | AM | ** | Praesepe |

**The Outer Planets**

| Planet | Date | Con. | R.A. | Dec. | Mag. | Diam. | Elon. | Vis. | Rat. | Close To |
|--------|------|------|------|------|------|-------|-------|------|------|----------|
| Jupiter | 25th | Tau | 5h 20m | 22° 25' | -2.2 | 45" | 130° W | AM | **** | |
| Saturn | 25th | Aqr | 23h 1m | -8° 38' | 1.0 | 19" | 135° E | PM | *** | |
| Uranus | 25th | Tau | 3h 35m | 19° 2' | 6.0 | 4" | 156° W | AM | *** | Pleiades |
| Neptune | 25th | Psc | 23h 53m | -2° 9' | 7.6 | 2" | 148° E | PM | ***** | |

## Highlights

| Date | Time (UT) | Event |
|------|-----------|-------|
| 21st | 09:46 | The waning gibbous Moon is north of Jupiter. (Morning sky.) |
| 22nd | N/A | The Orionid meteor shower is at its maximum. (ZHR: 15) |
| 23rd | 19:10 | The nearly last quarter Moon is north of Mars. (Morning sky.) |
| 24th | 08:04 | Last Quarter Moon. (Morning sky.) |
| | 17:47 | The last quarter Moon is north of the Praesepe open star cluster. (Morning sky.) |
| 25th | 11:49 | Venus is 3.1° north of the bright star Antares. (Evening sky.) |
| 26th | 15:14 | The waning crescent Moon is north of the bright star Regulus. (Morning sky.) |
| 29th | N/A | Good opportunity to see Earthshine on the waning crescent Moon. (Morning sky.) |

# November 1st to 10th, 2024

## The Moon

    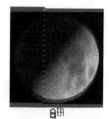

| 1st | 3rd | 5th | 7th | 9th |
|-----|-----|-----|-----|-----|

| Date | Con | R.A. | Dec | Mag | Diam | Ill. | Elon. | Phase | Close To |
|------|-----|------|-----|-----|------|------|-------|-------|----------|
| 1st | Vir | 14h 0m | -14° 48' | -4.6 | 30' | 0% | 6° W | NM | Spica |
| 2nd | Lib | 14h 46m | -19° 31' | -4.5 | 30' | 0% | 4° E | NM | |
| 3rd | Lib | 15h 36m | -23° 29' | -5.6 | 30' | 2% | 15° E | NM | Mercury |
| 4th | Sco | 16h 28m | -26° 28' | -6.6 | 30' | 6% | 27° E | NM | Mercury, Antares |
| 5th | Oph | 17h 23m | -28° 12' | -7.5 | 30' | 11% | 40° E | NM | Venus |
| 6th | Sgr | 18h 20m | -28° 32' | -8.3 | 31' | 18% | 54° E | +Cr | |
| 7th | Sgr | 19h 18m | -27° 19' | -8.9 | 31' | 27% | 67° E | +Cr | |
| 8th | Cap | 20h 15m | -24° 37' | -9.5 | 31' | 37% | 80° E | +Cr | |
| 9th | Cap | 21h 10m | -20° 32' | -10.0 | 32' | 47% | 93° E | FQ | |
| 10th | Aqr | 22h 4m | -15° 17' | -10.5 | 32' | 58% | 105° E | FQ | |

## Mercury and Venus

Mercury
5th

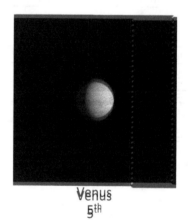

Venus
5th

## Mercury

| Date | Con. | R.A. | Dec. | Mag. | Diam. | Ill. | Elon. | Vis. | Rat. | Close To |
|------|------|------|------|------|-------|------|-------|------|------|----------|
| 1st | Lib | 15h 38m | -21° 32' | -0.2 | 5" | 86% | 18° E | PM | ** | |
| 3rd | Sco | 15h 50m | -22° 19' | -0.2 | 5" | 84% | 19° E | PM | ** | Moon, Antares |
| 5th | Sco | 16h 1m | -23° 1' | -0.2 | 6" | 82% | 20° E | PM | *** | Antares |
| 7th | Sco | 16h 13m | -23° 39' | -0.2 | 6" | 80% | 21° E | PM | *** | Antares |
| 9th | Sco | 16h 24m | -24° 12' | -0.2 | 6" | 77% | 22° E | PM | *** | Antares |

## Venus

| Date | Con. | R.A. | Dec. | Mag. | Diam. | Ill. | Elon. | Vis. | Rat. | Close To |
|------|------|------|------|------|-------|------|-------|------|------|----------|
| 1st | Oph | 17h 3m | -24° 35' | -3.5 | 14" | 77% | 39° E | PM | * | Antares |
| 3rd | Oph | 17h 13m | -24° 52' | -3.5 | 14" | 77% | 40° E | PM | * | |
| 5th | Oph | 17h 24m | -25° 7' | -3.5 | 15" | 76% | 40° E | PM | ** | Moon |
| 7th | Oph | 17h 34m | -25° 19' | -3.5 | 15" | 75% | 41° E | PM | ** | |
| 9th | Sgr | 17h 45m | -25° 28' | -3.5 | 15" | 75% | 42° E | PM | ** | |

## Mars and the Outer Planets

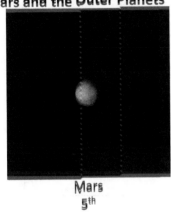

Mars
5th

Jupiter
5th

Saturn
5th

## Mars

| Date | Con. | R.A. | Dec. | Mag. | Diam. | Ill. | Elon. | Vis. | Rat. | Close To |
|------|------|------|------|------|-------|------|-------|------|------|----------|
| 1st | Cnc | 8h 5m | 21° 45' | 0.3 | 9" | 89% | 95° W | AM | ** | Praesepe |
| 5th | Cnc | 8h 11m | 21° 35' | 0.2 | 9" | 89% | 98° W | AM | ** | Praesepe |
| 10th | Cnc | 8h 18m | 21° 24' | 0.1 | 10" | 90% | 101° W | AM | ** | Praesepe |

## The Outer Planets

| Planet | Date | Con. | R.A. | Dec. | Mag. | Diam. | Elon. | Vis. | Rat. | Close To |
|--------|------|------|------|------|------|-------|-------|------|------|----------|
| Jupiter | 5th | Tau | 5h 17m | 22° 22' | -2.3 | 47" | 141° W | AM | **** | |
| Saturn | 5th | Aqr | 22h 59m | -8° 43' | 1.0 | 18" | 124° E | PM | *** | |
| Uranus | 5th | Tau | 3h 33m | 18° 56' | 6.0 | 4" | 167° W | AM | *** | Pleiades |
| Neptune | 5th | Psc | 23h 52m | -2° 14' | 7.7 | 2" | 138° E | PM | ***** | |

## Highlights

| Date | Time (UT) | Event |
|------|-----------|-------|
| 1st | 12:48 | New Moon. (Not visible.) |
| 3rd | 05:39 | The waxing crescent Moon is south of Mercury. (Evening sky.) |
| 4th | 00:48 | The waxing crescent Moon is north of the bright star Antares. (Evening sky.) |
| | N/A | Good opportunity to see Earthshine on the waxing crescent Moon. (Evening sky.) |
| 5th | 01:00 | The waxing crescent Moon is south of Venus. (Evening sky.) |
| 8th | N/A | The Andromedid meteor shower is at its maximum. (ZHR: Var.) |
| 9th | 05:56 | First Quarter Moon. (Evening sky.) |
| | 21:19 | Mercury is 2.0° north of the bright star Antares. (Evening sky.) |

# November 11ᵗʰ to 20ᵗʰ, 2024

## The Moon

| 11ᵗʰ | 13ᵗʰ | 15ᵗʰ | 17ᵗʰ | 19ᵗʰ |

| Date | Con | R.A. | Dec | Mag | Diam | Ill. | Elon. | Phase | Close To |
|------|-----|------|-----|-----|------|------|-------|-------|----------|
| 11th | Aqr | 22h 55m | -9° 7' | -10.9 | 32' | 69% | 117° E | +G | Saturn |
| 12th | Psc | 23h 47m | -2° 21' | -11.3 | 33' | 79% | 129° E | +G | Saturn, Neptune |
| 13th | Psc | 0h 38m | 4° 40' | -11.7 | 33' | 88% | 141° E | +G | Neptune |
| 14th | Psc | 1h 32m | 11° 33' | -12.0 | 33' | 95% | 153° E | +G | |
| 15th | Ari | 2h 28m | 17° 48' | -12.4 | 33' | 99% | 167° E | FM | |
| 16th | Ari | 3h 28m | 22° 58' | -12.6 | 33' | 100% | 180° W | FM | Uranus, Pleiades |
| 17th | Tau | 4h 31m | 26° 35' | -12.3 | 33' | 98% | 165° W | FM | Jupiter, Pleiades, Hyades, Aldeba |
| 18th | Tau | 5h 36m | 28° 22' | -12.0 | 32' | 94% | 150° W | -G | Jupiter |
| 19th | Aur | 6h 39m | 28° 14' | -11.6 | 32' | 87% | 135° W | -G | |
| 20th | Gem | 7h 40m | 26° 24' | -11.3 | 31' | 79% | 121° W | -G | |

## Mercury and Venus

Mercury
15ᵗʰ

Venus
15ᵗʰ

**Mercury**

| Date | Con. | R.A. | Dec. | Mag. | Diam. | Ill. | Elon. | Vis. | Rat. | Close To |
|------|------|------|------|------|-------|------|-------|------|------|----------|
| 11th | Oph | 16h 35m | -24° 39' | -0.1 | 6" | 74% | 22° E | PM | *** | Antares |
| 13th | Sco | 16h 45m | -25° 1' | -0.1 | 6" | 71% | 23° E | PM | *** | Antares |
| 15th | Oph | 16h 55m | -25° 17' | -0.1 | 6" | 66% | 23° E | PM | *** | Antares |
| 17th | Oph | 17h 4m | -25° 27' | -0.1 | 7" | 62% | 23° E | PM | *** | Antares |
| 19th | Oph | 17h 12m | -25° 31' | 0.0 | 7" | 56% | 23° E | PM | *** | |

## Venus

| Date | Con. | R.A. | Dec. | Mag. | Diam. | Ill. | Elon. | Vis. | Rat. | Close To |
|------|------|------|------|------|-------|------|-------|------|------|----------|
| 11th | Sgr | 17h 55m | -25° 34' | -3.6 | 15" | 74% | 42° E | PM | ** | |
| 13th | Sgr | 18h 6m | -25° 37' | -3.6 | 15" | 74% | 43° E | PM | ** | |
| 15th | Sgr | 18h 17m | -25° 37' | -3.6 | 15" | 73% | 44° E | PM | ** | |
| 17th | Sgr | 18h 27m | -25° 35' | -3.6 | 16" | 72% | 44° E | PM | ** | |
| 19th | Sgr | 18h 38m | -25° 29' | -3.6 | 16" | 72% | 45° E | PM | ** | |

## Mars and the Outer Planets

Mars
15th

Jupiter
15th

Saturn
15th

### Mars

| Date | Con. | R.A. | Dec. | Mag. | Diam. | Ill. | Elon. | Vis. | Rat. | Close To |
|------|------|------|------|------|-------|------|-------|------|------|----------|
| 11th | Cnc | 8h 19m | 21° 23' | 0.1 | 10" | 90% | 102° W | AM | ** | Praesepe |
| 15th | Cnc | 8h 23m | 21° 16' | 0.0 | 10" | 90% | 105° W | AM | ** | Praesepe |
| 20th | Cnc | 8h 28m | 21° 12' | -0.1 | 11" | 91% | 109° W | AM | *** | Praesepe |

### The Outer Planets

| Planet | Date | Con. | R.A. | Dec. | Mag. | Diam. | Elon. | Vis. | Rat. | Close To |
|--------|------|------|------|------|------|-------|-------|------|------|----------|
| Jupiter | 15th | Tau | 5h 13m | 22° 18' | -2.3 | 47" | 152° W | AM | **** | Aldebaran |
| Saturn | 15th | Aqr | 22h 59m | -8° 44' | 1.1 | 18" | 114° E | PM | *** | |
| Uranus | 15th | Tau | 3h 32m | 18° 50' | 6.0 | 4" | 178° W | AN | *** | Pleiades |
| Neptune | 15th | Psc | 23h 52m | -2° 17' | 7.7 | 2" | 127° E | PM | ***** | |

## Highlights

| Date | Time (UT) | Event |
|------|-----------|-------|
| 11th | 03:24 | The waxing gibbous Moon is north of Saturn. (Evening sky.) |
| | N/A | The Northern Taurid meteor shower is at its maximum. (ZHR: 5) |
| 12th | 04:07 | The waxing gibbous Moon is north of Neptune. (Evening sky.) |
| 15th | 21:29 | Full Moon. (Visible all night.) |
| 16th | 02:14 | The just-past full Moon is north of Uranus. (Visible all night.) |
| | 04:59 | Saturn is stationary prior to resuming prograde motion. (Evening sky.) |
| | 08:01 | Mercury is at greatest eastern elongation from the Sun. (Evening sky.) |
| 17th | 05:56 | Uranus is at opposition from the Sun. (Visible all night.) |
| | 14:27 | The waning gibbous Moon is north of Jupiter. (Morning sky.) |
| 18th | N/A | The Leonid meteor shower is at its maximum. (ZHR: 15) |
| 20th | 19:43 | The waning gibbous Moon is north of Mars. (Morning sky.) |

# November 21st to 30th, 2024

## The Moon

 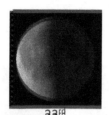   

| 21st | 23rd | 25th | 27th | 29th |

| Date | Con | R.A. | Dec | Mag | Diam | Ill. | Elon. | Phase | Close To |
|------|-----|------|-----|-----|------|------|-------|-------|----------|
| 21st | Cnc | 8h 36m | 23° 11' | -10.9 | 31' | 70% | 108° W | =G | Mars, Praesepe |
| 22nd | Leo | 9h 27m | 18° 56' | -10.6 | 30' | 60% | 96° W | LQ | Praesepe, Regulus |
| 23rd | Leo | 10h 14m | 14° 1' | -10.2 | 30' | 51% | 85° W | LQ | Regulus |
| 24th | Leo | 10h 58m | 8° 39' | -9.7 | 30' | 41% | 75° W | LQ | |
| 25th | Vir | 11h 41m | 3° 5' | -9.2 | 30' | 32% | 66° W | =Cr | |
| 26th | Vir | 12h 22m | -2° 34' | -8.7 | 29' | 24% | 57° W | =Cr | |
| 27th | Vir | 13h 4m | -8° 7' | -8.0 | 29' | 16% | 47° W | =Cr | Spica |
| 28th | Vir | 13h 47m | -13° 25' | -7.3 | 30' | 10% | 37° W | NM | Spica |
| 29th | Lib | 14h 33m | -18° 16' | -6.4 | 30' | 5% | 27° W | NM | |
| 30th | Lib | 15h 22m | -22° 28' | -5.5 | 30' | 2% | 16° W | NM | |

## Mercury and Venus

Mercury
25th

Venus
25th

### Mercury

| Date | Con. | R.A. | Dec. | Mag. | Diam. | Ill. | Elon. | Vis. | Rat. | Close To |
|------|------|------|------|------|-------|------|-------|------|------|----------|
| 21st | Oph | 17h 19m | -25° 29' | 0.1 | 7" | 50% | 23° E | PM | *** | |
| 23rd | Oph | 17h 24m | -25° 19' | 0.2 | 8" | 43% | 22° E | PM | *** | |
| 25th | Oph | 17h 27m | -25° 3' | 0.4 | 8" | 35% | 21° E | PM | *** | |
| 27th | Oph | 17h 27m | -24° 38' | 0.6 | 9" | 27% | 19° E | PM | *** | |
| 29th | Oph | 17h 24m | -24° 6' | 1.0 | 9" | 18% | 16° E | PM | ** | |

## Venus

| Date | Con. | R.A. | Dec. | Mag. | Diam. | Ill. | Elon. | Vis. | Rat. | Close To |
|---|---|---|---|---|---|---|---|---|---|---|
| 21st | Sgr | 18h 48m | -25° 21' | -3.6 | 16" | 71% | 45° E | PM | ** | |
| 23rd | Sgr | 18h 58m | -25° 9' | -3.6 | 16" | 70% | 46° E | PM | ** | |
| 25th | Sgr | 19h 9m | -24° 55' | -3.7 | 16" | 70% | 46° E | PM | ** | |
| 27th | Sgr | 19h 19m | -24° 38' | -3.7 | 17" | 69% | 47° E | PM | ** | |
| 29th | Sgr | 19h 29m | -24° 18' | -3.7 | 17" | 68% | 47° E | PM | ** | |

## Mars and the Outer Planets

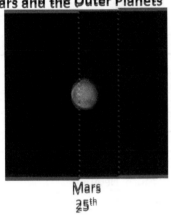

Mars
25th

Jupiter
25th

Saturn
25th

### Mars

| Date | Con. | R.A. | Dec. | Mag. | Diam. | Ill. | Elon. | Vis. | Rat. | Close To |
|---|---|---|---|---|---|---|---|---|---|---|
| 21st | Cnc | 8h 29m | 21° 11' | -0.1 | 11" | 91% | 109° W | AM | *** | Moon, Praesepe |
| 25th | Cnc | 8h 32m | 21° 11' | -0.2 | 11" | 92% | 113° W | AM | *** | Praesepe |
| 30th | Cnc | 8h 35m | 21° 14' | -0.3 | 12" | 92% | 118° W | AM | *** | Praesepe |

### The Outer Planets

| Planet | Date | Con. | R.A. | Dec. | Mag. | Diam. | Elon. | Vis. | Rat. | Close To |
|---|---|---|---|---|---|---|---|---|---|---|
| Jupiter | 25th | Tau | 5h 7m | 22° 12' | -2.3 | 48" | 164° W | AM | **** | Aldebaran |
| Saturn | 25th | Aqr | 22h 59m | -8° 41' | 1.1 | 18" | 104° E | PM | *** | |
| Uranus | 25th | Tau | 3h 30m | 18° 44' | 6.0 | 4" | 171° E | AN | *** | Pleiades |
| Neptune | 25th | Psc | 23h 51m | -2° 19' | 7.7 | 2" | 117° E | PM | ***** | |

## Highlights

| Date | Time (UT) | Event |
|---|---|---|
| 21st | N/A | The Alpha Monocerotid meteor shower is at its maximum. (ZHR: Var) |
| 22nd | 19:51 | The nearly last quarter Moon is north of the bright star Regulus. (Morning sky.) |
| 23rd | 01:29 | Last Quarter Moon. (Morning sky.) |
| 26th | 04:19 | Mercury is stationary prior to beginning retrograde motion. (Evening sky.) |
| 27th | N/A | Good opportunity to see Earthshine on the waning crescent Moon. (Morning sky.) |
| | 13:25 | The waning crescent Moon is south of the bright star Spica. (Morning sky.) |
| 29th | N/A | The November Orionid meteor shower is at its maximum. (ZHR: 3) |

# December 1st to 10th, 2024

## The Moon

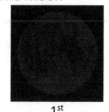  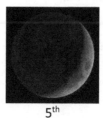  

1st        3rd        5th        7th        9th

| Date | Con | R.A. | Dec | Mag | Diam | Ill. | Elon. | Phase | Close To |
|------|-----|------|-----|-----|------|------|-------|-------|----------|
| 1st | Sco | 16h 14m | -25° 44' | -4.5 | 30' | 0% | 4° W | NM | Antares |
| 2nd | Oph | 17h 9m | -27° 48' | -4.9 | 30' | 1% | 9° E | NM | Mercury, Antares |
| 3rd | Sgr | 18h 6m | -28° 28' | -6.0 | 31' | 3% | 22° E | NM | |
| 4th | Sgr | 19h 5m | -27° 34' | -7.0 | 31' | 8% | 35° E | NM | |
| 5th | Sgr | 20h 2m | -25° 10' | -7.9 | 31' | 14% | 49° E | +Cr | Venus |
| 6th | Cap | 20h 58m | -21° 21' | -8.6 | 31' | 23% | 62° E | +Cr | |
| 7th | Cap | 21h 51m | -16° 23' | -9.2 | 32' | 32% | 74° E | +Cr | |
| 8th | Aqr | 22h 42m | -10° 32' | -9.8 | 32' | 43% | 85° E | FQ | Saturn |
| 9th | Aqr | 23h 32m | -4° 6' | -10.3 | 32' | 54% | 97° E | FQ | Saturn, Neptune |
| 10th | Psc | 0h 22m | 2° 39' | -10.7 | 32' | 65% | 108° E | FQ | Neptune |

## Mercury and Venus

Mercury
5th

Venus
5th

### Mercury

| Date | Con. | R.A. | Dec. | Mag. | Diam. | Ill. | Elon. | Vis. | Rat. | Close To |
|------|------|------|------|------|-------|------|-------|------|------|----------|
| 1st | Oph | 17h 18m | -23° 24' | 1.4 | 9" | 10% | 12° E | NV | N/A | |
| 3rd | Oph | 17h 10m | -22° 35' | 2.0 | 10" | 4% | 8° E | NV | N/A | |
| 5th | Oph | 16h 59m | -21° 40' | 2.7 | 10" | 1% | 3° E | NV | N/A | Antares |
| 7th | Oph | 16h 48m | -20° 44' | 2.7 | 10" | 1% | 2° W | NV | N/A | Antares |
| 9th | Oph | 16h 37m | -19° 53' | 2.0 | 10" | 4% | 7° W | NV | N/A | Antares |

## Venus

| Date | Con. | R.A. | Dec. | Mag. | Diam. | Ill. | Elon. | Vis. | Rat. | Close To |
|------|------|------|------|------|-------|------|-------|------|------|----------|
| 1st | Sgr | 19h 39m | -23° 56' | -3.7 | 17" | 68% | 47° E | PM | ** | |
| 3rd | Sgr | 19h 49m | -23° 31' | -3.7 | 17" | 67% | 48° E | PM | ** | |
| 5th | Sgr | 19h 59m | -23° 4' | -3.7 | 18" | 66% | 48° E | PM | ** | Moon |
| 7th | Cap | 20h 9m | -22° 34' | -3.7 | 18" | 66% | 48° E | PM | ** | |
| 9th | Cap | 20h 19m | -22° 2' | -3.8 | 18" | 65% | 49° E | PM | ** | |

## Mars and the Outer Planets

Mars
5th

Jupiter
5th

Saturn
5th

### Mars

| Date | Con. | R.A. | Dec. | Mag. | Diam. | Ill. | Elon. | Vis. | Rat. | Close To |
|------|------|------|------|------|-------|------|-------|------|------|----------|
| 1st | Cnc | 8h 35m | 21° 16' | -0.3 | 12" | 93% | 119° W | AM | *** | Praesepe |
| 5th | Cnc | 8h 36m | 21° 23' | -0.4 | 12" | 93% | 123° W | AM | *** | Praesepe |
| 10th | Cnc | 8h 37m | 21° 37' | -0.5 | 12" | 94% | 128° W | AM | *** | Praesepe |

### The Outer Planets

| Planet | Date | Con. | R.A. | Dec. | Mag. | Diam. | Elon. | Vis. | Rat. | Close To |
|--------|------|------|------|------|------|-------|-------|------|------|----------|
| Jupiter | 5th | Tau | 5h 2m | 22° 6' | -2.3 | 48" | 176° W | AN | ***** | Hyades, Aldebaran |
| Saturn | 5th | Aqr | 23h 0m | -8° 34' | 1.2 | 17" | 93° E | PM | ** | |
| Uranus | 5th | Tau | 3h 28m | 18° 39' | 6.0 | 4" | 160° E | PM | *** | Pleiades |
| Neptune | 5th | Psc | 23h 51m | -2° 20' | 7.7 | 2" | 106° E | PM | ***** | |

## Highlights

| Date | Time (UT) | Event |
|------|-----------|-------|
| 1st | 06:22 | New Moon. (Not visible.) |
| 4th | N/A | Good opportunity to see Earthshine on the waxing crescent Moon. (Evening sky.) |
| 5th | 00:02 | The waxing crescent Moon is south of Venus. (Evening sky.) |
| 6th | 02:11 | Mercury is at inferior conjunction with the Sun. (Not visible.) |
| | N/A | The Sigma Hydrid meteor shower is at its maximum. (ZHR: 3) |
| 7th | 21:06 | Mars is stationary prior to beginning retrograde motion. (Morning sky.) |
| | 22:05 | Jupiter is at opposition from the Sun. (Visible all night.) |
| 8th | 06:27 | Neptune is stationary prior to resuming prograde motion. (Evening sky.) |
| | 07:56 | The almost first quarter Moon is north of Saturn. (Evening sky.) |
| | 15:27 | First Quarter Moon. (Evening sky.) |
| 9th | 08:27 | The just-past first quarter Moon is north of Neptune. (Evening sky.) |

# December 11th to 20th, 2024

## The Moon

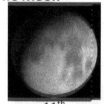
11th

13th

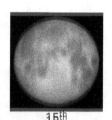
15th

17th

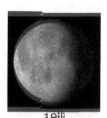
19th

| Date | Con | R.A. | Dec | Mag | Diam | Ill. | Elon. | Phase | Close To |
|------|-----|------|-----|-----|------|------|-------|-------|----------|
| 11th | Psc | 1h 13m | 9° 20' | -11.1 | 33' | 76% | 120° E | +G | |
| 12th | Ari | 2h 6m | 15° 38' | -11.5 | 33' | 85% | 132° E | +G | |
| 13th | Ari | 3h 3m | 21° 5' | -11.9 | 33' | 92% | 145° E | +G | Uranus, Pleiades |
| 14th | Tau | 4h 4m | 25° 15' | -12.2 | 33' | 97% | 159° E | FM | Uranus, Pleiades, Hyades, Aldeba |
| 15th | Tau | 5h 8m | 27° 46' | -12.5 | 32' | 100% | 174° E | FM | Jupiter, Hyades, Aldebaran |
| 16th | Aur | 6h 12m | 28° 25' | -12.5 | 32' | 99% | 171° W | FM | |
| 17th | Gem | 7h 14m | 27° 13' | -12.2 | 32' | 96% | 156° W | FM | |
| 18th | Cnc | 8h 13m | 24° 27' | -11.8 | 31' | 92% | 143° W | -G | Mars, Praesepe |
| 19th | Cnc | 9h 7m | 20° 27' | -11.5 | 31' | 85% | 130° W | -G | Mars, Praesepe |
| 20th | Leo | 9h 57m | 15° 38' | -11.2 | 30' | 77% | 119° W | -G | Regulus |

## Mercury and Venus

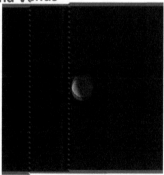
Mercury
15th

Venus
15th

### Mercury

| Date | Con. | R.A. | Dec. | Mag. | Diam. | Ill. | Elon. | Vis. | Rat. | Close To |
|------|------|------|------|------|-------|------|-------|------|------|----------|
| 11th | Oph | 16h 28m | -19° 12' | 1.4 | 9" | 10% | 11° W | NV | N/A | Antares |
| 13th | Sco | 16h 23m | -18° 46' | 0.9 | 9" | 18% | 15° W | NV | N/A | Antares |
| 15th | Sco | 16h 20m | -18° 34' | 0.6 | 9" | 27% | 18° W | AM | *** | Antares |
| 17th | Sco | 16h 20m | -18° 36' | 0.3 | 8" | 36% | 20° W | AM | *** | Antares |
| 19th | Sco | 16h 23m | -18° 49' | 0.1 | 8" | 44% | 22° W | AM | *** | Antares |

## Venus

| Date | Con. | R.A. | Dec. | Mag. | Diam. | Ill. | Elon. | Vis. | Rat. | Close To |
|------|------|------|------|------|-------|------|-------|------|------|----------|
| 11th | Cap | 20h 28m | -21° 27' | -3.8 | 18" | 64% | 49° E | PM | ** | |
| 13th | Cap | 20h 38m | -20° 50' | -3.8 | 19" | 63% | 49° E | PM | *** | |
| 15th | Cap | 20h 47m | -20° 12' | -3.8 | 19" | 63% | 49° E | PM | *** | |
| 17th | Cap | 20h 56m | -19° 31' | -3.8 | 19" | 62% | 49° E | PM | *** | |
| 19th | Cap | 21h 6m | -18° 49' | -3.8 | 20" | 61% | 49° E | PM | *** | |

## Mars and the Outer Planets

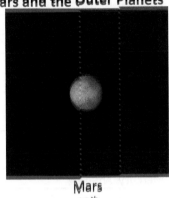

Mars
15th

Jupiter
15th

Saturn
15th

## Mars

| Date | Con. | R.A. | Dec. | Mag. | Diam. | Ill. | Elon. | Vis. | Rat. | Close To |
|------|------|------|------|------|-------|------|-------|------|------|----------|
| 11th | Cnc | 8h 36m | 21° 40' | -0.5 | 13" | 95% | 129° W | AM | *** | Praesepe |
| 15th | Cnc | 8h 35m | 21° 56' | -0.6 | 13" | 95% | 134° W | AM | *** | Praesepe |
| 20th | Cnc | 8h 33m | 22° 20' | -0.7 | 13" | 97% | 140° W | AM | *** | Praesepe |

## The Outer Planets

| Planet | Date | Con. | R.A. | Dec. | Mag. | Diam. | Elon. | Vis. | Rat. | Close To |
|--------|------|------|------|------|------|-------|-------|------|------|----------|
| Jupiter | 15th | Tau | 4h 56m | 21° 59' | -2.3 | 48" | 171° E | AN | **** | Moon, Hyades, Aldebaran |
| Saturn | 15th | Aqr | 23h 2m | -8° 22' | 1.2 | 17" | 83° E | PM | ** | |
| Uranus | 15th | Tau | 3h 27m | 18° 33' | 6.0 | 4" | 149° E | PM | *** | Pleiades |
| Neptune | 15th | Psc | 23h 51m | -2° 19' | 7.7 | 2" | 95° E | PM | **** | |

## Highlights

| Date | Time (UT) | Event |
|------|-----------|-------|
| 13th | 09:41 | The waxing gibbous Moon is north of Uranus. (Evening sky.) |
| | N/A | The Geminid meteor shower is at its maximum. (ZHR: 120) |
| 14th | 18:04 | The nearly full Moon is north of Jupiter. (Visible all night.) |
| 15th | 09:02 | Full Moon. (Visible all night.) |
| 18th | 10:24 | The waning gibbous Moon is north of Mars. (Morning sky.) |
| 20th | 06:55 | The waning gibbous Moon is north of the bright star Regulus. (Morning sky.) |
| | N/A | The December Leonis Minorid meteor shower is at its maximum. (ZHR: 5) |

# December 21<sup>st</sup> to 31<sup>st</sup>, 2024

## The Moon

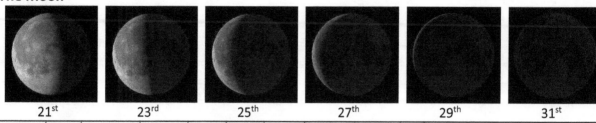

| 21<sup>st</sup> | 23<sup>rd</sup> | 25<sup>th</sup> | 27<sup>th</sup> | 29<sup>th</sup> | 31<sup>st</sup> |

| Date | Con | R.A. | Dec | Mag | Diam | Ill. | Elon. | Phase | Close To |
|------|-----|------|-----|-----|------|------|-------|-------|----------|
| 21st | Leo | 10h 43m | 10° 18' | -10.9 | 30' | 68% | 109° W | -G | Regulus |
| 22nd | Leo | 11h 26m | 4° 42' | -10.5 | 30' | 59% | 99° W | LQ | |
| 23rd | Vir | 12h 8m | 0° 59' | -10.1 | 30' | 49% | 90° W | LQ | |
| 24th | Vir | 12h 50m | -6° 35' | -9.7 | 30' | 40% | 80° W | LQ | Spica |
| 25th | Vir | 13h 33m | -11° 57' | -9.2 | 30' | 31% | 71° W | -Cr | Spica |
| 26th | Vir | 14h 18m | -16° 56' | -8.6 | 30' | 23% | 61° W | -Cr | |
| 27th | Lib | 15h 5m | -21° 19' | -8.0 | 30' | 15% | 50° W | -Cr | |
| 28th | Sco | 15h 56m | -24° 52' | -7.2 | 30' | 9% | 38° W | NM | Antares |
| 29th | Sco | 16h 51m | -27° 19' | -6.3 | 30' | 4% | 26° W | NM | Mercury, Antares |
| 30th | Sgr | 17h 48m | -28° 24' | -5.2 | 31' | 1% | 12° W | NM | Mercury |
| 31st | Sgr | 18h 47m | -27° 56' | -4.4 | 31' | 0% | 1° E | NM | |

## Mercury and Venus

Mercury
25<sup>th</sup>

Venus
25<sup>th</sup>

**Mercury**

| Date | Con. | R.A. | Dec. | Mag. | Diam. | Ill. | Elon. | Vis. | Rat. | Close To |
|------|------|------|------|------|-------|------|-------|------|------|----------|
| 21st | Oph | 16h 27m | -19° 11' | 0.0 | 7" | 51% | 23° W | AM | *** | Antares |
| 23rd | Oph | 16h 34m | -19° 38' | -0.1 | 7" | 58% | 23° W | AM | *** | Antares |
| 25th | Oph | 16h 41m | -20° 8' | -0.1 | 7" | 63% | 24° W | AM | *** | Antares |
| 27th | Oph | 16h 50m | -20° 40' | -0.2 | 6" | 68% | 24° W | AM | *** | Antares |
| 29th | Oph | 17h 0m | -21° 11' | -0.2 | 6" | 72% | 23° W | AM | *** | Moon, Antares |
| 31st | Oph | 17h 10m | -21° 42' | -0.2 | 6" | 76% | 23° W | AM | *** | |

**Venus**

| Date | Con. | R.A. | Dec. | Mag. | Diam. | Ill. | Elon. | Vis. | Rat. | Close To |
|------|------|------|------|------|-------|------|-------|------|------|----------|
| 21st | Cap | 21h 14m | -18° 4' | -3.9 | 20" | 60% | 49° E | PM | *** | |
| 23rd | Cap | 21h 23m | -17° 19' | -3.9 | 20" | 59% | 49° E | PM | *** | |
| 25th | Cap | 21h 32m | -16° 31' | -3.9 | 21" | 59% | 49° E | PM | *** | |
| 27th | Cap | 21h 40m | -15° 43' | -3.9 | 21" | 58% | 49° E | PM | *** | |
| 29th | Cap | 21h 49m | -14° 52' | -3.9 | 22" | 57% | 49° E | PM | *** | |
| 31st | Cap | 21h 57m | -14° 1' | -4.0 | 22" | 56% | 49° E | PM | *** | |

## Mars and the Outer Planets

Mars
25th

Jupiter
25th

Saturn
25th

**Mars**

| Date | Con. | R.A. | Dec. | Mag. | Diam. | Ill. | Elon. | Vis. | Rat. | Close To |
|------|------|------|------|------|-------|------|-------|------|------|----------|
| 21st | Cnc | 8h 32m | 22° 25' | -0.8 | 13" | 97% | 141° W | AM | *** | Praesepe |
| 25th | Cnc | 8h 28m | 22° 48' | -0.8 | 14" | 98% | 147° W | AM | **** | Praesepe |
| 31st | Cnc | 8h 21m | 23° 26' | -1.0 | 14" | 99% | 155° W | AM | **** | Praesepe |

**The Outer Planets**

| Planet | Date | Con. | R.A. | Dec. | Mag. | Diam. | Elon. | Vis. | Rat. | Close To |
|--------|------|------|------|------|------|-------|-------|------|------|----------|
| Jupiter | 25th | Tau | 4h 50m | 21° 52' | -2.3 | 48" | 159° E | PM | **** | Hyades, Aldebaran |
| Saturn | 25th | Aqr | 23h 4m | -8° 7' | 1.3 | 17" | 72° E | PM | ** | |
| Uranus | 25th | Tau | 3h 26m | 18° 29' | 6.0 | 4" | 138° E | PM | ** | Pleiades |
| Neptune | 25th | Psc | 23h 51m | -2° 17' | 7.7 | 2" | 84° E | PM | **** | |

## Highlights

| Date | Time (UT) | Event |
|------|-----------|-------|
| 21st | 09:21 | Hibernal solstice. |
| | 12:18 | Mercury is 7.1° north of the bright star Antares. (Morning sky.) |
| 22nd | 22:19 | Last Quarter Moon. (Morning sky.) |
| | N/A | The Ursid meteor shower is at its maximum. (ZHR: 10) |
| 24th | 19:21 | The waning crescent Moon is north of Spica. (Morning sky.) |
| 25th | 02:14 | Mercury is at greatest western elongation from the Sun. (Morning sky.) |
| 27th | N/A | Good opportunity to see Earthshine on the waning crescent Moon. (Morning sky.) |
| 28th | 16:26 | The waning crescent Moon is south of the bright star Antares. (Morning sky.) |
| 29th | 02:40 | The nearly new Moon is south of Mercury. (Morning sky.) |
| 30th | 22:28 | New Moon. (Not visible.) |
| 31st | N/A | The Coma Berenicid meteor shower is at its maximum. (ZHR: 5) |

# Planet Visibility Ratings

| Month | Day | Morning Sky | | | | | | | Evening Sky | | | | | | |
|---|---|---|---|---|---|---|---|---|---|---|---|---|---|---|---|
| | | Me | Ve | Ma | Ju | Sa | Ur | Ne | Me | Ve | Ma | Ju | Sa | Ur | Ne |
| Jan | 5th | *** | | **** | | | | | | **** | | ***** | ** | *** | ** |
| | 15th | ** | | | | | | | | **** | **** | **** | ** | *** | ** |
| | 25th | | | | | | | | | **** | **** | **** | ** | *** | ** |
| Feb | 5th | | | | | | | | | **** | **** | **** | * | *** | ** |
| | 15th | | | | | | | | | **** | *** | *** | * | ** | * |
| | 25th | | | | | | | | | *** | *** | *** | | ** | * |
| Mar | 5th | | | | | | | | ** | *** | *** | *** | | ** | |
| | 15th | | | | | | | | | | *** | *** | | ** | |
| | 25th | | | | | | | | | | ** | ** | | ** | |
| Apr | 5th | *** | | | | * | | | | | ** | ** | | * | |
| | 15th | *** | *** | | | * | | * | | | ** | ** | | * | |
| | 25th | *** | **** | | | * | | * | | | ** | ** | | * | |
| May | 5th | *** | **** | | | ** | | ** | | | ** | * | | | |
| | 15th | ** | *** | | | ** | | ** | | | ** | * | | | |
| | 25th | | *** | | | ** | | ** | | | * | * | | | |
| Jun | 5th | | *** | | | ** | * | ** | | | * | * | | | |
| | 15th | | *** | | | ** | * | *** | *** | | * | | | | |
| | 25th | | *** | | | *** | * | *** | **** | | * | | | | |
| Jul | 5th | | *** | | | *** | ** | *** | **** | | * | | | | |
| | 15th | | *** | | * | *** | ** | **** | *** | | * | | | | |
| | 25th | | ** | | * | *** | ** | **** | | | * | | | | |
| Aug | 5th | | ** | | * | **** | ** | **** | | | * | | | | |
| | 15th | *** | ** | | * | **** | ** | **** | | | * | | | | |
| | 25th | ** | ** | | ** | **** | *** | ***** | | | * | | | | |
| Sep | 5th | | ** | | ** | **** | *** | ***** | | | * | | | | |
| | 15th | | * | | ** | **** | *** | ***** | | | * | | | | |
| | 25th | | * | | ** | **** | *** | ***** | | | * | | | | |
| Oct | 5th | | * | | ** | **** | *** | ***** | | | * | | | | |
| | 15th | | * | | *** | **** | **** | ***** | ** | | * | | | | |
| | 25th | | * | | *** | **** | **** | **** | *** | | | | | | |
| Nov | 5th | | | | *** | **** | **** | **** | *** | | | | | | |
| | 15th | | | | **** | *** | **** | **** | | | | | | | |
| | 25th | | | | **** | *** | **** | **** | | | | | | | |
| Dec | 5th | *** | | | **** | *** | **** | *** | | | | | | | |
| | 15th | *** | | | **** | *** | **** | *** | | | | | | | |
| | 25th | ** | | | ***** | ** | **** | *** | | | | | | | |

# Solar and Lunar Eclipses

| Date | Time (UT) | Type | Visible From |
|---|---|---|---|
| Mat 14th | 06:59 | Total Lunar | Africa, east Asia, Australia, Europe, North America and South America |
| Mar 29th | 10:49 | Partial Solar | North-western Africa, northern Asia, Europe and north-eastern North America |
| Sep 7th | 18:11 | Total Lunar | Africa, Antarctica, Asia, Australia and Europe |
| Sep 21st | 19:42 | Partial Solar | Antarctica, south-eastern Australia and New Zealand |

# Planetary Highlights

| Date | Time (UT) | Elon. | Vis. | Description |
|------|-----------|-------|------|-------------|
| Jan 10th | 04:41 | | PM | Venus is at greatest eastern elongation from the Sun. |
| Jan 16th | 16:48 | | AN | Mars is at opposition. |
| Jan 20th | 05:22 | | PM | Venus is 2.5° north of Saturn. |
| Mar 8th | 05:57 | | PM | Mercury is at greatest eastern elongation from the Sun. |
| Apr 21st | 18:34 | | AM | Mercury is at greatest western elongation from the Sun. |
| May 4th | 04:45 | | PM | Mars is 0.7° north of the Praesepe star cluster. |
| Jun 1st | 03:14 | | AM | Venus is at greatest western elongation from the Sun. |
| Jun 16th | 12;56 | | PM | Mars is 0.8° north of the bright star Regulus |
| Jul 4th | 01:00 | | AM | Venus is 2.4° south of Uranus. |
| Jul 4th | 04:24 | | PM | Mercury is at greatest eastern elongation from the Sun. |
| Jul 12th | 17:22 | | AM | Uranus is 4.3° south of the Pleiades star cluster. |
| Aug 6th | 10:12 | | AM | Saturn is 1.1° south of Neptune. |
| Aug 12th | 07:41 | | AM | Venus is 0.9° south of Jupiter. |
| Aug 19th | 09:37 | | AM | Mercury is at greatest western elongation from the Sun. |
| Aug 31st | 23:51 | | AM | Venus is 1.2° south of the Praesepe star cluster. |
| Sep 11th | 18:20 | | PM | Mars is 2.4° north of the bright star Spica. |
| Sep 22nd | 06:28 | | AN | Saturn is at opposition. |
| Sep 24th | 04:36 | | AN | Neptune is at opposition. |
| Oct 21st | 06:20 | | PM | Mercury is 2.1° south of Mars. |
| Oct 29th | 21:46 | | PM | Mercury is at greatest eastern elongation from the Sun. |
| Nov 21st | 12:36 | | AN | Uranus is at opposition. |
| Dec 7th | 20:51 | | AM | Mercury is at greatest western elongation from the Sun. |

# Major Meteor Showers

| Shower Name | Start Date | End Date | Peak | ZHR | Speed | Brightness | Moon |
|-------------|-----------|----------|------|-----|-------|-----------|------|
| Quadrantids | Dec 28th | Jan 12th | Jan 4th | 120 | *** | ***** | 🌑 |
| Lyrids | Apr 14th | Apr 30th | Apr 23rd | 18 | *** | ***** | 🌒 |
| Eta Aquariids | Apr 19th | May 28th | May 6th | 40 | * | **** | 🌓 |
| Eta Lyrids | May 3rd | May 14th | May 9th | 3 | *** | ** | 🌕 |
| June Bootids | Jun 22nd | Jul 2nd | Jun 27th | Var | ***** | ***** | 🌑 |
| Alpha Capricornids | Jul 3rd | Aug 15th | Jul 30th | 5 | ***** | **** | 🌑 |
| Southern Delta Aquariids | Jul 12th | Aug 23rd | Jul 30th | 16 | *** | * | 🌑 |
| Perseids | Jul 17th | Aug 24th | Aug 13th | 100 | * | ***** | 🌕 |
| Kappa Cygnids | Aug 3rd | Aug 25th | Aug 18th | 3 | ***** | ** | 🌒 |
| Aurigids | Aug 28th | Sep 5th | Sep 1st | 6 | * | **** | 🌒 |
| September Epsilon Perseids | Sep 5th | Sep 21st | Sep 9th | 5 | * | ** | 🌕 |
| Draconids | Oct 6th | Oct 10th | Oct 9th | 10 | ***** | *** | 🌖 |
| Southern Taurids | Sep 10th | Nov 20th | Oct 10th | 5 | **** | **** | 🌖 |
| Epsilon Geminds | Oct 14th | Oct 27th | Oct 19th | 3 | * | ** | 🌑 |
| Orionids | Oct 2nd | Nov 7th | Oct 22nd | 20 | * | **** | 🌑 |
| Andromedids | Sep 25th | Dec 6th | Nov 9th | 3 | ***** | **** | 🌕 |
| Northern Taurids | Oct 20th | Dec 10th | Nov 13th | 5 | **** | **** | 🌒 |
| Leonids | Nov 6th | Nov 30th | Nov 18th | 15 | * | **** | 🌑 |
| Alpha Monocerotids | Nov 15th | Nov 25th | Nov 22nd | Var | * | **** | 🌒 |
| November Orionids | Nov 13th | Dec 6th | Nov 28th | 3 | *** | ** | 🌓 |
| Sigma Hydrids | Dec 3rd | Dec 15th | Dec 12th | 3 | * | ** | 🌗 |
| Geminids | Dec 4th | Dec 17th | Dec 14th | 140 | **** | *** | 🌘 |
| Coma Berenicids | Dec 12th | Dec 23rd | Dec 16th | 3 | * | ** | 🌑 |
| Ursids | Dec 17th | Dec 26th | Dec 23rd | 10 | **** | ** | 🌑 |

# January 1ˢᵗ to 10ᵗʰ, 2025

## The Moon

| | 1ˢᵗ | | 3ʳᵈ | | 5ᵗʰ | | 7ᵗʰ | | 9ᵗʰ |

| Date | Con | R.A. | Dec | Mag | Diam | Ill. | Elon. | Phase | Close To |
|------|-----|------|-----|-----|------|------|-------|-------|----------|
| 1st | Cap | 20h 15m | -24° 15' | -5.9 | 31' | 3% | 22° E | NM | |
| 2nd | Cap | 21h 11m | -20° 1' | -7.0 | 32' | 8% | 34° E | NM | |
| 3rd | Aqr | 22h 4m | -14° 41' | -7.9 | 32' | 14% | 47° E | +Cr | Venus |
| 4th | Aqr | 22h 55m | -8° 34' | -8.6 | 32' | 23% | 58° E | +Cr | Venus, Saturn |
| 5th | Psc | 23h 45m | -1° 60' | -9.3 | 32' | 33% | 70° E | +Cr | Saturn, Neptune |
| 6th | Psc | 0h 35m | 4° 42' | -9.9 | 32' | 44% | 81° E | FQ | Neptune |
| 7th | Psc | 1h 25m | 11° 12' | -10.4 | 32' | 56% | 93° E | FQ | |
| 8th | Ari | 2h 18m | 17° 9' | -10.8 | 32' | 67% | 105° E | +G | |
| 9th | Ari | 3h 15m | 22° 11' | -11.2 | 32' | 77% | 118° E | +G | Uranus, Pleiades |
| 10th | Tau | 4h 14m | 25° 55' | -11.6 | 32' | 86% | 132° E | +G | Jupiter, Pleiades, Hyades |

## Mercury and Venus

Mercury
5ᵗʰ

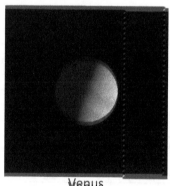

Venus
5ᵗʰ

### Mercury

| Date | Con. | R.A. | Dec. | Mag. | Diam. | Ill. | Elon. | Vis. | Rat. | Close To |
|------|------|------|------|------|-------|------|-------|------|------|----------|
| 1st | Oph | 17h 19m | -22° 4' | -0.4 | 6" | 78% | 23° W | AM | *** | |
| 3rd | Oph | 17h 30m | -22° 30' | -0.4 | 6" | 81% | 22° W | AM | *** | |
| 5th | Oph | 17h 42m | -22° 53' | -0.3 | 6" | 83% | 21° W | AM | *** | |
| 7th | Sgr | 17h 54m | -23° 13' | -0.4 | 5" | 85% | 20° W | AM | *** | |
| 9th | Sgr | 18h 6m | -23° 28' | -0.4 | 5" | 87% | 19° W | AM | ** | |

## Venus

| Date | Con. | R.A. | Dec. | Mag. | Diam. | Ill. | Elon. | Vis. | Rat. | Close To |
|------|------|------|------|------|-------|------|-------|------|------|----------|
| 1st | Aqr | 22h 3m | -13° 22' | -4.4 | 22" | 55% | 49° E | PM | *** | |
| 3rd | Aqr | 22h 11m | -12° 29' | -4.4 | 23" | 54% | 48° E | PM | **** | Moon |
| 5th | Aqr | 22h 19m | -11° 35' | -4.4 | 23" | 53% | 48° E | PM | **** | |
| 7th | Aqr | 22h 27m | -10° 40' | -4.4 | 24" | 52% | 48° E | PM | **** | |
| 9th | Aqr | 22h 34m | -9° 45' | -4.4 | 24" | 51% | 48° E | PM | **** | Saturn |

## Mars and the Outer Planets

Mars
5th

Jupiter
5th

Saturn
5th

## Mars

| Date | Con. | R.A. | Dec. | Mag. | Diam. | Ill. | Elon. | Vis. | Rat. | Close To |
|------|------|------|------|------|-------|------|-------|------|------|----------|
| 1st | Cnc | 8h 19m | 23° 36' | -1.2 | 14" | 99% | 157° W | AM | **** | Praesepe |
| 5th | Cnc | 8h 13m | 24° 2' | -1.3 | 14" | 99% | 163° W | AM | **** | Praesepe |
| 10th | Cnc | 8h 5m | 24° 35' | -1.3 | 15" | 100% | 171° W | AN | **** | Praesepe |

## The Outer Planets

| Planet | Date | Con. | R.A. | Dec. | Mag. | Diam. | Elon. | Vis. | Rat. | Close To |
|--------|------|------|------|------|------|-------|-------|------|------|----------|
| Jupiter | 5th | Tau | 4h 45m | 21° 45' | -2.7 | 47" | 145° E | PM | ***** | Hyades, Aldebaran |
| Saturn | 5th | Aqr | 23h 7m | -7° 46' | 1.1 | 16" | 60° E | PM | ** | Moon |
| Uranus | 5th | Ari | 3h 24m | 18° 25' | 5.7 | 4" | 125° E | PM | *** | Pleiades |
| Neptune | 5th | Psc | 23h 52m | -2° 13' | 7.9 | 2" | 71° E | PM | ** | Moon |

## Highlights

| Date | Time (UT) | Event |
|------|-----------|-------|
| 3rd | 15:24 | The waxing crescent Moon is south of Venus. (Evening sky.) |
| 4th | N/A | The Quadrantid meteor shower is at its maximum (ZHR: 120) |
| | 17:25 | The waxing crescent Moon is south of Saturn. (Evening sky.) |
| 5th | 16:18 | The waxing crescent Moon is north of Neptune. (Evening sky.) |
| 6th | 23:57 | First Quarter Moon. (Evening sky.) |
| 9th | 15:54 | The waxing gibbous Moon is north of Uranus (Evening sky.) |
| 10th | 04:41 | Venus is at greatest eastern elongation from the Sun. (Evening sky.) |
| | 20:20 | The waxing gibbous Moon is north of the bright star Aldebaran. (Evening sky.) |
| | 23:14 | The waxing gibbous Moon is north of Jupiter. (Evening sky.) |

# January 11ᵗʰ to 20ᵗʰ, 2025

## The Moon

| 11ᵗʰ | 13ᵗʰ | 15ᵗʰ | 17ᵗʰ | 19ᵗʰ |

| Date | Con | R.A. | Dec | Mag | Diam | Ill. | Elon. | Phase | | Close To |
|------|-----|------|-----|-----|------|------|-------|-------|--|----------|
| 11th | Tau | 5h 16m | 28° 3' | -11.9 | 32' | 93% | 146° E | +G | | Jupiter, Aldebaran |
| 12th | Aur | 6h 19m | 28° 23' | -12.2 | 32' | 97% | 161° E | FM | | |
| 13th | Gem | 7h 21m | 26° 57' | -12.5 | 31' | 100% | 175° E | FM | | Mars |
| 14th | Cnc | 8h 19m | 23° 59' | -12.5 | 31' | 99% | 172° W | FM | | Mars, Praesepe |
| 15th | Cnc | 9h 13m | 19° 50' | -12.2 | 31' | 97% | 159° W | FM | | Praesepe |
| 16th | Leo | 10h 2m | 14° 51' | -11.9 | 30' | 93% | 148° W | -G | | Regulus |
| 17th | Leo | 10h 48m | 9° 22' | -11.6 | 30' | 87% | 138° W | -G | | Regulus |
| 18th | Leo | 11h 32m | 3° 40' | -11.3 | 30' | 80% | 128° W | -G | | |
| 19th | Vir | 12h 14m | -2° 5' | -11.0 | 30' | 71% | 118° W | -G | | |
| 20th | Vir | 12h 56m | -7° 42' | -10.6 | 30' | 63% | 109° W | LQ | | Spica |

## Mercury and Venus

Mercury
15ᵗʰ

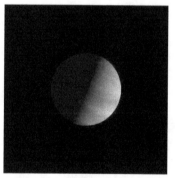

Venus
15ᵗʰ

**Mercury**

| Date | Con. | R.A. | Dec. | Mag. | Diam. | Ill. | Elon. | Vis. | Rat. | Close To |
|------|------|------|------|------|-------|------|-------|------|------|----------|
| 11th | Sgr | 18h 19m | -23° 39' | -0.4 | 5" | 89% | 18° W | AM | ** | |
| 13th | Sgr | 18h 32m | -23° 45' | -0.4 | 5" | 90% | 17° W | AM | ** | |
| 15th | Sgr | 18h 45m | -23° 47' | -0.4 | 5" | 91% | 16° W | AM | ** | |
| 17th | Sgr | 18h 58m | -23° 43' | -0.4 | 5" | 93% | 15° W | AM | ** | |
| 19th | Sgr | 19h 11m | -23° 35' | -0.5 | 5" | 94% | 14° W | NV | N/A | |

## Venus

| Date | Con. | R.A. | Dec. | Mag. | Diam. | Ill. | Elon. | Vis. | Rat. | Close To |
|---|---|---|---|---|---|---|---|---|---|---|
| 11th | Aqr | 22h 42m | -8° 49' | -4.4 | 25" | 50% | 47° E | PM | **** | Saturn |
| 13th | Aqr | 22h 49m | -7° 53' | -4.5 | 25" | 49% | 47° E | PM | **** | Saturn |
| 15th | Aqr | 22h 56m | -6° 56' | -4.5 | 26" | 48% | 47° E | PM | **** | Saturn |
| 17th | Aqr | 23h 3m | -5° 60' | -4.5 | 27" | 47% | 46° E | PM | **** | Saturn |
| 19th | Aqr | 23h 9m | -5° 3' | -4.5 | 27" | 46% | 46° E | PM | **** | Saturn |

## Mars and the Outer Planets

Mars
15th

Jupiter
15th

Saturn
15th

### Mars

| Date | Con. | R.A. | Dec. | Mag. | Diam. | Ill. | Elon. | Vis. | Rat. | Close To |
|---|---|---|---|---|---|---|---|---|---|---|
| 11th | Cnc | 8h 4m | 24° 41' | -1.4 | 15" | 100% | 172° W | AN | **** | Praesepe |
| 15th | Gem | 7h 57m | 25° 4' | -1.4 | 15" | 100% | 178° W | AN | **** | |
| 20th | Gem | 7h 48m | 25° 29' | -1.3 | 14" | 100% | 174° E | AN | **** | |

### The Outer Planets

| Planet | Date | Con. | R.A. | Dec. | Mag. | Diam. | Elon. | Vis. | Rat. | Close To |
|---|---|---|---|---|---|---|---|---|---|---|
| Jupiter | 15th | Tau | 4h 42m | 21° 40' | -2.6 | 46" | 133° E | PM | **** | Hyades, Aldebaran |
| Saturn | 15th | Aqr | 23h 10m | -7° 25' | 1.1 | 16" | 50° E | PM | ** | Venus |
| Uranus | 15th | Ari | 3h 24m | 18° 22' | 5.7 | 4" | 113° E | PM | *** | Pleiades |
| Neptune | 15th | Psc | 23h 53m | -2° 9' | 7.9 | 2" | 61° E | PM | ** | |

## Highlights

| Date | Time (UT) | Event |
|---|---|---|
| 13th | 22:28 | Full Moon (Visible all night.) |
| 14th | 03:42 | The just-past full Moon appears south of Mars (Visible all night.) |
| 16th | 14:56 | The waning gibbous Moon appears north of the bright star Regulus. (Morning sky.) |
| | 16:48 | Mars is at opposition. (Visible all night.) |
| 20th | 05:22 | Venus is 2.5° north of Saturn. (Evening sky.) |

# January 21st to 31st, 2025

## The Moon

| 21st | 23rd | 25th | 27th | 29th | 31st |
|------|------|------|------|------|------|

| Date | Con | R.A. | Dec | Mag | Diam | Ill. | Elon. | Phase | Close To |
|------|-----|------|-----|-----|------|------|-------|-------|----------|
| 21st | Vir | 13h 39m | -13° 1' | -10.3 | 30' | 53% | 99° W | LQ | Spica |
| 22nd | Lib | 14h 24m | -17° 53' | -9.9 | 30' | 44% | 89° W | LQ | |
| 23rd | Lib | 15h 12m | -22° 6' | -9.4 | 30' | 35% | 78° W | Cr | |
| 24th | Sco | 16h 3m | -25° 28' | -8.8 | 30' | 26% | 66° W | Cr | Antares |
| 25th | Oph | 16h 58m | -27° 41' | -8.2 | 30' | 18% | 54° W | Cr | Antares |
| 26th | Sgr | 17h 55m | -28° 33' | -7.4 | 31' | 11% | 40° W | NM | |
| 27th | Sgr | 18h 55m | -27° 49' | -6.5 | 31' | 5% | 26° W | NM | |
| 28th | Sgr | 19h 54m | -25° 28' | -5.4 | 32' | 1% | 13° W | NM | Mercury |
| 29th | Cap | 20h 52m | -21° 35' | -4.3 | 32' | 0% | 1° E | NM | Mercury |
| 30th | Cap | 21h 47m | -16° 25' | -5.3 | 32' | 1% | 14° E | NM | |
| 31st | Aqr | 22h 40m | -10° 19' | -6.5 | 32' | 5% | 26° E | NM | Saturn |

## Mercury and Venus

Mercury
25th

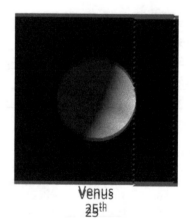

Venus
25th

## Mercury

| Date | Con. | R.A. | Dec. | Mag. | Diam. | Ill. | Elon. | Vis. | Rat. | Close To |
|------|------|------|------|------|-------|------|-------|------|------|----------|
| 21st | Sgr | 19h 25m | -23° 21' | -0.5 | 5" | 95% | 13° W | NV | N/A | |
| 23rd | Sgr | 19h 38m | -23° 1' | -0.6 | 5" | 96% | 11° W | NV | N/A | |
| 25th | Sgr | 19h 52m | -22° 37' | -0.6 | 5" | 96% | 10° W | NV | N/A | |
| 27th | Sgr | 20h 6m | -22° 6' | -0.7 | 5" | 97% | 9° W | NV | N/A | |
| 29th | Cap | 20h 19m | -21° 30' | -0.8 | 5" | 98% | 7° W | NV | N/A | Moon |
| 31st | Cap | 20h 33m | -20° 48' | -0.9 | 5" | 98% | 6° W | NV | N/A | |

## Venus

| Date | Con. | R.A. | Dec. | Mag. | Diam. | Ill. | Elon. | Vis. | Rat. | Close To |
|------|------|------|------|------|-------|------|-------|------|------|----------|
| 21st | Aqr | 23h 16m | -4° 6' | -4.5 | 28" | 45% | 45° E | PM | **** | Saturn |
| 23rd | Psc | 23h 22m | -3° 10' | -4.5 | 29" | 44% | 45° E | PM | **** | Saturn |
| 25th | Psc | 23h 28m | -2° 13' | -4.5 | 29" | 42% | 44° E | PM | **** | Saturn |
| 27th | Psc | 23h 34m | -1° 18' | -4.6 | 30" | 41% | 44° E | PM | **** | Saturn, Neptune |
| 29th | Psc | 23h 40m | 0° 22' | -4.6 | 31" | 40% | 43° E | PM | **** | Saturn, Neptune |
| 31st | Psc | 23h 46m | 0° 32' | -4.6 | 32" | 38% | 42° E | PM | **** | Saturn, Neptune |

## Mars and the Outer Planets

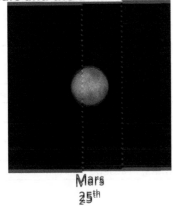

Mars
25th

Jupiter
25th

Saturn
25th

## Mars

| Date | Con. | R.A. | Dec. | Mag. | Diam. | Ill. | Elon. | Vis. | Rat. | Close To |
|------|------|------|------|------|-------|------|-------|------|------|----------|
| 21st | Gem | 7h 47m | 25° 33' | -1.3 | 14" | 100% | 173° E | AN | **** | |
| 25th | Gem | 7h 40m | 25° 48' | -1.2 | 14" | 99% | 167° E | PM | **** | |
| 31st | Gem | 7h 32m | 26° 4' | -1.1 | 14" | 99% | 159° E | PM | **** | |

## The Outer Planets

| Planet | Date | Con. | R.A. | Dec. | Mag. | Diam. | Elon. | Vis. | Rat. | Close To |
|--------|------|------|------|------|------|-------|-------|------|------|----------|
| Jupiter | 25th | Tau | 4h 39m | 21° 39' | -2.6 | 44" | 122° E | PM | **** | Hyades, Aldebaran |
| Saturn | 25th | Aqr | 23h 14m | -7° 1' | 1.1 | 16" | 40° E | PM | ** | Venus |
| Uranus | 25th | Ari | 3h 23m | 18° 21' | 5.7 | 4" | 103° E | PM | *** | Pleiades |
| Neptune | 25th | Psc | 23h 53m | -2° 3' | 7.9 | 2" | 50° E | PM | ** | Venus, Saturn |

## Highlights

| Date | Time (UT) | Event |
|------|-----------|-------|
| 21st | 03:53 | The almost last quarter Moon is south of the bright star Spica. (Morning sky.) |
| | 20:31 | Last Quarter Moon. (Morning sky.) |
| | 22:31 | Dwarf planet Pluto is in conjunction with the Sun. (Not visible.) |
| 24th | 23:34 | The waning crescent Moon is south of the bright star Antares. (Morning sky.) |
| 29th | 12:37 | New Moon. (Not visible.) |
| 30th | 16:03 | Uranus is stationary, prior to resuming prograde motion. (Evening sky.) |

# February 1ˢᵗ to 10ᵗʰ, 2025

## The Moon

| | 1ˢᵗ | | 3ʳᵈ | | 5ᵗʰ | | 7ᵗʰ | | 9ᵗʰ |

| Date | Con | R.A. | Dec | Mag | Diam | Ill. | Elon. | Phase | Close To |
|------|-----|------|-----|-----|------|------|-------|-------|----------|
| 1st | Aqr | 23h 31m | -3° 37' | -7.5 | 33' | 11% | 38° E | NM | Venus, Saturn, Neptune |
| 2nd | Psc | 0h 22m | 3° 16' | -8.4 | 33' | 20% | 49° E | +Cr | Venus, Neptune |
| 3rd | Psc | 1h 13m | 9° 57' | -9.1 | 32' | 30% | 61° E | +Cr | |
| 4th | Ari | 2h 6m | 16° 6' | -9.7 | 32' | 41% | 73° E | FQ | |
| 5th | Ari | 3h 1m | 21° 21' | -10.2 | 32' | 52% | 86° E | FQ | Uranus, Pleiades |
| 6th | Tau | 4h 0m | 25° 22' | -10.7 | 32' | 63% | 100° E | FQ | Jupiter, Uranus, Pleiades, Hyades, |
| 7th | Tau | 5h 0m | 27° 50' | -11.1 | 32' | 73% | 114° E | +G | Jupiter, Hyades, Aldebaran |
| 8th | Aur | 6h 2m | 28° 36' | -11.4 | 31' | 82% | 128° E | +G | |
| 9th | Gem | 7h 3m | 27° 39' | -11.8 | 31' | 90% | 142° E | +G | Mars |
| 10th | Gem | 8h 1m | 25° 8' | -12.1 | 31' | 95% | 156° E | FM | Mars, Praesepe |

## Mercury and Venus

Mercury
5ᵗʰ

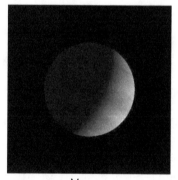

Venus
5ᵗʰ

### Mercury

| Date | Con. | R.A. | Dec. | Mag. | Diam. | Ill. | Elon. | Vis. | Rat. | Close To |
|------|------|------|------|------|-------|------|-------|------|------|----------|
| 1st | Cap | 20h 40m | -20° 25' | -1.0 | 5" | 99% | 5° W | NV | N/A | |
| 3rd | Cap | 20h 54m | -19° 34' | -1.1 | 5" | 99% | 4° W | NV | N/A | |
| 5th | Cap | 21h 8m | -18° 37' | -1.2 | 5" | 99% | 2° W | NV | N/A | |
| 7th | Cap | 21h 22m | -17° 35' | -1.4 | 5" | 100% | 1° W | NV | N/A | |
| 9th | Cap | 21h 36m | -16° 27' | -1.5 | 5" | 100% | 1° E | NV | N/A | |

## Venus

| Date | Con. | R.A. | Dec. | Mag. | Diam. | Ill. | Elon. | Vis. | Rat. | Close To |
|------|------|------|------|------|-------|------|-------|------|------|----------|
| 1st | Psc | 23h 48m | 0° 59' | -4.6 | 32" | 38% | 42° E | PM | **** | Moon, Saturn, Neptune |
| 3rd | Psc | 23h 54m | 1° 53' | -4.6 | 33" | 36% | 41° E | PM | **** | Saturn, Neptune |
| 5th | Psc | 23h 58m | 2° 45' | -4.6 | 34" | 35% | 40° E | PM | **** | Neptune |
| 7th | Psc | 0h 3m | 3° 36' | -4.6 | 35" | 33% | 40° E | PM | **** | Neptune |
| 9th | Psc | 0h 7m | 4° 25' | -4.6 | 36" | 32% | 39° E | PM | **** | Neptune |

## Mars and the Outer Planets

Mars
5th

Jupiter
5th

Saturn
5th

### Mars

| Date | Con. | R.A. | Dec. | Mag. | Diam. | Ill. | Elon. | Vis. | Rat. | Close To |
|------|------|------|------|------|-------|------|-------|------|------|----------|
| 1st | Gem | 7h 30m | 26° 6' | -1.0 | 14" | 99% | 157° E | PM | **** | |
| 5th | Gem | 7h 25m | 26° 12' | -0.9 | 13" | 98% | 152° E | PM | **** | |
| 10th | Gem | 7h 21m | 26° 14' | -0.8 | 13" | 97% | 146° E | PM | *** | Moon |

### The Outer Planets

| Planet | Date | Con. | R.A. | Dec. | Mag. | Diam. | Elon. | Vis. | Rat. | Close To |
|--------|------|------|------|------|------|-------|-------|------|------|----------|
| Jupiter | 5th | Tau | 4h 39m | 21° 40' | -2.5 | 43" | 110° E | PM | **** | Hyades, Aldebaran |
| Saturn | 5th | Aqr | 23h 18m | -6° 33' | 1.1 | 16" | 30° E | PM | * | |
| Uranus | 5th | Ari | 3h 23m | 18° 22' | 5.7 | 4" | 92° E | PM | *** | Moon, Pleiades |
| Neptune | 5th | Psc | 23h 55m | -1° 55' | 7.9 | 2" | 39° E | PM | ** | Venus, Saturn |

## Highlights

| Date | Time (UT) | Event |
|------|-----------|-------|
| 1st | 04:50 | The waxing crescent Moon is north of Saturn. (Evening sky.) |
| | 20:24 | The waxing crescent Moon is south of Venus. (Evening sky.) |
| | 22:43 | The waxing crescent Moon is north of Aldebaran. (Evening sky.) |
| 3rd | 20:00 | Venus is 4.0° north of Neptune. (Evening sky.) |
| 4th | 12:36 | Jupiter is stationary prior to resuming prograde motion. (Evening sky.) |
| 5th | 08:03 | First Quarter Moon. (Evening sky.) |
| | 21:13 | The just-past first quarter Moon is north of Uranus. (Evening sky.) |
| 6th | 06:45 | The just-past first quarter Moon is south of the Pleiades star cluster. (Evening sky.) |
| 7th | 02:19 | The waxing gibbous Moon is north of the bright star Aldebaran. (Evening sky.) |
| | 03:38 | The waxing gibbous Moon is north of Jupiter. (Evening sky.) |
| 9th | 11:53 | Mercury is at superior conjunction with the Sun. (Not visible.) |
| | 19:38 | The waxing gibbous Moon is south of Mars. (Evening sky.) |

## The Moon

   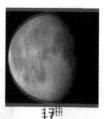 

11th     13th     15th     17th     19th

| Date | Con | R.A. | Dec | Mag | Diam | Ill. | Elon. | Phase | Close To |
|------|-----|------|-----|-----|------|------|-------|-------|----------|
| 11th | Cnc | 8h 55m | 21° 20' | -12.4 | 31' | 99% | 169° E | FM | Praesepe |
| 12th | Leo | 9h 45m | 16° 36' | -12.6 | 30' | 100% | 180° W | FM | Regulus |
| 13th | Leo | 10h 32m | 11° 15' | -12.4 | 30' | 99% | 169° W | FM | Regulus |
| 14th | Leo | 11h 17m | 5° 33' | -12.2 | 30' | 96% | 159° W | FM | |
| 15th | Vir | 12h 0m | 0° 16' | -11.9 | 30' | 92% | 149° W | =G | |
| 16th | Vir | 12h 42m | -5° 59' | -11.6 | 30' | 86% | 140° W | =G | Spica |
| 17th | Vir | 13h 25m | -11° 27' | -11.3 | 30' | 79% | 130° W | =G | Spica |
| 18th | Vir | 14h 9m | -16° 30' | -10.9 | 30' | 71% | 120° W | =G | Spica |
| 19th | Lib | 14h 55m | -20° 56' | -10.6 | 30' | 62% | 109° W | LQ | |
| 20th | Lib | 15h 45m | -24° 34' | -10.2 | 30' | 52% | 98° W | LQ | Antares |

## Mercury and Venus

Mercury
15th

Venus
15th

### Mercury

| Date | Con. | R.A. | Dec. | Mag. | Diam. | Ill. | Elon. | Vis. | Rat. | Close To |
|------|------|------|------|------|-------|------|-------|------|------|----------|
| 11th | Cap | 21h 50m | -15° 12' | -1.5 | 5" | 100% | 2° E | NV | N/A | |
| 13th | Aqr | 22h 4m | -13° 53' | -1.5 | 5" | 99% | 4° E | NV | N/A | |
| 15th | Aqr | 22h 17m | -12° 28' | -1.4 | 5" | 99% | 5° E | NV | N/A | |
| 17th | Aqr | 22h 31m | -10° 57' | -1.4 | 5" | 97% | 7° E | NV | N/A | |
| 19th | Aqr | 22h 45m | -9° 22' | -1.3 | 5" | 96% | 8° E | NV | N/A | Saturn |

## Venus

| Date | Con. | R.A. | Dec. | Mag. | Diam. | Ill. | Elon. | Vis. | Rat. | Close To |
|------|------|------|------|------|-------|------|-------|------|------|----------|
| 11th | Psc | 0h 11m | 5° 13' | -4.6 | 37" | 30% | 38° E | PM | **** | Neptune |
| 13th | Psc | 0h 15m | 5° 59' | -4.6 | 38" | 28% | 37° E | PM | **** | Neptune |
| 15th | Psc | 0h 18m | 6° 43' | -4.6 | 40" | 27% | 35° E | PM | **** | |
| 17th | Psc | 0h 21m | 7° 25' | -4.6 | 41" | 25% | 34° E | PM | **** | |
| 19th | Psc | 0h 24m | 8° 4' | -4.6 | 42" | 23% | 33° E | PM | **** | |

## Mars and the Outer Planets

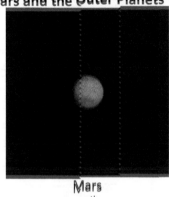

Mars
15th

Jupiter
15th

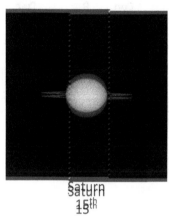

Saturn
15th

## Mars

| Date | Con. | R.A. | Dec. | Mag. | Diam. | Ill. | Elon. | Vis. | Rat. | Close To |
|------|------|------|------|------|-------|------|-------|------|------|----------|
| 11th | Gem | 7h 20m | 26° 14' | -0.8 | 13" | 97% | 145° E | PM | *** | |
| 15th | Gem | 7h 18m | 26° 12' | -0.7 | 12" | 96% | 140° E | PM | *** | |
| 20th | Gem | 7h 16m | 26° 7' | -0.5 | 12" | 95% | 135° E | PM | *** | |

## The Outer Planets

| Planet | Date | Con. | R.A. | Dec. | Mag. | Diam. | Elon. | Vis. | Rat. | Close To |
|--------|------|------|------|------|------|-------|-------|------|------|----------|
| Jupiter | 15th | Tau | 4h 40m | 21° 43' | -2.4 | 41" | 101° E | PM | *** | Hyades, Aldebaran |
| Saturn | 15th | Aqr | 23h 22m | -6° 5' | 1.2 | 16" | 21° E | PM | * | |
| Uranus | 15th | Ari | 3h 24m | 18° 23' | 5.7 | 4" | 82° E | PM | ** | Pleiades |
| Neptune | 15th | Psc | 23h 56m | -1° 47' | 7.9 | 2" | 30° E | PM | * | Venus, Saturn |

## Highlights

| Date | Time (UT) | Event |
|------|-----------|-------|
| 12th | 13:54 | Full Moon. (Visible all night.) |
| | 23:21 | The full Moon is north of the bright star Regulus. (Visible all night.) |
| 17th | 12:01 | The waxing gibbous Moon is south of the bright star Spica. (Morning sky.) |
| 18th | 15:51 | Dwarf planet Ceres is in conjunction with the Sun. (Not visible.) |
| 20th | 17:33 | Last Quarter Moon. (Morning sky.) |

# February 21st to 28th, 2025

## The Moon

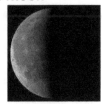
21st

23rd

25th

27th

| Date | Con | R.A. | Dec | Mag | Diam | Ill. | Elon. | Phase | Close To |
|------|-----|------|-----|-----|------|------|-------|-------|----------|
| 21st | Sco | 16h 37m | -27° 11' | -9.8 | 30' | 43% | 86° W | LQ | Antares |
| 22nd | Oph | 17h 33m | -28° 32' | -9.3 | 30' | 33% | 73° W | -Cr | |
| 23rd | Sgr | 18h 31m | -28° 26' | -8.7 | 31' | 24% | 59° W | -Cr | |
| 24th | Sgr | 19h 29m | -26° 44' | -8.0 | 31' | 15% | 45° W | -Cr | |
| 25th | Cap | 20h 28m | -23° 28' | -7.1 | 32' | 8% | 32° W | NM | |
| 26th | Cap | 21h 24m | -18° 46' | -6.0 | 32' | 3% | 19° W | NM | |
| 27th | Aqr | 22h 19m | -12° 54' | -4.7 | 33' | 0% | 6° W | NM | |
| 28th | Aqr | 23h 12m | -6° 13' | -4.6 | 33' | 0% | 6° E | NM | Mercury, Saturn, Neptune |

## Mercury and Venus

Mercury
24th

Venus
24th

**Mercury**

| Date | Con. | R.A. | Dec. | Mag. | Diam. | Ill. | Elon. | Vis. | Rat. | Close To |
|------|------|------|------|------|-------|------|-------|------|------|----------|
| 21st | Aqr | 22h 58m | -7° 43' | -1.3 | 5" | 93% | 10° E | NV | N/A | Saturn |
| 23rd | Aqr | 23h 12m | -6° 1' | -1.2 | 5" | 90% | 11° E | NV | N/A | Saturn |
| 25th | Aqr | 23h 24m | -4° 17' | -1.2 | 6" | 86% | 12° E | NV | N/A | Saturn |
| 27th | Psc | 23h 37m | -2° 33' | -1.1 | 6" | 80% | 14° E | NV | N/A | Saturn |

**Venus**

| Date | Con. | R.A. | Dec. | Mag. | Diam. | Ill. | Elon. | Vis. | Rat. | Close To |
|------|------|------|------|------|-------|------|-------|------|------|----------|
| 21st | Psc | 0h 26m | 8° 41' | -4.6 | 43" | 22% | 31° E | PM | **** | |
| 23rd | Psc | 0h 27m | 9° 14' | -4.6 | 45" | 20% | 30° E | PM | **** | |
| 25th | Psc | 0h 28m | 9° 44' | -4.6 | 46" | 18% | 28° E | PM | *** | |
| 27th | Psc | 0h 29m | 10° 10' | -4.6 | 48" | 16% | 27° E | PM | *** | |

## Mars and the Outer Planets

Mars
24th

Jupiter
24th

Saturn
24th

**Mars**

| Date | Con. | R.A. | Dec. | Mag. | Diam. | Ill. | Elon. | Vis. | Rat. | Close To |
|------|------|------|------|------|-------|------|-------|------|------|----------|
| 21st | Gem | 7h 16m | 26° 5' | -0.5 | 12" | 95% | 134° E | PM | *** | |
| 24th | Gem | 7h 15m | 26° 0' | -0.4 | 11" | 95% | 131° E | PM | *** | |
| 28th | Gem | 7h 16m | 25° 52' | -0.3 | 11" | 94% | 127° E | PM | *** | |

**The Outer Planets**

| Planet | Date | Con. | R.A. | Dec. | Mag. | Diam. | Elon. | Vis. | Rat. | Close To |
|--------|------|------|------|------|------|-------|-------|------|------|----------|
| Jupiter | 24th | Tau | 4h 42m | 21° 49' | -2.4 | 40" | 93° E | PM | *** | Hyades, Aldebaran |
| Saturn | 24th | Aqr | 23h 26m | -5° 40' | 1.1 | 16" | 14° E | NV | N/A | Mercury |
| Uranus | 24th | Ari | 3h 24m | 18° 26' | 5.7 | 4" | 73° E | PM | ** | Pleiades |
| Neptune | 24th | Psc | 23h 57m | -1° 40' | 8.0 | 2" | 21° E | PM | * | Mercury, Venus, Saturn |

## Highlights

| Date | Time (UT) | Event |
|------|-----------|-------|
| 21st | 08:21 | The just-past last quarter Moon is south of the bright star Antares. (Morning sky.) |
| 24th | 09:34 | Mars is stationary prior to resuming prograde motion. (Evening sky.) |
| 28th | 00:45 | New Moon. (Not visible.) |
| | 03:42 | Venus is stationary prior to beginning retrograde motion. (Evening sky.) |

# March 1st to 10th, 2025

## The Moon

  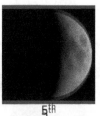 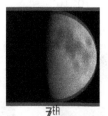 

| 1st | 3rd | 5th | 7th | 9th |

| Date | Con | R.A. | Dec | Mag | Diam | Ill: | Elon: | Phase | Close To |
|------|-----|------|-----|-----|------|------|-------|-------|----------|
| 1st | Psc | 0h 4m | 0° 51' | -6.0 | 33' | 3% | 18° E | NM | Mercury, Venus, Saturn, Neptune |
| 2nd | Psc | 0h 56m | 7° 54' | -7.1 | 33' | 9% | 31° E | NM | Venus |
| 3rd | Ari | 1h 50m | 14° 30' | -8.1 | 33' | 16% | 43° E | +Cr | |
| 4th | Ari | 2h 46m | 20° 12' | -8.9 | 33' | 26% | 56° E | +Cr | Uranus |
| 5th | Tau | 3h 45m | 24° 39' | -9.5 | 32' | 37% | 70° E | +Cr | Uranus, Pleiades, Hyades |
| 6th | Tau | 4h 46m | 27° 33' | -10.0 | 32' | 48% | 84° E | FQ | Jupiter, Hyades, Aldebaran |
| 7th | Aur | 5h 48m | 28° 42' | -10.5 | 32' | 59% | 99° E | FQ | |
| 8th | Gem | 6h 49m | 28° 7' | -10.9 | 31' | 69% | 113° E | +G | Mars |
| 9th | Gem | 7h 47m | 25° 57' | -11.3 | 31' | 79% | 127° E | +G | Mars |
| 10th | Cnc | 8h 41m | 22° 28' | -11.6 | 31' | 86% | 140° E | +G | Praesepe |

## Mercury and Venus

Mercury
5th

Venus
5th

### Mercury

| Date | Con: | R.A. | Dec: | Mag: | Diam: | Ill: | Elon: | Vis: | Rat: | Close To |
|------|------|------|------|------|-------|------|-------|------|------|----------|
| 1st | Psc | 23h 48m | 0° 51' | -1.0 | 6" | 74% | 15° E | NV | N/A | Moon, Saturn, Neptune |
| 3rd | Psc | 23h 59m | 0° 47' | -0.8 | 6" | 67% | 15° E | PM | ** | Venus, Saturn, Neptune |
| 5th | Psc | 0h 8m | 2° 19' | -0.7 | 7" | 58% | 16° E | PM | ** | Venus, Saturn, Neptune |
| 7th | Psc | 0h 16m | 3° 40' | -0.4 | 7" | 49% | 16° E | PM | ** | Venus, Neptune |
| 9th | Psc | 0h 22m | 4° 50' | -0.1 | 8" | 41% | 16° E | PM | ** | Venus |

## Venus

| Date | Con. | R.A. | Dec. | Mag. | Diam. | Ill. | Elon. | Vis. | Rat. | Close To |
|------|------|------|------|------|-------|------|-------|------|------|----------|
| 1st | Psc | 0h 29m | 10° 32' | -4.6 | 49" | 14% | 25° E | PM | *** | Moon |
| 3rd | Psc | 0h 28m | 10° 49' | -4.5 | 51" | 13% | 23° E | PM | *** | Mercury |
| 5th | Psc | 0h 27m | 11° 1' | -4.5 | 52" | 11% | 20° E | PM | *** | Mercury |
| 7th | Psc | 0h 25m | 11° 8' | -4.5 | 53" | 9% | 18° E | PM | *** | Mercury |
| 9th | Psc | 0h 22m | 11° 9' | -4.4 | 55" | 7% | 16° E | PM | *** | Mercury |

## Mars and the Outer Planets

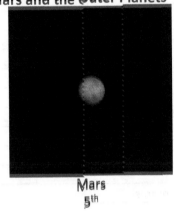

Mars
5th

Jupiter
5th

Saturn
5th

## Mars

| Date | Con. | R.A. | Dec. | Mag. | Diam. | Ill. | Elon. | Vis. | Rat. | Close To |
|------|------|------|------|------|-------|------|-------|------|------|----------|
| 1st | Gem | 7h 16m | 25° 50' | -0.3 | 11" | 94% | 127° E | PM | *** | |
| 5th | Gem | 7h 18m | 25° 40' | -0.2 | 10" | 93% | 123° E | PM | *** | |
| 10th | Gem | 7h 20m | 25° 26' | 0.0 | 10" | 92% | 119° E | PM | *** | |

## The Outer Planets

| Planet | Date | Con. | R.A. | Dec. | Mag. | Diam. | Elon. | Vis. | Rat. | Close To |
|--------|------|------|------|------|------|-------|-------|------|------|----------|
| Jupiter | 5th | Tau | 4h 45m | 21° 56' | -2.3 | 39" | 85° E | PM | *** | Hyades, Aldebaran |
| Saturn | 5th | Aqr | 23h 30m | -5° 14' | 1.1 | 16" | 6° E | NV | N/A | Mercury |
| Uranus | 5th | Tau | 3h 25m | 18° 29' | 5.8 | 4" | 65° E | PM | ** | Moon, Pleiades |
| Neptune | 5th | Psc | 23h 58m | -1° 32' | 8.0 | 2" | 13° E | NV | N/A | Mercury, Venus, Saturn |

## Highlights

| Date | Time (UT) | Event |
|------|-----------|-------|
| 1st | 23:18 | The just-past new Moon is south of Venus. (Evening sky.) |
| 3rd | N/A | Good opportunity to see Earthshine on the waxing crescent Moon. (Evening sky.) |
| 5th | 03:57 | The nearly first quarter Moon is north of Uranus. (Evening sky.) |
| | 12:31 | The nearly first quarter Moon is south of the Pleiades star cluster. (Evening sky.) |
| 6th | 11:31 | The almost first quarter Moon is north of Jupiter. (Evening sky.) |
| | 16:32 | First Quarter Moon. (Evening sky.) |
| 8th | 05:57 | Mercury is at greatest eastern elongation from the Sun. (Evening sky.) |
| 9th | 00:27 | The waxing gibbous Moon is north of Mars. (Evening sky.) |
| | 10:46 | Mercury is 6.3° south of Venus. (Evening sky.) |

# March 11<sup>th</sup> to 20<sup>th</sup>, 2025

## The Moon

| 11<sup>th</sup> | 13<sup>th</sup> | 15<sup>th</sup> | 17<sup>th</sup> | 19<sup>th</sup> |

| Date | Con | R.A. | Dec | Mag | Diam | Ill. | Elon. | Phase | Close To |
|------|-----|------|-----|-----|------|------|-------|-------|----------|
| 11th | Leo | 9h 32m | 18° 0' | -11.9 | 30' | 93% | 151° E | +G | Regulus |
| 12th | Leo | 10h 19m | 12° 50' | -12.2 | 30' | 97% | 162° E | FM | Regulus |
| 13th | Leo | 11h 4m | 7° 15' | -12.5 | 30' | 99% | 172° E | FM | |
| 14th | Vir | 11h 47m | 1° 28' | -12.7 | 30' | 100% | 178° W | FM | |
| 15th | Vir | 12h 29m | -4° 18' | -12.4 | 30' | 99% | 168° W | FM | |
| 16th | Vir | 13h 12m | -9° 53' | -12.1 | 29' | 96% | 158° W | FM | Spica |
| 17th | Vir | 13h 55m | -15° 5' | -11.8 | 29' | 91% | 148° W | -G | Spica |
| 18th | Lib | 14h 41m | -19° 43' | -11.5 | 29' | 85% | 138° W | -G | |
| 19th | Lib | 15h 29m | -23° 37' | -11.2 | 30' | 77% | 127° W | -G | |
| 20th | Sco | 16h 20m | -26° 32' | -10.9 | 30' | 69% | 115° W | -G | Antares |

## Mercury and Venus

Mercury
15<sup>th</sup>

Venus
15<sup>th</sup>

### Mercury

| Date | Con. | R.A. | Dec. | Mag. | Diam. | Ill. | Elon. | Vis. | Rat. | Close To |
|------|------|------|------|------|-------|------|-------|------|------|----------|
| 11th | Psc | 0h 27m | 5° 46' | 0.3 | 8" | 32% | 15° E | NV | N/A | Venus |
| 13th | Psc | 0h 29m | 6° 25' | 0.8 | 9" | 24% | 14° E | NV | N/A | Venus |
| 15th | Psc | 0h 30m | 6° 46' | 1.4 | 9" | 17% | 12° E | NV | N/A | Venus |
| 17th | Psc | 0h 28m | 6° 49' | 2.1 | 10" | 11% | 10° E | NV | N/A | Venus |
| 19th | Psc | 0h 25m | 6° 34' | 3.0 | 10" | 6% | 7° E | NV | N/A | Venus |

**Venus**

| Date | Con. | R.A. | Dec. | Mag. | Diam. | Ill. | Elon. | Vis. | Rat. | Close To |
|------|------|------|------|------|-------|------|-------|------|------|----------|
| 11th | Psc | 0h 19m | 11° 5' | -4.4 | 56" | 6% | 13° E | NV | N/A | Mercury |
| 13th | Psc | 0h 16m | 10° 54' | -4.3 | 57" | 4% | 10° E | NV | N/A | Mercury, Neptune |
| 15th | Psc | 0h 12m | 10° 37' | -4.2 | 58" | 3% | 8° E | NV | N/A | Mercury, Saturn, Neptune |
| 17th | Psc | 0h 8m | 10° 15' | -4.2 | 59" | 2% | 5° E | NV | N/A | Mercury, Saturn, Neptune |
| 19th | Psc | 0h 4m | 9° 47' | -4.1 | 59" | 2% | 2° E | NV | N/A | Mercury, Saturn, Neptune |

## Mars and the Outer Planets

Mars
15th

Jupiter
15th

Saturn
15th

**Mars**

| Date | Con. | R.A. | Dec. | Mag. | Diam. | Ill. | Elon. | Vis. | Rat. | Close To |
|------|------|------|------|------|-------|------|-------|------|------|----------|
| 11th | Gem | 7h 21m | 25° 22' | 0.0 | 10" | 92% | 119° E | PM | *** | |
| 15th | Gem | 7h 24m | 25° 9' | 0.1 | 9" | 92% | 116° E | PM | *** | |
| 20th | Gem | 7h 29m | 24° 51' | 0.2 | 9" | 91% | 112° E | PM | ** | |

**The Outer Planets**

| Planet | Date | Con. | R.A. | Dec. | Mag. | Diam. | Elon. | Vis. | Rat. | Close To |
|--------|------|------|------|------|------|-------|-------|------|------|----------|
| Jupiter | 15th | Tau | 4h 49m | 22° 6' | -2.2 | 38" | 77° E | PM | *** | Hyades, Aldebaran |
| Saturn | 15th | Aqr | 23h 35m | -4° 45' | 1.2 | 16" | 2° W | NV | N/A | Venus |
| Uranus | 15th | Tau | 3h 27m | 18° 34' | 5.8 | 4" | 56° E | PM | ** | Pleiades |
| Neptune | 15th | Psc | 23h 59m | -1° 23' | 8.0 | 2" | 4° E | NV | N/A | Mercury, Venus, Saturn |

## Highlights

| Date | Time (UT) | Event |
|------|-----------|-------|
| 12th | 06:08 | The waxing gibbous Moon is north of the bright star Regulus. (Evening sky.) |
| 13th | 09:07 | Saturn is in conjunction with the Sun. (Not visible.) |
| 14th | 06:55 | Full Moon. (Visible all night.) |
| | 06:59 | Total lunar eclipse. (Visible from Africa, east Asia, the Atlantic ocean, Australia, Europe, North America, the Pacific ocean and South America.) |
| | 20:40 | Mercury is stationary prior to beginning retrograde motion. (Not visible.) |
| 16th | 19:17 | The waning gibbous Moon is south of the bright star Spica. (Morning sky.) |
| 20th | 09:02 | Spring equinox. |
| | 13:10 | Neptune is in conjunction with the Sun. (Not visible.) |

# March 21st to 31st, 2025

## The Moon

| 21st | 23rd | 25th | 27th | 29th | 31st |

| Date | Con | R.A. | Dec | Mag | Diam | Ill. | Elon. | Phase | Close To |
|------|-----|------|-----|-----|------|------|-------|-------|----------|
| 21st | Oph | 17h 14m | -28° 18' | -10.5 | 30' | 60% | 102° W | LQ | Antares |
| 22nd | Sgr | 18h 10m | -28° 41' | -10.1 | 30' | 50% | 89° W | LQ | |
| 23rd | Sgr | 19h 7m | -27° 36' | -9.6 | 31' | 40% | 76° W | LQ | |
| 24th | Sgr | 20h 4m | -24° 60' | -9.1 | 31' | 30% | 63° W | Cr | |
| 25th | Cap | 21h 0m | -20° 57' | -8.4 | 32' | 20% | 49° W | Cr | |
| 26th | Cap | 21h 55m | -15° 38' | -7.6 | 32' | 12% | 37° W | NM | |
| 27th | Aqr | 22h 48m | -9° 20' | -6.6 | 33' | 6% | 24° W | NM | |
| 28th | Psc | 23h 40m | -2° 22' | -5.3 | 33' | 1% | 12° W | NM | Mercury, Venus, Saturn, Neptune |
| 29th | Psc | 0h 33m | 4° 51' | -4.0 | 33' | 0% | 0° E | NM | Mercury, Neptune |
| 30th | Psc | 1h 28m | 11° 52' | -5.5 | 33' | 2% | 13° E | NM | |
| 31st | Ari | 2h 25m | 18° 10' | -6.7 | 33' | 6% | 26° E | NM | |

## Mercury and Venus

Mercury
25th

Venus
25th

### Mercury

| Date | Con. | R.A. | Dec. | Mag. | Diam. | Ill. | Elon. | Vis. | Rat. | Close To |
|------|------|------|------|------|-------|------|-------|------|------|----------|
| 21st | Psc | 0h 20m | 6° 3' | 3.9 | 11" | 3% | 4° E | NV | N/A | Venus, Neptune |
| 23rd | Psc | 0h 15m | 5° 17' | 4.8 | 11" | 1% | 1° E | NV | N/A | Venus, Saturn, Neptune |
| 25th | Psc | 0h 9m | 4° 20' | 5.1 | 11" | 1% | 2° W | NV | N/A | Venus, Saturn, Neptune |
| 27th | Psc | 0h 3m | 3° 18' | 4.4 | 11" | 1% | 6° W | NV | N/A | Venus, Saturn, Neptune |
| 29th | Psc | 23h 58m | 2° 14' | 3.7 | 11" | 4% | 9° W | NV | N/A | Moon, Venus, Saturn |
| 31st | Psc | 23h 53m | 1° 13' | 3.0 | 11" | 6% | 12° W | NV | N/A | Venus, Saturn, Neptune |

## Venus

| Date | Con. | R.A. | Dec. | Mag. | Diam. | Ill. | Elon. | Vis. | Rat. | Close To |
|------|------|------|------|------|-------|------|-------|------|------|----------|
| 21st | Psc | 23h 59m | 9° 14' | -4.1 | 59" | 1% | 1° W | NV | N/A | Mercury, Saturn, Neptune |
| 23rd | Peg | 23h 55m | 8° 37' | -4.0 | 59" | 1% | 4° W | NV | N/A | Mercury, Saturn, Neptune |
| 25th | Psc | 23h 50m | 7° 58' | -4.1 | 59" | 1% | 7° W | NV | N/A | Mercury, Saturn, Neptune |
| 27th | Psc | 23h 46m | 7° 16' | -4.1 | 59" | 2% | 10° W | NV | N/A | Mercury, Saturn, Neptune |
| 29th | Psc | 23h 43m | 6° 33' | -4.2 | 58" | 2% | 12° W | NV | N/A | Mercury, Saturn, Neptune |
| 31st | Psc | 23h 40m | 5° 50' | -4.2 | 57" | 3% | 15° W | AM | *** | Mercury, Saturn |

## Mars and the Outer Planets

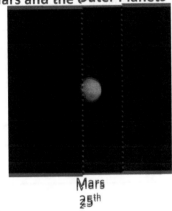

Mars
25th

Jupiter
25th

Saturn
25th

## Mars

| Date | Con. | R.A. | Dec. | Mag. | Diam. | Ill. | Elon. | Vis. | Rat. | Close To |
|------|------|------|------|------|-------|------|-------|------|------|----------|
| 21st | Gem | 7h 30m | 24° 47' | 0.2 | 9" | 91% | 112° E | PM | ** | |
| 25th | Gem | 7h 35m | 24° 30' | 0.3 | 9" | 91% | 109° E | PM | ** | |
| 31st | Gem | 7h 43m | 24° 3' | 0.4 | 8" | 91% | 106° E | PM | ** | |

## The Outer Planets

| Planet | Date | Con. | R.A. | Dec. | Mag. | Diam. | Elon. | Vis. | Rat. | Close To |
|--------|------|------|------|------|------|-------|-------|------|------|----------|
| Jupiter | 25th | Tau | 4h 55m | 22° 16' | -2.2 | 37" | 69° E | PM | ** | Hyades, Aldebaran |
| Saturn | 25th | Aqr | 23h 39m | -4° 16' | 1.2 | 16" | 10° W | NV | N/A | Mercury, Venus |
| Uranus | 25th | Tau | 3h 28m | 18° 40' | 5.8 | 3" | 47° E | PM | ** | Pleiades |
| Neptune | 25th | Psc | 0h 1m | -1° 14' | 8.0 | 2" | 4° W | NV | N/A | Mercury, Venus, Saturn |

## Highlights

| Date | Time (UT) | Event |
|------|-----------|-------|
| 22nd | 11:30 | Last Quarter Moon. (Morning sky.) |
| 23rd | 01:01 | Venus is at inferior conjunction with the Sun. (Not visible.) |
| 24th | 19:41 | Mercury is at inferior conjunction with the Sun. (Not visible.) |
| 29th | 10:49 | Partial solar eclipse. (Visible from north-western Africa, northern Asia, northern Atlantic ocean, Europe and north-eastern North America.) |
| | 10:58 | New Moon. (Not visible.) |

# April 1st to 10th, 2025

## The Moon

 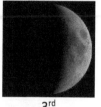   

| | 1st | | 3rd | | 5th | | 7th | | 9th |

| Date | Con | R.A. | Dec | Mag | Diam | Ill. | Elon. | Phase | | Close To |
|------|-----|------|-----|-----|------|------|-------|-------|--|----------|
| **1st** | Ari | 3h 25m | 23° 17' | -7.8 | 33' | 13% | 40° E | +Cr | | Uranus, Pleiades |
| **2nd** | Tau | 4h 27m | 26° 50' | -8.6 | 33' | 22% | 55° E | +Cr | | Jupiter, Pleiades, Hyades |
| **3rd** | Tau | 5h 31m | 28° 33' | -9.3 | 32' | 33% | 70° E | +Cr | | Jupiter |
| **4th** | Aur | 6h 34m | 28° 24' | -9.8 | 32' | 44% | 85° E | FQ | | |
| **5th** | Gem | 7h 33m | 26° 34' | -10.3 | 31' | 54% | 99° E | FQ | | Mars |
| **6th** | Cnc | 8h 29m | 23° 22' | -10.7 | 31' | 65% | 112° E | FQ | | Mars, Praesepe |
| **7th** | Cnc | 9h 21m | 19° 6' | -11.1 | 30' | 74% | 124° E | +G | | Praesepe, Regulus |
| **8th** | Leo | 10h 8m | 14° 6' | -11.4 | 30' | 82% | 135° E | +G | | Regulus |
| **9th** | Leo | 10h 53m | 8° 39' | -11.7 | 30' | 89% | 145° E | +G | | Regulus |
| **10th** | Leo | 11h 36m | 2° 57' | -12.0 | 30' | 94% | 155° E | +G | | |

## Mercury and Venus

Mercury
5th

Venus
5th

### Mercury

| Date | Con. | R.A. | Dec. | Mag. | Diam. | Ill. | Elon. | Vis. | Rat. | | Close To |
|------|------|------|------|------|-------|------|-------|------|------|--|----------|
| **1st** | Psc | 23h 52m | 0° 45' | 2.7 | 11" | 8% | 13° W | NV | N/A | | Venus, Saturn, Neptune |
| **3rd** | Psc | 23h 49m | 0° 6' | 2.2 | 11" | 12% | 15° W | AM | ** | | Venus, Saturn, Neptune |
| **5th** | Psc | 23h 48m | 0° 48' | 1.8 | 11" | 16% | 18° W | AM | *** | | Venus, Saturn, Neptune |
| **7th** | Psc | 23h 48m | -1° 20' | 1.5 | 10" | 20% | 19° W | AM | *** | | Venus, Saturn, Neptune |
| **9th** | Psc | 23h 49m | -1° 42' | 1.2 | 10" | 24% | 21° W | AM | *** | | Venus, Saturn, Neptune |

## Venus

| Date | Con. | R.A. | Dec. | Mag. | Diam. | Ill. | Elon. | Vis. | Rat. | Close To |
|------|------|------|------|------|-------|------|-------|------|------|----------|
| 1st | Psc | 23h 38m | 5° 29' | -4.2 | 57" | 4% | 16° W | AM | *** | Mercury, Saturn |
| 3rd | Psc | 23h 36m | 4° 47' | -4.3 | 55" | 5% | 19° W | AM | *** | Mercury, Saturn |
| 5th | Psc | 23h 34m | 4° 8' | -4.3 | 54" | 7% | 21° W | AM | *** | Mercury, Saturn |
| 7th | Psc | 23h 33m | 3° 31' | -4.4 | 53" | 8% | 23° W | AM | *** | Mercury, Saturn |
| 9th | Psc | 23h 32m | 2° 58' | -4.4 | 51" | 10% | 25° W | AM | *** | Mercury, Saturn |

## Mars and the Outer Planets

Mars
5th

Jupiter
5th

Saturn
5th

## Mars

| Date | Con. | R.A. | Dec. | Mag. | Diam. | Ill. | Elon. | Vis. | Rat. | Close To |
|------|------|------|------|------|-------|------|-------|------|------|----------|
| 1st | Gem | 7h 44m | 23° 58' | 0.5 | 8" | 90% | 105° E | PM | ** | |
| 5th | Gem | 7h 50m | 23° 37' | 0.5 | 8" | 90% | 103° E | PM | ** | Moon |
| 10th | Gem | 7h 58m | 23° 9' | 0.6 | 8" | 90% | 100° E | PM | ** | |

## The Outer Planets

| Planet | Date | Con. | R.A. | Dec. | Mag. | Diam. | Elon. | Vis. | Rat. | Close To |
|--------|------|------|------|------|------|-------|-------|------|------|----------|
| Jupiter | 5th | Tau | 5h 2m | 22° 29' | -2.1 | 36" | 61° E | PM | ** | Hyades, Aldebaran |
| Saturn | 5th | Aqr | 23h 44m | -3° 46' | 1.2 | 16" | 18° W | AM | * | Mercury, Venus, Neptune |
| Uranus | 5th | Tau | 3h 30m | 18° 48' | 5.8 | 3" | 38° E | PM | * | Pleiades |
| Neptune | 5th | Psc | 0h 2m | -1° 4' | 8.0 | 2" | 14° W | NV | N/A | Mercury, Venus, Saturn |

## Highlights

| Date | Time (UT) | Event |
|------|-----------|-------|
| 1st | 13:53 | The waxing crescent Moon is north of Uranus. (Evening sky.) |
| | 20:29 | The waxing crescent Moon is south of the Pleiades star cluster. (Evening sky.) |
| 3rd | 00:24 | The waxing crescent Moon is north of Jupiter. (Evening sky.) |
| 5th | 02:15 | First Quarter Moon. (Evening sky.) |
| | 19:02 | The first quarter Moon is north of Mars. (Evening sky.) |
| 6th | 06:16 | Mercury is stationary prior to resuming prograde motion. (Morning sky.) |
| 8th | 11:51 | The waxing gibbous Moon is north of Regulus. (Evening sky.) |
| 10th | 14:57 | Venus is stationary prior to resuming prograde motion. (Morning sky.) |

# April 11th to 20th, 2025

## The Moon

 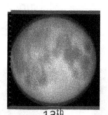 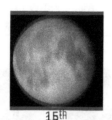  

| 11th | 13th | 15th | 17th | 19th |

| Date | Con. | R.A. | Dec. | Mag. | Diam. | Ill. | Elon. | Phase | Close To |
|------|------|------|------|------|-------|------|-------|-------|----------|
| 11th | Vir | 12h 18m | -2° 48' | -12.3 | 30' | 98% | 164° E | FM | |
| 12th | Vir | 13h 0m | -8° 25' | -12.6 | 29' | 100% | 174° E | FM | Spica |
| 13th | Vir | 13h 44m | -13° 43' | -12.6 | 29' | 100% | 176° W | FM | Spica |
| 14th | Lib | 14h 29m | -18° 31' | -12.3 | 29' | 98% | 166° W | FM | |
| 15th | Lib | 15h 16m | -22° 37' | -12.0 | 29' | 94% | 155° W | -G | |
| 16th | Sco | 16h 6m | -25° 48' | -11.7 | 30' | 89% | 143° W | -G | Antares |
| 17th | Oph | 16h 59m | -27° 52' | -11.4 | 30' | 83% | 131° W | -G | Antares |
| 18th | Sgr | 17h 54m | -28° 38' | -11.1 | 30' | 75% | 118° W | -G | |
| 19th | Sgr | 18h 50m | -27° 59' | -10.8 | 30' | 66% | 105° W | LQ | |
| 20th | Sgr | 19h 46m | -25° 53' | -10.4 | 31' | 56% | 92° W | LQ | |

## Mercury and Venus

Mercury
15th

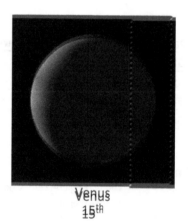

Venus
15th

### Mercury

| Date | Con. | R.A. | Dec. | Mag. | Diam. | Ill. | Elon. | Vis. | Rat. | Close To |
|------|------|------|------|------|-------|------|-------|------|------|----------|
| 11th | Psc | 23h 52m | -1° 53' | 1.0 | 10" | 28% | 22° W | AM | *** | Venus, Saturn, Neptune |
| 13th | Psc | 23h 56m | -1° 54' | 0.9 | 9" | 31% | 23° W | AM | *** | Venus, Saturn, Neptune |
| 15th | Psc | 0h 0m | -1° 46' | 0.7 | 9" | 35% | 24° W | AM | *** | Venus, Saturn, Neptune |
| 17th | Psc | 0h 6m | -1° 30' | 0.6 | 9" | 39% | 24° W | AM | *** | Venus, Saturn, Neptune |
| 19th | Psc | 0h 12m | -1° 5' | 0.5 | 8" | 42% | 24° W | AM | *** | Venus, Saturn, Neptune |

## Venus

| Date | Con. | R.A. | Dec. | Mag. | Diam. | Ill. | Elon. | Vis. | Rat. | Close To |
|------|------|------|------|------|-------|------|-------|------|------|----------|
| 11th | Psc | 23h 32m | 2° 29' | -4.5 | 50" | 12% | 27° W | AM | *** | Mercury, Saturn |
| 13th | Psc | 23h 33m | 2° 3' | -4.5 | 48" | 13% | 29° W | AM | *** | Mercury, Saturn |
| 15th | Psc | 23h 34m | 1° 41' | -4.5 | 47" | 15% | 30° W | AM | *** | Mercury, Saturn |
| 17th | Psc | 23h 35m | 1° 23' | -4.5 | 45" | 17% | 32° W | AM | **** | Mercury, Saturn |
| 19th | Psc | 23h 38m | 1° 10' | -4.5 | 44" | 19% | 33° W | AM | **** | Mercury, Saturn |

## Mars and the Outer Planets

Mars
15th

Jupiter
15th

Saturn
15th

## Mars

| Date | Con. | R.A. | Dec. | Mag. | Diam. | Ill. | Elon. | Vis. | Rat. | Close To |
|------|------|------|------|------|-------|------|-------|------|------|----------|
| 11th | Gem | 7h 59m | 23° 3' | 0.6 | 8" | 90% | 100° E | PM | ** | |
| 15th | Cnc | 8h 6m | 22° 39' | 0.7 | 7" | 90% | 98° E | PM | ** | Praesepe |
| 20th | Cnc | 8h 14m | 22° 6' | 0.8 | 7" | 90% | 95° E | PM | ** | Praesepe |

## The Outer Planets

| Planet | Date | Con. | R.A. | Dec. | Mag. | Diam. | Elon. | Vis. | Rat. | Close To |
|--------|------|------|------|------|------|-------|-------|------|------|----------|
| Jupiter | 15th | Tau | 5h 9m | 22° 40' | -2.0 | 35" | 54° E | PM | ** | Aldebaran |
| Saturn | 15th | Aqr | 23h 49m | -3° 19' | 1.2 | 16" | 27° W | AM | * | Mercury, Venus, Neptune |
| Uranus | 15th | Tau | 3h 32m | 18° 55' | 5.8 | 3" | 29° E | PM | * | Pleiades |
| Neptune | 15th | Psc | 0h 4m | 0° 56' | 8.0 | 2" | 23° W | AM | * | Mercury, Venus, Saturn |

## Highlights

| Date | Time (UT) | Event |
|------|-----------|-------|
| 13th | 00:23 | Full Moon. (Visible all night.) |
| | 01:40 | The full Moon is south of Spica. (Visible all night.) |
| 16th | 22:20 | The waning gibbous Moon is south of the bright star Antares. (Morning sky.) |

# April 21st to 30th, 2025

## The Moon

| 21st | 23rd | 25th | 27th | 29th |

| Date | Con | R.A. | Dec | Mag | Diam | Ill. | Elon. | Phase | Close To |
|------|-----|------|-----|-----|------|------|-------|-------|----------|
| 21st | Cap | 20h 41m | -22° 25' | -9.9 | 31' | 45% | 79° W | LQ | |
| 22nd | Cap | 21h 34m | -17° 41' | -9.4 | 32' | 35% | 67° W | -Cr | |
| 23rd | Aqr | 22h 26m | -11° 56' | -8.8 | 32' | 25% | 55° W | -Cr | |
| 24th | Aqr | 23h 17m | -5° 23' | -8.0 | 33' | 15% | 43° W | -Cr | Venus, Saturn, Neptune |
| 25th | Psc | 0h 9m | 1° 37' | -7.0 | 33' | 8% | 31° W | NM | Mercury, Venus, Saturn, Neptune |
| 26th | Psc | 1h 2m | 8° 42' | -5.8 | 33' | 3% | 18° W | NM | Mercury |
| 27th | Ari | 1h 58m | 15° 23' | -4.5 | 33' | 0% | 5° W | NM | |
| 28th | Ari | 2h 58m | 21° 9' | -5.0 | 33' | 1% | 9° E | NM | Uranus |
| 29th | Tau | 4h 1m | 25° 29' | -6.3 | 33' | 4% | 23° E | NM | Pleiades, Hyades, Aldebaran |
| 30th | Tau | 5h 7m | 27° 60' | -7.4 | 33' | 11% | 39° E | NM | Jupiter, Hyades, Aldebaran |

## Mercury and Venus

Mercury
25th

Venus
25th

**Mercury**

| Date | Con. | R.A. | Dec. | Mag. | Diam. | Ill. | Elon. | Vis. | Rat. | Close To |
|------|------|------|------|------|-------|------|-------|------|------|----------|
| 21st | Psc | 0h 19m | 0° 32' | 0.4 | 8" | 45% | 25° W | AM | *** | Venus, Saturn, Neptune |
| 23rd | Psc | 0h 27m | 0° 7' | 0.4 | 8" | 48% | 24° W | AM | *** | Saturn |
| 25th | Cet | 0h 35m | 0° 52' | 0.3 | 7" | 51% | 24° W | AM | *** | Moon |
| 27th | Cet | 0h 44m | 1° 44' | 0.2 | 7" | 54% | 24° W | AM | *** | |
| 29th | Cet | 0h 54m | 2° 41' | 0.1 | 7" | 57% | 23° W | AM | *** | |

**Venus**

| Date | Con. | R.A. | Dec. | Mag. | Diam. | Ill. | Elon. | Vis. | Rat. | Close To |
|------|------|------|------|------|-------|------|-------|------|------|----------|
| 21st | Psc | 23h 40m | 0° 60' | -4.5 | 43" | 21% | 34° W | AM | **** | Mercury, Saturn |
| 23rd | Psc | 23h 43m | 0° 54' | -4.5 | 41" | 23% | 35° W | AM | **** | Saturn |
| 25th | Psc | 23h 47m | 0° 52' | -4.5 | 40" | 24% | 36° W | AM | **** | Moon, Saturn, Neptune |
| 27th | Psc | 23h 51m | 0° 54' | -4.5 | 39" | 26% | 37° W | AM | **** | Saturn, Neptune |
| 29th | Psc | 23h 55m | 0° 58' | -4.5 | 37" | 28% | 38° W | AM | **** | Saturn, Neptune |

## Mars and the Outer Planets

Mars
25th

Jupiter
25th

Saturn
25th

**Mars**

| Date | Con. | R.A. | Dec. | Mag. | Diam. | Ill. | Elon. | Vis. | Rat. | Close To |
|------|------|------|------|------|-------|------|-------|------|------|----------|
| 21st | Cnc | 8h 16m | 21° 59' | 0.8 | 7" | 90% | 95° E | PM | ** | Praesepe |
| 25th | Cnc | 8h 23m | 21° 30' | 0.9 | 7" | 90% | 93° E | PM | ** | Praesepe |
| 30th | Cnc | 8h 32m | 20° 52' | 0.9 | 7" | 90% | 90° E | PM | ** | Praesepe |

**The Outer Planets**

| Planet | Date | Con. | R.A. | Dec. | Mag. | Diam. | Elon. | Vis. | Rat. | Close To |
|--------|------|------|------|------|------|-------|-------|------|------|----------|
| Jupiter | 25th | Tau | 5h 17m | 22° 50' | -2.0 | 34" | 46° E | PM | ** | |
| Saturn | 25th | Psc | 23h 53m | -2° 54' | 1.2 | 16" | 35° W | AM | * | Moon, Venus, Neptune |
| Uranus | 25th | Tau | 3h 34m | 19° 3' | 5.8 | 3" | 21° E | PM | * | Pleiades |
| Neptune | 25th | Psc | 0h 5m | 0° 48' | 7.9 | 2" | 32° W | AM | * | Moon, Mercury, Venus, Saturn |

## Highlights

| Date | Time (UT) | Event |
|------|-----------|-------|
| 21st | 01:36 | Last Quarter Moon. (Morning sky.) |
| | 18:34 | Mercury is at greatest western elongation from the Sun. (Morning sky.) |
| 23rd | N/A | The Lyrid meteor shower is at its maximum. (ZHR: 18) |
| 25th | 01:19 | The waning crescent Moon is south of Venus. (Morning sky.) |
| | 04:22 | The waning crescent Moon is north of Saturn. (Morning sky.) |
| 26th | 01:02 | The nearly new Moon is north of Mercury. (Morning sky.) |
| 27th | 19:32 | New Moon. (Not visible.) |
| 29th | 02:34 | Venus is 3.7° north of Saturn. (Morning sky.) |
| | 06:36 | The waxing crescent Moon is south of the Pleiades star cluster. (Evening sky.) |
| 30th | 00:42 | The waxing crescent Moon is north of the bright star Aldebaran. (Evening sky.) |
| | 17:37 | The waxing crescent Moon is north of Jupiter. (Evening sky.) |

# May 1ˢᵗ to 10ᵗʰ, 2025

## The Moon

  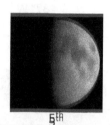  

| 1ˢᵗ | 3ʳᵈ | 5ᵗʰ | 7ᵗʰ | 9ᵗʰ |

| Date | Con | R.A. | Dec | Mag | Diam | Ill. | Elon. | Phase | Close To |
|------|-----|------|-----|-----|------|------|-------|-------|----------|
| 1st | Aur | 6h 12m | 28° 32' | -8.3 | 32' | 19% | 54° E | +Cr | |
| 2nd | Gem | 7h 15m | 27° 11' | -9.0 | 32' | 28% | 69° E | +Cr | |
| 3rd | Cnc | 8h 13m | 24° 16' | -9.6 | 31' | 39% | 83° E | FQ | Mars, Praesepe |
| 4th | Cnc | 9h 7m | 20° 11' | -10.1 | 31' | 49% | 95° E | FQ | Mars, Praesepe |
| 5th | Leo | 9h 56m | 15° 17' | -10.5 | 30' | 59% | 106° E | FQ | Regulus |
| 6th | Leo | 10h 42m | 9° 54' | -10.9 | 30' | 69% | 117° E | +G | Regulus |
| 7th | Leo | 11h 25m | 4° 15' | -11.2 | 30' | 77% | 127° E | +G | |
| 8th | Vir | 12h 8m | -1° 28' | -11.5 | 30' | 85% | 136° E | +G | |
| 9th | Vir | 12h 50m | -7° 5' | -11.8 | 29' | 91% | 146° E | +G | Spica |
| 10th | Vir | 13h 32m | -12° 26' | -12.1 | 29' | 96% | 156° E | FM | Spica |

## Mercury and Venus

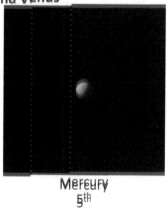

Mercury
5ᵗʰ

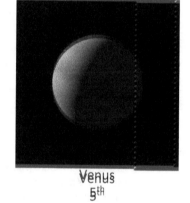

Venus
5ᵗʰ

### Mercury

| Date | Con. | R.A. | Dec. | Mag. | Diam. | Ill. | Elon. | Vis. | Rat. | Close To |
|------|------|------|------|------|-------|------|-------|------|------|----------|
| 1st | Psc | 1h 4m | 3° 43' | 0.1 | 7" | 60% | 23° W | AM | *** | |
| 3rd | Psc | 1h 14m | 4° 49' | 0.0 | 7" | 63% | 22° W | AM | *** | |
| 5th | Psc | 1h 25m | 6° 0' | -0.1 | 6" | 66% | 21° W | AM | *** | |
| 7th | Psc | 1h 36m | 7° 15' | -0.2 | 6" | 69% | 20° W | AM | *** | |
| 9th | Psc | 1h 48m | 8° 33' | -0.3 | 6" | 72% | 19° W | AM | *** | |

## Venus

| Date | Con. | R.A. | Dec. | Mag. | Diam. | Ill. | Elon. | Vis. | Rat. | Close To |
|---|---|---|---|---|---|---|---|---|---|---|
| 1st | Psc | 23h 59m | 1° 6' | -4.5 | 36" | 29% | 39° W | AM | **** | Saturn, Neptune |
| 3rd | Psc | 0h 4m | 1° 17' | -4.5 | 35" | 31% | 40° W | AM | **** | Saturn, Neptune |
| 5th | Psc | 0h 9m | 1° 31' | -4.5 | 34" | 32% | 40° W | AM | **** | Saturn, Neptune |
| 7th | Psc | 0h 15m | 1° 47' | -4.5 | 33" | 34% | 41° W | AM | **** | Saturn, Neptune |
| 9th | Psc | 0h 20m | 2° 6' | -4.5 | 32" | 35% | 41° W | AM | **** | Saturn, Neptune |

## Mars and the Outer Planets

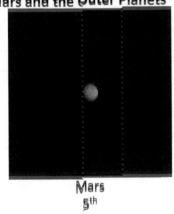

Mars
5th

Jupiter
5th

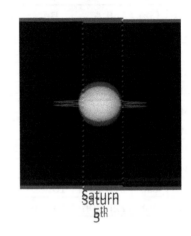

Saturn
5th

## Mars

| Date | Con. | R.A. | Dec. | Mag. | Diam. | Ill. | Elon. | Vis. | Rat. | Close To |
|---|---|---|---|---|---|---|---|---|---|---|
| 1st | Cnc | 8h 34m | 20° 44' | 1.0 | 7" | 90% | 90° E | PM | ** | Praesepe |
| 5th | Cnc | 8h 42m | 20° 11' | 1.0 | 6" | 90% | 88° E | PM | ** | Praesepe |
| 10th | Cnc | 8h 52m | 19° 28' | 1.1 | 6" | 90% | 85° E | PM | ** | Praesepe |

## The Outer Planets

| Planet | Date | Con. | R.A. | Dec. | Mag. | Diam. | Elon. | Vis. | Rat. | Close To |
|---|---|---|---|---|---|---|---|---|---|---|---|
| Jupiter | 5th | Tau | 5h 26m | 22° 59' | -2.0 | 33" | 39° E | PM | * | |
| Saturn | 5th | Psc | 23h 56m | -2° 32' | 1.2 | 16" | 44° W | AM | ** | Venus, Neptune |
| Uranus | 5th | Tau | 3h 37m | 19° 11' | 5.8 | 3" | 12° E | NV | N/A | Pleiades |
| Neptune | 5th | Psc | 0h 6m | 0° 41' | 7.9 | 2" | 41° W | AM | ** | Venus, Saturn |

## Highlights

| Date | Time (UT) | Event |
|---|---|---|
| 3rd | 23:12 | The nearly first quarter Moon is north of Mars. (Evening sky.) |
| 4th | 02:56 | Venus is 2.0° north of Neptune. (Morning sky.) |
| | 04:45 | Mars is 0.7° north of the Praesepe star cluster. (Evening sky.) |
| | 13:52 | First Quarter Moon. (Evening sky.) |
| 5th | 05:04 | Asteroid Vesta is at opposition. (Visible all night.) |
| | 15:03 | Dwarf planet Pluto is stationary prior to beginning retrograde motion. (Morning sky.) |
| | 17:58 | The just-past first quarter Moon is north of the bright star Regulus. (Evening sky.) |
| 6th | N/A | The Eta Aquariid meteor shower is at its maximum. (ZHR: 40) |
| 10th | 07:45 | The waxing gibbous Moon is south of the bright star Spica. (Evening sky.) |

# May 11<sup>th</sup> to 20<sup>th</sup>, 2025

## The Moon

| | 11<sup>th</sup> | 13<sup>th</sup> | 15<sup>th</sup> | 17<sup>th</sup> | 19<sup>th</sup> |

| Date | Con | R.A. | Dec | Mag | Diam | Ill. | Elon. | Phase | Close To |
|------|-----|------|-----|-----|------|------|-------|-------|----------|
| 11th | Vir | 14h 17m | -17° 21' | -12.4 | 29' | 99% | 166° E | FM | |
| 12th | Lib | 15h 4m | -21° 37' | -12.6 | 29' | 100% | 177° E | FM | |
| 13th | Sco | 15h 54m | -25° 2' | -12.5 | 30' | 99% | 172° W | FM | Antares |
| 14th | Sco | 16h 46m | -27° 22' | -12.2 | 30' | 97% | 160° W | FM | Antares |
| 15th | Oph | 17h 41m | -28° 26' | -11.9 | 30' | 93% | 147° W | -G | |
| 16th | Sgr | 18h 37m | -28° 7' | -11.6 | 30' | 87% | 134° W | -G | |
| 17th | Sgr | 19h 32m | -26° 21' | -11.3 | 30' | 79% | 121° W | -G | |
| 18th | Cap | 20h 27m | -23° 15' | -10.9 | 31' | 71% | 109° W | -G | |
| 19th | Cap | 21h 19m | -18° 56' | -10.6 | 31' | 61% | 97° W | LQ | |
| 20th | Aqr | 22h 10m | -13° 36' | -10.1 | 32' | 50% | 85° W | LQ | |

## Mercury and Venus

Mercury
15<sup>th</sup>

Venus
15<sup>th</sup>

**Mercury**

| Date | Con. | R.A. | Dec. | Mag. | Diam. | Ill. | Elon. | Vis. | Rat. | Close To |
|------|------|------|------|------|-------|------|-------|------|------|----------|
| 11th | Psc | 2h 1m | 9° 53' | -0.4 | 6" | 76% | 18° W | AM | ** | |
| 13th | Ari | 2h 14m | 11° 16' | -0.6 | 6" | 79% | 17° W | AM | ** | |
| 15th | Ari | 2h 28m | 12° 41' | -0.7 | 6" | 82% | 16° W | AM | ** | |
| 17th | Ari | 2h 42m | 14° 7' | -0.9 | 5" | 86% | 14° W | NV | N/A | |
| 19th | Ari | 2h 57m | 15° 33' | -1.0 | 5" | 89% | 12° W | NV | N/A | |

**Venus**

| Date | Con. | R.A. | Dec. | Mag. | Diam. | Ill. | Elon. | Vis. | Rat. | Close To |
|------|------|------|------|------|-------|------|-------|------|------|----------|
| 11th | Psc | 0h 26m | 2° 27' | -4.5 | 31" | 37% | 42° W | AM | **** | Saturn, Neptune |
| 13th | Psc | 0h 32m | 2° 51' | -4.4 | 30" | 38% | 42° W | AM | **** | Saturn |
| 15th | Psc | 0h 38m | 3° 16' | -4.4 | 29" | 40% | 43° W | AM | *** | Saturn |
| 17th | Psc | 0h 45m | 3° 43' | -4.4 | 29" | 41% | 43° W | AM | *** | |
| 19th | Psc | 0h 51m | 4° 12' | -4.4 | 28" | 42% | 44° W | AM | *** | |

## Mars and the Outer Planets

Mars
15th

Jupiter
15th

Saturn
15th

**Mars**

| Date | Con. | R.A. | Dec. | Mag. | Diam. | Ill. | Elon. | Vis. | Rat. | Close To |
|------|------|------|------|------|-------|------|-------|------|------|----------|
| 11th | Cnc | 8h 54m | 19° 19' | 1.1 | 6" | 90% | 85° E | PM | ** | Praesepe |
| 15th | Cnc | 9h 2m | 18° 42' | 1.1 | 6" | 90% | 83° E | PM | ** | Praesepe |
| 20th | Cnc | 9h 12m | 17° 53' | 1.2 | 6" | 90% | 81° E | PM | * | Praesepe |

**The Outer Planets**

| Planet | Date | Con. | R.A. | Dec. | Mag. | Diam. | Elon. | Vis. | Rat. | Close To |
|--------|------|------|------|------|------|-------|-------|------|------|----------|
| Jupiter | 15th | Tau | 5h 35m | 23° 7' | -1.9 | 33" | 31° E | PM | * | |
| Saturn | 15th | Psc | 0h 0m | -2° 11' | 1.2 | 16" | 52° W | AM | ** | Venus, Neptune |
| Uranus | 15th | Tau | 3h 39m | 19° 19' | 5.8 | 3" | 2° E | NV | N/A | Pleiades |
| Neptune | 15th | Psc | 0h 7m | 0° 35' | 7.9 | 2" | 51° W | AM | ** | Venus, Saturn |

## Highlights

| Date | Time (UT) | Event |
|------|-----------|-------|
| 12th | 16:57 | Full Moon. (Visible all night.) |
| 14th | 04:11 | The waning gibbous Moon is south of the bright star Antares. (Morning sky.) |
| 18th | 00:50 | Uranus is in conjunction with the Sun. (Not visible.) |
| 20th | 11:59 | Last Quarter Moon. (Morning sky.) |

# May 21st to 31st, 2025

## The Moon

| | | | | | |
|---|---|---|---|---|---|
| 21st | 23rd | 25th | 27th | 29th | 31st |

| Date | Con | R.A. | Dec | Mag | Diam | Ill. | Elon. | Phase | Close To |
|---|---|---|---|---|---|---|---|---|---|
| 21st | Aqr | 23h 0m | -7° 29' | -9.6 | 32' | 39% | 73° W | LQ | |
| 22nd | Psc | 23h 50m | 0° 51' | -9.0 | 32' | 28% | 62° W | -Cr | Saturn, Neptune |
| 23rd | Psc | 0h 40m | 5° 60' | -8.3 | 33' | 18% | 50° W | -Cr | Venus, Saturn, Neptune |
| 24th | Psc | 1h 34m | 12° 41' | -7.4 | 33' | 10% | 38° W | NM | Venus |
| 25th | Ari | 2h 31m | 18° 46' | -6.2 | 33' | 4% | 25° W | NM | |
| 26th | Tau | 3h 32m | 23° 43' | -5.0 | 33' | 1% | 10° W | NM | Mercury, Uranus, Pleiades |
| 27th | Tau | 4h 37m | 27° 3' | -4.7 | 33' | 0% | 5° E | NM | Mercury, Hyades, Aldebaran |
| 28th | Tau | 5h 44m | 28° 25' | -5.9 | 33' | 3% | 21° E | NM | Jupiter |
| 29th | Gem | 6h 50m | 27° 46' | -7.0 | 32' | 8% | 36° E | NM | |
| 30th | Gem | 7h 52m | 25° 20' | -8.0 | 32' | 15% | 50° E | +Cr | Praesepe |
| 31st | Cnc | 8h 49m | 21° 30' | -8.7 | 31' | 24% | 64° E | +Cr | Mars, Praesepe |

## Mercury and Venus

Mercury
25th

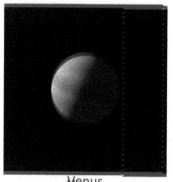

Venus
25th

### Mercury

| Date | Con. | R.A. | Dec. | Mag. | Diam. | Ill. | Elon. | Vis. | Rat. | Close To |
|---|---|---|---|---|---|---|---|---|---|---|
| 21st | Ari | 3h 13m | 16° 58' | -1.2 | 5" | 92% | 10° W | NV | N/A | Pleiades |
| 23rd | Tau | 3h 29m | 18° 20' | -1.5 | 5" | 95% | 8° W | NV | N/A | Uranus, Pleiades |
| 25th | Tau | 3h 46m | 19° 40' | -1.7 | 5" | 97% | 6° W | NV | N/A | Uranus, Pleiades |
| 27th | Tau | 4h 4m | 20° 54' | -2.0 | 5" | 99% | 3° W | NV | N/A | Moon, Pleiades, Hyades |
| 29th | Tau | 4h 22m | 22° 1' | -2.3 | 5" | 100% | 1° W | NV | N/A | Pleiades, Hyades, Aldebaran |
| 31st | Tau | 4h 41m | 23° 0' | -2.2 | 5" | 100% | 2° E | NV | N/A | Hyades, Aldebaran |

## Venus

| Date | Con. | R.A. | Dec. | Mag. | Diam. | Ill. | Elon. | Vis. | Rat. | Close To |
|------|------|------|------|------|-------|------|-------|------|------|----------|
| 21st | Psc | 0h 58m | 4° 42' | -4.4 | 27" | 43% | 44° W | AM | *** | |
| 23rd | Psc | 1h 5m | 5° 13' | -4.4 | 26" | 45% | 44° W | AM | *** | Moon |
| 25th | Psc | 1h 12m | 5° 46' | -4.4 | 26" | 46% | 45° W | AM | *** | |
| 27th | Psc | 1h 19m | 6° 20' | -4.3 | 25" | 47% | 45° W | AM | *** | |
| 29th | Psc | 1h 26m | 6° 55' | -4.3 | 25" | 48% | 45° W | AM | *** | |
| 31st | Psc | 1h 33m | 7° 30' | -4.3 | 24" | 49% | 45° W | AM | *** | |

## Mars and the Outer Planets

Mars
25th

Jupiter
25th

Saturn
25th

## Mars

| Date | Con. | R.A. | Dec. | Mag. | Diam. | Ill. | Elon. | Vis. | Rat. | Close To |
|------|------|------|------|------|-------|------|-------|------|------|----------|
| 21st | Cnc | 9h 14m | 17° 43' | 1.2 | 6" | 90% | 80° E | PM | * | Praesepe |
| 25th | Cnc | 9h 22m | 17° 3' | 1.2 | 6" | 91% | 78° E | PM | * | |
| 31st | Leo | 9h 34m | 15° 57' | 1.3 | 6" | 91% | 75° E | PM | * | Moon, Regulus |

## The Outer Planets

| Planet | Date | Con. | R.A. | Dec. | Mag. | Diam. | Elon. | Vis. | Rat. | Close To |
|--------|------|------|------|------|------|-------|-------|------|------|----------|
| Jupiter | 25th | Tau | 5h 45m | 23° 12' | -1.9 | 33" | 24° E | PM | * | |
| Saturn | 25th | Psc | 0h 3m | -1° 54' | 1.1 | 17" | 62° W | AM | ** | Neptune |
| Uranus | 25th | Tau | 3h 42m | 19° 27' | 5.8 | 3" | 7° W | NV | N/A | Mercury, Pleiades |
| Neptune | 25th | Psc | 0h 8m | 0° 30' | 7.9 | 2" | 60° W | AM | ** | Saturn |

## Highlights

| Date | Time (UT) | Event |
|------|-----------|-------|
| 22nd | 18:01 | The waning crescent Moon is north of Saturn. (Morning sky.) |
| | 20:40 | The waning crescent Moon is north of Neptune. (Morning sky.) |
| 23rd | 23:49 | The waning crescent Moon is north of Venus. (Morning sky.) |
| 27th | 03:03 | New Moon. (Not visible.) |
| 28th | 13:14 | The just-past new Moon is north of Jupiter. (Evening sky.) |
| 30th | 04:00 | Mercury is at superior conjunction with the Sun. (Not visible.) |
| 31st | 08:00 | The waxing crescent Moon is north of the Praesepe star cluster. (Evening sky.) |

# June 1ˢᵗ to 10ᵗʰ, 2025

## The Moon

1ˢᵗ

3ʳᵈ

5ᵗʰ

7ᵗʰ

9ᵗʰ

| Date | Con | R.A. | Dec | Mag | Diam | Ill. | Elon. | Phase | Close To |
|------|-----|------|-----|-----|------|------|-------|-------|----------|
| 1ˢᵗ | Leo | 9h 41m | 16° 42' | -9.3 | 31' | 34% | 76° E | +Cr | Mars, Regulus |
| 2nd | Leo | 10h 28m | 11° 20' | -9.8 | 30' | 43% | 86° E | FQ | Regulus |
| 3rd | Leo | 11h 13m | 5° 39' | -10.3 | 30' | 53% | 97° E | FQ | |
| 4th | Vir | 11h 56m | 0° 6' | -10.7 | 30' | 63% | 106° E | FQ | |
| 5th | Vir | 12h 38m | -5° 46' | -11.0 | 30' | 72% | 116° E | +G | Spica |
| 6th | Vir | 13h 21m | -11° 11' | -11.3 | 29' | 80% | 125° E | +G | Spica |
| 7th | Vir | 14h 5m | -16° 12' | -11.6 | 29' | 87% | 135° E | +G | Spica |
| 8th | Lib | 14h 51m | -20° 38' | -11.9 | 30' | 93% | 146° E | +G | |
| 9th | Lib | 15h 40m | -24° 15' | -12.2 | 30' | 97% | 157° E | FM | |
| 10th | Sco | 16h 32m | -26° 52' | -12.5 | 30' | 99% | 169° E | FM | Antares |

## Mercury and Venus

Mercury
5ᵗʰ

Venus
5ᵗʰ

### Mercury

| Date | Con. | R.A. | Dec. | Mag. | Diam. | Ill. | Elon. | Vis. | Rat. | Close To |
|------|------|------|------|------|-------|------|-------|------|------|----------|
| 1ˢᵗ | Tau | 4h 50m | 23° 26' | -2.0 | 5" | 99% | 3° E | NV | N/A | Hyades, Aldebaran |
| 3rd | Tau | 5h 9m | 24° 10' | -1.8 | 5" | 97% | 6° E | NV | N/A | Aldebaran |
| 5th | Tau | 5h 28m | 24° 43' | -1.5 | 5" | 95% | 8° E | NV | N/A | Jupiter |
| 7th | Tau | 5h 46m | 25° 6' | -1.3 | 5" | 92% | 11° E | NV | N/A | Jupiter |
| 9th | Gem | 6h 4m | 25° 17' | -1.1 | 5" | 88% | 13° E | NV | N/A | Jupiter |

**Venus**

| Date | Con. | R.A. | Dec. | Mag. | Diam. | Ill. | Elon. | Vis. | Rat. | Close To |
|------|------|------|------|------|-------|------|-------|------|------|----------|
| 1st | Psc | 1h 37m | 7° 48' | -4.3 | 24" | 50% | 45° W | AM | *** | |
| 3rd | Psc | 1h 44m | 8° 24' | -4.3 | 23" | 51% | 46° W | AM | *** | |
| 5th | Psc | 1h 52m | 9° 1' | -4.3 | 23" | 52% | 46° W | AM | *** | |
| 7th | Psc | 2h 0m | 9° 38' | -4.3 | 22" | 53% | 46° W | AM | *** | |
| 9th | Psc | 2h 7m | 10° 16' | -4.2 | 22" | 54% | 46° W | AM | *** | |

## Mars and the Outer Planets

Mars
5th

Jupiter
5th

Saturn
5th

**Mars**

| Date | Con. | R.A. | Dec. | Mag. | Diam. | Ill. | Elon. | Vis. | Rat. | Close To |
|------|------|------|------|------|-------|------|-------|------|------|----------|
| 1st | Leo | 9h 36m | 15° 46' | 1.3 | 6" | 91% | 74° E | PM | * | Moon, Regulus |
| 5th | Leo | 9h 45m | 14° 60' | 1.3 | 5" | 91% | 72° E | PM | * | Regulus |
| 10th | Leo | 9h 55m | 14° 1' | 1.4 | 5" | 91% | 70° E | PM | * | Regulus |

**The Outer Planets**

| Planet | Date | Con. | R.A. | Dec. | Mag. | Diam. | Elon. | Vis. | Rat. | Close To |
|--------|------|------|------|------|------|-------|-------|------|------|----------|
| Jupiter | 5th | Tau | 5h 55m | 23° 16' | -1.9 | 32" | 15° E | PM | * | Mercury |
| Saturn | 5th | Psc | 0h 6m | -1° 39' | 1.1 | 17" | 72° W | AM | ** | Neptune |
| Uranus | 5th | Tau | 3h 44m | 19° 35' | 5.8 | 3" | 18° W | AM | * | Pleiades |
| Neptune | 5th | Psc | 0h 9m | 0° 25' | 7.9 | 2" | 72° W | AM | ** | Saturn |

## Highlights

| Date | Time (UT) | Event |
|------|-----------|-------|
| 1st | 03:14 | Venus is at greatest western elongation from the Sun. (Morning sky.) |
| | 09:52 | The waxing crescent Moon is north of Mars. (Evening sky.) |
| 2nd | 01:33 | The nearly first quarter Moon is north of the bright star Regulus. (Evening sky.) |
| 3rd | 03:42 | First Quarter Moon. (Evening sky.) |
| 6th | 14:18 | The waxing gibbous Moon is south of the bright star Spica. (Evening sky.) |
| 10th | 10:27 | The nearly full Moon is south of the bright star Antares. (Evening sky.) |

# June 11th to 20th, 2025

## The Moon

| 11th | 13th | 15th | 17th | 19th |

| Date | Con | R.A. | Dec | Mag | Diam | Ill. | Elon. | Phase | Close To |
|------|-----|------|-----|-----|------|------|-------|-------|----------|
| 11th | Oph | 17h 27m | -28° 14' | -12.6 | 30' | 100% | 178° W | FM | |
| 12th | Sgr | 18h 23m | -28° 13' | -12.4 | 30' | 98% | 165° W | FM | |
| 13th | Sgr | 19h 19m | -26° 45' | -12.1 | 30' | 95% | 152° W | FM | |
| 14th | Cap | 20h 14m | -23° 53' | -11.8 | 31' | 90% | 139° W | -G | |
| 15th | Cap | 21h 7m | -19° 48' | -11.4 | 31' | 83% | 127° W | -G | |
| 16th | Cap | 21h 58m | -14° 42' | -11.1 | 31' | 74% | 116° W | -G | |
| 17th | Aqr | 22h 47m | -8° 50' | -10.7 | 32' | 64% | 104° W | LQ | |
| 18th | Psc | 23h 36m | -2° 26' | -10.3 | 32' | 53% | 93° W | LQ | Saturn, Neptune |
| 19th | Psc | 0h 25m | 4° 11' | -9.8 | 32' | 42% | 82° W | LQ | Saturn, Neptune |
| 20th | Psc | 1h 16m | 10° 45' | -9.2 | 33' | 31% | 70° W | -Cr | |

## Mercury and Venus

 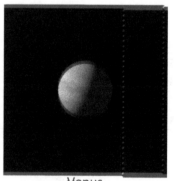

Mercury
15th

Venus
15th

### Mercury

| Date | Con. | R.A. | Dec. | Mag. | Diam. | Ill. | Elon. | Vis. | Rat. | Close To |
|------|------|------|------|------|-------|------|-------|------|------|----------|
| 11th | Gem | 6h 22m | 25° 18' | -0.9 | 6" | 84% | 16° E | PM | ** | Jupiter |
| 13th | Gem | 6h 39m | 25° 9' | -0.8 | 6" | 79% | 18° E | PM | ** | Jupiter |
| 15th | Gem | 6h 55m | 24° 52' | -0.6 | 6" | 75% | 20° E | PM | *** | |
| 17th | Gem | 7h 10m | 24° 27' | -0.4 | 6" | 71% | 21° E | PM | *** | |
| 19th | Gem | 7h 24m | 23° 56' | -0.3 | 6" | 67% | 23° E | PM | *** | |

## Venus

| Date | Con. | R.A. | Dec. | Mag. | Diam. | Ill. | Elon. | Vis. | Rat. | Close To |
|------|------|------|------|------|-------|------|-------|------|------|----------|
| 11th | Ari | 2h 15m | 10° 53' | -4.2 | 21" | 55% | 46° W | AM | *** | |
| 13th | Ari | 2h 23m | 11° 31' | -4.2 | 21" | 56% | 46° W | AM | *** | |
| 15th | Ari | 2h 31m | 12° 8' | -4.2 | 20" | 57% | 46° W | AM | *** | |
| 17th | Ari | 2h 39m | 12° 46' | -4.2 | 20" | 58% | 46° W | AM | *** | |
| 19th | Ari | 2h 48m | 13° 22' | -4.2 | 20" | 59% | 46° W | AM | *** | |

## Mars and the Outer Planets

Mars
15th

Jupiter
15th

Saturn
15th

### Mars

| Date | Con. | R.A. | Dec. | Mag. | Diam. | Ill. | Elon. | Vis. | Rat. | Close To |
|------|------|------|------|------|-------|------|-------|------|------|----------|
| 11th | Leo | 9h 57m | 13° 49' | 1.4 | 5" | 91% | 69° E | PM | * | Regulus |
| 15th | Leo | 10h 6m | 12° 59' | 1.4 | 5" | 92% | 67° E | PM | * | Regulus |
| 20th | Leo | 10h 16m | 11° 56' | 1.4 | 5" | 92% | 65° E | PM | * | Regulus |

### The Outer Planets

| Planet | Date | Con. | R.A. | Dec. | Mag. | Diam. | Elon. | Vis. | Rat. | Close To |
|--------|------|------|------|------|------|-------|-------|------|------|----------|
| Jupiter | 15th | Gem | 6h 5m | 23° 17' | -1.9 | 32" | 7° E | NV | N/A | |
| Saturn | 15th | Psc | 0h 8m | -1° 28' | 1.1 | 17" | 82° W | AM | ** | Neptune |
| Uranus | 15th | Tau | 3h 46m | 19° 43' | 5.8 | 3" | 27° W | AM | * | Pleiades |
| Neptune | 15th | Psc | 0h 9m | 0° 23' | 7.9 | 2" | 82° W | AM | *** | Saturn |

## Highlights

| Date | Time (UT) | Event |
|------|-----------|-------|
| 11th | 07:45 | Full Moon. (Visible all night.) |
| 16th | 12:56 | Mars is 0.8° north of the bright star Regulus. (Evening sky.) |
| 18th | 19:20 | Last Quarter Moon. (Evening sky.) |
| 19th | 03:59 | The just-past last quarter Moon is north of Saturn. (Morning sky.) |
| | 04:29 | The just-past last quarter Moon is north of Neptune. (Morning sky.) |

# June 21st to 30th, 2025

## The Moon

| 21st | 23rd | 25th | 27th | 29th |

| Date | Con | R.A. | Dec | Mag | Diam | Ill. | Elon. | Phase | Close To |
|------|-----|------|-----|-----|------|------|-------|-------|----------|
| 21st | Ari | 2h 10m | 16° 51' | -8.5 | 33' | 21% | 58° W | -Cr | Venus |
| 22nd | Ari | 3h 8m | 22° 5' | -7.6 | 33' | 12% | 44° W | NM | Venus, Uranus, Pleiades |
| 23rd | Tau | 4h 10m | 25° 58' | -6.6 | 33' | 5% | 30° W | NM | Pleiades, Hyades, Aldebaran |
| 24th | Tau | 5h 16m | 28° 5' | -5.4 | 33' | 1% | 14° W | NM | Aldebaran |
| 25th | Aur | 6h 22m | 28° 13' | -4.4 | 33' | 0% | 1° E | NM | Jupiter |
| 26th | Gem | 7h 26m | 26° 25' | -5.5 | 32' | 2% | 16° E | NM | Mercury |
| 27th | Cnc | 8h 26m | 23° 1' | -6.6 | 32' | 6% | 30° E | NM | Mercury, Praesepe |
| 28th | Cnc | 9h 21m | 18° 26' | -7.6 | 31' | 12% | 43° E | NM | Praesepe, Regulus |
| 29th | Leo | 10h 11m | 13° 7' | -8.3 | 31' | 19% | 54° E | +Cr | Mars, Regulus |
| 30th | Leo | 10h 58m | 7° 23' | -9.0 | 30' | 28% | 65° E | +Cr | Mars |

## Mercury and Venus

Mercury
25th

Venus
25th

**Mercury**

| Date | Con. | R.A. | Dec. | Mag. | Diam. | Ill. | Elon. | Vis. | Rat. | Close To |
|------|------|------|------|------|-------|------|-------|------|------|----------|
| 21st | Gem | 7h 38m | 23° 19' | -0.2 | 6" | 63% | 24° E | PM | *** | |
| 23rd | Gem | 7h 50m | 22° 37' | -0.1 | 7" | 60% | 25° E | PM | *** | |
| 25th | Cnc | 8h 2m | 21° 51' | 0.1 | 7" | 56% | 26° E | PM | **** | Praesepe |
| 27th | Cnc | 8h 13m | 21° 2' | 0.2 | 7" | 52% | 27° E | PM | **** | Moon, Praesepe |
| 29th | Cnc | 8h 23m | 20° 12' | 0.3 | 7" | 49% | 27° E | PM | **** | Praesepe |

**Venus**

| Date | Con. | R.A. | Dec. | Mag. | Diam. | Ill. | Elon. | Vis. | Rat. | Close To |
|------|------|------|------|------|-------|------|-------|------|------|----------|
| **21st** | Ari | 2h 56m | 13° 59' | -4.2 | 19" | 60% | 46° W | AM | *** | Moon |
| **23rd** | Ari | 3h 4m | 14° 35' | -4.2 | 19" | 60% | 46° W | AM | *** | |
| **25th** | Ari | 3h 13m | 15° 10' | -4.1 | 19" | 61% | 46° W | AM | *** | Pleiades |
| **27th** | Ari | 3h 21m | 15° 44' | -4.1 | 18" | 62% | 46° W | AM | *** | Pleiades |

## Mars and the Outer Planets

Mars
25th

Jupiter
25th

Saturn
25th

**Mars**

| Date | Con. | R.A. | Dec. | Mag. | Diam. | Ill. | Elon. | Vis. | Rat. | Close To |
|------|------|------|------|------|-------|------|-------|------|------|----------|
| **21st** | Leo | 10h 18m | 11° 43' | 1.4 | 5" | 92% | 64° E | PM | * | Regulus |
| **25th** | Leo | 10h 27m | 10° 50' | 1.5 | 5" | 92% | 62° E | PM | * | Regulus |
| **30th** | Leo | 10h 38m | 9° 43' | 1.5 | 5" | 92% | 60° E | PM | * | Moon, Regulus |

**The Outer Planets**

| Planet | Date | Con. | R.A. | Dec. | Mag. | Diam. | Elon. | Vis. | Rat. | Close To |
|--------|------|------|------|------|------|-------|-------|------|------|----------|
| Jupiter | 25th | Gem | 6h 15m | 23° 15' | -1.9 | 32" | 1° W | NV | N/A | Moon |
| Saturn | 25th | Psc | 0h 9m | -1° 22' | 1.0 | 18" | 92° W | AM | *** | Neptune |
| Uranus | 25th | Tau | 3h 49m | 19° 50' | 5.8 | 3" | 37° W | AM | * | Venus, Pleiades, Hyades |
| Neptune | 25th | Psc | 0h 9m | 0° 21' | 7.9 | 2" | 92° W | AM | *** | Saturn |

## Highlights

| Date | Time (UT) | Event |
|------|-----------|-------|
| 21st | 02:43 | Summer solstice. |
| 22nd | 08:34 | The waning crescent Moon is north of Venus. (Morning sky.) |
| 24th | 15:16 | Jupiter is in conjunction with the Sun. (Not visible.) |
| 25th | 10:32 | New Moon. (Not visible.) |
| 27th | N/A | The June Bootid meteor shower is at its maximum. (ZHR: Variable) |
| | 06:03 | The waxing crescent Moon is north of Mercury. (Evening sky.) |
| 29th | 08:05 | Saturn is 1.0° south of Neptune. (Morning sky.) |
| | 10:28 | The waxing crescent Moon is north of the bright star Regulus. (Evening sky.) |
| 30th | 01:09 | The waxing crescent Moon is south of Mars. (Evening sky.) |

# July 1st to 10th, 2025

## The Moon

    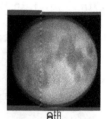

| 1st | | 3rd | | 5th | | 7th | | 9th |

| Date | Con | R.A. | Dec | Mag | Diam | Ill. | Elon. | Phase | Close To |
|------|-----|------|-----|-----|------|------|-------|-------|----------|
| 1st | Vir | 11h 42m | 1° 32' | -9.5 | 30' | 37% | 75° E | +Cr | |
| 2nd | Vir | 12h 24m | -4° 15' | -10.0 | 30' | 47% | 84° E | FQ | |
| 3rd | Vir | 13h 7m | -9° 48' | -10.4 | 30' | 57% | 94° E | FQ | Spica |
| 4th | Vir | 13h 51m | -14° 58' | -10.8 | 30' | 66% | 104° E | FQ | Spica |
| 5th | Lib | 14h 37m | -19° 34' | -11.1 | 30' | 74% | 114° E | +G | |
| 6th | Lib | 15h 25m | -23° 25' | -11.4 | 30' | 82% | 125° E | +G | |
| 7th | Sco | 16h 16m | -26° 19' | -11.7 | 30' | 89% | 137° E | +G | Antares |
| 8th | Oph | 17h 11m | -28° 2' | -12.0 | 30' | 94% | 150° E | +G | Antares |
| 9th | Sgr | 18h 7m | -28° 23' | -12.3 | 30' | 98% | 163° E | FM | |
| 10th | Sgr | 19h 4m | -27° 16' | -12.6 | 30' | 100% | 176° E | FM | |

## Mercury and Venus

Mercury
5th

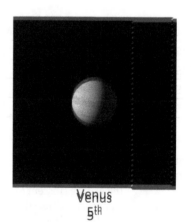

Venus
5th

### Mercury

| Date | Con. | R.A. | Dec. | Mag. | Diam. | Ill. | Elon. | Vis. | Rat. | Close To |
|------|------|------|------|------|-------|------|-------|------|------|----------|
| 1st | Cnc | 8h 32m | 19° 20' | 0.4 | 8" | 46% | 27° E | PM | **** | Praesepe |
| 3rd | Cnc | 8h 40m | 18° 27' | 0.5 | 8" | 42% | 27° E | PM | **** | Praesepe |
| 5th | Cnc | 8h 47m | 17° 35' | 0.6 | 8" | 39% | 27° E | PM | **** | Praesepe |
| 7th | Cnc | 8h 54m | 16° 44' | 0.7 | 9" | 35% | 27° E | PM | **** | Praesepe |
| 9th | Cnc | 8h 59m | 15° 55' | 0.9 | 9" | 32% | 26° E | PM | **** | Praesepe |

## Venus

| Date | Con. | R.A. | Dec. | Mag. | Diam. | Ill. | Elon. | Vis. | Rat. | Close To |
|------|------|------|------|------|-------|------|-------|------|------|----------|
| 1st | Tau | 3h 39m | 16° 50' | −4.1 | 18" | 64% | 46° W | AM | *** | Uranus, Pleiades |
| 3rd | Tau | 3h 48m | 17° 21' | −4.1 | 17" | 65% | 46° W | AM | *** | Uranus, Pleiades, Hyades |
| 5th | Tau | 3h 57m | 17° 52' | −4.1 | 17" | 66% | 46° W | AM | *** | Uranus, Pleiades, Hyades |
| 7th | Tau | 4h 6m | 18° 20' | −4.1 | 17" | 66% | 45° W | AM | *** | Pleiades, Hyades, Aldebaran |
| 9th | Tau | 4h 15m | 18° 48' | −4.1 | 17" | 67% | 45° W | AM | *** | Pleiades, Hyades, Aldebaran |

## Mars and the Outer Planets

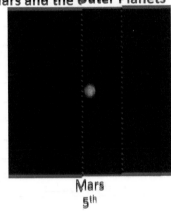

Mars
5th

Jupiter
5th

Saturn
5th

## Mars

| Date | Con. | R.A. | Dec. | Mag. | Diam. | Ill. | Elon. | Vis. | Rat. | Close To |
|------|------|------|------|------|-------|------|-------|------|------|----------|
| 1st | Leo | 10h 40m | 9° 29' | 1.5 | 5" | 92% | 59° E | PM | * | Regulus |
| 5th | Leo | 10h 49m | 8° 33' | 1.5 | 5" | 93% | 57° E | PM | * | |
| 10th | Leo | 10h 59m | 7° 22' | 1.5 | 5" | 93% | 55° E | PM | * | |

## The Outer Planets

| Planet | Date | Con. | R.A. | Dec. | Mag. | Diam. | Elon. | Vis. | Rat. | Close To |
|--------|------|------|------|------|------|-------|-------|------|------|----------|
| Jupiter | 5th | Gem | 6h 25m | 23° 11' | −1.9 | 32" | 9° W | NV | N/A | |
| Saturn | 5th | Psc | 0h 10m | −1° 19' | 1.0 | 18" | 102° W | AM | *** | Neptune |
| Uranus | 5th | Tau | 3h 51m | 19° 56' | 5.8 | 3" | 47° W | AM | ** | Venus, Pleiades, Hyades |
| Neptune | 5th | Psc | 0h 10m | 0° 21' | 7.9 | 2" | 102° W | AM | *** | Saturn |

## Highlights

| Date | Time (UT) | Event |
|------|-----------|-------|
| 2nd | 19:31 | First Quarter Moon. (Evening sky.) |
| 3rd | 04:19 | Venus is 6.8° south of the Pleiades open star cluster. (Morning sky.) |
| | 21:43 | The just-past first quarter Moon is south of the bright star Spica. (Evening sky.) |
| 4th | 01:00 | Venus is 2.4° south of Uranus. (Morning sky.) |
| | 04:24 | Mercury is at greatest eastern elongation from the Sun. (Evening sky.) |
| 5th | 10:01 | Neptune is stationary prior to beginning retrograde motion. (Morning sky.) |
| 7th | 17:38 | The waxing gibbous Moon is south of the bright star Antares. (Evening sky.) |
| 10th | 20:37 | Full Moon. (Visible all night.) |

# July 11th to 20th, 2025

## The Moon

11th

13th

15th

17th

19th

| Date | Con | R.A. | Dec | Mag | Diam | Ill. | Elon. | Phase | Close To |
|------|-----|------|-----|-----|------|------|-------|-------|----------|
| 11th | Sgr | 20h 0m | -24° 42' | -12.5 | 31' | 99% | 171° W | FM | |
| 12th | Cap | 20h 54m | -20° 49' | -12.2 | 31' | 97% | 159° W | FM | |
| 13th | Cap | 21h 46m | -15° 51' | -11.9 | 31' | 92% | 147° W | -G | |
| 14th | Aqr | 22h 36m | -10° 2' | -11.6 | 32' | 86% | 135° W | -G | |
| 15th | Aqr | 23h 25m | -3° 42' | -11.2 | 32' | 77% | 124° W | -G | Saturn, Neptune |
| 16th | Psc | 0h 13m | 2° 54' | -10.8 | 32' | 67% | 113° W | -G | Saturn, Neptune |
| 17th | Psc | 1h 3m | 9° 25' | -10.4 | 32' | 56% | 101° W | LQ | |
| 18th | Ari | 1h 55m | 15° 33' | -9.9 | 32' | 45% | 89° W | LQ | |
| 19th | Ari | 2h 51m | 20° 55' | -9.3 | 32' | 33% | 76° W | -Cr | |
| 20th | Tau | 3h 50m | 25° 5' | -8.6 | 32' | 23% | 62° W | -Cr | Uranus, Pleiades, Hyades |

## Mercury and Venus

Mercury
15th

Venus
15th

**Mercury**

| Date | Con. | R.A. | Dec. | Mag. | Diam. | Ill. | Elon. | Vis. | Rat. | Close To |
|------|------|------|------|------|-------|------|-------|------|------|----------|
| 11th | Cnc | 9h 3m | 15° 9' | 1.0 | 9" | 29% | 25° E | PM | **** | Praesepe |
| 13th | Cnc | 9h 6m | 14° 26' | 1.2 | 10" | 25% | 23° E | PM | *** | Praesepe |
| 15th | Cnc | 9h 7m | 13° 49' | 1.4 | 10" | 22% | 22° E | PM | *** | Praesepe |
| 17th | Cnc | 9h 8m | 13° 17' | 1.7 | 10" | 18% | 20° E | PM | *** | Praesepe |
| 19th | Cnc | 9h 7m | 12° 51' | 2.0 | 11" | 15% | 18° E | PM | *** | Praesepe |

## Venus

| Date | Con. | R.A. | Dec. | Mag. | Diam. | Ill. | Elon. | Vis. | Rat. | Close To |
|------|------|------|------|------|-------|------|-------|------|------|----------|
| 11th | Tau | 4h 24m | 19° 14' | -4.1 | 16" | 68% | 45° W | AM | *** | Pleiades, Hyades, Aldebaran |
| 13th | Tau | 4h 34m | 19° 38' | -4.1 | 16" | 69% | 45° W | AM | *** | Hyades, Aldebaran |
| 15th | Tau | 4h 43m | 20° 1' | -4.1 | 16" | 69% | 44° W | AM | *** | Hyades, Aldebaran |
| 17th | Tau | 4h 53m | 20° 22' | -4.0 | 16" | 70% | 44° W | AM | *** | Hyades, Aldebaran |

## Mars and the Outer Planets

Mars
15th

Jupiter
15th

Saturn
15th

### Mars

| Date | Con. | R.A. | Dec. | Mag. | Diam. | Ill. | Elon. | Vis. | Rat. | Close To |
|------|------|------|------|------|-------|------|-------|------|------|----------|
| 11th | Leo | 11h 2m | 7° 8' | 1.5 | 5" | 93% | 54° E | PM | * | |
| 15th | Leo | 11h 10m | 6° 10' | 1.5 | 5" | 93% | 53° E | PM | * | |
| 20th | Leo | 11h 21m | 4° 56' | 1.6 | 5" | 94% | 50° E | PM | * | |

### The Outer Planets

| Planet | Date | Con. | R.A. | Dec. | Mag. | Diam. | Elon. | Vis. | Rat. | Close To |
|--------|------|------|------|------|------|-------|-------|------|------|----------|
| Jupiter | 15th | Gem | 6h 35m | 23° 5' | -1.9 | 32" | 16° W | AM | * | |
| Saturn | 15th | Psc | 0h 10m | -1° 21' | 0.9 | 18" | 112° W | AM | *** | Moon, Neptune |
| Uranus | 15th | Tau | 3h 52m | 20° 1' | 5.8 | 4" | 57° W | AM | ** | Pleiades, Hyades |
| Neptune | 15th | Psc | 0h 9m | 0° 22' | 7.9 | 2" | 113° W | AM | **** | Moon, Saturn |

## Highlights

| Date | Time (UT) | Event |
|------|-----------|-------|
| 12th | 17:22 | Uranus is 4.3° south of the Pleiades star cluster. (Morning sky.) |
| 13th | 20:21 | Venus is 3.2° north of the bright star Aldebaran. (Morning sky.) |
| 14th | 06:52 | Saturn is stationary prior to beginning retrograde motion. (Morning sky.) |
| 16th | 10:08 | The waning gibbous Moon is north of Neptune. (Morning sky.) |
| | 10:30 | The waning gibbous Moon is north of Saturn. (Morning sky.) |
| 17th | 07:16 | Mercury is stationary prior to beginning retrograde motion. (Evening sky.) |
| 18th | 00:38 | Last Quarter Moon. (Morning sky.) |
| 20th | 10:28 | The waning crescent Moon is south of the Pleiades star cluster. (Morning sky.) |
| | 13:06 | The waning crescent Moon is north of Uranus. (Morning sky.) |

# July 21st to 31st, 2025

## The Moon

| 21st | 23rd | 25th | 27th | 29th | 31st |
|------|------|------|------|------|------|

| Date | Con | R.A. | Dec | Mag | Diam | Ill. | Elon. | Phase | Close To |
|------|-----|------|-----|-----|------|------|-------|-------|----------|
| 21st | Tau | 4h 54m | 27° 42' | -7.8 | 32' | 14% | 48° W | -Cr | Venus, Hyades, Aldebaran |
| 22nd | Tau | 5h 58m | 28° 28' | -6.9 | 32' | 7% | 33° W | NM | Venus, Jupiter |
| 23rd | Gem | 7h 2m | 27° 20' | -5.7 | 32' | 2% | 17° W | NM | Jupiter |
| 24th | Cnc | 8h 3m | 24° 29' | -4.5 | 32' | 0% | 3° W | NM | Praesepe |
| 25th | Cnc | 9h 0m | 20° 16' | -4.9 | 32' | 1% | 10° E | NM | Mercury, Praesepe |
| 26th | Leo | 9h 52m | 15° 7' | -6.1 | 31' | 3% | 22° E | NM | Regulus |
| 27th | Leo | 10h 40m | 9° 25' | -7.1 | 31' | 8% | 33° E | NM | Regulus |
| 28th | Leo | 11h 26m | 3° 29' | -7.9 | 30' | 15% | 43° E | +Cr | Mars |
| 29th | Vir | 12h 9m | -2° 27' | -8.6 | 30' | 22% | 53° E | +Cr | Mars |
| 30th | Vir | 12h 52m | -8° 10' | -9.2 | 30' | 31% | 63° E | +Cr | Spica |
| 31st | Vir | 13h 36m | -13° 31' | -9.7 | 30' | 40% | 73° E | FQ | Spica |

## Mercury and Venus

Mercury
25th

Venus
25th

### Mercury

| Date | Con. | R.A. | Dec. | Mag. | Diam. | Ill. | Elon. | Vis. | Rat. | Close To |
|------|------|------|------|------|-------|------|-------|------|------|----------|
| 21st | Cnc | 9h 5m | 12° 33' | 2.4 | 11" | 11% | 15° E | PM | ** | Praesepe |
| 23rd | Cnc | 9h 2m | 12° 23' | 2.8 | 11" | 8% | 13° E | NV | N/A | Praesepe |
| 25th | Cnc | 8h 58m | 12° 22' | 3.3 | 11" | 5% | 9° E | NV | N/A | Moon, Praesepe |
| 27th | Cnc | 8h 53m | 12° 29' | 3.9 | 11" | 3% | 6° E | NV | N/A | Praesepe |
| 29th | Cnc | 8h 47m | 12° 45' | 4.4 | 11" | 2% | 3° E | NV | N/A | Praesepe |
| 31st | Cnc | 8h 41m | 13° 8' | 4.8 | 11" | 1% | 1° W | NV | N/A | Praesepe |

## Venus

| Date | Con. | R.A. | Dec. | Mag. | Diam. | Ill. | Elon. | Vis. | Rat. | Close To |
|------|------|------|------|------|-------|------|-------|------|------|----------|
| 21st | Tau | 5h 12m | 20° 59' | -4.0 | 15" | 72% | 43° W | AM | ** | Moon, Aldebaran |
| 23rd | Tau | 5h 22m | 21° 14' | -4.0 | 15" | 72% | 43° W | AM | ** | |
| 25th | Tau | 5h 31m | 21° 27' | -4.0 | 15" | 73% | 42° W | AM | ** | |
| 27th | Tau | 5h 41m | 21° 38' | -4.0 | 15" | 74% | 42° W | AM | ** | |
| 29th | Ori | 5h 51m | 21° 47' | -4.0 | 15" | 74% | 41° W | AM | ** | |
| 31st | Ori | 6h 1m | 21° 54' | -4.0 | 14" | 75% | 41° W | AM | ** | |

## Mars and the Outer Planets

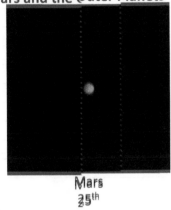

Mars
25th

Jupiter
25th

Saturn
25th

### Mars

| Date | Con. | R.A. | Dec. | Mag. | Diam. | Ill. | Elon. | Vis. | Rat. | Close To |
|------|------|------|------|------|-------|------|-------|------|------|----------|
| 21st | Leo | 11h 24m | 4° 41' | 1.6 | 5" | 94% | 50° E | PM | * | |
| 25th | Leo | 11h 32m | 3° 41' | 1.6 | 5" | 94% | 48° E | PM | * | |
| 31st | Vir | 11h 46m | 2° 10' | 1.6 | 4" | 94% | 46° E | PM | * | |

### The Outer Planets

| Planet | Date | Con. | R.A. | Dec. | Mag. | Diam. | Elon. | Vis. | Rat. | Close To |
|--------|------|------|------|------|------|-------|-------|------|------|----------|
| Jupiter | 25th | Gem | 6h 44m | 22° 57' | -1.9 | 32" | 24° W | AM | * | |
| Saturn | 25th | Psc | 0h 10m | -1° 26' | 0.9 | 18" | 123° W | AM | *** | Neptune |
| Uranus | 25th | Tau | 3h 54m | 20° 6' | 5.8 | 4" | 66° W | AM | ** | Pleiades, Hyades |
| Neptune | 25th | Psc | 0h 9m | 0° 25' | 7.8 | 2" | 123° W | AM | **** | Saturn |

## Highlights

| Date | Time (UT) | Event |
|------|-----------|-------|
| 21st | 05:14 | The waning crescent Moon is north of the bright star Aldebaran. (Morning sky.) |
| | 19:27 | The waning crescent Moon is north of Venus. (Morning sky.) |
| 23rd | 04:19 | The nearly new Moon is north of Jupiter. (Morning sky.) |
| 24th | 19:12 | New Moon. (Not visible.) |
| 26th | 19:43 | The waxing crescent Moon is north of the bright star Regulus. (Evening sky.) |
| 28th | 19:49 | The waxing crescent Moon is south of Mars. (Evening sky.) |
| 30th | N/A | The Alpha Capricornid meteor shower is at its maximum. (ZHR: 5) |
| | N/A | The Southern Delta Aquariid meteor shower is at its maximum. (ZHR: 16) |
| 31st | 05:50 | The nearly first quarter Moon is south of the bright star Spica. (Evening sky.) |
| | 23:36 | Mercury is at inferior conjunction with the Sun. (Not visible.) |

# August 1ˢᵗ to 10ᵗʰ, 2025

## The Moon

1ˢᵗ

3ʳᵈ

5ᵗʰ

7ᵗʰ

9ᵗʰ

| Date | Con | R.A. | Dec | Mag | Diam | Ill. | Elon. | Phase | Close To |
|------|-----|------|-----|-----|------|------|-------|-------|----------|
| 1st | Vir | 14h 21m | -18° 20' | -10.1 | 30' | 50% | 84° E | FQ | |
| 2nd | Lib | 15h 9m | -22° 26' | -10.5 | 30' | 59% | 94° E | FQ | |
| 3rd | Sco | 15h 59m | -25° 38' | -10.9 | 30' | 68% | 106° E | +G | Antares |
| 4th | Sco | 16h 52m | -27° 44' | -11.2 | 30' | 77% | 118° E | +G | Antares |
| 5th | Sgr | 17h 48m | -28° 32' | -11.5 | 30' | 85% | 131° E | +G | |
| 6th | Sgr | 18h 44m | -27° 53' | -11.8 | 30' | 91% | 144° E | +G | |
| 7th | Sgr | 19h 41m | -25° 45' | -12.1 | 31' | 96% | 158° E | FM | |
| 8th | Cap | 20h 36m | -22° 13' | -12.4 | 31' | 99% | 171° E | FM | |
| 9th | Cap | 21h 30m | -17° 27' | -12.6 | 31' | 100% | 177° W | FM | |
| 10th | Aqr | 22h 21m | -11° 43' | -12.3 | 32' | 98% | 165° W | FM | |

## Mercury and Venus

Mercury
5ᵗʰ

Venus
5ᵗʰ

**Mercury**

| Date | Con. | R.A. | Dec. | Mag. | Diam. | Ill. | Elon. | Vis. | Rat. | Close To |
|------|------|------|------|------|-------|------|-------|------|------|----------|
| 1st | Cnc | 8h 38m | 13° 21' | 4.8 | 11" | 1% | 2° W | NV | N/A | Praesepe |
| 3rd | Cnc | 8h 33m | 13° 53' | 4.3 | 11" | 2% | 5° W | NV | N/A | Praesepe |
| 5th | Cnc | 8h 28m | 14° 27' | 3.7 | 11" | 4% | 9° W | NV | N/A | Praesepe |
| 7th | Cnc | 8h 25m | 15° 3' | 3.0 | 10" | 7% | 11° W | NV | N/A | Praesepe |
| 9th | Cnc | 8h 23m | 15° 39' | 2.3 | 10" | 10% | 14° W | NV | N/A | Praesepe |

## Venus

| Date | Con. | R.A. | Dec. | Mag. | Diam. | Ill. | Elon. | Vis. | Rat. | Close To |
|------|------|------|------|------|-------|------|-------|------|------|----------|
| 1st | Gem | 6h 6m | 21° 56' | -4.0 | 14" | 75% | 40° W | AM | ** | |
| 3rd | Gem | 6h 16m | 21° 60' | -4.0 | 14" | 76% | 40° W | AM | ** | Jupiter |
| 5th | Gem | 6h 26m | 22° 1' | -4.0 | 14" | 77% | 39° W | AM | ** | Jupiter |
| 7th | Gem | 6h 36m | 21° 59' | -4.0 | 14" | 77% | 39° W | AM | ** | Jupiter |
| 9th | Gem | 6h 46m | 21° 55' | -4.0 | 14" | 78% | 38° W | AM | ** | Jupiter |

## Mars and the Outer Planets

Mars
5th

Jupiter
5th

Saturn
5th

### Mars

| Date | Con. | R.A. | Dec. | Mag. | Diam. | Ill. | Elon. | Vis. | Rat. | Close To |
|------|------|------|------|------|-------|------|-------|------|------|----------|
| 1st | Vir | 11h 48m | 1° 54' | 1.6 | 4" | 94% | 45° E | PM | * | |
| 5th | Vir | 11h 57m | 0° 53' | 1.6 | 4" | 95% | 44° E | PM | * | |
| 10th | Vir | 12h 8m | 0° 25' | 1.6 | 4" | 95% | 42° E | PM | * | |

### The Outer Planets

| Planet | Date | Con. | R.A. | Dec. | Mag. | Diam. | Elon. | Vis. | Rat. | Close To |
|--------|------|------|------|------|------|-------|-------|------|------|----------|
| Jupiter | 5th | Gem | 6h 55m | 22° 46' | -1.9 | 33" | 32° W | AM | * | Venus |
| Saturn | 5th | Psc | 0h 9m | -1° 36' | 0.8 | 19" | 133° W | AM | **** | Neptune |
| Uranus | 5th | Tau | 3h 55m | 20° 10' | 5.7 | 4" | 77° W | AM | ** | Pleiades, Hyades, Aldebaran |
| Neptune | 5th | Psc | 0h 9m | 0° 29' | 7.8 | 2" | 134° W | AM | **** | Saturn |

## Highlights

| Date | Time (UT) | Event |
|------|-----------|-------|
| 1st | 12:42 | First Quarter Moon. (Evening sky.) |
| 4th | 01:40 | The waxing gibbous Moon is south of the bright star Antares. (Evening sky.) |
| 6th | 10:12 | Saturn is 1.1° south of Neptune. (Morning sky.) |
| 9th | 07:56 | Full Moon. (Visible all night.) |
| 10th | 18:03 | Mercury is stationary prior to resuming prograde motion. (Not visible.) |

# August 11th to 20th, 2025

## The Moon

| 11th | 13th | 15th | 17th | 19th |

| Date | Con | R.A. | Dec | Mag | Diam | Ill. | Elon. | Phase | Close To |
|------|-----|------|-----|-----|------|------|-------|-------|----------|
| 11th | Aqr | 23h 11m | -5° 20' | -12.0 | 32' | 94% | 154° W | =G | |
| 12th | Psc | 0h 1m | 1° 22' | -11.7 | 32' | 88% | 142° W | =G | Saturn, Neptune |
| 13th | Psc | 0h 51m | 8° 4' | -11.3 | 32' | 79% | 130° W | =G | Saturn, Neptune |
| 14th | Psc | 1h 43m | 14° 23' | -10.9 | 32' | 69% | 118° W | =G | |
| 15th | Ari | 2h 38m | 19° 57' | -10.5 | 32' | 58% | 106° W | LQ | |
| 16th | Tau | 3h 36m | 24° 24' | -10.0 | 32' | 47% | 92° W | LQ | Uranus, Pleiades |
| 17th | Tau | 4h 37m | 27° 21' | -9.4 | 32' | 35% | 78° W | =Cr | Uranus, Hyades, Aldebaran |
| 18th | Tau | 5h 41m | 28° 34' | -8.8 | 32' | 25% | 63° W | =Cr | |
| 19th | Gem | 6h 44m | 27° 56' | -8.0 | 32' | 16% | 48° W | =Cr | Jupiter |
| 20th | Gem | 7h 45m | 25° 36' | -7.1 | 32' | 9% | 34° W | NM | Venus, Jupiter |

## Mercury and Venus

 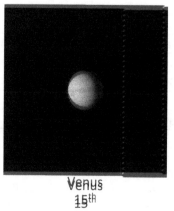

Mercury
15th

Venus
15th

## Mercury

| Date | Con. | R.A. | Dec. | Mag. | Diam. | Ill. | Elon. | Vis. | Rat. | Close To |
|------|------|------|------|------|-------|------|-------|------|------|----------|
| 11th | Cnc | 8h 23m | 16° 12' | 1.7 | 9" | 15% | 16° W | AM | ** | Praesepe |
| 13th | Cnc | 8h 24m | 16° 40' | 1.2 | 9" | 21% | 17° W | AM | *** | Praesepe |
| 15th | Cnc | 8h 28m | 17° 3' | 0.7 | 8" | 27% | 18° W | AM | *** | Praesepe |
| 17th | Cnc | 8h 34m | 17° 18' | 0.3 | 8" | 35% | 19° W | AM | *** | Praesepe |
| 19th | Cnc | 8h 41m | 17° 24' | 0.0 | 7" | 42% | 19° W | AM | *** | Praesepe |

## Venus

| Date | Con. | R.A. | Dec. | Mag. | Diam. | Ill. | Elon. | Vis. | Rat. | Close To |
|------|------|------|------|------|-------|------|-------|------|------|----------|
| 11th | Gem | 6h 56m | 21° 49' | -4.0 | 14" | 78% | 37° W | AM | ** | Jupiter |
| 13th | Gem | 7h 7m | 21° 41' | -4.0 | 13" | 79% | 37° W | AM | ** | Jupiter |
| 15th | Gem | 7h 17m | 21° 30' | -4.0 | 13" | 80% | 36° W | AM | ** | Jupiter |
| 17th | Gem | 7h 27m | 21° 17' | -4.0 | 13" | 80% | 35° W | AM | ** | Jupiter |
| 19th | Gem | 7h 37m | 21° 1' | -4.0 | 13" | 81% | 35° W | AM | ** | Jupiter |

## Mars and the Outer Planets

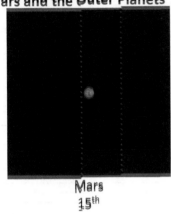

Mars
15th

Jupiter
15th

Saturn
15th

## Mars

| Date | Con. | R.A. | Dec. | Mag. | Diam. | Ill. | Elon. | Vis. | Rat. | Close To |
|------|------|------|------|------|-------|------|-------|------|------|----------|
| 11th | Vir | 12h 11m | 0° 41' | 1.6 | 4" | 95% | 41° E | PM | * | |
| 15th | Vir | 12h 20m | -1° 43' | 1.6 | 4" | 95% | 40° E | PM | * | |
| 20th | Vir | 12h 31m | -3° 2' | 1.6 | 4" | 95% | 38° E | PM | * | |

## The Outer Planets

| Planet | Date | Con. | R.A. | Dec. | Mag. | Diam. | Elon. | Vis. | Rat. | Close To |
|--------|------|------|------|------|------|-------|-------|------|------|----------|
| Jupiter | 15th | Gem | 7h 3m | 22° 34' | -1.9 | 33" | 39° W | AM | * | Venus |
| Saturn | 15th | Psc | 0h 7m | -1° 49' | 0.8 | 19" | 143° W | AM | **** | Neptune |
| Uranus | 15th | Tau | 3h 56m | 20° 12' | 5.7 | 4" | 86° W | AM | ** | Pleiades, Hyades, Aldebaran |
| Neptune | 15th | Psc | 0h 8m | 0° 33' | 7.8 | 2" | 143° W | AM | **** | Saturn |

## Highlights

| Date | Time (UT) | Event |
|------|-----------|-------|
| 12th | 07:41 | Venus is 0.9° south of Jupiter. (Morning sky.) |
| | 15:16 | The waning gibbous Moon is north of Saturn. (Morning sky.) |
| | 15:32 | The waning gibbous Moon is north of Neptune. (Morning sky.) |
| 13th | N/A | The Perseid meteor shower is at its maximum. (ZHR: 100) |
| 16th | 05:13 | Last Quarter Moon. (Morning sky.) |
| | 16:10 | The last quarter Moon is south of the Pleiades star cluster. (Morning sky.) |
| | 20:03 | The last quarter Moon is north of Uranus. (Morning sky.) |
| 18th | N/A | The Kappa Cygnid meteor shower is at its maximum. (ZHR: 3) |
| 19th | 09:37 | Mercury is at greatest western elongation from the Sun. (Morning sky.) |
| | 21:06 | The waning crescent Moon is north of Jupiter. (Morning sky.) |
| 20th | 10:52 | The waning crescent Moon is north of Venus. (Morning sky.) |

## The Moon

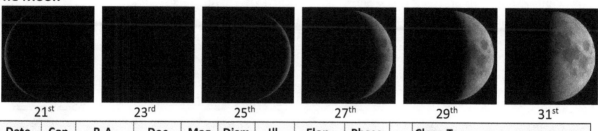

| 21st | 23rd | 25th | 27th | 29th | 31st |

| Date | Con | R.A. | Dec | Mag | Diam | Ill. | Elon. | Phase | Close To |
|------|-----|------|-----|-----|------|------|-------|-------|----------|
| 21st | Cnc | 8h 42m | 21° 49' | -6.1 | 31' | 4% | 20° W | NM | Mercury, Praesepe |
| 22nd | Leo | 9h 34m | 16° 59' | -4.9 | 31' | 1% | 8° W | NM | Mercury, Regulus |
| 23rd | Leo | 10h 23m | 11° 27' | -4.2 | 31' | 0% | 3° E | NM | Regulus |
| 24th | Leo | 11h 10m | 5° 33' | -5.5 | 30' | 2% | 14° E | NM | |
| 25th | Vir | 11h 54m | 0° 27' | -6.5 | 30' | 5% | 24° E | NM | |
| 26th | Vir | 12h 37m | -6° 19' | -7.4 | 30' | 10% | 34° E | NM | Mars, Spica |
| 27th | Vir | 13h 21m | -11° 51' | -8.1 | 30' | 17% | 44° E | +Cr | Mars, Spica |
| 28th | Vir | 14h 6m | -16° 54' | -8.8 | 30' | 25% | 54° E | +Cr | Spica |
| 29th | Lib | 14h 53m | -21° 16' | -9.3 | 30' | 34% | 65° E | +Cr | |
| 30th | Lib | 15h 42m | -24° 47' | -9.8 | 30' | 43% | 76° E | FQ | Antares |
| 31st | Sco | 16h 34m | -27° 15' | -10.2 | 30' | 52% | 89° E | FQ | Antares |

## Mercury and Venus

Mercury
25ᵗʰ

Venus
25ᵗʰ

**Mercury**

| Date | Con. | R.A. | Dec. | Mag. | Diam. | Ill. | Elon. | Vis. | Rat. | Close To |
|------|------|------|------|------|-------|------|-------|------|------|----------|
| 21st | Cnc | 8h 50m | 17° 20' | -0.3 | 7" | 50% | 18° W | AM | *** | Moon, Praesepe |
| 23rd | Cnc | 9h 1m | 17° 5' | -0.6 | 7" | 58% | 17° W | AM | ** | Praesepe |
| 25th | Cnc | 9h 13m | 16° 39' | -0.8 | 6" | 66% | 16° W | AM | ** | Praesepe |
| 27th | Leo | 9h 26m | 16° 1' | -1.0 | 6" | 74% | 15° W | NV | N/A | |
| 29th | Leo | 9h 40m | 15° 11' | -1.1 | 6" | 81% | 13° W | NV | N/A | Regulus |
| 31st | Leo | 9h 55m | 14° 11' | -1.2 | 6" | 86% | 11° W | NV | N/A | Regulus |

**Venus**

| Date | Con. | R.A. | Dec. | Mag. | Diam. | Ill. | Elon. | Vis. | Rat. | Close To |
|------|------|------|------|------|-------|------|-------|------|------|----------|
| 21st | Gem | 7h 47m | 20° 43' | -4.0 | 13" | 81% | 34° W | AM | ** | Jupiter |
| 23rd | Gem | 7h 57m | 20° 22' | -4.0 | 13" | 82% | 33° W | AM | ** | |
| 25th | Cnc | 8h 7m | 19° 59' | -4.0 | 13" | 83% | 33° W | AM | ** | Praesepe |
| 27th | Cnc | 8h 17m | 19° 34' | -4.0 | 13" | 83% | 32° W | AM | ** | Praesepe |
| 29th | Cnc | 8h 27m | 19° 7' | -4.0 | 12" | 84% | 31° W | AM | ** | Praesepe |
| 31st | Cnc | 8h 37m | 18° 38' | -4.0 | 12" | 84% | 31° W | AM | ** | Praesepe |

## Mars and the Outer Planets

Mars
25th

Jupiter
25th

Saturn
25th

**Mars**

| Date | Con. | R.A. | Dec. | Mag. | Diam. | Ill. | Elon. | Vis. | Rat. | Close To |
|------|------|------|------|------|-------|------|-------|------|------|----------|
| 21st | Vir | 12h 34m | -3° 17' | 1.6 | 4" | 96% | 38° E | PM | * | |
| 25th | Vir | 12h 43m | -4° 20' | 1.6 | 4" | 96% | 36° E | PM | * | |
| 31st | Vir | 12h 57m | -5° 54' | 1.6 | 4" | 96% | 34° E | PM | * | Spica |

**The Outer Planets**

| Planet | Date | Con. | R.A. | Dec. | Mag. | Diam. | Elon. | Vis. | Rat. | Close To |
|--------|------|------|------|------|------|-------|-------|------|------|----------|
| Jupiter | 25th | Gem | 7h 12m | 22° 21' | -2.0 | 34" | 46° W | AM | ** | |
| Saturn | 25th | Psc | 0h 5m | -2° 5' | 0.7 | 19" | 153° W | AM | **** | Neptune |
| Uranus | 25th | Tau | 3h 57m | 20° 14' | 5.7 | 4" | 95° W | AM | *** | Pleiades, Hyades, Aldebaran |
| Neptune | 25th | Psc | 0h 7m | 0° 39' | 7.8 | 2" | 153° W | AM | ***** | Saturn |

## Highlights

| Date | Time (UT) | Event |
|------|-----------|-------|
| 21st | 16:15 | The waning crescent Moon is north of Mercury. (Morning sky.) |
| 23rd | 06:07 | New Moon. (Not visible.) |
| 26th | 16:45 | The waxing crescent Moon is south of Mars. (Evening sky.) |
| 27th | 14:00 | The waxing crescent Moon is south of the bright star Spica. (Evening sky.) |
| 31st | 06:26 | First Quarter Moon. (Morning sky.) |
| | 09:55 | The first quarter Moon is south of the bright star Antares. (Evening sky.) |
| | 23:51 | Venus is 1.2° south of the Praesepe star cluster. (Morning sky.) |

# September 1st to 10th, 2025

## The Moon

   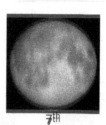 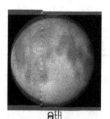

1st    3rd    5th    7th    9th

| Date | Con | R.A. | Dec | Mag | Diam | Ill. | Elon. | Phase | Close To |
|------|-----|------|-----|-----|------|------|-------|-------|----------|
| 1st | Oph | 17h 28m | -28° 30' | -10.6 | 30' | 62% | 101° E | FQ | |
| 2nd | Sgr | 18h 23m | -28° 23' | -11.0 | 30' | 71% | 114° E | +G | |
| 3rd | Sgr | 19h 20m | -26° 48' | -11.3 | 31' | 80% | 127° E | +G | |
| 4th | Cap | 20h 15m | -23° 47' | -11.6 | 31' | 87% | 140° E | +G | |
| 5th | Cap | 21h 9m | -19° 27' | -12.0 | 31' | 94% | 153° E | +G | |
| 6th | Aqr | 22h 2m | -14° 0' | -12.3 | 32' | 98% | 165° E | FM | |
| 7th | Aqr | 22h 53m | -7° 43' | -12.6 | 32' | 100% | 177° E | FM | |
| 8th | Psc | 23h 44m | 0° 56' | -12.5 | 33' | 99% | 171° W | FM | Saturn, Neptune |
| 9th | Psc | 0h 35m | 5° 60' | -12.1 | 33' | 96% | 159° W | FM | Saturn, Neptune |
| 10th | Psc | 1h 28m | 12° 40' | -11.8 | 33' | 90% | 147° W | -G | |

## Mercury and Venus

 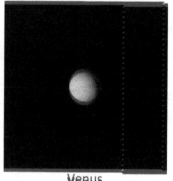

Mercury          Venus
5th              5th

### Mercury

| Date | Con. | R.A. | Dec. | Mag. | Diam. | Ill. | Elon. | Vis. | Rat. | Close To |
|------|------|------|------|------|-------|------|-------|------|------|----------|
| 1st | Leo | 10h 2m | 13° 38' | -1.3 | 5" | 89% | 10° W | NV | N/A | Regulus |
| 3rd | Leo | 10h 17m | 12° 24' | -1.4 | 5" | 93% | 8° W | NV | N/A | Regulus |
| 5th | Leo | 10h 32m | 11° 4' | -1.5 | 5" | 96% | 6° W | NV | N/A | Regulus |
| 7th | Leo | 10h 47m | 9° 39' | -1.5 | 5" | 98% | 5° W | NV | N/A | Regulus |
| 9th | Leo | 11h 1m | 8° 9' | -1.6 | 5" | 99% | 3° W | NV | N/A | |

## Venus

| Date | Con. | R.A. | Dec. | Mag. | Diam. | Ill. | Elon. | Vis. | Rat. | Close To |
|------|------|------|------|------|-------|------|-------|------|------|----------|
| 1st | Cnc | 8h 42m | 18° 22' | -4.0 | 12" | 84% | 38° W | AM | ** | Praesepe |
| 3rd | Cnc | 8h 52m | 17° 50' | -4.0 | 12" | 85% | 38° W | AM | ** | Praesepe |
| 5th | Cnc | 9h 2m | 17° 15' | -4.0 | 12" | 85% | 29° W | AM | ** | Praesepe |
| 7th | Cnc | 9h 11m | 16° 38' | -3.9 | 12" | 86% | 28° W | AM | * | Praesepe |
| 9th | Cnc | 9h 21m | 15° 60' | -3.9 | 12" | 86% | 28° W | AM | * | |

## Mars and the Outer Planets

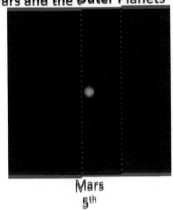

Mars
5th

Jupiter
5th

Saturn
5th

## Mars

| Date | Con. | R.A. | Dec. | Mag. | Diam. | Ill. | Elon. | Vis. | Rat. | Close To |
|------|------|------|------|------|-------|------|-------|------|------|----------|
| 1st | Vir | 13h 0m | -6° 10' | 1.6 | 4" | 96% | 34° E | PM | * | Spica |
| 5th | Vir | 13h 9m | -7° 12' | 1.6 | 4" | 96% | 33° E | PM | * | Spica |
| 10th | Vir | 13h 22m | -8° 29' | 1.6 | 4" | 97% | 32° E | PM | * | Spica |

## The Outer Planets

| Planet | Date | Con. | R.A. | Dec. | Mag. | Diam. | Elon. | Vis. | Rat. | Close To |
|--------|------|------|------|------|------|-------|-------|------|------|----------|
| Jupiter | 5th | Gem | 7h 30m | 22° 7' | -2.0 | 35" | 54° W | AM | ** | |
| Saturn | 5th | Psc | 0h 3m | -2° 24' | 0.7 | 19" | 164° W | AM | **** | Neptune |
| Uranus | 5th | Tau | 3h 57m | 20° 15' | 5.7 | 4" | 105° W | AM | *** | Pleiades, Hyades, Aldebaran |
| Neptune | 5th | Psc | 0h 6m | 0° 46' | 7.8 | 2" | 163° W | AM | ***** | Saturn |

## Highlights

| Date | Time (UT) | Event |
|------|-----------|-------|
| 1st | N/A | The Aurigid meteor shower reaches its maximum. (ZHR: 6) |
| 6th | 02:48 | Uranus is stationary prior to beginning retrograde motion. (Morning sky.) |
| 7th | 18:10 | Full Moon. (Visible all night) |
| | 18:11 | Total lunar eclipse. (Visible from Africa, Antarctica, Asia, eastern Atlantic ocean, Australia, Europe, Indian ocean, western Pacific ocean.) |
| 8th | 20:23 | The just-past full Moon is north of Saturn. (Visible all night.) |
| | 22:28 | The just-past full Moon is north of Neptune. (Visible all night.) |
| 9th | N/A | The September Epsilon Perseid meteor shower reaches its maximum. (ZHR: 5) |

# September 11<sup>th</sup> to 20<sup>th</sup>, 2025

## The Moon

  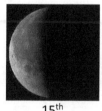  

| 11<sup>th</sup> | 13<sup>th</sup> | 15<sup>th</sup> | 17<sup>th</sup> | 19<sup>th</sup> |

| Date | Con | R.A. | Dec | Mag | Diam | Ill. | Elon. | Phase | Close To |
|------|-----|------|-----|-----|------|------|-------|-------|----------|
| 11th | Ari | 2h 23m | 18° 38' | -11.4 | 33' | 82% | 134° W | -G | |
| 12th | Ari | 3h 22m | 23° 30' | -11.0 | 33' | 72% | 120° W | -G | Uranus, Pleiades |
| 13th | Tau | 4h 23m | 26° 52' | -10.6 | 32' | 61% | 106° W | LQ | Uranus, Pleiades, Hyades, Aldeba |
| 14th | Tau | 5h 26m | 28° 30' | -10.1 | 32' | 49% | 91° W | LQ | |
| 15th | Aur | 6h 30m | 28° 17' | -9.6 | 32' | 38% | 76° W | LQ | |
| 16th | Gem | 7h 30m | 26° 20' | -9.0 | 32' | 28% | 62° W | -Cr | Jupiter |
| 17th | Cnc | 8h 27m | 22° 56' | -8.3 | 31' | 19% | 48° W | -Cr | Praesepe |
| 18th | Cnc | 9h 20m | 18° 25' | -7.5 | 31' | 11% | 36° W | NM | Venus, Praesepe, Regulus |
| 19th | Leo | 10h 10m | 13° 7' | -6.6 | 31' | 5% | 25° W | NM | Venus, Regulus |
| 20th | Leo | 10h 56m | 7° 22' | -5.5 | 30' | 2% | 14° W | NM | Venus |

## Mercury and Venus

Mercury
15<sup>th</sup>

Venus
15<sup>th</sup>

**Mercury**

| Date | Con. | R.A. | Dec. | Mag. | Diam. | Ill. | Elon. | Vis. | Rat. | Close To |
|------|------|------|------|------|-------|------|-------|------|------|----------|
| 11th | Leo | 11h 15m | 6° 37' | -1.7 | 5" | 100% | 1° W | NV | N/A | |
| 13th | Leo | 11h 29m | 5° 3' | -1.7 | 5" | 100% | 1° E | NV | N/A | |
| 15th | Vir | 11h 42m | 3° 28' | -1.5 | 5" | 100% | 2° E | NV | N/A | |
| 17th | Vir | 11h 56m | 1° 52' | -1.4 | 5" | 99% | 4° E | NV | N/A | |
| 19th | Vir | 12h 8m | 0° 17' | -1.2 | 5" | 99% | 5° E | NV | N/A | |

**Venus**

| Date | Con. | R.A. | Dec. | Mag. | Diam. | Ill. | Elon. | Vis. | Rat. | Close To |
|------|------|------|------|------|-------|------|-------|------|------|----------|
| 11th | Leo | 9h 31m | 15° 20' | -3.9 | 12" | 87% | 27° W | AM | * | Regulus |
| 13th | Leo | 9h 40m | 14° 38' | -3.9 | 12" | 87% | 26° W | AM | * | Regulus |
| 15th | Leo | 9h 50m | 13° 54' | -3.9 | 12" | 88% | 26° W | AM | * | Regulus |
| 17th | Leo | 10h 0m | 13° 8' | -3.9 | 12" | 88% | 25° W | AM | * | Regulus |
| 19th | Leo | 10h 9m | 12° 22' | -3.9 | 11" | 89% | 25° W | AM | * | Moon, Regulus |

## Mars and the Outer Planets

Mars
15th

Jupiter
15th

Saturn
15th

**Mars**

| Date | Con. | R.A. | Dec. | Mag. | Diam. | Ill. | Elon. | Vis. | Rat. | Close To |
|------|------|------|------|------|-------|------|-------|------|------|----------|
| 11th | Vir | 13h 24m | -8° 44' | 1.6 | 4" | 97% | 31° E | PM | * | Spica |
| 15th | Vir | 13h 34m | -9° 45' | 1.6 | 4" | 97% | 30° E | PM | * | Spica |
| 20th | Vir | 13h 47m | -11° 0' | 1.6 | 4" | 97% | 29° E | PM | * | Spica |

**The Outer Planets**

| Planet | Date | Con. | R.A. | Dec. | Mag. | Diam. | Elon. | Vis. | Rat. | Close To |
|--------|------|------|------|------|------|-------|-------|------|------|----------|
| Jupiter | 15th | Gem | 7h 27m | 21° 54' | -2.1 | 35" | 62° W | AM | ** | |
| Saturn | 15th | Psc | 0h 0m | -2° 43' | 0.6 | 19" | 173° W | AN | **** | Neptune |
| Uranus | 15th | Tau | 3h 57m | 20° 14' | 5.7 | 4" | 114° W | AM | *** | Pleiades, Hyades, Aldebaran |
| Neptune | 15th | Psc | 0h 5m | 0° 52' | 7.8 | 2" | 172° W | AN | ***** | Saturn |

## Highlights

| Date | Time (UT) | Event |
|------|-----------|-------|
| 11th | 18:20 | Mars is 2.4° north of the bright star Spica. (Evening sky.) |
| 12th | 21:47 | The waning gibbous Moon is north of the Pleaides star cluster. (Morning sky.) |
| 13th | 01:52 | The nearly last quarter Moon is north of Uranus. (Morning sky.) |
| | 10:35 | Mercury is at superior conjunction with the Sun. (Not visible.) |
| 14th | 10:34 | Last Quarter Moon. (Morning sky.) |
| 16th | 11:05 | The waning crescent Moon is north of Jupiter. (Morning sky.) |
| 17th | 17:22 | The waning crescent Moon is north of the Praesepe star cluster. (Morning sky.) |
| 19th | 05:44 | Venus is 0.5° north of the bright star Regulus. (Morning sky.) |
| | 11:11 | The waning crescent Moon is north of the bright star Regulus. (Morning sky.) |
| | 11:47 | The waning crescent Moon is south of Venus. (Morning sky.) |

# September 21st to 30th, 2025

## The Moon

| 21st | 23rd | 25th | 27th | 29th |

| Date | Con | R.A. | Dec | Mag | Diam | Ill. | Elon. | Phase | Close To |
|------|-----|------|-----|-----|------|------|-------|-------|----------|
| 21st | Vir | 11h 40m | 1° 25' | -4.3 | 30' | 0% | 4° W | NM | Mercury |
| 22nd | Vir | 12h 24m | -4° 29' | -4.7 | 30' | 0% | 6° E | NM | Mercury |
| 23rd | Vir | 13h 7m | -10° 9' | -5.9 | 30' | 3% | 16° E | NM | Mercury, Mars, Spica |
| 24th | Vir | 13h 52m | -15° 22' | -6.8 | 30' | 7% | 26° E | NM | Mars, Spica |
| 25th | Lib | 14h 38m | -19° 59' | -7.6 | 29' | 12% | 37° E | NM | Mars |
| 26th | Lib | 15h 26m | -23° 46' | -8.3 | 29' | 19% | 48° E | ±Cr | |
| 27th | Sco | 16h 17m | -26° 35' | -8.9 | 30' | 27% | 60° E | ±Cr | Antares |
| 28th | Oph | 17h 10m | -28° 13' | -9.4 | 30' | 36% | 72° E | ±Cr | Antares |
| 29th | Sgr | 18h 4m | -28° 33' | -9.9 | 30' | 45% | 85° E | FQ | |
| 30th | Sgr | 18h 59m | -27° 31' | -10.3 | 30' | 55% | 98° E | FQ | |

## Mercury and Venus

| Mercury 25th | Venus 25th |

### Mercury

| Date | Con. | R.A. | Dec. | Mag. | Diam. | Ill. | Elon. | Vis. | Rat. | Close To |
|------|------|------|------|------|-------|------|-------|------|------|----------|
| 21st | Vir | 12h 21m | -1° 17' | -1.0 | 5" | 98% | 6° E | NV | N/A | Moon |
| 23rd | Vir | 12h 33m | -2° 49' | -0.9 | 5" | 97% | 8° E | NV | N/A | Moon |
| 25th | Vir | 12h 45m | -4° 21' | -0.8 | 5" | 96% | 9° E | NV | N/A | Spica |
| 27th | Vir | 12h 57m | -5° 50' | -0.7 | 5" | 95% | 10° E | NV | N/A | Spica |
| 29th | Vir | 13h 9m | -7° 18' | -0.6 | 5" | 94% | 11° E | NV | N/A | Spica |

## Venus

| Date | Con. | R.A. | Dec. | Mag. | Diam. | Ill. | Elon. | Vis. | Rat. | Close To |
|------|------|------|------|------|-------|------|-------|------|------|----------|
| 21st | Leo | 10h 19m | 11° 33' | -3.9 | 11" | 89% | 24° W | AM | * | Regulus |
| 23rd | Leo | 10h 28m | 10° 44' | -3.9 | 11" | 90% | 24° W | AM | * | Regulus |
| 25th | Leo | 10h 37m | 9° 53' | -3.9 | 11" | 90% | 23° W | AM | * | Regulus |
| 27th | Leo | 10h 47m | 9° 1' | -3.9 | 11" | 90% | 22° W | AM | * | Regulus |
| 29th | Leo | 10h 56m | 8° 8' | -3.9 | 11" | 91% | 22° W | AM | * | |

## Mars and the Outer Planets

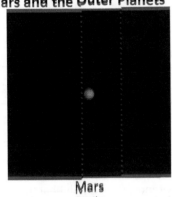

Mars
25th

Jupiter
25th

Saturn
25th

## Mars

| Date | Con. | R.A. | Dec. | Mag. | Diam. | Ill. | Elon. | Vis. | Rat. | Close To |
|------|------|------|------|------|-------|------|-------|------|------|----------|
| 21st | Vir | 13h 49m | -11° 15' | 1.6 | 4" | 97% | 29° E | PM | * | Spica |
| 25th | Vir | 13h 59m | -12° 14' | 1.6 | 4" | 97% | 27° E | PM | * | Moon, Spica |
| 30th | Vir | 14h 12m | -13° 26' | 1.6 | 4" | 98% | 26° E | PM | * | |

## The Outer Planets

| Planet | Date | Con. | R.A. | Dec. | Mag. | Diam. | Elon. | Vis. | Rat. | Close To |
|--------|------|------|------|------|------|-------|-------|------|------|----------|
| Jupiter | 25th | Gem | 7h 34m | 21° 41' | -2.1 | 36" | 69° W | AM | ** | |
| Saturn | 25th | Psc | 23h 57m | -3° 2' | 0.6 | 19" | 177° E | AN | **** | Neptune |
| Uranus | 25th | Tau | 3h 56m | 20° 13' | 5.7 | 4" | 123° W | AM | *** | Pleiades, Hyades, Aldebaran |
| Neptune | 25th | Psc | 0h 4m | 0° 59' | 7.8 | 2" | 179° E | AN | ***** | Saturn |

## Highlights

| Date | Time (UT) | Event |
|------|-----------|-------|
| 21st | 19:42 | Partial solar eclipse. (Visible from Antarctica, south-eastern Australia, New Zealand and the southern Pacific ocean.) |
| | 19:55 | New Moon. (Not visible.) |
| 22nd | 06:28 | Saturn is at opposition. (Visible all night.) |
| | 18:20 | Autumn equinox. |
| 23rd | 21:34 | The waxing crescent Moon is south of the bright star Spica. (Evening sky.) |
| 24th | 04:36 | Neptune is at opposition. (Visible all night.) |
| | 14:55 | The waxing crescent Moon is south of Mars. (Evening sky.) |
| 27th | 17:35 | The waxing crescent Moon is south of the bright star Antares. (Evening sky.) |
| 29th | 23:54 | First Quarter Moon. (Evening sky.) |

# October 1ˢᵗ to 10ᵗʰ, 2025

## The Moon

| | 1ˢᵗ | | 3ʳᵈ | | 5ᵗʰ | | 7ᵗʰ | | 9ᵗʰ |

| Date | Con | R.A. | Dec | Mag | Diam | Ill. | Elon. | Phase | Close To |
|------|-----|------|-----|-----|------|------|-------|-------|----------|
| 1st | Sgr | 19h 54m | -25° 4' | -10.7 | 31' | 65% | 111° E | FQ | |
| 2nd | Cap | 20h 48m | -21° 19' | -11.1 | 31' | 74% | 123° E | +G | |
| 3rd | Cap | 21h 40m | -16° 23' | -11.4 | 32' | 83% | 135° E | +G | |
| 4th | Aqr | 22h 31m | -10° 29' | -11.8 | 32' | 91% | 147° E | +G | |
| 5th | Aqr | 23h 22m | -3° 53' | -12.1 | 33' | 96% | 159° E | FM | Saturn, Neptune |
| 6th | Psc | 0h 13m | 3° 5' | -12.5 | 33' | 99% | 171° E | FM | Saturn, Neptune |
| 7th | Psc | 1h 6m | 10° 2' | -12.6 | 33' | 100% | 177° W | FM | |
| 8th | Ari | 2h 2m | 16° 29' | -12.2 | 33' | 97% | 164° W | FM | |
| 9th | Ari | 3h 2m | 21° 57' | -11.9 | 33' | 92% | 150° W | -G | Pleiades |
| 10th | Tau | 4h 4m | 25° 57' | -11.5 | 33' | 84% | 135° W | -G | Uranus, Hyades, Aldebaran |

## Mercury and Venus

Mercury
5ᵗʰ

Venus
5ᵗʰ

**Mercury**

| Date | Con. | R.A. | Dec. | Mag. | Diam. | Ill. | Elon. | Vis. | Rat. | Close To |
|------|------|------|------|------|-------|------|-------|------|------|----------|
| 1st | Vir | 13h 21m | -8° 43' | -0.5 | 5" | 93% | 12° E | NV | N/A | Spica |
| 3rd | Vir | 13h 32m | -10° 6' | -0.4 | 5" | 92% | 13° E | NV | N/A | Spica |
| 5th | Vir | 13h 43m | -11° 26' | -0.4 | 5" | 91% | 14° E | NV | N/A | Spica |
| 7th | Vir | 13h 55m | -12° 43' | -0.3 | 5" | 89% | 15° E | PM | ** | Mars, Spica |
| 9th | Vir | 14h 6m | -13° 58' | -0.3 | 5" | 88% | 16° E | PM | ** | Mars |

## Venus

| Date | Con. | R.A. | Dec. | Mag. | Diam. | Ill. | Elon. | Vis. | Rat. | Close To |
|------|------|------|------|------|-------|------|-------|------|------|----------|
| 1st | Leo | 11h 5m | 7° 14' | -3.9 | 11" | 91% | 21° W | AM | * | |
| 3rd | Leo | 11h 14m | 6° 19' | -3.9 | 11" | 92% | 21° W | AM | * | |
| 5th | Leo | 11h 24m | 5° 23' | -3.9 | 11" | 92% | 21° W | AM | * | |
| 7th | Leo | 11h 33m | 4° 27' | -3.9 | 11" | 92% | 20° W | AM | * | |

## Mars and the Outer Planets

Mars
5th

Jupiter
5th

Saturn
5th

### Mars

| Date | Con. | R.A. | Dec. | Mag. | Diam. | Ill. | Elon. | Vis. | Rat. | Close To |
|------|------|------|------|------|-------|------|-------|------|------|----------|
| 1st | Vir | 14h 15m | -13° 40' | 1.6 | 4" | 98% | 26° E | PM | * | |
| 5th | Lib | 14h 26m | -14° 36' | 1.6 | 4" | 98% | 25° E | PM | * | |
| 10th | Lib | 14h 39m | -15° 43' | 1.5 | 4" | 98% | 24° E | PM | * | Mercury |

### The Outer Planets

| Planet | Date | Con. | R.A. | Dec. | Mag. | Diam. | Elon. | Vis. | Rat. | Close To |
|--------|------|------|------|------|------|-------|-------|------|------|----------|
| Jupiter | 5th | Gem | 7h 39m | 21° 31' | -2.2 | 37" | 77° W | AM | ** | |
| Saturn | 5th | Aqr | 23h 54m | -3° 20' | 0.7 | 19" | 167° E | PM | **** | Moon, Neptune |
| Uranus | 5th | Tau | 3h 56m | 20° 11' | 5.6 | 4" | 132° W | AM | *** | Pleiades, Hyades, Aldebaran |
| Neptune | 5th | Psc | 0h 3m | -1° 5' | 7.8 | 2" | 169° E | PM | ***** | Moon, Saturn |

## Highlights

| Date | Time (UT) | Event |
|------|-----------|-------|
| 6th | 02:59 | The nearly full Moon is north of Saturn. (Visible all night.) |
| | 07:19 | The nearly full Moon is north of Neptune. (Visible all night.) |
| 7th | 03:48 | Full Moon. (Visible all night.) |
| | 12:35 | Dwarf planet Ceres is at opposition. (Visible all night.) |
| 9th | N/A | The Draconid meteor shower reaches its maximum. (ZHR: 10) |
| 10th | N/A | The Southern Taurid meteor shower reaches its maximum. (ZHR: 5) |
| | 05:24 | The waning gibbous Moon is south of the Pleiades star cluster. (Morning sky.) |
| | 08:38 | The waning gibbous Moon is north of Uranus. (Morning sky.) |

# October 11<sup>th</sup> to 20<sup>th</sup>, 2025

## The Moon

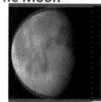

| | 11<sup>th</sup> | | 13<sup>th</sup> | | 15<sup>th</sup> | | 17<sup>th</sup> | | 19<sup>th</sup> |

| Date | Con | R.A. | Dec | Mag | Diam | Ill. | Elon. | Phase | Close To |
|---|---|---|---|---|---|---|---|---|---|
| 11th | Tau | 5h 9m | 28° 10' | -11.1 | 33' | 75% | 120° W | -G | Hyades, Aldebaran |
| 12th | Aur | 6h 14m | 28° 26' | -10.7 | 32' | 64% | 104° W | LQ | |
| 13th | Gem | 7h 16m | 26° 51' | -10.2 | 32' | 53% | 90° W | LQ | Jupiter |
| 14th | Cnc | 8h 15m | 23° 45' | -9.8 | 31' | 42% | 76° W | LQ | Jupiter, Praesepe |
| 15th | Cnc | 9h 8m | 19° 27' | -9.2 | 31' | 32% | 63° W | -Cr | Praesepe |
| 16th | Leo | 9h 58m | 14° 22' | -8.6 | 31' | 23% | 52° W | -Cr | Regulus |
| 17th | Leo | 10h 45m | 8° 46' | -7.9 | 30' | 15% | 41° W | -Cr | Regulus |
| 18th | Leo | 11h 29m | 2° 56' | -7.1 | 30' | 8% | 31° W | NM | |
| 19th | Vir | 12h 12m | -2° 56' | -6.2 | 30' | 4% | 21° W | NM | Venus |
| 20th | Vir | 12h 55m | -8° 37' | -5.1 | 30' | 1% | 11° W | NM | Venus, Spica |

## Mercury and Venus

Mercury
15<sup>th</sup>

Venus
15<sup>th</sup>

### Mercury

| Date | Con. | R.A. | Dec. | Mag. | Diam. | Ill. | Elon. | Vis. | Rat. | Close To |
|---|---|---|---|---|---|---|---|---|---|---|
| 11th | Vir | 14h 17m | -15° 9' | -0.2 | 5" | 86% | 17° E | PM | ** | Mars |
| 13th | Lib | 14h 28m | -16° 17' | -0.2 | 5" | 85% | 18° E | PM | ** | Mars |
| 15th | Lib | 14h 38m | -17° 22' | -0.2 | 5" | 83% | 19° E | PM | ** | Mars |
| 17th | Lib | 14h 49m | -18° 23' | -0.2 | 6" | 81% | 20° E | PM | *** | Mars |
| 19th | Lib | 15h 0m | -19° 20' | -0.2 | 6" | 79% | 21° E | PM | *** | Mars |

## Venus

| Date | Con. | R.A. | Dec. | Mag. | Diam. | Ill. | Elon. | Vis. | Rat. | Close To |
|------|------|------|------|------|-------|------|-------|------|------|----------|
| 11th | Vir | 11h 51m | 2° 32' | -3.9 | 11" | 93% | 19° W | AM | * | |
| 13th | Vir | 12h 0m | 1° 34' | -3.9 | 11" | 93% | 19° W | AM | * | |
| 15th | Vir | 12h 9m | 0° 36' | -3.9 | 11" | 94% | 18° W | AM | * | |
| 17th | Vir | 12h 18m | 0° 23' | -3.9 | 11" | 94% | 18° W | AM | * | |
| 19th | Vir | 12h 28m | -1° 22' | -3.9 | 11" | 94% | 17° W | AM | * | Moon |

## Mars and the Outer Planets

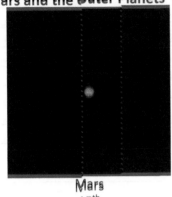

Mars
15th

Jupiter
15th

Saturn
15th

### Mars

| Date | Con. | R.A. | Dec. | Mag. | Diam. | Ill. | Elon. | Vis. | Rat. | Close To |
|------|------|------|------|------|-------|------|-------|------|------|----------|
| 11th | Lib | 14h 42m | -15° 56' | 1.5 | 4" | 98% | 24° E | PM | * | Mercury |
| 15th | Lib | 14h 53m | -16° 48' | 1.5 | 4" | 98% | 23° E | PM | * | Mercury |
| 20th | Lib | 15h 7m | -17° 50' | 1.5 | 4" | 98% | 21° E | PM | * | Mercury |

### The Outer Planets

| Planet | Date | Con. | R.A. | Dec. | Mag. | Diam. | Elon. | Vis. | Rat. | Close To |
|--------|------|------|------|------|------|-------|-------|------|------|----------|
| Jupiter | 15th | Gem | 7h 43m | 21° 22' | -2.2 | 39" | 85° W | AM | *** | |
| Saturn | 15th | Aqr | 23h 51m | -3° 36' | 0.8 | 19" | 157° E | PM | **** | Neptune |
| Uranus | 15th | Tau | 3h 54m | 20° 7' | 5.6 | 4" | 142° W | AM | **** | Pleiades, Hyades |
| Neptune | 15th | Psc | 0h 2m | -1° 12' | 7.8 | 2" | 160° E | PM | ***** | Saturn |

## Highlights

| Date | Time (UT) | Event |
|------|-----------|-------|
| 13th | 18:13 | Last Quarter Moon. (Morning sky.) |
| | 20:12 | Dwarf planet Pluto is stationary prior to resuming prograde motion. (Evening sky.) |
| | 22:27 | The last quarter Moon is north of Jupiter. |
| 14th | 22:47 | The just-past last quarter Moon is north of the Praesepe star cluster. (Morning sky.) |
| 16th | 16:54 | The waning crescent Moon is north of the bright star Regulus. (Morning sky.) |
| 17th | N/A | Good opportunity to see Earthshine on the waning crescent Moon. (Morning sky.) |
| 19th | N/A | The Epsilon Geminid meteor shower is at its maximum. (ZHR: 3) |
| | 21:43 | The waning crescent Moon is south of Venus. (Morning sky.) |

# October 21st to 31st, 2025

## The Moon

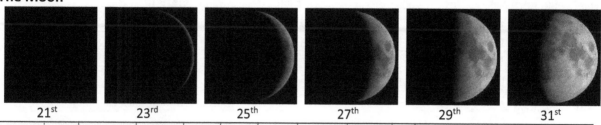

| 21st | 23rd | 25th | 27th | 29th | 31st |

| Date | Con | R.A. | Dec | Mag | Diam | Ill. | Elon. | Phase | Close To |
|------|-----|------|-----|-----|------|------|-------|-------|----------|
| 21st | Vir | 13h 39m | -13° 56' | -4.2 | 30' | 0% | 1° W | NM | Spica |
| 22nd | Lib | 14h 25m | -18° 42' | -5.1 | 29' | 1% | 9° E | NM | Mars |
| 23rd | Lib | 15h 12m | -22° 43' | -6.1 | 29' | 4% | 20° E | NM | Mercury, Mars |
| 24th | Sco | 16h 2m | -25° 48' | -7.0 | 29' | 8% | 32° E | NM | Mercury, Mars, Antares |
| 25th | Oph | 16h 55m | -27° 46' | -7.8 | 29' | 14% | 44° E | +Cr | Antares |
| 26th | Sgr | 17h 48m | -28° 27' | -8.5 | 30' | 21% | 56° E | +Cr | |
| 27th | Sgr | 18h 43m | -27° 49' | -9.1 | 30' | 29% | 69° E | +Cr | |
| 28th | Sgr | 19h 37m | -25° 50' | -9.6 | 30' | 38% | 81° E | FQ | |
| 29th | Cap | 20h 29m | -22° 36' | -10.0 | 30' | 48% | 93° E | FQ | |
| 30th | Cap | 21h 20m | -18° 12' | -10.5 | 31' | 58% | 105° E | FQ | |
| 31st | Aqr | 22h 11m | -12° 49' | -10.9 | 31' | 69% | 117° E | +G | |

## Mercury and Venus

Mercury
25th

Venus
25th

### Mercury

| Date | Con. | R.A. | Dec. | Mag. | Diam. | Ill. | Elon. | Vis. | Rat. | Close To |
|------|------|------|------|------|-------|------|-------|------|------|----------|
| 21st | Lib | 15h 10m | -20° 13' | -0.2 | 6" | 77% | 21° E | PM | *** | Mars |
| 23rd | Lib | 15h 20m | -21° 1' | -0.2 | 6" | 74% | 22° E | PM | *** | Moon, Mars |
| 25th | Lib | 15h 30m | -21° 45' | -0.2 | 6" | 71% | 22° E | PM | *** | Mars |
| 27th | Lib | 15h 39m | -22° 24' | -0.1 | 6" | 68% | 23° E | PM | *** | Mars |
| 29th | Lib | 15h 48m | -22° 56' | -0.1 | 7" | 64% | 23° E | PM | *** | Mars |
| 31st | Sco | 15h 56m | -23° 23' | -0.1 | 7" | 59% | 23° E | PM | *** | Mars, Antares |

## Venus

| Date | Con. | R.A. | Dec. | Mag. | Diam. | Ill. | Elon. | Vis. | Rat. | Close To |
|------|------|------|------|------|-------|------|-------|------|------|----------|
| 21st | Vir | 12h 37m | -2° 20' | -3.9 | 11" | 95% | 17° W | AM | * | |
| 23rd | Vir | 12h 46m | -3° 19' | -3.9 | 10" | 95% | 17° W | AM | * | Spica |
| 25th | Vir | 12h 55m | -4° 17' | -3.9 | 10" | 95% | 16° W | AM | * | Spica |
| 27th | Vir | 13h 4m | -5° 15' | -3.9 | 10" | 96% | 16° W | AM | * | Spica |
| 29th | Vir | 13h 14m | -6° 13' | -3.9 | 10" | 96% | 15° W | AM | * | Spica |
| 31st | Vir | 13h 23m | -7° 10' | -3.9 | 10" | 96% | 15° W | AM | * | Spica |

## Mars and the Outer Planets

Mars
25th

Jupiter
25th

Saturn
25th

### Mars

| Date | Con. | R.A. | Dec. | Mag. | Diam. | Ill. | Elon. | Vis. | Rat. | Close To |
|------|------|------|------|------|-------|------|-------|------|------|----------|
| 21st | Lib | 15h 10m | -18° 2' | 1.5 | 4" | 99% | 21° E | PM | * | Mercury |
| 25th | Lib | 15h 21m | -18° 48' | 1.5 | 4" | 99% | 20° E | PM | * | Mercury |
| 31st | Lib | 15h 38m | -19° 53' | 1.5 | 4" | 99% | 19° E | PM | * | Mercury |

### The Outer Planets

| Planet | Date | Con. | R.A. | Dec. | Mag. | Diam. | Elon. | Vis. | Rat. | Close To |
|--------|------|------|------|------|------|-------|-------|------|------|----------|
| Jupiter | 25th | Gem | 7h 46m | 21° 16' | -2.3 | 40" | 94° W | AM | *** | |
| Saturn | 25th | Aqr | 23h 49m | -3° 49' | 0.8 | 19" | 147° E | PM | **** | Neptune |
| Uranus | 25th | Tau | 3h 53m | 20° 3' | 5.6 | 4" | 152° W | AM | **** | Pleiades, Hyades |
| Neptune | 25th | Psc | 0h 1m | -1° 17' | 7.8 | 2" | 150° E | PM | **** | Saturn |

## Highlights

| Date | Time (UT) | Event |
|------|-----------|-------|
| 21st | 06:20 | Mercury is 2.1° south of Mars. (Evening sky.) |
| | 12:26 | New Moon. (Not visible.) |
| 22nd | N/A | The Orionid meteor shower is at its maximum. (ZHR: 20) |
| 23rd | 13:31 | The waxing crescent Moon is south of Mars. (Evening sky.) |
| | 16:18 | The waxing crescent Moon is south of Mercury. (Evening sky.) |
| 25th | 00:16 | The waxing crescent Moon is south of the bright star Antares. (Evening sky.) |
| 29th | 16:21 | First Quarter Moon. (Evening sky.) |
| | 21:46 | Mercury is at greatest eastern elongation from the Sun. (Evening sky.) |
| 31st | 21:16 | Venus is 3.8° north of the bright star Spica. (Morning sky.) |

# November 1st to 10th, 2025

## The Moon

    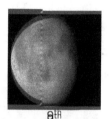

| | 1st | | 3rd | | 5th | | 7th | | 9th |
|---|---|---|---|---|---|---|---|---|---|

| Date | Con | R.A. | Dec | Mag | Diam | Ill. | Elon | Phase | Close To |
|---|---|---|---|---|---|---|---|---|---|
| 1st | Aqr | 23h 0m | -6° 40' | -11.2 | 32' | 78% | 128° E | +G | Saturn |
| 2nd | Psc | 23h 50m | 0° 1' | -11.6 | 33' | 87% | 140° E | +G | Saturn, Neptune |
| 3rd | Psc | 0h 42m | 6° 54' | -12.0 | 33' | 94% | 152° E | +G | Neptune |
| 4th | Psc | 1h 36m | 13° 36' | -12.3 | 33' | 98% | 164° E | FM | |
| 5th | Ari | 2h 34m | 19° 36' | -12.6 | 33' | 100% | 178° E | FM | |
| 6th | Tau | 3h 37m | 24° 22' | -12.4 | 33' | 98% | 167° W | FM | Uranus, Pleiades |
| 7th | Tau | 4h 43m | 27° 24' | -12.0 | 33' | 94% | 152° W | -G | Hyades, Aldebaran |
| 8th | Tau | 5h 51m | 28° 23' | -11.6 | 33' | 87% | 136° W | -G | |
| 9th | Gem | 6h 57m | 27° 22' | -11.3 | 32' | 78% | 121° W | -G | |
| 10th | Gem | 7h 58m | 24° 35' | -10.9 | 32' | 68% | 106° W | -G | Jupiter, Praesepe |

## Mercury and Venus

Mercury
5th

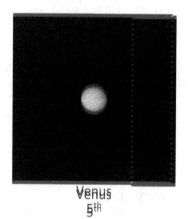

Venus
5th

### Mercury

| Date | Con. | R.A. | Dec. | Mag. | Diam. | Ill. | Elon. | Vis. | Rat. | Close To |
|---|---|---|---|---|---|---|---|---|---|---|
| 1st | Sco | 16h 0m | -23° 34' | -0.1 | 7" | 57% | 23° E | PM | *** | Mars, Antares |
| 3rd | Sco | 16h 7m | -23° 51' | 0.0 | 7" | 52% | 23° E | PM | *** | Mars, Antares |
| 5th | Sco | 16h 12m | -23° 60' | 0.1 | 8" | 46% | 22° E | PM | *** | Mars, Antares |
| 7th | Sco | 16h 16m | -23° 59' | 0.2 | 8" | 39% | 21° E | PM | *** | Mars, Antares |
| 9th | Sco | 16h 18m | -23° 49' | 0.5 | 8" | 32% | 20° E | PM | *** | Mars, Antares |

## Venus

| Date | Con. | R.A. | Dec. | Mag. | Diam. | Ill. | Elon. | Vis. | Rat. | Close To |
|------|------|------|------|------|-------|------|-------|------|------|----------|
| 1st | Vir | 13h 28m | -7° 39' | -3.9 | 10" | 96% | 15° W | NV | N/A | Spica |
| 3rd | Vir | 13h 37m | -8° 35' | -3.9 | 10" | 96% | 15° W | NV | N/A | Spica |
| 5th | Vir | 13h 46m | -9° 31' | -3.9 | 10" | 97% | 14° W | NV | N/A | Spica |
| 7th | Vir | 13h 56m | -10° 26' | -3.9 | 10" | 97% | 14° W | NV | N/A | Spica |

## Mars and the Outer Planets

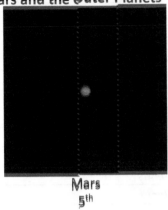

Mars
5th

Jupiter
5th

Saturn
5th

### Mars

| Date | Con. | R.A. | Dec. | Mag. | Diam. | Ill. | Elon. | Vis. | Rat. | Close To |
|------|------|------|------|------|-------|------|-------|------|------|----------|
| 1st | Lib | 15h 41m | -20° 4' | 1.5 | 4" | 99% | 19° E | PM | * | Mercury |
| 5th | Sco | 15h 53m | -20° 43' | 1.5 | 4" | 99% | 18° E | PM | * | Mercury, Antares |
| 10th | Sco | 16h 8m | -21° 28' | 1.4 | 4" | 99% | 16° E | PM | * | Mercury, Antares |

### The Outer Planets

| Planet | Date | Con. | R.A. | Dec. | Mag. | Diam. | Elon. | Vis. | Rat. | Close To |
|--------|------|------|------|------|------|-------|-------|------|------|----------|
| Jupiter | 5th | Gem | 7h 48m | 21° 13' | -2.4 | 41" | 104° W | AM | *** | |
| Saturn | 5th | Aqr | 23h 47m | -3° 60' | 0.9 | 19" | 136° E | PM | **** | Neptune |
| Uranus | 5th | Tau | 3h 51m | 19° 58' | 5.6 | 4" | 163° W | AM | **** | Pleiades, Hyades |
| Neptune | 5th | Psc | 0h 1m | -1° 22' | 7.8 | 2" | 139° E | PM | **** | Saturn |

## Highlights

| Date | Time (UT) | Event |
|------|-----------|-------|
| 3rd | 10:57 | The waxing gibbous Moon is north of Saturn. (Evening sky.) |
| | 17:07 | The waxing gibbous Moon is north of Neptune. (Evening sky.) |
| 5th | 13:20 | Full Moon. (Visible all night.) |
| 6th | 17:15 | The just-past full Moon is north of Uranus. (Visible all night.) |
| 9th | N/A | The Andromedid meteor shower reaches its maximum. (ZHR: 3) |
| | 23:13 | Mercury is stationary prior to beginning retrograde motion. (Evening sky.) |
| 10th | 07:56 | The waning gibbous Moon is north of Jupiter. (Morning sky.) |

# November 11<sup>th</sup> to 20<sup>th</sup>, 2025

## The Moon

11<sup>th</sup>

13<sup>th</sup>

15<sup>th</sup>

17<sup>th</sup>

19<sup>th</sup>

| Date | Con | R.A. | Dec | Mag | Diam | Ill. | Elon. | Phase | Close To |
|------|-----|------|-----|-----|------|------|-------|-------|----------|
| 11th | Cnc | 8h 54m | 20° 29' | -10.4 | 31' | 58% | 93° W | LQ | Praesepe |
| 12th | Leo | 9h 46m | 15° 29' | -10.0 | 31' | 47% | 81° W | LQ | Regulus |
| 13th | Leo | 10h 34m | 9° 57' | -9.5 | 31' | 37% | 70° W | -Cr | Regulus |
| 14th | Leo | 11h 18m | 4° 10' | -9.0 | 30' | 28% | 60° W | -Cr | |
| 15th | Vir | 12h 2m | -1° 40' | -8.3 | 30' | 19% | 50° W | -Cr | |
| 16th | Vir | 12h 45m | -7° 21' | -7.6 | 30' | 12% | 41° W | NM | Spica |
| 17th | Vir | 13h 28m | -12° 42' | -6.8 | 30' | 7% | 31° W | NM | Spica |
| 18th | Vir | 14h 13m | -17° 34' | -5.9 | 29' | 3% | 21° W | NM | Venus |
| 19th | Lib | 15h 0m | -21° 45' | -4.9 | 29' | 1% | 10° W | NM | Venus |
| 20th | Sco | 15h 50m | -25° 3' | -4.5 | 29' | 0% | 1° E | NM | Mercury, Antares |

## Mercury and Venus

Mercury
15<sup>th</sup>

Venus
15<sup>th</sup>

### Mercury

| Date | Con. | R.A. | Dec. | Mag. | Diam. | Ill. | Elon. | Vis. | Rat. | Close To |
|------|------|------|------|------|-------|------|-------|------|------|----------|
| 11th | Sco | 16h 17m | -23° 27' | 0.9 | 9" | 24% | 18° E | PM | *** | Mars, Antares |
| 13th | Sco | 16h 14m | -22° 52' | 1.4 | 9" | 16% | 15° E | NV | N/A | Mars, Antares |
| 15th | Sco | 16h 8m | -22° 3' | 2.3 | 10" | 9% | 11° E | NV | N/A | Mars, Antares |
| 17th | Sco | 15h 59m | -21° 1' | 3.4 | 10" | 3% | 7° E | NV | N/A | Mars, Antares |
| 19th | Lib | 15h 49m | -19° 49' | 5.0 | 10" | 0% | 2° E | NV | N/A | Antares |

## Venus

| Date | Con. | R.A. | Dec. | Mag. | Diam. | Ill. | Elon. | Vis. | Rat. | Close To |
|------|------|------|------|------|-------|------|-------|------|------|----------|
| 11th | Vir | 14h 15m | -12° 12' | -3.9 | 10" | 97% | 13° W | NV | N/A | |
| 13th | Lib | 14h 25m | -13° 4' | -3.9 | 10" | 97% | 13° W | NV | N/A | |
| 15th | Lib | 14h 34m | -13° 54' | -3.9 | 10" | 98% | 12° W | NV | N/A | |
| 17th | Lib | 14h 44m | -14° 43' | -3.9 | 10" | 98% | 12° W | NV | N/A | |

## Mars and the Outer Planets

Mars
15th

Jupiter
15th

Saturn
15th

### Mars

| Date | Con. | R.A. | Dec. | Mag. | Diam. | Ill. | Elon. | Vis. | Rat. | Close To |
|------|------|------|------|------|-------|------|-------|------|------|----------|
| 11th | Sco | 16h 11m | -21° 37' | 1.4 | 4" | 99% | 16° E | PM | * | Mercury, Antares |
| 15th | Sco | 16h 24m | -22° 9' | 1.4 | 4" | 99% | 15° E | PM | * | Mercury, Antares |
| 20th | Oph | 16h 39m | -22° 44' | 1.4 | 4" | 99% | 14° E | NV | N/A | Antares |

### The Outer Planets

| Planet | Date | Con. | R.A. | Dec. | Mag. | Diam. | Elon. | Vis. | Rat. | Close To |
|--------|------|------|------|------|------|-------|-------|------|------|----------|
| Jupiter | 15th | Gem | 7h 48m | 21° 14' | -2.4 | 42" | 114° W | AM | **** | |
| Saturn | 15th | Aqr | 23h 46m | -4° 5' | 1.0 | 19" | 126° E | PM | *** | Neptune |
| Uranus | 15th | Tau | 3h 50m | 19° 53' | 5.6 | 4" | 173° W | AN | **** | Pleiades, Hyades |
| Neptune | 15th | Psc | 0h 0m | -1° 26' | 7.8 | 2" | 129° E | PM | **** | Saturn |

## Highlights

| Date | Time (UT) | Event |
|------|-----------|-------|
| 11th | 19:03 | Jupiter is stationary prior to beginning retrograde motion. (Morning sky.) |
| 12th | 05:29 | Last Quarter Moon. (Morning sky.) |
| | 22:47 | The last quarter Moon is north of the bright star Regulus. (Morning sky.) |
| 13th | N/A | The Northern Taurid meteor shower is at its maximum. (ZHR: 5) |
| 17th | 10:13 | The waning crescent Moon is south of the bright star Spica. (Morning sky.) |
| 18th | N/A | The Leonid meteor shower is at its maximum. (ZHR: 15) |
| 20th | 06:48 | New Moon. (Not visible.) |
| | 09:17 | Mercury is at inferior conjunction with the Sun. (Not visible.) |

# November 21st to 30th, 2025

## The Moon

    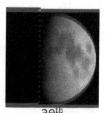

| 21st | 23rd | 25th | 27th | 29th |

| Date | Con | R.A. | Dec | Mag | Diam | Ill. | Elon. | Phase | Close To |
|------|-----|------|-----|-----|------|------|-------|-------|----------|
| 21st | Sco | 16h 41m | -27° 16' | -5.4 | 29' | 2% | 13° E | NM | Mars, Antares |
| 22nd | Oph | 17h 35m | -28° 15' | -6.4 | 30' | 5% | 26° E | NM | |
| 23rd | Sgr | 18h 29m | -27° 55' | -7.2 | 30' | 9% | 38° E | NM | |
| 24th | Sgr | 19h 23m | -26° 15' | -8.0 | 30' | 15% | 50° E | +Cr | |
| 25th | Cap | 20h 15m | -23° 20' | -8.6 | 30' | 23% | 62° E | +Cr | |
| 26th | Cap | 21h 6m | -19° 18' | -9.2 | 30' | 32% | 74° E | +Cr | |
| 27th | Cap | 21h 55m | -14° 20' | -9.8 | 31' | 42% | 85° E | FQ | |
| 28th | Aqr | 22h 43m | -8° 36' | -10.2 | 31' | 52% | 96° E | FQ | |
| 29th | Psc | 23h 31m | -2° 20' | -10.7 | 32' | 63% | 107° E | FQ | Saturn, Neptune |
| 30th | Psc | 0h 20m | 4° 15' | -11.1 | 32' | 73% | 118° E | +G | Saturn, Neptune |

## Mercury and Venus

Mercury
25th

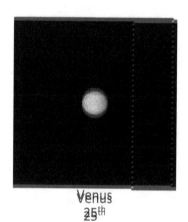

Venus
25th

### Mercury

| Date | Con. | R.A. | Dec. | Mag. | Diam. | Ill. | Elon. | Vis. | Rat. | Close To |
|------|------|------|------|------|-------|------|-------|------|------|----------|
| 21st | Lib | 15h 38m | -18° 34' | 4.7 | 10" | 1% | 2° W | NV | N/A | Venus |
| 23rd | Lib | 15h 29m | -17° 24' | 3.1 | 10" | 4% | 7° W | NV | N/A | Venus |
| 25th | Lib | 15h 22m | -16° 27' | 1.9 | 9" | 11% | 11° W | NV | N/A | Venus |
| 27th | Lib | 15h 17m | -15° 48' | 1.0 | 9" | 19% | 14° W | NV | N/A | Venus |
| 29th | Lib | 15h 15m | -15° 29' | 0.4 | 8" | 29% | 17° W | AM | *** | Venus |

## Venus

| Date | Con. | R.A. | Dec. | Mag. | Diam. | Ill. | Elon. | Vis. | Rat. | Close To |
|------|------|------|------|------|-------|------|-------|------|------|----------|
| 21st | Lib | 15h 4m | -16° 17' | -3.9 | 10" | 98% | 11° W | NV | N/A | Mercury |
| 23rd | Lib | 15h 14m | -17° 1' | -3.9 | 10" | 98% | 11° W | NV | N/A | Mercury |
| 25th | Lib | 15h 24m | -17° 44' | -3.9 | 10" | 99% | 10° W | NV | N/A | Mercury |
| 27th | Lib | 15h 34m | -18° 24' | -3.9 | 10" | 99% | 10° W | NV | N/A | Mercury |

## Mars and the Outer Planets

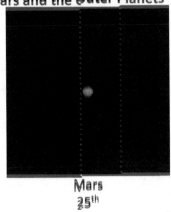

Mars
25th

Jupiter
25th

Saturn
25th

## Mars

| Date | Con. | R.A. | Dec. | Mag. | Diam. | Ill. | Elon. | Vis. | Rat. | Close To |
|------|------|------|------|------|-------|------|-------|------|------|----------|
| 21st | Oph | 16h 42m | -22° 51' | 1.4 | 4" | 99% | 14° E | NV | N/A | Moon, Antares |
| 25th | Oph | 16h 55m | -23° 14' | 1.4 | 4" | 100% | 12° E | NV | N/A | Antares |
| 30th | Oph | 17h 11m | -23° 38' | 1.3 | 4" | 100% | 11° E | NV | N/A | |

## The Outer Planets

| Planet | Date | Con. | R.A. | Dec. | Mag. | Diam. | Elon. | Vis. | Rat. | Close To |
|--------|------|------|------|------|------|-------|-------|------|------|----------|
| Jupiter | 25th | Gem | 7h 47m | 21° 18' | -2.5 | 44" | 125° W | AM | **** | |
| Saturn | 25th | Aqr | 23h 46m | -4° 7' | 1.0 | 18" | 115° E | PM | *** | Neptune |
| Uranus | 25th | Tau | 3h 48m | 19° 48' | 5.6 | 4" | 176° E | AN | **** | Pleiades, Hyades |
| Neptune | 25th | Psc | 0h 0m | -1° 28' | 7.9 | 2" | 119° E | PM | **** | Saturn |

## Highlights

| Date | Time (UT) | Event |
|------|-----------|-------|
| 21st | 12:36 | Uranus is at opposition. (Visible all night.) |
| 22nd | N/A | The Alpha Monocerotid meteor shower is at its maximum. (ZHR: Var) |
| 27th | 21:15 | Dwarf planet Ceres is stationary prior to resuming prograde motion. (Evening sky.) |
| 28th | N/A | The November Orionid meteor shower is at its maximum. (ZHR: 3) |
| | 07:00 | First Quarter Moon. (Evening sky.) |
| | 23:59 | Saturn is stationary prior to resuming prograde motion. (Evening sky.) |
| 29th | 15:06 | Mercury is stationary prior to resuming prograde motion. (Morning sky.) |
| | 19:20 | The just-past first quarter Moon is north of Saturn. (Evening sky.) |
| 30th | 02:10 | The waxing gibbous Moon is north of Neptune. (Evening sky.) |

# December 1st to 10th, 2025

## The Moon

| 1st | 3rd | 5th | 7th | 9th |

| Date | Con | R.A. | Dec | Mag | Diam | Ill. | Elon. | Phase | Close To |
|------|-----|------|-----|-----|------|------|-------|-------|----------|
| 1st | Psc | 1h 11m | 10° 50' | -11.4 | 33' | 83% | 130° E | +G | |
| 2nd | Ari | 2h 7m | 17° 2' | -11.8 | 33' | 91% | 143° E | +G | |
| 3rd | Ari | 3h 7m | 22° 19' | -12.2 | 33' | 97% | 157° E | FM | Uranus, Pleiades |
| 4th | Tau | 4h 12m | 26° 9' | -12.5 | 33' | 99% | 172° E | FM | Uranus, Pleiades, Hyades |
| 5th | Tau | 5h 20m | 28° 5' | -12.5 | 33' | 99% | 172° W | FM | Aldebaran |
| 6th | Gem | 6h 28m | 27° 53' | -12.2 | 33' | 96% | 156° W | FM | |
| 7th | Gem | 7h 33m | 25° 42' | -11.8 | 33' | 91% | 141° W | -G | Jupiter |
| 8th | Cnc | 8h 34m | 21° 54' | -11.4 | 32' | 83% | 127° W | -G | Praesepe |
| 9th | Leo | 9h 29m | 16° 59' | -11.1 | 32' | 74% | 114° W | -G | Regulus |
| 10th | Leo | 10h 19m | 11° 26' | -10.7 | 31' | 64% | 103° W | LQ | Regulus |

## Mercury and Venus

Mercury
5th

Venus
5th

### Mercury

| Date | Con. | R.A. | Dec. | Mag. | Diam. | Ill. | Elon. | Vis. | Rat. | Close To |
|------|------|------|------|------|-------|------|-------|------|------|----------|
| 1st | Lib | 15h 17m | -15° 29' | 0.1 | 8" | 38% | 19° W | AM | *** | Venus |
| 3rd | Lib | 15h 20m | -15° 43' | -0.2 | 7" | 47% | 20° W | AM | *** | |
| 5th | Lib | 15h 26m | -16° 9' | -0.3 | 7" | 54% | 21° W | AM | *** | |
| 7th | Lib | 15h 33m | -16° 44' | -0.4 | 7" | 61% | 21° W | AM | *** | |
| 9th | Lib | 15h 41m | -17° 24' | -0.5 | 6" | 67% | 21° W | AM | *** | |

**Venus**

| Date | Con. | R.A. | Dec. | Mag. | Diam. | Ill. | Elon. | Vis. | Rat. | Close To |
|------|------|------|------|------|-------|------|-------|------|------|----------|
| 1st | Lib | 15h 55m | -19° 39' | -3.9 | 10" | 99% | 9° W | NV | N/A | Mercury, Antares |
| 3rd | Sco | 16h 5m | -20° 13' | -3.9 | 10" | 99% | 9° W | NV | N/A | Antares |
| 5th | Sco | 16h 16m | -20° 45' | -3.9 | 10" | 99% | 8° W | NV | N/A | Antares |
| 7th | Oph | 16h 26m | -21° 15' | -3.9 | 10" | 99% | 8° W | NV | N/A | Antares |

## Mars and the Outer Planets

Mars
5th

Jupiter
5th

Saturn
5th

**Mars**

| Date | Con. | R.A. | Dec. | Mag. | Diam. | Ill. | Elon. | Vis. | Rat. | Close To |
|------|------|------|------|------|-------|------|-------|------|------|----------|
| 1st | Oph | 17h 14m | -23° 42' | 1.3 | 4" | 100% | 11° E | NV | N/A | |
| 5th | Oph | 17h 27m | -23° 56' | 1.3 | 4" | 100% | 10° E | NV | N/A | |
| 10th | Oph | 17h 44m | -24° 8' | 1.3 | 4" | 100% | 8° E | NV | N/A | |

**The Outer Planets**

| Planet | Date | Con. | R.A. | Dec. | Mag. | Diam. | Elon. | Vis. | Rat. | Close To |
|--------|------|------|------|------|------|-------|-------|------|------|----------|
| Jupiter | 5th | Gem | 7h 44m | 21° 26' | -2.6 | 45" | 136° W | AM | **** | |
| Saturn | 5th | Aqr | 23h 46m | -4° 4' | 1.1 | 18" | 104° E | PM | *** | Neptune |
| Uranus | 5th | Tau | 3h 46m | 19° 42' | 5.6 | 4" | 164° E | PM | **** | Pleiades |
| Neptune | 5th | Psc | 23h 59m | -1° 29' | 7.9 | 2" | 108° E | PM | *** | Saturn |

## Highlights

| Date | Time (UT) | Event |
|------|-----------|-------|
| 4th | 02:53 | The almost full Moon is north of Uranus. (Visible all night.) |
| | 20:31 | The almost full Moon is north of the bright star Aldebaran. (Visible all night.) |
| | 23:15 | Full Moon. (Visible all night.) |
| 7th | 15:51 | The waning gibbous Moon is north of Jupiter. (Morning sky.) |
| | 20:51 | Mercury is at greatest western elongation from the Sun. (Morning sky.) |
| 10th | 06:31 | The nearly last quarter Moon is close to the bright star Regulus. (Morning sky.) |
| | 19:49 | Neptune is stationary prior to resuming prograde motion. (Evening sky.) |

# December 11ᵗʰ to 20ᵗʰ, 2025

## The Moon

11ᵗʰ

13ᵗʰ

15ᵗʰ

17ᵗʰ

19ᵗʰ

| Date | Con. | R.A. | Dec. | Mag. | Diam. | Ill. | Elon. | Phase | Close To |
|------|------|------|------|------|-------|------|-------|-------|----------|
| 11th | Leo | 11h 5m | 5° 33' | -10.3 | 31' | 54% | 92° W | LQ | |
| 12th | Vir | 11h 50m | 0° 21' | -9.8 | 30' | 44% | 82° W | LQ | |
| 13th | Vir | 12h 33m | -6° 7' | -9.4 | 30' | 34% | 73° W | -Cr | |
| 14th | Vir | 13h 17m | -11° 34' | -8.8 | 30' | 26% | 63° W | -Cr | Spica |
| 15th | Vir | 14h 1m | -16° 32' | -8.2 | 30' | 18% | 53° W | -Cr | Spica |
| 16th | Lib | 14h 48m | -20° 52' | -7.5 | 29' | 11% | 42° W | NM | |
| 17th | Lib | 15h 36m | -24° 21' | -6.7 | 29' | 6% | 31° W | NM | Mercury |
| 18th | Sco | 16h 28m | -26° 49' | -5.8 | 29' | 2% | 19° W | NM | Mercury, Antares |
| 19th | Oph | 17h 21m | -28° 5' | -4.8 | 30' | 0% | 7° W | NM | Mercury, Venus |
| 20th | Sgr | 18h 16m | -28° 2' | -4.6 | 30' | 0% | 5° E | NM | Venus, Mars |

## Mercury and Venus

Mercury
15ᵗʰ

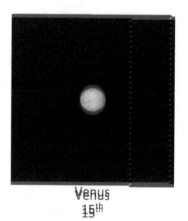
Venus
15ᵗʰ

### Mercury

| Date | Con. | R.A. | Dec. | Mag. | Diam. | Ill. | Elon. | Vis. | Rat. | Close To |
|------|------|------|------|------|-------|------|-------|------|------|----------|
| 11th | Lib | 15h 50m | -18° 7' | -0.5 | 6" | 72% | 21° W | AM | *** | Antares |
| 13th | Lib | 16h 0m | -18° 52' | -0.5 | 6" | 76% | 21° W | AM | *** | Antares |
| 15th | Sco | 16h 11m | -19° 36' | -0.5 | 6" | 80% | 20° W | AM | *** | Antares |
| 17th | Sco | 16h 22m | -20° 20' | -0.5 | 6" | 83% | 20° W | AM | *** | Moon, Antares |
| 19th | Oph | 16h 34m | -21° 1' | -0.5 | 5" | 85% | 19° W | AM | ** | Moon, Antares |

## Venus

| Date | Con. | R.A. | Dec. | Mag. | Diam. | Ill. | Elon. | Vis. | Rat. | Close To |
|------|------|------|------|------|-------|------|-------|------|------|----------|
| 11th | Oph | 16h 48m | -22° 6' | -3.9 | 10" | 99% | 7° W | NV | N/A | Antares |
| 13th | Oph | 16h 58m | -22° 28' | -3.9 | 10" | 100% | 6° W | NV | N/A | Antares |
| 15th | Oph | 17h 9m | -22° 48' | -3.9 | 10" | 100% | 6° W | NV | N/A | |
| 17th | Oph | 17h 20m | -23° 4' | -3.9 | 10" | 100% | 5° W | NV | N/A | |

## Mars and the Outer Planets

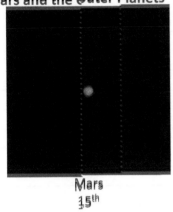

Mars
15th

Jupiter
15th

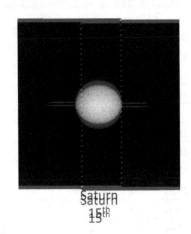

Saturn
15th

## Mars

| Date | Con. | R.A. | Dec. | Mag. | Diam. | Ill. | Elon. | Vis. | Rat. | Close To |
|------|------|------|------|------|-------|------|-------|------|------|----------|
| 11th | Sgr | 17h 47m | -24° 9' | 1.3 | 4" | 100% | 8° E | NV | N/A | |
| 15th | Sgr | 18h 0m | -24° 13' | 1.3 | 4" | 100% | 7° E | NV | N/A | |
| 20th | Sgr | 18h 17m | -24° 12' | 1.2 | 4" | 100% | 6° E | NV | N/A | Moon |

## The Outer Planets

| Planet | Date | Con. | R.A. | Dec. | Mag. | Diam. | Elon. | Vis. | Rat. | Close To |
|--------|------|------|------|------|------|-------|-------|------|------|----------|
| Jupiter | 15th | Gem | 7h 40m | 21° 37' | -2.6 | 46" | 148° W | AM | **** | |
| Saturn | 15th | Aqr | 23h 46m | -3° 57' | 1.1 | 18" | 94° E | PM | *** | Neptune |
| Uranus | 15th | Tau | 3h 45m | 19° 37' | 5.6 | 4" | 153° E | PM | **** | Pleiades |
| Neptune | 15th | Psc | 23h 59m | -1° 29' | 7.9 | 2" | 97° E | PM | *** | Saturn |

## Highlights

| Date | Time (UT) | Event |
|------|-----------|-------|
| 11th | 20:52 | Last Quarter Moon. (Morning sky.) |
| 12th | N/A | The Sigma Hydrid meteor shower is at its maximum. (ZHR: 3) |
| 14th | N/A | The Geminid meteor shower is at its maximum. (ZHR: 140) |
| 16th | N/A | The Coma Berenicid meteor shower is at its maximum. (ZHR: 3) |
| 18th | 12:13 | The waning crescent Moon is south of Mercury. (Morning sky.) |
| | 12:29 | The waning crescent Moon is south of the bright star Antares. (Morning sky.) |
| | 14:36 | Mercury is 5.7° north of the bright star Antares. (Morning sky.) |
| 20th | 01:44 | New Moon. (Not visible.) |

# December 21ˢᵗ to 31ˢᵗ, 2025

## The Moon

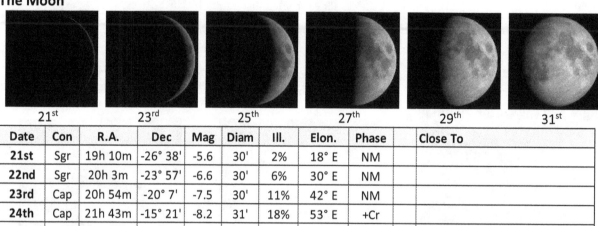

| 21ˢᵗ | 23ʳᵈ | 25ᵗʰ | 27ᵗʰ | 29ᵗʰ | 31ˢᵗ |

| Date | Con | R.A. | Dec | Mag | Diam | Ill. | Elon. | Phase | Close To |
|------|-----|------|-----|-----|------|------|-------|-------|----------|
| 21st | Sgr | 19h 10m | -26° 38' | -5.6 | 30' | 2% | 18° E | NM | |
| 22nd | Sgr | 20h 3m | -23° 57' | -6.6 | 30' | 6% | 30° E | NM | |
| 23rd | Cap | 20h 54m | -20° 7' | -7.5 | 30' | 11% | 42° E | NM | |
| 24th | Cap | 21h 43m | -15° 21' | -8.2 | 31' | 18% | 53° E | +Cr | |
| 25th | Aqr | 22h 31m | -9° 49' | -8.9 | 31' | 26% | 63° E | +Cr | |
| 26th | Aqr | 23h 18m | -3° 47' | -9.5 | 31' | 36% | 74° E | +Cr | Saturn, Neptune |
| 27th | Psc | 0h 5m | 2° 34' | -10.0 | 32' | 47% | 85° E | FQ | Saturn, Neptune |
| 28th | Psc | 0h 54m | 8° 57' | -10.4 | 32' | 58% | 96° E | FQ | |
| 29th | Psc | 1h 46m | 15° 3' | -10.9 | 32' | 69% | 108° E | +G | |
| 30th | Ari | 2h 42m | 20° 30' | -11.3 | 33' | 79% | 121° E | +G | |
| 31st | Tau | 3h 43m | 24° 49' | -11.7 | 33' | 88% | 135° E | +G | Uranus, Pleiades, Hyades |

## Mercury and Venus

Mercury
25ᵗʰ

Venus
25ᵗʰ

**Mercury**

| Date | Con. | R.A. | Dec. | Mag. | Diam. | Ill. | Elon. | Vis. | Rat. | Close To |
|------|------|------|------|------|-------|------|-------|------|------|----------|
| 21st | Oph | 16h 46m | -21° 40' | -0.5 | 5" | 88% | 18° W | AM | ** | Antares |
| 23rd | Oph | 16h 59m | -22° 15' | -0.5 | 5" | 89% | 17° W | AM | ** | Antares |
| 25th | Oph | 17h 11m | -22° 47' | -0.5 | 5" | 91% | 16° W | AM | ** | |
| 27th | Oph | 17h 24m | -23° 14' | -0.5 | 5" | 92% | 15° W | AM | ** | |
| 29th | Oph | 17h 37m | -23° 37' | -0.5 | 5" | 94% | 14° W | NV | N/A | |
| 31st | Sgr | 17h 50m | -23° 56' | -0.6 | 5" | 95% | 13° W | NV | N/A | |

**Venus**

| Date | Con. | R.A. | Dec. | Mag. | Diam. | Ill. | Elon. | Vis. | Rat. | Close To |
|------|------|------|------|------|-------|------|-------|------|------|----------|
| **21st** | Oph | 17h 42m | -23° 28' | -3.9 | 10" | 100% | 4° W | NV | N/A | Mars |
| **23rd** | Sgr | 17h 53m | -23° 36' | -3.9 | 10" | 100% | 4° W | NV | N/A | Mars |
| **25th** | Sgr | 18h 4m | -23° 41' | -3.9 | 10" | 100% | 3° W | NV | N/A | Mars |
| **27th** | Sgr | 18h 15m | -23° 43' | -3.9 | 10" | 100% | 3° W | NV | N/A | Mars |
| **29th** | Sgr | 18h 26m | -23° 43' | -3.9 | 10" | 100% | 2° W | NV | N/A | Mars |
| **31st** | Sgr | 18h 37m | -23° 39' | -3.9 | 10" | 100% | 2° W | NV | N/A | Mars |

## Mars and the Outer Planets

Mars
25th

Jupiter
25th

Saturn
25th

**Mars**

| Date | Con. | R.A. | Dec. | Mag. | Diam. | Ill. | Elon. | Vis. | Rat. | Close To |
|------|------|------|------|------|-------|------|-------|------|------|----------|
| **21st** | Sgr | 18h 20m | -24° 11' | 1.2 | 4" | 100% | 5° E | NV | N/A | Venus |
| **25th** | Sgr | 18h 33m | -24° 4' | 1.2 | 4" | 100% | 4° E | NV | N/A | Venus |
| **31st** | Sgr | 18h 53m | -23° 45' | 1.2 | 4" | 100% | 3° E | NV | N/A | Venus |

**The Outer Planets**

| Planet | Date | Con. | R.A. | Dec. | Mag. | Diam. | Elon. | Vis. | Rat. | Close To |
|--------|------|------|------|------|------|-------|-------|------|------|----------|
| Jupiter | 25th | Gem | 7h 35m | 21° 50' | -2.6 | 46" | 160° W | AM | ***** | |
| Saturn | 25th | Aqr | 23h 48m | -3° 45' | 1.1 | 17" | 83° E | PM | ** | Neptune |
| Uranus | 25th | Tau | 3h 43m | 19° 33' | 5.6 | 4" | 142° E | PM | **** | Pleiades |
| Neptune | 25th | Psc | 0h 0m | -1° 27' | 7.9 | 2" | 86° E | PM | *** | Saturn |

## Highlights

| Date | Time (UT) | Event |
|------|-----------|-------|
| 21st | 15:04 | Winter solstice. |
| 23rd | N/A | The Ursid meteor shower is at its maximum. (ZHR: 10) |
| 27th | 03:37 | The almost first quarter Moon ` is north of Saturn. (Evening sky.) |
| | 09:27 | The almost first quarter Moon is north of Neptune. (Evening sky.) |
| | 19:11 | First Quarter Moon. (Evening sky.) |
| 31st | 11:48 | The waxing gibbous Moon is north of Uranus. (Evening sky.) |

# Glossary

### Aphelion

The point at which an object is farthest from the Sun. (See also *perihelion*.)

### Apogee

The point at which an object is farthest from the Earth. (See also *perigee*.)

### Apparent Diameter

Apparent Diameter is the size an object appears in the sky and is measured in degrees, arc-minutes and arc-seconds. If you were to stand facing due north and slowly turn toward east, south, west and then north again, you would be turning 360° (degrees.)

You might therefore think that you can see 360° of sky overhead – in fact, you can only see 180° because the sky is only the visible half a sphere and since you can't see the entire sphere (the ground is in the way,) you can't look 360° in every direction at the sky.

If, however, you were to face due north and look directly overhead at the zenith, this would be 90°. Look down from the zenith to the southern horizon and you would see another 90°, making 180° total. (Incidentally, how high an object appears in the sky is called its *altitude*, but it isn't necessary to know that to use this book.)

To put this into perspective, the Sun and Moon both appear to be about half a degree in diameter but because these are the largest astronomical objects in the sky and everything else appears to be smaller, we need a more convenient (and accurate) measurement.

A degree then is broken up into sixty arc-minutes. Therefore, because the Sun and Moon both appear to be about half a degree, we say their apparent diameter is 30' (arc-minutes.)

The planets and asteroids are even smaller, so we break each arc-minute up into sixty again, thereby creating arc-seconds. A planets' apparent diameter will greatly depend on its actual size and its distance from Earth.

For example, although the planet Venus is slightly smaller than the Earth, it is also the closest world to our own. So at its closest (inferior conjunction) it can have an apparent diameter of 66.01" – in other words, 1' 01" (one arc-minute and one arc-second.) Despite Jupiter being large enough to swallow all the other planets within it, it is much further away – therefore, at its best, it is only able to reach 50.12" (fifty arc-seconds). Neptune is the fourth largest planet but is also the most distant and barely manages to reach 2.37" (arc-seconds.)

Even through a small telescope observers can easily see all of the planets as discs; however, how large the planet appears and the details seen will vary greatly, depending upon the equipment used and the apparent diameter of the planet itself. The dark bands of Jupiter's' atmosphere are easily visible in almost any sized 'scope, whereas Uranus and Neptune typically only show tiny discs in a small to medium sized telescope. Under low power those distant worlds can easily be mistaken for stars.

Only the apparent diameters of the Moon and planets are noted in this book; the dwarf planets Pluto and Ceres, along with the asteroids Juno, Pallas and Vesta and all the bright stars mentioned, only appear as points of light through amateur instruments and their apparent diameter is therefore negligible.

To help put this into perspective, the image below depicts the average apparent size of the planets in comparison with one another.

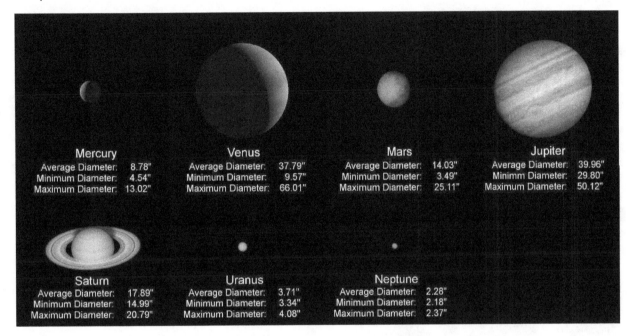

| Mercury | | Venus | | Mars | | Jupiter | |
|---|---|---|---|---|---|---|---|
| Average Diameter: | 8.78" | Average Diameter: | 37.79" | Average Diameter: | 14.03" | Average Diameter: | 39.96" |
| Minimum Diameter: | 4.54" | Minimum Diameter: | 9.57" | Minimum Diameter: | 3.49" | Minimm Diameter: | 29.80" |
| Maximum Diameter: | 13.02" | Maximum Diameter: | 66.01" | Maximum Diameter: | 25.11" | Maximum Diameter: | 50.12" |

| Saturn | | Uranus | | Neptune | |
|---|---|---|---|---|---|
| Average Diameter: | 17.89" | Average Diameter: | 3.71" | Average Diameter: | 2.28" |
| Minimum Diameter: | 14.99" | Minimum Diameter: | 3.34" | Minimum Diameter: | 2.18" |
| Maximum Diameter: | 20.79" | Maximum Diameter: | 4.08" | Maximum Diameter: | 2.37" |

## Asterism

An asterism is a recognizable pattern of stars within a constellation. For example, the "backwards question mark" depicting the head of Leo the Lion, or the seven stars of the Big Dipper (aka, the Plough) in the much-larger constellation of Ursa Major, the Great Bear.

## AU – Astronomical Unit

An astronomical unit is basically the mean distance of the Earth to the Sun and is the standard measurement of distance within the solar system. One astronomical unit is equivalent to almost 150 million kilometers (specifically, 149,597,870 km) or roughly 92.96 million miles.

## Conjunction

Astronomically, this is a fairly vague term. It basically means any situation when two bodies appear close to one another in the sky. However, there is no officially recognized separation limit that would clearly define when a conjunction is taking place. An object in conjunction with the Sun is never visible because the light from the Sun is too over-powering and the object will be lost in the glare. (See also *inferior conjunction, opposition* and *superior conjunction.*)

## Culminate

Culmination occurs when an object is at its highest point in the sky. For most objects, this means it is due south, but this will depend on the object and your latitude. For example, some stars or objects may be directly overhead when they culminate. (In fact, if you faced due south and followed an invisible line between due south and due north, your gaze would pass overhead and any object culminating could also be in sight.)

## Earthshine

Earthshine is when the Moon is a crescent but you can see the "dark side" of the Moon too – so you can see the whole Moon in the sky. This happens when light is reflected from the daylight side of the Earth and illuminates the unlit portion of the Moons' surface. It can make for a beautiful sight in the twilight, especially when the Moon is close to a bright star or planet.

*Left: An example of Earthshine on the waxing crescent Moon. Light is reflected from the Earth, causing the "dark side" of the Moon to be visible. In this image, the unlit portion of the Moon has been lightened to be more apparent. Image by Steve Jurvetson and used under the Creative Commons Attribution 2.0 Generic license.*

## Ecliptic

The approximate path the Sun, Moon and planets appear to follow across the sky. The ecliptic crosses the traditional twelve signs of the zodiac as well as briefly passing through other constellations, such as Ophiuchus and Orion. The ecliptic is depicted as a pale blue line in the images used throughout this book.

## Elongation

Elongation is how far to the east or west an object appears in relation to another object (most usually in relation to the Sun.) If Mercury or Venus is at eastern elongation, it will appear in the evening sky. If Mercury or Venus is at western elongation, it will appear in the pre-dawn sky. (It is worth noting that there is no guarantee the planet will be visible – it will also depend upon the time of year and the observers' latitude.)

## Gibbous

The Moon is said to be gibbous between the half phases (first and last quarter) and full Moon. It's hard to describe the shape – it's not a half Moon, but it's not completely circular either. The inner planets Mercury and Venus can also show a gibbous phase. (See also *illumination*.)

## Globular Star Cluster

A globular star cluster is, quite literally and simply, a sphere of thousands of stars. Globular clusters appear as faint, misty balls of grey light against the night, and although they all require at least a pair of binoculars to be seen, many can be resolved into their individual stars through a telescope. The best (and most famous) example in the northern hemisphere is M13, the Great Hercules Cluster (see image below and also *open star cluster*.)

*Left: M13, the Great Hercules Cluster. Photo taken by the author using Slooh.*

## Illumination

Simply how much of an object's visible surface is lit. For example, when the Moon is new, none of the lit surface is visible, so the Moon is 0% illuminated. At half phase (first or last quarter) half the lit surface is visible, so the Moon is 50% illuminated. At full Moon, the entire lit surface is visible, so the Moon is 100% illuminated. The planets Mercury, Venus and Mars can also show phases (Mars, being an outer planet, is more limited) and so their illumination will also change over time. (See also *gibbous*.)

## Inferior Conjunction

Inferior conjunction occurs when either Mercury or Venus are directly between the Earth and the Sun. Because the planet will appear so close to the Sun in the sky, it will not be visible from Earth. Mercury and Venus are the only two planets that can go through inferior conjunction because only these two worlds orbit closer to the Sun than the Earth. (See also *conjunction* and *superior conjunction*.)

## Magnitude

An object's magnitude is simply a measurement of its brightness. The ancient Greeks created a system where the brightest stars were given a magnitude of 1 and the faintest were magnitude 6. Since that time, astronomers have refined the system and increased its accuracy, but as a result, the magnitude range has increased dramatically.

For example, there are some objects that are brighter than zero and therefore have a negative magnitude. Sirius, the brightest star in the sky, has a magnitude of -1.47. All of the naked eye planets – Mercury, Venus, Mars, Jupiter and Saturn – can all have negative magnitudes. The other two planets, Uranus and Neptune, dwarf planets and asteroids are all more than magnitude five.

(The bright star Vega, in the constellation Lyra, is used as the standard reference point. It has a magnitude of 0.0.)

The naked eye can, theoretically, see objects up to magnitude six, but this greatly depends upon the observer's vision and the conditions of the night sky. Most people can see up to around magnitude five under clear, dark, rural skies. Light pollution is so bad in many towns and cities that, even in the suburbs, it is often difficult to see anything fainter than magnitude 3 or 4 at best.

However, the Moon, Mercury, Venus, Mars, Jupiter and Saturn should easily be visible to anyone, anywhere, assuming that the object is not too close to the Sun. A planets' magnitude will vary, depending upon how close it is to the Earth, how large it appears in the sky and how much of its lit surface is visible. (See also *apparent diameter* and *illumination*.)

Constellations can be problematic, depending upon the brightness of the stars that form the constellation itself.

Meteors are best observed from rural skies but the bright stars mentioned, as well as the Pleiades and Hyades star clusters can be seen from the suburbs.

Almost everything else is best observed under rural skies and/or with binoculars or a telescope.

The magnitude ranges of the solar system objects mentioned in this book are detailed below. (The Sun is, on average, about magnitude -26.74)

| Object | Minimum Magnitude (Faintest) | Maximum Magnitude (Brightest) |
| --- | --- | --- |
| Moon | -2.5 (New) | -12.9 (Full) |
| Mercury | 5.73 | -2.45 |
| Venus | -3.82 | -4.89 |
| Mars | 1.84 | -2.91 |
| Jupiter | -1.61 | -2.94 |
| Saturn | 1.47 | -0.49 |
| Uranus | 5.95 | 5.32 |
| Neptune | 8.02 | 7.78 |
| Pluto | 16.3 | 13.65 |
| Dwarf planet Ceres | 9.34 | 6.64 |
| Asteroid 2 Pallas | 10.65 | 6.49 |
| Asteroid 3 Juno | 11.55 | 7.4 |
| Asteroid 4 Vesta | 8.48 | 5.1 |

## Occultation

When one object completely covers another. Many occultations involve the Moon occulting a star or, sometimes, a planet, but on occasion, a planet may be seen to occult a star. On very rare occasions, one planet may occult another.

## Open Star Cluster

An open star cluster is one where the stars appear to be loosely scattered against the night. Unlike globular clusters, their member stars are usually quite young and only number a couple of hundred at most. Several open clusters can be seen with the naked eye (for example, M44 - the Praesepe - in Cancer and the Hyades, which forms the V shaped asterism in the constellation Taurus.) The most famous example of an open cluster is M45, the Pleiades, a naked eye cluster (also in Taurus) that is easily seen throughout the winter. (See image below and also *globular star cluster*.)

*Left: M45, the famous Pleiades open star cluster in Taurus. Easily visible with the naked eye throughout the winter, this image shows the deep blue nebulosity that surround the stars. This nebulosity is the remains of the cloud that gave birth to the stars themselves, but unfortunately this is not visible to the vast majority of observers. Photo by the author using Slooh.*

## Opposition

An object is said to be at opposition when it is directly opposite the Sun in the sky. On that date, it is visible throughout the night as it will rise at sunset, culminate at midnight and set at sunrise. For that reason, this is the best opportunity to observe that object. (See also *conjunction*.)

## Perihelion

The point at which an object is closest to the Sun. (See also *aphelion*.)

## Perigee

The point at which an object is closest to the Earth. (See also *apogee*.)

## Prograde Motion

Prograde motion is when a body appears to move forwards through the constellations from west to east. It's the normal motion of the Sun, Moon, planets and asteroids across the sky. (See also *retrograde motion*.)


## Retrograde Motion

Retrograde motion is when a body appears to move *backwards* through the constellations, from east to west. For the inferior planets Mercury and Venus, this happens for a time after greatest eastern elongation (when the planet appears in the evening sky) and before greatest western elongation (when it appears in the pre-dawn sky) as the planet catches up to and then passes the Earth in its orbit. For all the other planets and asteroids, it happens for a time before and after opposition when the Earth catches up to that world and then passes it. This YouTube video from 2009 does a good job of graphically depicting how this happens. (See also *prograde motion*.)


## Superior Conjunction

Superior conjunction occurs when either Mercury or Venus are on the opposite side of the Sun from the Earth. For example, if Mercury is at superior conjunction, the Sun would be directly between the Earth and the Mercury. Like *inferior conjunction*, the planet appears very close to the Sun in the sky and is not visible from Earth.

Again, like *inferior conjunction*, Mercury and Venus are the only two planets that can go through inferior conjunction because only these two worlds orbit closer to the Sun than the Earth. (See also *conjunction* and *inferior conjunction*.)


## Universal Time

Universal Time is the standard method of notating when an astronomical event takes place. It is based upon Greenwich Mean Time and requires adjustment for other time zones:

Greenwich Mean Time = no change. (Summer Time = add one hour.)

Eastern Time = deduct five hours. (Summer Time = deduct four hours.)

Central Time = deduct six hours. (Summer Time = deduct five hours.)

Mountain Time = deduct seven hours. (Summer Time = deduct six hours.)

Pacific Time = deduct eight hours. (Summer Time = deduct seven hours.)

A useful website that will convert Universal Time to other time zones can be found at the following address: http://www.worldtimeserver.com/convert_time_in_UTC.aspx

Bear in mind that if an event takes place during the day at your location, it may still be visible in the evening or pre-dawn sky. For example, two planets may be at their closest at 1pm local time but because they don't move quickly, they'll still be very close together during the night. The only exception is the Moon – it *does* move relatively quickly, but may still appear fairly close to the object when it next becomes visible. It just won't be as close as it was at the time listed in the book.

## Waning

The Moon is said to be "waning" between the full and new Moon. When the Moon is full, the Earth lies between the Moon and the Sun and the lit surface is completely visible to us. When the Moon wanes, the visible, lit portion of the Moon appears to decrease until it is completely invisible at new Moon. A waning Moon is best seen in the pre-dawn sky. (See also *waxing*.)

## Waxing

The Moon is said to be "waxing" between the new and full Moon. When the Moon is new, it lies between the Earth and the Sun and the lit surface is not visible to us. When the Moon waxes, the visible, lit portion of the Moon appears to be increasing until it is completely lit at full Moon. A waxing Moon is best seen in the evening sky. (See also *waning*.)

## Zenith

The point directly overhead in the sky. (See also *zenith hourly rate*.)

## Zenith Hourly Rate

The number of meteors an observer can expect to see each hour at the zenith (directly overhead) on the shower's peak date. It's worth remembering that meteor showers can be somewhat unpredictable and, hence, the maximum zenith hourly rate is only an estimate at best. (See also *zenith*.)

# Also by the Author

## Signposts to the Stars

## Easy Things to See With a Small Telescope

Aimed at absolute beginners, this book will help you to locate and learn the constellations using the brightest stars of Ursa Major and Orion as signposts.

More than that, the book also details:
*Key astronomical terms and phrases
*The brightest stars and constellations for each season
*The myths and legends of the stars
*Fascinating stars, star clusters, nebulae and galaxies, many of which can be seen with just your eyes or binoculars
*An introduction to the planets, comets and meteor showers

If you've ever stopped and stared at the stars but didn't know where to begin, these signposts will get you started on your journey!

Specifically written with the beginner in mind, this book highlights over sixty objects easily found and observed in the night sky. Objects such as:
* Stunning multiple stars
* Star clusters
* Nebulae
* And the Andromeda Galaxy!

Each object has its own page which includes a map, a view of the area through your finderscope and a depiction of the object through the eyepiece.

There's also a realistic description of every object based upon the author's own notes written over years of observations.

Additionally, there are useful tips and tricks designed to make your start in astronomy easier and pages to record your observations.

If you're new to astronomy and own a small telescope, this book is an invaluable introduction to the night sky.

## The Easy Guide to the Night Sky

## The Deep Sky Observer's Guide

Written for the amateur astronomer who wants to discover more in the night sky, this book explores the constellations and reveals many of the highlights visible with just your eyes or binoculars.

The highlights include:
* The myths and legends associated with the stars
* Bright stars and multiple stars
* Star clusters
* Nebulae
* Galaxies

Each constellation has its own star chart and almost all are accompanied by graphics depicting the highlights and binocular views of the best objects.

Whether you're new to astronomy or are an experienced stargazer simply looking to learn more about the constellations, this book is an invaluable guide to the night sky and the stars to be found there.

The Deep Sky Observer's Guide offers you the night sky at your fingertips. As an amateur astronomer, you want to know what's up tonight and you don't always have the time to plan ahead. The Deep Sky Observer's Guide can solve this problem in a conveniently sized paperback that easily fits in your back pocket. Take it outside and let the guide suggest any one of over 1,300 deep sky objects, all visible with a small telescope and many accessible via binoculars.

* Multiple stars with 2" or more of separation
* Open clusters up to magnitude 9
* Nebulae up to magnitude 10
* Globular clusters up to magnitude 10
* Planetary nebulae up to magnitude 12
* Galaxies up to magnitude 12
* Includes lists of deep sky objects for the entire sky with R.A. and declination for each and accompanying images for many

Whether you use a GoTo or prefer to star hop, no matter where you live in the world and no matter what time of year or night, the Deep Sky Observer's Guide is the indispensable companion for every adventure among the stars.

## The Wonder of it All

## The Daily Astronomical and Space Quiz Book

From our home here on Earth, past the Sun, Moon and planets, this is a journey out to the stars and beyond.

A journey of discovery that shows us the beauty and wonder of the cosmos and our special and unique place within it.

Written by an amateur astronomer with a life-long love of the stars, The Wonder of It All will open your child's eyes to the universe and includes notes for parents to help develop an interest in astronomy.

Are you an expert on space or is your knowledge like a black hole? Can you find Polaris or can't tell the difference between Ursa Major and Orion?

With hundreds of daily questions covering everything from astronomy, cosmology, history, the planets and their moons, you'll journey across the very length and breadth of time and space with over 700 questions and more than 350 illustrations.

Great for pub and club quiz nights or simply to test yourself and expand your knowledge, the Daily Astronomical and Space Quiz Book makes learning more about the universe fun!

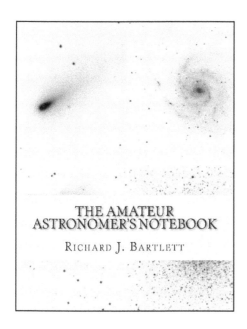

## *The Amateur Astronomer's Notebook*

The Amateur Astronomer's Notebook is the perfect way to log your observations of the Moon, stars, planets and deep sky objects.

With an additional appendix with hundreds of suggested deep sky objects, this 8.5" by 11" notebook allows you to record everything you need for 150 observing sessions under the stars:
*Date
*Time
*Lunar Phase
*Limiting Magnitude
*Transparency
*Seeing
*Equipment
*Eyepieces
*Additional Notes
*Pre-drawn circles to sketch your observations
*Plenty of room to record your notes and impressions

Whether you're an experienced astronomer or just beginning to discover the universe around us, you'll find the notebook to be an invaluable tool and record of your exploration of the cosmos.